GAS TABLES

BOOKS BY J. H. KEENAN
    Thermodynamics
BY J. H. KEENAN, J. CHAO, AND J. KAYE
    Gas Tables
BY J. H. KEENAN AND F. G. KEYES
    Mollier Chart
BY J. H. KEENAN, F. G. KEYES, P. G. HILL, AND J. G. MOORE
    Steam Tables
BY G. N. HATSOPOULOS AND J. H. KEENAN
    Principles of General Thermodynamics

GAS TABLES

THERMODYNAMIC PROPERTIES OF AIR
PRODUCTS OF COMBUSTION
AND
COMPONENT GASES

COMPRESSIBLE FLOW FUNCTIONS

Including those of Ascher H. Shapiro
and Gilbert M. Edelman

SECOND EDITION
(ENGLISH UNITS)

BY

JOSEPH H. KEENAN

Professor of Mechanical Engineering
Massachusetts Institute of Technology

JING CHAO

Research Scientist
Thermodynamics Research Center
Texas A & M University

AND

JOSEPH KAYE

Professor of Mechanical Engineering
Massachusetts Institute of Technology

A WILEY-INTERSCIENCE PUBLICATION

JOHN WILEY AND SONS

NEW YORK   CHICHESTER   BRISBANE   TORONTO

Published by John Wiley & Sons, Inc.
Copyright, 1945 by Joseph H. Keenan and Joseph Kaye under the title
Thermodynamic Properties of Air, Including Polytropic Functions
Copyright, 1948 by Joseph H. Keenan and Joseph Kaye
Copyright © 1980 by John Wiley & Sons, Inc.

Library of Congress Cataloging in Publication Data:

Keenan, Joseph Henry, 1900–1977
    Gas tables.

    "A Wiley-Interscience publication."
    Bibliography: p.
    1. Air — Tables, calculations, etc. 2. Thermodynamics —
Tables, calculations, etc.   I. Chao, Jing, joint
author. II. Kaye, Joseph, 1912-1961, joint author.
III. Title.

QC161.5.K43 1979   621.43'021'2   79-15098
ISBN 0-471-02207-1

Printed in the United States of America

10 9 8 7 6 5 4 3 2 1

IN MEMORIAM
Joseph H. Keenan

Professor Joseph H. Keenan, who was the motivating force behind these tables from the beginning, died at the age of 77 on July 17, 1977. Although his health had gradually declined during the last two years of his life, he continued to engage in preparations leading to the publication of this revision.

With the project seriously jeopardized by the loss of Professor Keenan's experience and guidance, coauthor Dr. Jing Chao and his son, William C. H. Chao, who were already actively involved, greatly expanded their participation. Dr. Chao, for example, completely rewrote the chapter now entitled, "Sources of Data and Calculation Methods." Other individuals, including many who had been very close to Professor Keenan over the years, were more than generous with their assistance. Dr. George N. Hatsopoulos in particular never failed to respond enthusiastically whenever his judgment and technical talents were sought, despite the fact that responsibilities for managing a burgeoning business organization were placing heavy demands upon his time.

The origins of these tables date back more than 30 years to a collaboration between Professor Keenan and Professor Joseph Kaye, who died in 1961. They developed the basic concepts and methodology used in formulating much of the data for the early editions. For this reason it was considered appropriate here to include their acknowledgments and the contents of each preface from the 1945 and 1948 printings. The wives of these men deserve recognition as well—Mrs. Ida Kaye for her understanding cooperation at the outset of the present revision and Mrs. Isabel M. Keenan for her encouragement and support during its later phases.

I know that my father would have wanted me to thank on his behalf all of the people who contributed their time and effort in making the revision a success. I know too how deeply he would have appreciated the readiness and willingness shown by those among them who knew him and were close to him.

MATTHEW A. KEENAN

Belmont, Massachusetts
January 1980

PREFACE TO THE SECOND EDITION
OF THE GAS TABLES

This revision of the *Gas Tables* not only endows their contents with a higher and more uniform degree of accuracy, but also extends and enhances their practical application.

The most recent physical, molecular, and spectroscopic constants were employed in reevaluating the thermodynamic properties of air at low pressures and of each of its component gases as well as the combustion products of hydrocarbon fuels and several other gaseous substances. Because of relevance to the energy industries, the thermodynamic properties for combustion products of hydrocarbon fuels with 100 percent theoretical air are included in the tables for the first time.

The values for viscosity and thermal conductivity of air in Table 2 were revised and extended to cover a wider temperature range. In column seven of that table, the values for maximum flow per unit area (G_{max}/p_i) were recalculated utilizing a special computer program to produce results with the highest degree of precision. Recent atomic weights for carbon (C) and hydrogen (H) have been used to evaluate the molecular weights and gas constants for combustion products of hydrocarbon fuels of composition $(CH_x)_n$ with air as shown in Table 9. In Table 10, the enthalpies of combustion for six more paraffins ($C_{11}H_{24}$—$C_{16}H_{34}$), olefins ($C_{11}H_{22}$—$C_{16}H_{32}$), and one more alkylbenzene ($C_{12}H_{18}$) were added to the three existing homologous series. Furthermore the enthalpies of combustion for both liquid and gaseous hydrocarbons are now presented as gross values and net values. The gross value refers to the enthalpy change for the combustion of a given hydrocarbon with air, at 77°F (25°C) and constant pressure, to produce combustion products of liquid water and gaseous carbon dioxide. The net value also represents the enthalpy change for the same combustion process, but with products of water vapor and carbon dioxide.

The thermodynamic properties for the gaseous species N_2, O_2, H_2O, CO_2, H_2, and CO have been recalculated using standard statistical mechanical methods and employing a nonrigid-rotor and anharmonic-oscillator molecular model. The thermodynamic functions for water vapor, carbon dioxide, and hydrogen were obtained by appropriate interpolation from the most reliable sources. For the monatomic gaseous argon, electronic energy levels were used to derive the respective properties.

Tables 32 to 53 have been rechecked, while Table 61, "Physical Constants," is included for the first time. Table 62 has been expanded to include the International Metric System of Units (SI) conversions.

The revised chapter on "Sources of Data and Calculation Methods" now explains how statistical mechanics is applied in calculating the ideal gas thermodynamic functions for simple gases. Also set forth are the working formulas for evaluating the thermodynamic properties of N_2, O_2, CO, and Ar.

My involvement in this project began four years ago when Professor Joseph H. Keenan invited me to coauthor the revision. It was with profound regret that I learned of his death during the summer of 1977. While work on the book was temporarily disrupted as a result, his son, Matthew A. Keenan, agreed to lend his aid by coordinating the various phases of the work and by editing the manuscript. Many others in addition to the son of Professor Keenan were instrumental in the success of the revision. Foremost among them was his close friend and colleague, Dr. George N. Hatsopoulos, whose counsel was frequently sought and always appreciated. Another friend and colleague, Dr. Elias P. Giftopoulos assisted with the chapter on "Examples." All of the illustrative problems in this chapter were reworked in detail by Raymond A. Kach, whose suggestions were extremely useful.

Another key contributor was my son, William C. H. Chao, who devoted part of his time to this project as an undergraduate at Massachusetts Institute of Technology. He developed the comprehensive computer program which recalculated the revised numerical tables and automatically generated photocomposition tapes, greatly facilitating the printing of the tables and eliminating the potential for typesetting errors. Sincere gratitude also belongs to the people at Inforonics, Inc., with special thanks to Kurt Lanza, Candy Verhulst, Joe Du Pré, and Paul Cesario for their kind services and efficient handling of the photocomposition galleys.

The author particularly wishes to thank Dr. Lester Haar at the National Bureau of Standards for recommending that Dr. Harold W. Woolley's recently reevaluated values of ideal gas thermodynamic properties for steam be adopted in this book.

A debt of gratitude is also owed to Pauline Northern, who consented upon short notice to type and proofread much of the text as preparations drew to a close.

On behalf of all the participants, it is hoped that Professor Keenan, who embarked upon this task more than four years ago, would have been pleased with the outcome and have considered it worthy of the high standards he maintained throughout a long and productive career in the field of thermodynamics.

Finally, any comments or observations that are forwarded by the reader of this revision will be greatly appreciated.

JING CHAO

Bryan, Texas
January 1980

PREFACE TO THE FIRST EDITION
OF THE GAS TABLES

THIS book supersedes a previous one entitled *Thermodynamic Properties of Air*. The properties of air have been reexamined and recalculated, the properties of products of combustion of hydrocarbons and of the constituent gases have been added, and tables of functions useful in analysis of the flow of compressible fluids have been greatly extended.

All values of the thermodynamic properties in this book are based on the examination of data from spectroscopic sources which was published by F. D. Rossini and his coworkers of the National Bureau of Standards in 1945. This was a revision of the calculated values of Johnston, Giauque, Gordon, and Kassel, and of the tabular values of Heck. The revision consisted mainly of application of new values of the fundamental constants.

The present status of these thermodynamic properties is relatively satisfactory. The uncertainties remaining in the interpretation of spectra appear to be small for monatomic and diatomic gases, and of no great order for CO_2 and H_2O. For engineering application to the design of power apparatus the precision of the present tables is of the same order as that of modern steam tables. The convenience is of a higher order.

The base temperature, for which enthalpy is zero, is not the same here as in *Thermodynamic Properties of Air*. It is taken to be zero on the absolute Fahrenheit scale in accordance with the data of Rossini. Negative values of properties are thereby avoided.

The molal unit is employed for all tables of thermodynamic properties except air. As compared with the mass unit, the molal unit makes the range of values between tables very much smaller. It also permits the use of a single table for the products of combustion corresponding to a wide range of carbon-hydrogen ratios, provided only that the "percentage of theoretical air" is held fixed. Certain other interpretations of the same table in terms of mixtures of air and fuel vapor and of air and water vapor are also valid to good precision.

Although the major tables of thermodynamic properties can be used in the analysis of the flow of compressible fluids, several excellent tabulations which have recently appeared are of greater specific convenience. The more generally useful of these have been included here. Some of the most valuable portions of this material prepared by Ascher H. Shapiro and Gilbert M. Edelman have not been previously published except in reports of limited circulation.

Dependable values of viscosity and thermal conductivity for air cover a limited range of temperature—for other gases an entirely inadequate range. Values of these properties are tabulated here only for air. Correlation of existing data and some measurements to obtain new data are in progress for other gases. As the results of this work become available they will be incorporated into the tables. The scope of the investigations in progress, however, is entirely too limited to fill the urgent need for knowledge of these two properties.

JOSEPH H. KEENAN
JOSEPH KAYE

Cambridge, May 1948

PREFACE TO
THERMODYNAMIC PROPERTIES OF
AIR

THE need for a working table of the thermodynamic properties of air has been emphasized recently by the rapidly growing interest in the gas turbine. The compression of atmospheric air which occurs in a gas turbine can be computed with reasonable convenience without the aid of a table. Computations of certain other processes, however—such as the heating of air in a regenerator or the expansion of air and similar gases from states of high temperature—involve laborious integrations if tabulated properties are not available.

For such computations air may be considered to be a relatively simple substance because the expression

$$pv = RT$$

is an adequate equation of state. By virtue of this fact the table of properties for air is simpler than the corresponding tables for vapors in that a single independent argument serves in place of two. Within the same space, therefore, a far more detailed table is possible for air than for steam. In the present table, as contrasted with existing tables for vapors, interpolation can often be dispensed with, and where it must be used it is a single interpolation which may be done by inspection.

Since the value of the gas constant R is known with great precision, the degree of uncertainty in the properties tabulated here depends only upon the state of knowledge of specific heats. The development of spectroscopic methods of determining the specific heats of the simpler gases at zero pressure has reduced this degree of uncertainty to such an extent that there is now little room for disagreement. In general, the effects of departures in composition and in pressure from those stipulated for this table will be greater than the effects of uncertainties in the data upon which the table is based. For this good fortune we are indebted to those who have labored to perfect our knowledge of the specific heats of gases.

Table 1 appeared in an abbreviated form in the *Journal of Applied Mechanics*. Since that publication the composition assumed and the values of the specific heat for air have been revised, and the entire table has been recomputed, interpolated to smaller intervals, and extended to higher temperatures.

In addition to the usual thermodynamic properties, values are given for viscosity, thermal conductivity, and Prandtl number because of their great utility in engineering computations. Certain commonly used functions of the pressure ratio are also given in some detail for a range of values of the polytropic exponent.

The present data will satisfy all but the most exacting requirements for calculations of gas-turbine processes. In order to satisfy certain other requirements work is in progress on the extension of the present data to cover the properties of air-fuel mixtures for hydrocarbon fuels and their products of combustion.

JOSEPH H. KEENAN
JOSEPH KAYE

Cambridge, March 1945

CONTENTS

SYMBOLS AND UNITS FOR TABLES 1 TO 8

a velocity of sound $= \sqrt{gkRT}$, ft/sec

c_p specific heat at constant pressure, Btu/°F lb

c_v specific heat at constant volume, Btu/°F lb

G flow per unit area or mass velocity, lbm/ft^{2}sec

h enthalpy per unit mass, Btu/lb*

$\bar{h}$ enthalpy per mole, Btu/lb-mole

k c_p/c_v

p pressure, lbf/in.^{2}

Pr Prandtl number $= c_p \mu/\lambda$

p_r relative pressure, lbf/in.^{2}†

R gas constant for air

T temperature, R

t temperature, °F

u internal energy per unit mass, Btu/lb*

$\bar{u}$ internal energy per mole, Btu/lb-mole

v_r relative volume, ft^3†

λ thermal conductivity, Btu/ft hr °F

μ viscosity, lbm/ft sec

ϕ $\int_{T_0}^{T} \frac{c_p}{T} dT$, Btu/R lb

$\bar{\phi}$ $\int_{T_0}^{T} \frac{\bar{c}_p}{T} dT$, Btu/R lb-mole‡

*To obtain values per pound-mole multiply tabulated values by molecular weight of air, namely 28.9669.

†The ratio of the pressures p_a and p_b corresponding to the temperatures T_a and T_b, respectively, along a given isentropic is equal to the ratio of the relative pressures p_{ra} and p_{rb} as tabulated for T_a and T_b, respectively. Thus

$$\left(\frac{p_a}{p_b} \right)_{s=\text{constant}} = \frac{p_{ra}}{p_{rb}}$$

Similarly

$$\left(\frac{v_a}{v_b} \right)_{s=\text{constant}} = \frac{v_{ra}}{v_{rb}}$$

‡For interpolation between Table 1 and Table 3 the value of ϕ from Table 1 should be augmented by unity and multiplied by the molecular weight of air, namely 28.9669.

Table 1 Air at Low Pressures (for One Pound)

T	t	h	p$_r$	u	v$_r$	φ	T	t	h	p$_r$	u	v$_r$	φ
100	−359.67	23.82	.003826	16.96	9684	.19698	150	−309.67	35.78	.015753	25.50	3528	.29401
101		24.06	.003961	17.13	9447	.19936	151		36.02	.016123	25.67	3470	.29560
102		24.30	.004100	17.30	9218	.20172	152		36.26	.016499	25.84	3413	.29718
103		24.54	.004242	17.47	8996	.20405	153		36.50	.016881	26.01	3358	.29875
104		24.77	.004387	17.65	8782	.20637	154		36.74	.017269	26.18	3304	.30031
105	−354.67	25.01	.004536	17.82	8576	.20866	155	−304.67	36.98	.017664	26.35	3251	.30186
106		25.25	.004689	17.99	8376	.21092	156		37.22	.018065	26.52	3199	.30339
107		25.49	.004845	18.16	8182	.21317	157		37.46	.018472	26.69	3149	.30492
108		25.73	.005005	18.33	7995	.21540	158		37.70	.018886	26.87	3099	.30644
109		25.97	.005168	18.50	7813	.21760	159		37.94	.019307	27.04	3051	.30795
110	−349.67	26.21	.005336	18.67	7637	.21979	160	−299.67	38.18	.019734	27.21	3004	.30945
111		26.45	.005507	18.84	7467	.22195	161		38.42	.020168	27.38	2958	.31094
112		26.69	.005682	19.01	7302	.22410	162		38.65	.020608	27.55	2912	.31243
113		26.93	.005861	19.18	7142	.22623	163		38.89	.021056	27.72	2868	.31390
114		27.17	.006044	19.35	6987	.22834	164		39.13	.021510	27.89	2825	.31536
115	−344.67	27.41	.006231	19.52	6837	.23043	165	−294.67	39.37	.021971	28.06	2782	.31682
116		27.65	.006423	19.69	6691	.23250	166		39.61	.022440	28.23	2741	.31826
117		27.89	.006618	19.86	6550	.23455	167		39.85	.022915	28.40	2700	.31970
118		28.13	.006817	20.04	6412	.23659	168		40.09	.023398	28.57	2660	.32113
119		28.36	.007021	20.21	6279	.23861	169		40.33	.023888	28.74	2621	.32255
120	−339.67	28.60	.007229	20.38	6149	.24061	170	−289.67	40.57	.024385	28.91	2583	.32396
121		28.84	.007442	20.55	6024	.24260	171		40.81	.024889	29.08	2545	.32536
122		29.08	.007659	20.72	5901	.24457	172		41.05	.025401	29.26	2509	.32676
123		29.32	.007880	20.89	5783	.24652	173		41.29	.025920	29.43	2473	.32815
124		29.56	.008106	21.06	5667	.24846	174		41.53	.026447	29.60	2437	.32953
125	−334.67	29.80	.008337	21.23	5555	.25038	175	−284.67	41.77	.026981	29.77	2403	.33090
126		30.04	.008572	21.40	5446	.25229	176		42.00	.027523	29.94	2369	.33226
127		30.28	.008811	21.57	5340	.25418	177		42.24	.028073	30.11	2336	.33362
128		30.52	.009056	21.74	5236	.25605	178		42.48	.028630	30.28	2303	.33496
129		30.76	.009305	21.91	5136	.25792	179		42.72	.029196	30.45	2271	.33631
130	−329.67	31.00	.009560	22.08	5038	.25976	180	−279.67	42.96	.029769	30.62	2240	.33764
131		31.24	.009819	22.26	4943	.26160	181		43.20	.030350	30.79	2209	.33896
132		31.48	.010083	22.43	4850	.26342	182		43.44	.030940	30.96	2179	.34028
133		31.71	.010352	22.60	4760	.26522	183		43.68	.031537	31.13	2150	.34159
134		31.95	.010626	22.77	4672	.26702	184		43.92	.032143	31.30	2121	.34290
135	−324.67	32.19	.010906	22.94	4586	.26880	185	−274.67	44.16	.032757	31.48	2092	.34420
136		32.43	.011190	23.11	4503	.27056	186		44.40	.033379	31.65	2064	.34549
137		32.67	.011480	23.28	4421	.27231	187		44.64	.034010	31.82	2037	.34677
138		32.91	.011775	23.45	4342	.27406	188		44.88	.034649	31.99	2010	.34805
139		33.15	.012076	23.62	4264	.27578	189		45.12	.035296	32.16	1984	.34931
140	−319.67	33.39	.012382	23.79	4189	.27750	190	−269.67	45.35	.035953	32.33	1958	.35058
141		33.63	.012693	23.96	4115	.27920	191		45.59	.036618	32.50	1932	.35183
142		33.87	.013010	24.13	4044	.28089	192		45.83	.037291	32.67	1907	.35308
143		34.11	.013333	24.30	3973	.28257	193		46.07	.037974	32.84	1883	.35433
144		34.35	.013661	24.47	3905	.28424	194		46.31	.038665	33.01	1859	.35556
145	−314.67	34.59	.013995	24.65	3838	.28590	195	−264.67	46.55	.039365	33.18	1835	.35679
146		34.83	.014335	24.82	3773	.28754	196		46.79	.040074	33.35	1812	.35802
147		35.06	.014681	24.99	3710	.28917	197		47.03	.040792	33.52	1789	.35924
148		35.30	.015032	25.16	3647	.29080	198		47.27	.041520	33.70	1767	.36045
149		35.54	.015390	25.33	3587	.29241	199		47.51	.042256	33.87	1745	.36165

Table 1 Air at Low Pressures (for One Pound)

T	t	h	p_r	u	v_r	ϕ	T	t	h	p_r	u	v_r	ϕ
200	−259.67	47.75	.04300	34.04	1723.0	.36285	250	−209.67	59.71	.09371	42.57	988.4	.41625
201		47.99	.04376	34.21	1701.8	.36405	251		59.95	.09502	42.75	978.6	.41721
202		48.23	.04452	34.38	1680.9	.36523	252		60.19	.09635	42.92	968.9	.41816
203		48.47	.04530	34.55	1660.3	.36642	253		60.43	.09769	43.09	959.4	.41911
204		48.71	.04608	34.72	1640.1	.36759	254		60.67	.09905	43.26	950.0	.42005
205	−254.67	48.94	.04687	34.89	1620.3	.36876	255	−204.67	60.91	.10042	43.43	940.8	.42099
206		49.18	.04768	35.06	1600.7	.36993	256		61.15	.10180	43.60	931.7	.42193
207		49.42	.04849	35.23	1581.6	.37109	257		61.39	.10319	43.77	922.7	.42286
208		49.66	.04931	35.40	1562.7	.37224	258		61.63	.10460	43.94	913.8	.42379
209		49.90	.05014	35.57	1544.1	.37339	259		61.87	.10602	44.11	905.0	.42472
210	−249.67	50.14	.05099	35.74	1525.9	.37453	260	−199.67	62.11	.10746	44.28	896.4	.42564
211		50.38	.05184	35.92	1507.9	.37567	261		62.35	.10891	44.45	887.8	.42656
212		50.62	.05270	36.09	1490.3	.37680	262		62.59	.11037	44.62	879.4	.42747
213		50.86	.05357	36.26	1472.9	.37792	263		62.82	.11185	44.79	871.1	.42839
214		51.10	.05446	36.43	1455.8	.37904	264		63.06	.11334	44.97	862.9	.42929
215	−244.67	51.34	.05535	36.60	1439.0	.38016	265	−194.67	63.30	.11485	45.14	854.8	.43020
216		51.58	.05626	36.77	1422.5	.38127	266		63.54	.11637	45.31	846.9	.43110
217		51.82	.05717	36.94	1406.2	.38238	267		63.78	.11790	45.48	839.0	.43200
218		52.06	.05809	37.11	1390.2	.38348	268		64.02	.11945	45.65	831.2	.43289
219		52.29	.05903	37.28	1374.4	.38457	269		64.26	.12101	45.82	823.5	.43378
220	−239.67	52.53	.05998	37.45	1358.9	.38566	270	−189.67	64.50	.12259	45.99	815.9	.43467
221		52.77	.06093	37.62	1343.7	.38675	271		64.74	.12418	46.16	808.5	.43556
222		53.01	.06190	37.79	1328.6	.38783	272		64.98	.12579	46.33	801.1	.43644
223		53.25	.06288	37.96	1313.9	.38890	273		65.22	.12741	46.50	793.8	.43732
224		53.49	.06387	38.13	1299.3	.38997	274		65.46	.12905	46.67	786.6	.43819
225	−234.67	53.73	.06487	38.31	1285.0	.39104	275	−184.67	65.70	.13070	46.84	779.5	.43906
226		53.97	.06588	38.48	1270.8	.39210	276		65.94	.13237	47.01	772.5	.43993
227		54.21	.06691	38.65	1256.9	.39316	277		66.18	.13405	47.19	765.5	.44080
228		54.45	.06794	38.82	1243.3	.39421	278		66.41	.13575	47.36	758.7	.44166
229		54.69	.06899	38.99	1229.8	.39526	279		66.65	.13746	47.53	752.0	.44252
230	−229.67	54.93	.07004	39.16	1216.5	.39630	280	−179.67	66.89	.13919	47.70	745.3	.44338
231		55.17	.07111	39.33	1203.4	.39734	281		67.13	.14093	47.87	738.7	.44423
232		55.41	.07219	39.50	1190.6	.39837	282		67.37	.14269	48.04	732.2	.44508
233		55.65	.07329	39.67	1177.9	.39940	283		67.61	.14446	48.21	725.8	.44593
234		55.88	.07439	39.84	1165.4	.40043	284		67.85	.14625	48.38	719.4	.44677
235	−224.67	56.12	.07551	40.01	1153.1	.40145	285	−174.67	68.09	.14806	48.55	713.1	.44761
236		56.36	.07663	40.18	1140.9	.40246	286		68.33	.14988	48.72	706.9	.44845
237		56.60	.07777	40.35	1129.0	.40347	287		68.57	.15172	48.89	700.8	.44929
238		56.84	.07892	40.53	1117.2	.40448	288		68.81	.15357	49.06	694.8	.45012
239		57.08	.08009	40.70	1105.6	.40549	289		69.05	.15544	49.23	688.8	.45095
240	−219.67	57.32	.08126	40.87	1094.1	.40648	290	−169.67	69.29	.15733	49.41	682.9	.45177
241		57.56	.08245	41.04	1082.9	.40748	291		69.53	.15923	49.58	677.1	.45260
242		57.80	.08365	41.21	1071.8	.40847	292		69.77	.16115	49.75	671.3	.45342
243		58.04	.08486	41.38	1060.8	.40946	293		70.00	.16308	49.92	665.6	.45424
244		58.28	.08609	41.55	1050.0	.41044	294		70.24	.16503	50.09	660.0	.45505
245	−214.67	58.52	.08733	41.72	1039.4	.41142	295	−164.67	70.48	.16700	50.26	654.4	.45587
246		58.76	.08858	41.89	1028.9	.41239	296		70.72	.16898	50.43	648.9	.45668
247		59.00	.08984	42.06	1018.5	.41336	297		70.96	.17099	50.60	643.5	.45748
248		59.24	.09112	42.23	1008.3	.41433	298		71.20	.17300	50.77	638.1	.45829
249		59.47	.09241	42.40	998.3	.41529	299		71.44	.17504	50.94	632.8	.45909

Table 1 Air at Low Pressures (for One Pound)

T	t	h	p_r	u	v_r	ϕ	T	t	h	p_r	u	v_r	ϕ
300	−159.67	71.68	.17709	51.11	627.6	.45989	350	−109.67	83.65	.3033	59.65	427.5	.49678
301		71.92	.17916	51.28	622.4	.46068	351		83.89	.3064	59.82	424.4	.49747
302		72.16	.18125	51.45	617.3	.46148	352		84.13	.3094	59.99	421.4	.49815
303		72.40	.18335	51.63	612.2	.46227	353		84.37	.3125	60.16	418.5	.49883
304		72.64	.18547	51.80	607.2	.46306	354		84.60	.3156	60.34	415.5	.49950
305	−154.67	72.88	.18761	51.97	602.3	.46384	355	−104.67	84.84	.3187	60.51	412.6	.50018
306		73.12	.18977	52.14	597.4	.46463	356		85.08	.3219	60.68	409.7	.50085
307		73.36	.19194	52.31	592.6	.46541	357		85.32	.3250	60.85	406.9	.50152
308		73.59	.19413	52.48	587.8	.46619	358		85.56	.3282	61.02	404.1	.50219
309		73.83	.19634	52.65	583.0	.46696	359		85.80	.3315	61.19	401.3	.50286
310	−149.67	74.07	.19857	52.82	578.4	.46774	360	−99.67	86.04	.3347	61.36	398.5	.50353
311		74.31	.20081	52.99	573.8	.46851	361		86.28	.3379	61.53	395.8	.50419
312		74.55	.20308	53.16	569.2	.46927	362		86.52	.3412	61.70	393.0	.50485
313		74.79	.20536	53.33	564.7	.47004	363		86.76	.3445	61.87	390.3	.50551
314		75.03	.20766	53.50	560.2	.47080	364		87.00	.3478	62.04	387.7	.50617
315	−144.67	75.27	.20998	53.67	555.8	.47156	365	−94.67	87.24	.3512	62.21	385.0	.50683
316		75.51	.21231	53.85	551.4	.47232	366		87.48	.3546	62.39	382.4	.50748
317		75.75	.21467	54.02	547.1	.47308	367		87.72	.3580	62.56	379.8	.50814
318		75.99	.21704	54.19	542.8	.47383	368		87.96	.3614	62.73	377.3	.50879
319		76.23	.21943	54.36	538.6	.47459	369		88.19	.3648	62.90	374.7	.50944
320	−139.67	76.47	.22184	54.53	534.4	.47533	370	−89.67	88.43	.3683	63.07	372.2	.51008
321		76.71	.22427	54.70	530.3	.47608	371		88.67	.3718	63.24	369.7	.51073
322		76.95	.22672	54.87	526.2	.47683	372		88.91	.3753	63.41	367.2	.51137
323		77.18	.22919	55.04	522.1	.47757	373		89.15	.3788	63.58	364.8	.51202
324		77.42	.23168	55.21	518.1	.47831	374		89.39	.3824	63.75	362.4	.51266
325	−134.67	77.66	.23418	55.38	514.1	.47904	375	−84.67	89.63	.3860	63.92	360.0	.51330
326		77.90	.23671	55.55	510.2	.47978	376		89.87	.3896	64.09	357.6	.51393
327		78.14	.23925	55.72	506.3	.48051	377		90.11	.3932	64.26	355.2	.51457
328		78.38	.24182	55.89	502.5	.48124	378		90.35	.3968	64.44	352.9	.51520
329		78.62	.24440	56.07	498.7	.48197	379		90.59	.4005	64.61	350.6	.51584
330	−129.67	78.86	.24700	56.24	495.0	.48270	380	−79.67	90.83	.4042	64.78	348.3	.51647
331		79.10	.24963	56.41	491.2	.48342	381		91.07	.4080	64.95	346.0	.51710
332		79.34	.25227	56.58	487.6	.48415	382		91.31	.4117	65.12	343.7	.51772
333		79.58	.25493	56.75	483.9	.48487	383		91.55	.4155	65.29	341.5	.51835
334		79.82	.25762	56.92	480.3	.48558	384		91.79	.4193	65.46	339.3	.51897
335	−124.67	80.06	.26032	57.09	476.8	.48630	385	−74.67	92.02	.4231	65.63	337.1	.51960
336		80.30	.26304	57.26	473.2	.48701	386		92.26	.4270	65.80	334.9	.52022
337		80.54	.26578	57.43	469.7	.48772	387		92.50	.4308	65.97	332.8	.52084
338		80.77	.26855	57.60	466.3	.48843	388		92.74	.4347	66.14	330.7	.52146
339		81.01	.27133	57.77	462.9	.48914	389		92.98	.4387	66.31	328.5	.52207
340	−119.67	81.25	.27414	57.94	459.5	.48984	390	−69.67	93.22	.4426	66.49	326.4	.52269
341		81.49	.27696	58.12	456.1	.49055	391		93.46	.4466	66.66	324.4	.52330
342		81.73	.27981	58.29	452.8	.49125	392		93.70	.4506	66.83	322.3	.52391
343		81.97	.28268	58.46	449.5	.49195	393		93.94	.4546	67.00	320.3	.52452
344		82.21	.28556	58.63	446.3	.49264	394		94.18	.4587	67.17	318.2	.52513
345	−114.67	82.45	.28847	58.80	443.1	.49334	395	−64.67	94.42	.4627	67.34	316.2	.52574
346		82.69	.29140	58.97	439.9	.49403	396		94.66	.4668	67.51	314.3	.52634
347		82.93	.29435	59.14	436.7	.49472	397		94.90	.4710	67.68	312.3	.52694
348		83.17	.29732	59.31	433.6	.49541	398		95.14	.4751	67.85	310.3	.52755
349		83.41	.30032	59.48	430.5	.49610	399		95.38	.4793	68.02	308.4	.52815

Table 1   Air at Low Pressures (for One Pound)

T	t	h	p_r	u	v_r	ϕ	T	t	h	p_r	u	v_r	ϕ
400	−59.67	95.62	.4835	68.19	306.5	.52875	450	−9.67	107.59	.7296	76.74	228.50	.55695
401		95.86	.4878	68.36	304.6	.52934	451		107.83	.7353	76.91	227.24	.55748
402		96.09	.4920	68.54	302.7	.52994	452		108.07	.7410	77.08	225.99	.55801
403		96.33	.4963	68.71	300.8	.53054	453		108.31	.7467	77.25	224.75	.55854
404		96.57	.5006	68.88	299.0	.53113	454		108.55	.7525	77.42	223.51	.55907
405	−54.67	96.81	.5050	69.05	297.1	.53172	455	−4.67	108.79	.7583	77.59	222.29	.55960
406		97.05	.5093	69.22	295.3	.53231	456		109.03	.7641	77.76	221.08	.56012
407		97.29	.5137	69.39	293.5	.53290	457		109.27	.7700	77.94	219.87	.56065
408		97.53	.5181	69.56	291.7	.53349	458		109.51	.7759	78.11	218.68	.56117
409		97.77	.5226	69.73	290.0	.53407	459		109.74	.7819	78.28	217.49	.56169
410	−49.67	98.01	.5271	69.90	288.2	.53466	460	.33	109.98	.7878	78.45	216.31	.56222
411		98.25	.5316	70.07	286.4	.53524	461		110.22	.7938	78.62	215.15	.56274
412		98.49	.5361	70.24	284.7	.53582	462		110.46	.7999	78.79	213.99	.56325
413		98.73	.5407	70.41	283.0	.53640	463		110.70	.8059	78.96	212.84	.56377
414		98.97	.5452	70.59	281.3	.53698	464		110.94	.8120	79.13	211.69	.56429
415	−44.67	99.21	.5499	70.76	279.6	.53756	465	5.33	111.18	.8182	79.30	210.56	.56481
416		99.45	.5545	70.93	277.9	.53814	466		111.42	.8243	79.47	209.43	.56532
417		99.69	.5592	71.10	276.3	.53871	467		111.66	.8305	79.65	208.32	.56583
418		99.93	.5639	71.27	274.6	.53929	468		111.90	.8368	79.82	207.21	.56635
419		100.17	.5686	71.44	273.0	.53986	469		112.14	.8430	79.99	206.11	.56686
420	−39.67	100.40	.5733	71.61	271.4	.54043	470	10.33	112.38	.8493	80.16	205.02	.56737
421		100.64	.5781	71.78	269.8	.54100	471		112.62	.8556	80.33	203.93	.56788
422		100.88	.5829	71.95	268.2	.54157	472		112.86	.8620	80.50	202.86	.56839
423		101.12	.5878	72.12	266.6	.54213	473		113.10	.8684	80.67	201.79	.56889
424		101.36	.5926	72.29	265.1	.54270	474		113.34	.8748	80.84	200.73	.56940
425	−34.67	101.60	.5975	72.47	263.5	.54326	475	15.33	113.58	.8813	81.01	199.67	.56990
426		101.84	.6025	72.64	262.0	.54383	476		113.82	.8878	81.18	198.63	.57041
427		102.08	.6074	72.81	260.4	.54439	477		114.06	.8944	81.36	197.59	.57091
428		102.32	.6124	72.98	258.9	.54495	478		114.30	.9009	81.53	196.56	.57141
429		102.56	.6174	73.15	257.4	.54551	479		114.54	.9075	81.70	195.54	.57191
430	−29.67	102.80	.6225	73.32	255.9	.54606	480	20.33	114.78	.9142	81.87	194.52	.57241
431		103.04	.6275	73.49	254.5	.54662	481		115.02	.9208	82.04	193.52	.57291
432		103.28	.6326	73.66	253.0	.54717	482		115.26	.9275	82.21	192.52	.57341
433		103.52	.6378	73.83	251.5	.54773	483		115.49	.9343	82.38	191.52	.57391
434		103.76	.6429	74.00	250.1	.54828	484		115.73	.9411	82.55	190.54	.57440
435	−24.67	104.00	.6481	74.17	248.7	.54883	485	25.33	115.97	.9479	82.72	189.56	.57490
436		104.24	.6533	74.35	247.2	.54938	486		116.21	.9547	82.90	188.59	.57539
437		104.48	.6586	74.52	245.8	.54993	487		116.45	.9616	83.07	187.62	.57588
438		104.71	.6639	74.69	244.4	.55048	488		116.69	.9685	83.24	186.66	.57637
439		104.95	.6692	74.86	243.0	.55102	489		116.93	.9755	83.41	185.71	.57686
440	−19.67	105.19	.6745	75.03	241.7	.55157	490	30.33	117.17	.9825	83.58	184.77	.57735
441		105.43	.6799	75.20	240.3	.55211	491		117.41	.9895	83.75	183.83	.57784
442		105.67	.6853	75.37	239.0	.55265	492		117.65	.9966	83.92	182.90	.57833
443		105.91	.6907	75.54	237.6	.55320	493		117.89	1.0037	84.09	181.98	.57882
444		106.15	.6962	75.71	236.3	.55374	494		118.13	1.0108	84.26	181.06	.57930
445	−14.67	106.39	.7017	75.88	235.0	.55427	495	35.33	118.37	1.0180	84.44	180.15	.57979
446		106.63	.7072	76.05	233.6	.55481	496		118.61	1.0252	84.61	179.24	.58027
447		106.87	.7127	76.23	232.3	.55535	497		118.85	1.0324	84.78	178.34	.58075
448		107.11	.7183	76.40	231.1	.55588	498		119.09	1.0397	84.95	177.45	.58123
449		107.35	.7239	76.57	229.8	.55642	499		119.33	1.0470	85.12	176.56	.58172

Table 1 Air at Low Pressures (for One Pound)

T	t	h	p_r	u	v_r	φ	T	t	h	p_r	u	v_r	φ
500	40.33	119.57	1.0544	85.29	175.68	.58220	550	90.33	131.56	1.4717	93.86	138.46	.60505
501		119.81	1.0618	85.46	174.81	.58267	551		131.80	1.4811	94.03	137.83	.60549
502		120.05	1.0692	85.63	173.94	.58315	552		132.04	1.4905	94.20	137.20	.60593
503		120.29	1.0767	85.80	173.08	.58363	553		132.28	1.5000	94.37	136.58	.60636
504		120.53	1.0842	85.98	172.22	.58411	554		132.52	1.5095	94.54	135.97	.60679
505	45.33	120.77	1.0917	86.15	171.37	.58458	555	95.33	132.76	1.5190	94.71	135.36	.60723
506		121.01	1.0993	86.32	170.53	.58506	556		133.00	1.5286	94.88	134.75	.60766
507		121.25	1.1069	86.49	169.69	.58553	557		133.24	1.5383	95.06	134.14	.60809
508		121.49	1.1146	86.66	168.86	.58600	558		133.48	1.5480	95.23	133.54	.60852
509		121.73	1.1223	86.83	168.03	.58647	559		133.72	1.5577	95.40	132.95	.60895
510	50.33	121.97	1.1300	87.00	167.21	.58694	560	100.33	133.96	1.5675	95.57	132.35	.60938
511		122.21	1.1378	87.17	166.39	.58741	561		134.20	1.5773	95.74	131.76	.60981
512		122.45	1.1456	87.35	165.58	.58788	562		134.44	1.5872	95.91	131.18	.61024
513		122.69	1.1534	87.52	164.78	.58835	563		134.68	1.5971	96.08	130.60	.61066
514		122.93	1.1613	87.69	163.98	.58882	564		134.92	1.6071	96.26	130.02	.61109
515	55.33	123.17	1.1692	87.86	163.18	.58928	565	105.33	135.16	1.6171	96.43	129.44	.61151
516		123.40	1.1772	88.03	162.39	.58975	566		135.40	1.6271	96.60	128.87	.61194
517		123.64	1.1852	88.20	161.61	.59021	567		135.64	1.6372	96.77	128.30	.61236
518		123.88	1.1932	88.37	160.83	.59067	568		135.88	1.6473	96.94	127.74	.61279
519		124.12	1.2013	88.54	160.06	.59114	569		136.12	1.6575	97.11	127.18	.61321
520	60.33	124.36	1.2094	88.71	159.29	.59160	570	110.33	136.36	1.6677	97.29	126.62	.61363
521		124.60	1.2176	88.89	158.53	.59206	571		136.60	1.6780	97.46	126.07	.61405
522		124.84	1.2258	89.06	157.77	.59252	572		136.84	1.6883	97.63	125.51	.61447
523		125.08	1.2340	89.23	157.02	.59298	573		137.08	1.6987	97.80	124.97	.61489
524		125.32	1.2423	89.40	156.27	.59344	574		137.32	1.7091	97.97	124.42	.61531
525	65.33	125.56	1.2506	89.57	155.53	.59389	575	115.33	137.56	1.7196	98.14	123.88	.61573
526		125.80	1.2589	89.74	154.79	.59435	576		137.80	1.7301	98.32	123.34	.61614
527		126.04	1.2673	89.91	154.06	.59481	577		138.04	1.7406	98.49	122.81	.61656
528		126.28	1.2758	90.09	153.33	.59526	578		138.28	1.7512	98.66	122.28	.61698
529		126.52	1.2842	90.26	152.61	.59571	579		138.52	1.7618	98.83	121.75	.61739
530	70.33	126.76	1.2927	90.43	151.89	.59617	580	120.33	138.76	1.7725	99.00	121.23	.61781
531		127.00	1.3013	90.60	151.17	.59662	581		139.00	1.7833	99.17	120.70	.61822
532		127.24	1.3099	90.77	150.47	.59707	582		139.24	1.7940	99.35	120.18	.61863
533		127.48	1.3185	90.94	149.76	.59752	583		139.49	1.8049	99.52	119.67	.61905
534		127.72	1.3272	91.11	149.06	.59797	584		139.73	1.8157	99.69	119.16	.61946
535	75.33	127.96	1.3359	91.28	148.37	.59842	585	125.33	139.97	1.8267	99.86	118.65	.61987
536		128.20	1.3447	91.46	147.67	.59887	586		140.21	1.8376	100.03	118.14	.62028
537		128.44	1.3535	91.63	146.99	.59931	587		140.45	1.8486	100.20	117.64	.62069
538		128.68	1.3623	91.80	146.31	.59976	588		140.69	1.8597	100.38	117.14	.62110
539		128.92	1.3712	91.97	145.63	.60021	589		140.93	1.8708	100.55	116.64	.62151
540	80.33	129.16	1.3801	92.14	144.96	.60065	590	130.33	141.17	1.8820	100.72	116.14	.62191
541		129.40	1.3891	92.31	144.29	.60110	591		141.41	1.8932	100.89	115.65	.62232
542		129.64	1.3981	92.48	143.62	.60154	592		141.65	1.9044	101.06	115.16	.62273
543		129.88	1.4071	92.66	142.96	.60198	593		141.89	1.9157	101.23	114.68	.62313
544		130.12	1.4162	92.83	142.31	.60242	594		142.13	1.9271	101.41	114.19	.62354
545	85.33	130.36	1.4254	93.00	141.65	.60286	595	135.33	142.37	1.9385	101.58	113.71	.62394
546		130.60	1.4345	93.17	141.01	.60330	596		142.61	1.9499	101.75	113.24	.62435
547		130.84	1.4438	93.34	140.36	.60374	597		142.85	1.9614	101.92	112.76	.62475
548		131.08	1.4530	93.51	139.72	.60418	598		143.09	1.9730	102.09	112.29	.62515
549		131.32	1.4623	93.68	139.09	.60462	599		143.33	1.9846	102.27	111.82	.62555

Table 1   Air at Low Pressures (for One Pound)

T	t	h	p_r	u	v_r	ϕ	T	t	h	p_r	u	v_r	ϕ
600	140.33	143.57	1.996	102.44	111.35	.62595	650	190.33	155.60	2.644	111.04	91.08	.64522
601		143.81	2.008	102.61	110.89	.62635	651		155.85	2.658	111.22	90.73	.64559
602		144.05	2.020	102.78	110.43	.62675	652		156.09	2.673	111.39	90.38	.64596
603		144.29	2.031	102.95	109.97	.62715	653		156.33	2.687	111.56	90.03	.64633
604		144.53	2.043	103.12	109.51	.62755	654		156.57	2.701	111.73	89.69	.64670
605	145.33	144.77	2.055	103.30	109.06	.62795	655	195.33	156.81	2.716	111.90	89.34	.64706
606		145.01	2.067	103.47	108.61	.62835	656		157.05	2.731	112.08	89.00	.64743
607		145.25	2.079	103.64	108.16	.62874	657		157.29	2.745	112.25	88.66	.64780
608		145.49	2.091	103.81	107.72	.62914	658		157.53	2.760	112.42	88.32	.64817
609		145.73	2.103	103.98	107.27	.62953	659		157.77	2.775	112.59	87.99	.64853
610	150.33	145.98	2.115	104.16	106.83	.62993	660	200.33	158.01	2.790	112.77	87.65	.64890
611		146.22	2.128	104.33	106.40	.63032	661		158.26	2.805	112.94	87.32	.64926
612		146.46	2.140	104.50	105.96	.63072	662		158.50	2.819	113.11	86.99	.64963
613		146.70	2.152	104.67	105.53	.63111	663		158.74	2.834	113.29	86.66	.64999
614		146.94	2.164	104.84	105.10	.63150	664		158.98	2.850	113.46	86.33	.65035
615	155.33	147.18	2.177	105.02	104.67	.63189	665	205.33	159.22	2.865	113.63	86.00	.65072
616		147.42	2.189	105.19	104.24	.63228	666		159.46	2.880	113.80	85.68	.65108
617		147.66	2.202	105.36	103.82	.63267	667		159.70	2.895	113.98	85.35	.65144
618		147.90	2.214	105.53	103.40	.63306	668		159.94	2.910	114.15	85.03	.65180
619		148.14	2.227	105.70	102.98	.63345	669		160.18	2.926	114.32	84.71	.65216
620	160.33	148.38	2.240	105.88	102.56	.63384	670	210.33	160.43	2.941	114.49	84.39	.65252
621		148.62	2.252	106.05	102.15	.63423	671		160.67	2.957	114.67	84.08	.65288
622		148.86	2.265	106.22	101.74	.63461	672		160.91	2.972	114.84	83.76	.65324
623		149.10	2.278	106.39	101.33	.63500	673		161.15	2.988	115.01	83.45	.65360
624		149.34	2.291	106.56	100.92	.63539	674		161.39	3.003	115.18	83.14	.65396
625	165.33	149.58	2.304	106.74	100.52	.63577	675	215.33	161.63	3.019	115.36	82.83	.65432
626		149.82	2.317	106.91	100.11	.63616	676		161.87	3.035	115.53	82.52	.65467
627		150.07	2.330	107.08	99.71	.63654	677		162.11	3.051	115.70	82.21	.65503
628		150.31	2.343	107.25	99.31	.63693	678		162.36	3.067	115.88	81.91	.65539
629		150.55	2.356	107.42	98.92	.63731	679		162.60	3.083	116.05	81.60	.65574
630	170.33	150.79	2.369	107.60	98.52	.63769	680	220.33	162.84	3.099	116.22	81.30	.65610
631		151.03	2.382	107.77	98.13	.63807	681		163.08	3.115	116.39	81.00	.65645
632		151.27	2.395	107.94	97.74	.63845	682		163.32	3.131	116.57	80.70	.65681
633		151.51	2.409	108.11	97.36	.63883	683		163.56	3.147	116.74	80.40	.65716
634		151.75	2.422	108.29	96.97	.63921	684		163.80	3.163	116.91	80.11	.65751
635	175.33	151.99	2.436	108.46	96.59	.63959	685	225.33	164.05	3.180	117.08	79.81	.65787
636		152.23	2.449	108.63	96.21	.63997	686		164.29	3.196	117.26	79.52	.65822
637		152.47	2.463	108.80	95.83	.64035	687		164.53	3.212	117.43	79.23	.65857
638		152.71	2.476	108.97	95.45	.64073	688		164.77	3.229	117.60	78.94	.65892
639		152.95	2.490	109.15	95.08	.64111	689		165.01	3.245	117.78	78.65	.65927
640	180.33	153.20	2.504	109.32	94.70	.64148	690	230.33	165.25	3.262	117.95	78.36	.65962
641		153.44	2.517	109.49	94.33	.64186	691		165.49	3.279	118.12	78.08	.65997
642		153.68	2.531	109.66	93.96	.64223	692		165.74	3.296	118.30	77.79	.66032
643		153.92	2.545	109.84	93.60	.64261	693		165.98	3.312	118.47	77.51	.66067
644		154.16	2.559	110.01	93.23	.64298	694		166.22	3.329	118.64	77.23	.66102
645	185.33	154.40	2.573	110.18	92.87	.64336	695	235.33	166.46	3.346	118.81	76.95	.66137
646		154.64	2.587	110.35	92.51	.64373	696		166.70	3.363	118.99	76.67	.66171
647		154.88	2.601	110.53	92.15	.64410	697		166.94	3.380	119.16	76.39	.66206
648		155.12	2.615	110.70	91.79	.64447	698		167.19	3.397	119.33	76.12	.66241
649		155.36	2.630	110.87	91.44	.64485	699		167.43	3.414	119.51	75.84	.66275

Table 1   Air at Low Pressures (for One Pound)

T	t	h	p_r	u	v_r	ϕ	T	t	h	p_r	u	v_r	ϕ
700	240.33	167.67	3.432	119.68	75.57	.66310	750	290.33	179.77	4.378	128.36	63.46	.67980
701		167.91	3.449	119.85	75.30	.66344	751		180.01	4.399	128.53	63.25	.68012
702		168.15	3.466	120.03	75.03	.66379	752		180.26	4.420	128.70	63.03	.68045
703		168.39	3.484	120.20	74.76	.66413	753		180.50	4.441	128.88	62.82	.68077
704		168.64	3.501	120.37	74.49	.66448	754		180.74	4.461	129.05	62.61	.68109
705	245.33	168.88	3.519	120.55	74.22	.66482	755	295.33	180.98	4.482	129.22	62.40	.68141
706		169.12	3.537	120.72	73.96	.66516	756		181.23	4.504	129.40	62.19	.68173
707		169.36	3.554	120.89	73.69	.66550	757		181.47	4.525	129.57	61.98	.68205
708		169.60	3.572	121.06	73.43	.66585	758		181.71	4.546	129.75	61.78	.68237
709		169.84	3.590	121.24	73.17	.66619	759		181.96	4.567	129.92	61.57	.68269
710	250.33	170.09	3.608	121.41	72.91	.66653	760	300.33	182.20	4.588	130.10	61.36	.68301
711		170.33	3.626	121.58	72.65	.66687	761		182.44	4.610	130.27	61.16	.68333
712		170.57	3.644	121.76	72.39	.66721	762		182.68	4.631	130.44	60.96	.68365
713		170.81	3.662	121.93	72.14	.66755	763		182.93	4.653	130.62	60.75	.68397
714		171.05	3.680	122.10	71.88	.66789	764		183.17	4.674	130.79	60.55	.68429
715	255.33	171.30	3.698	122.28	71.63	.66823	765	305.33	183.41	4.696	130.97	60.35	.68460
716		171.54	3.716	122.45	71.37	.66856	766		183.65	4.718	131.14	60.15	.68492
717		171.78	3.735	122.62	71.12	.66890	767		183.90	4.740	131.31	59.95	.68524
718		172.02	3.753	122.80	70.87	.66924	768		184.14	4.762	131.49	59.75	.68555
719		172.26	3.772	122.97	70.62	.66957	769		184.38	4.784	131.66	59.55	.68587
720	260.33	172.51	3.790	123.14	70.38	.66991	770	310.33	184.63	4.806	131.84	59.36	.68619
721		172.75	3.809	123.32	70.13	.67025	771		184.87	4.828	132.01	59.16	.68650
722		172.99	3.828	123.49	69.88	.67058	772		185.11	4.850	132.19	58.97	.68682
723		173.23	3.846	123.67	69.64	.67092	773		185.35	4.872	132.36	58.77	.68713
724		173.47	3.865	123.84	69.40	.67125	774		185.60	4.895	132.53	58.58	.68744
725	265.33	173.72	3.884	124.01	69.15	.67159	775	315.33	185.84	4.917	132.71	58.39	.68776
726		173.96	3.903	124.19	68.91	.67192	776		186.08	4.940	132.88	58.20	.68807
727		174.20	3.922	124.36	68.67	.67225	777		186.33	4.962	133.06	58.01	.68838
728		174.44	3.941	124.53	68.44	.67259	778		186.57	4.985	133.23	57.82	.68870
729		174.68	3.960	124.71	68.20	.67292	779		186.81	5.008	133.41	57.63	.68901
730	270.33	174.93	3.979	124.88	67.96	.67325	780	320.33	187.06	5.031	133.58	57.44	.68932
731		175.17	3.999	125.05	67.73	.67358	781		187.30	5.054	133.76	57.25	.68963
732		175.41	4.018	125.23	67.49	.67391	782		187.54	5.077	133.93	57.07	.68994
733		175.65	4.037	125.40	67.26	.67424	783		187.78	5.100	134.11	56.88	.69025
734		175.89	4.057	125.57	67.03	.67457	784		188.03	5.123	134.28	56.70	.69056
735	275.33	176.14	4.077	125.75	66.80	.67490	785	325.33	188.27	5.146	134.45	56.51	.69087
736		176.38	4.096	125.92	66.57	.67523	786		188.51	5.169	134.63	56.33	.69118
737		176.62	4.116	126.10	66.34	.67556	787		188.76	5.193	134.80	56.15	.69149
738		176.86	4.136	126.27	66.11	.67589	788		189.00	5.216	134.98	55.97	.69180
739		177.11	4.155	126.44	65.88	.67622	789		189.24	5.240	135.15	55.79	.69211
740	280.33	177.35	4.175	126.62	65.66	.67655	790	330.33	189.49	5.263	135.33	55.61	.69242
741		177.59	4.195	126.79	65.43	.67687	791		189.73	5.287	135.50	55.43	.69273
742		177.83	4.215	126.96	65.21	.67720	792		189.97	5.311	135.68	55.25	.69303
743		178.08	4.236	127.14	64.99	.67753	793		190.22	5.334	135.85	55.07	.69334
744		178.32	4.256	127.31	64.77	.67785	794		190.46	5.358	136.03	54.90	.69365
745	285.33	178.56	4.276	127.49	64.55	.67818	795	335.33	190.70	5.382	136.20	54.72	.69395
746		178.80	4.296	127.66	64.33	.67850	796		190.95	5.406	136.38	54.55	.69426
747		179.04	4.317	127.83	64.11	.67883	797		191.19	5.431	136.55	54.37	.69457
748		179.29	4.337	128.01	63.89	.67915	798		191.43	5.455	136.73	54.20	.69487
749		179.53	4.358	128.18	63.68	.67948	799		191.68	5.479	136.90	54.02	.69518

Table 1   Air at Low Pressures (for One Pound)

T	t	h	p_r	u	v_r	φ	T	t	h	p_r	u	v_r	φ
800	340.33	191.92	5.504	137.08	53.85	.69548	850	390.33	204.12	6.829	145.85	46.11	.71027
801		192.16	5.528	137.25	53.68	.69578	851		204.37	6.858	146.02	45.97	.71056
802		192.41	5.553	137.43	53.51	.69609	852		204.61	6.886	146.20	45.84	.71085
803		192.65	5.577	137.60	53.34	.69639	853		204.86	6.915	146.38	45.70	.71114
804		192.89	5.602	137.78	53.17	.69669	854		205.10	6.944	146.55	45.56	.71142
805	345.33	193.14	5.627	137.95	53.00	.69700	855	395.33	205.34	6.973	146.73	45.42	.71171
806		193.38	5.652	138.13	52.84	.69730	856		205.59	7.003	146.91	45.29	.71199
807		193.63	5.677	138.30	52.67	.69760	857		205.83	7.032	147.08	45.15	.71228
808		193.87	5.702	138.48	52.50	.69790	858		206.08	7.061	147.26	45.02	.71257
809		194.11	5.727	138.65	52.34	.69820	859		206.32	7.091	147.43	44.88	.71285
810	350.33	194.36	5.752	138.83	52.17	.69851	860	400.33	206.57	7.120	147.61	44.75	.71314
811		194.60	5.777	139.00	52.01	.69881	861		206.81	7.150	147.79	44.61	.71342
812		194.84	5.803	139.18	51.84	.69911	862		207.06	7.179	147.96	44.48	.71370
813		195.09	5.828	139.35	51.68	.69941	863		207.30	7.209	148.14	44.35	.71399
814		195.33	5.854	139.53	51.52	.69971	864		207.55	7.239	148.32	44.22	.71427
815	355.33	195.58	5.879	139.70	51.36	.70001	865	405.33	207.79	7.269	148.49	44.09	.71455
816		195.82	5.905	139.88	51.20	.70030	866		208.04	7.299	148.67	43.95	.71484
817		196.06	5.931	140.05	51.04	.70060	867		208.28	7.329	148.84	43.82	.71512
818		196.31	5.956	140.23	50.88	.70090	868		208.53	7.360	149.02	43.69	.71540
819		196.55	5.982	140.40	50.72	.70120	869		208.77	7.390	149.20	43.57	.71569
820	360.33	196.79	6.008	140.58	50.56	.70150	870	410.33	209.02	7.420	149.37	43.44	.71597
821		197.04	6.035	140.75	50.40	.70179	871		209.26	7.451	149.55	43.31	.71625
822		197.28	6.061	140.93	50.25	.70209	872		209.51	7.482	149.73	43.18	.71653
823		197.53	6.087	141.10	50.09	.70239	873		209.75	7.512	149.90	43.05	.71681
824		197.77	6.113	141.28	49.93	.70268	874		210.00	7.543	150.08	42.93	.71709
825	365.33	198.01	6.140	141.46	49.78	.70298	875	415.33	210.24	7.574	150.26	42.80	.71737
826		198.26	6.166	141.63	49.63	.70328	876		210.49	7.605	150.43	42.67	.71765
827		198.50	6.193	141.81	49.47	.70357	877		210.73	7.636	150.61	42.55	.71793
828		198.75	6.220	141.98	49.32	.70387	878		210.98	7.667	150.79	42.42	.71821
829		198.99	6.246	142.16	49.17	.70416	879		211.22	7.699	150.96	42.30	.71849
830	370.33	199.23	6.273	142.33	49.02	.70446	880	420.33	211.47	7.730	151.14	42.18	.71877
831		199.48	6.300	142.51	48.87	.70475	881		211.72	7.761	151.32	42.05	.71905
832		199.72	6.327	142.68	48.71	.70504	882		211.96	7.793	151.49	41.93	.71933
833		199.97	6.354	142.86	48.57	.70534	883		212.21	7.825	151.67	41.81	.71960
834		200.21	6.382	143.04	48.42	.70563	884		212.45	7.856	151.85	41.69	.71988
835	375.33	200.46	6.409	143.21	48.27	.70592	885	425.33	212.70	7.888	152.02	41.56	.72016
836		200.70	6.436	143.39	48.12	.70621	886		212.94	7.920	152.20	41.44	.72044
837		200.94	6.464	143.56	47.97	.70651	887		213.19	7.952	152.38	41.32	.72071
838		201.19	6.491	143.74	47.83	.70680	888		213.43	7.984	152.56	41.20	.72099
839		201.43	6.519	143.91	47.68	.70709	889		213.68	8.017	152.73	41.08	.72127
840	380.33	201.68	6.547	144.09	47.53	.70738	890	430.33	213.92	8.049	152.91	40.96	.72154
841		201.92	6.575	144.27	47.39	.70767	891		214.17	8.081	153.09	40.85	.72182
842		202.17	6.603	144.44	47.25	.70796	892		214.42	8.114	153.26	40.73	.72209
843		202.41	6.631	144.62	47.10	.70825	893		214.66	8.147	153.44	40.61	.72237
844		202.65	6.659	144.79	46.96	.70854	894		214.91	8.179	153.62	40.49	.72264
845	385.33	202.90	6.687	144.97	46.82	.70883	895	435.33	215.15	8.212	153.79	40.38	.72292
846		203.14	6.715	145.14	46.67	.70912	896		215.40	8.245	153.97	40.26	.72319
847		203.39	6.743	145.32	46.53	.70941	897		215.64	8.278	154.15	40.14	.72347
848		203.63	6.772	145.50	46.39	.70970	898		215.89	8.311	154.33	40.03	.72374
849		203.88	6.800	145.67	46.25	.70999	899		216.14	8.345	154.50	39.91	.72401

Table 1 Air at Low Pressures (for One Pound)

T	t	h	p_r	u	v_r	φ	T	t	h	p_r	u	v_r	φ
900	440.33	216.38	8.378	154.68	39.80	.72429	950	490.33	228.70	10.175	163.58	34.59	.73761
901		216.63	8.411	154.86	39.68	.72456	951		228.95	10.214	163.75	34.49	.73787
902		216.87	8.445	155.04	39.57	.72483	952		229.20	10.253	163.93	34.40	.73813
903		217.12	8.478	155.21	39.46	.72511	953		229.45	10.292	164.11	34.31	.73839
904		217.36	8.512	155.39	39.34	.72538	954		229.69	10.331	164.29	34.21	.73865
905	445.33	217.61	8.546	155.57	39.23	.72565	955	495.33	229.94	10.370	164.47	34.12	.73891
906		217.86	8.580	155.74	39.12	.72592	956		230.19	10.409	164.65	34.03	.73917
907		218.10	8.614	155.92	39.01	.72619	957		230.44	10.448	164.83	33.93	.73943
908		218.35	8.648	156.10	38.90	.72646	958		230.68	10.488	165.01	33.84	.73969
909		218.59	8.682	156.28	38.79	.72673	959		230.93	10.527	165.18	33.75	.73994
910	450.33	218.84	8.717	156.45	38.68	.72700	960	500.33	231.18	10.567	165.36	33.66	.74020
911		219.09	8.751	156.63	38.57	.72728	961		231.42	10.607	165.54	33.57	.74046
912		219.33	8.786	156.81	38.46	.72755	962		231.67	10.647	165.72	33.47	.74072
913		219.58	8.820	156.99	38.35	.72781	963		231.92	10.687	165.90	33.38	.74097
914		219.82	8.855	157.16	38.24	.72808	964		232.17	10.727	166.08	33.29	.74123
915	455.33	220.07	8.890	157.34	38.13	.72835	965	505.33	232.41	10.767	166.26	33.20	.74149
916		220.32	8.925	157.52	38.02	.72862	966		232.66	10.807	166.44	33.11	.74174
917		220.56	8.960	157.70	37.92	.72889	967		232.91	10.848	166.62	33.02	.74200
918		220.81	8.995	157.88	37.81	.72916	968		233.16	10.888	166.79	32.94	.74226
919		221.06	9.030	158.05	37.70	.72943	969		233.40	10.929	166.97	32.85	.74251
920	460.33	221.30	9.066	158.23	37.60	.72970	970	510.33	233.65	10.970	167.15	32.76	.74277
921		221.55	9.101	158.41	37.49	.72996	971		233.90	11.011	167.33	32.67	.74302
922		221.80	9.137	158.59	37.39	.73023	972		234.15	11.052	167.51	32.58	.74328
923		222.04	9.172	158.76	37.28	.73050	973		234.40	11.093	167.69	32.50	.74353
924		222.29	9.208	158.94	37.18	.73076	974		234.64	11.134	167.87	32.41	.74379
925	465.33	222.53	9.244	159.12	37.07	.73103	975	515.33	234.89	11.176	168.05	32.32	.74404
926		222.78	9.280	159.30	36.97	.73130	976		235.14	11.217	168.23	32.23	.74430
927		223.03	9.316	159.48	36.86	.73156	977		235.39	11.259	168.41	32.15	.74455
928		223.27	9.352	159.65	36.76	.73183	978		235.64	11.300	168.59	32.06	.74480
929		223.52	9.388	159.83	36.66	.73209	979		235.88	11.342	168.77	31.98	.74506
930	470.33	223.77	9.425	160.01	36.56	.73236	980	520.33	236.13	11.384	168.95	31.89	.74531
931		224.01	9.461	160.19	36.45	.73263	981		236.38	11.426	169.13	31.81	.74556
932		224.26	9.498	160.37	36.35	.73289	982		236.63	11.469	169.30	31.72	.74582
933		224.51	9.535	160.54	36.25	.73315	983		236.88	11.511	169.48	31.64	.74607
934		224.75	9.572	160.72	36.15	.73342	984		237.12	11.553	169.66	31.55	.74632
935	475.33	225.00	9.608	160.90	36.05	.73368	985	525.33	237.37	11.596	169.84	31.47	.74657
936		225.25	9.645	161.08	35.95	.73395	986		237.62	11.639	170.02	31.39	.74682
937		225.49	9.683	161.26	35.85	.73421	987		237.87	11.681	170.20	31.30	.74708
938		225.74	9.720	161.43	35.75	.73447	988		238.12	11.724	170.38	31.22	.74733
939		225.99	9.757	161.61	35.65	.73474	989		238.36	11.767	170.56	31.14	.74758
940	480.33	226.23	9.795	161.79	35.55	.73500	990	530.33	238.61	11.810	170.74	31.05	.74783
941		226.48	9.832	161.97	35.46	.73526	991		238.86	11.854	170.92	30.97	.74808
942		226.73	9.870	162.15	35.36	.73552	992		239.11	11.897	171.10	30.89	.74833
943		226.98	9.908	162.33	35.26	.73579	993		239.36	11.941	171.28	30.81	.74858
944		227.22	9.946	162.51	35.16	.73605	994		239.61	11.984	171.46	30.73	.74883
945	485.33	227.47	9.984	162.68	35.07	.73631	995	535.33	239.85	12.028	171.64	30.65	.74908
946		227.72	10.022	162.86	34.97	.73657	996		240.10	12.072	171.82	30.57	.74933
947		227.96	10.060	163.04	34.87	.73683	997		240.35	12.116	172.00	30.49	.74958
948		228.21	10.098	163.22	34.78	.73709	998		240.60	12.160	172.18	30.41	.74983
949		228.46	10.137	163.40	34.68	.73735	999		240.85	12.204	172.36	30.33	.75008

Table 1 Air at Low Pressures (for One Pound)

T	t	h	p_r	u	v_r	ϕ	T	t	h	p_r	u	v_r	ϕ
1000	540.33	241.10	12.248	172.54	30.25	.75032	1050	590.33	253.56	14.626	181.58	26.60	.76249
1001		241.34	12.293	172.72	30.17	.75057	1051		253.81	14.677	181.76	26.53	.76273
1002		241.59	12.337	172.90	30.09	.75082	1052		254.06	14.728	181.94	26.46	.76296
1003		241.84	12.382	173.08	30.01	.75107	1053		254.31	14.779	182.12	26.40	.76320
1004		242.09	12.427	173.26	29.93	.75132	1054		254.56	14.830	182.30	26.33	.76344
1005	545.33	242.34	12.472	173.44	29.85	.75156	1055	595.33	254.81	14.882	182.48	26.26	.76368
1006		242.59	12.517	173.62	29.78	.75181	1056		255.06	14.933	182.67	26.20	.76391
1007		242.84	12.562	173.80	29.70	.75206	1057		255.31	14.985	182.85	26.13	.76415
1008		243.09	12.608	173.98	29.62	.75231	1058		255.56	15.037	183.03	26.07	.76439
1009		243.33	12.653	174.16	29.54	.75255	1059		255.81	15.089	183.21	26.00	.76462
1010	550.33	243.58	12.699	174.34	29.47	.75280	1060	600.33	256.06	15.141	183.39	25.94	.76486
1011		243.83	12.744	174.52	29.39	.75305	1061		256.31	15.193	183.57	25.87	.76509
1012		244.08	12.790	174.70	29.31	.75329	1062		256.56	15.245	183.76	25.81	.76533
1013		244.33	12.836	174.88	29.24	.75354	1063		256.81	15.298	183.94	25.74	.76557
1014		244.58	12.882	175.06	29.16	.75378	1064		257.06	15.351	184.12	25.68	.76580
1015	555.33	244.83	12.928	175.24	29.09	.75403	1065	605.33	257.31	15.403	184.30	25.62	.76604
1016		245.08	12.975	175.42	29.01	.75427	1066		257.56	15.456	184.48	25.55	.76627
1017		245.33	13.021	175.60	28.94	.75452	1067		257.82	15.509	184.67	25.49	.76651
1018		245.57	13.068	175.78	28.86	.75476	1068		258.07	15.562	184.85	25.42	.76674
1019		245.82	13.114	175.97	28.79	.75501	1069		258.32	15.616	185.03	25.36	.76698
1020	560.33	246.07	13.161	176.15	28.71	.75525	1070	610.33	258.57	15.669	185.21	25.30	.76721
1021		246.32	13.208	176.33	28.64	.75550	1071		258.82	15.723	185.39	25.24	.76744
1022		246.57	13.255	176.51	28.56	.75574	1072		259.07	15.777	185.58	25.17	.76768
1023		246.82	13.302	176.69	28.49	.75598	1073		259.32	15.830	185.76	25.11	.76791
1024		247.07	13.350	176.87	28.42	.75623	1074		259.57	15.885	185.94	25.05	.76815
1025	565.33	247.32	13.397	177.05	28.34	.75647	1075	615.33	259.82	15.939	186.12	24.99	.76838
1026		247.57	13.445	177.23	28.27	.75671	1076		260.07	15.993	186.31	24.93	.76861
1027		247.82	13.493	177.41	28.20	.75696	1077		260.32	16.047	186.49	24.86	.76885
1028		248.07	13.540	177.59	28.13	.75720	1078		260.57	16.102	186.67	24.80	.76908
1029		248.32	13.588	177.77	28.05	.75744	1079		260.82	16.157	186.85	24.74	.76931
1030	570.33	248.57	13.636	177.95	27.98	.75768	1080	620.33	261.08	16.212	187.03	24.68	.76954
1031		248.82	13.685	178.13	27.91	.75793	1081		261.33	16.267	187.22	24.62	.76978
1032		249.06	13.733	178.31	27.84	.75817	1082		261.58	16.322	187.40	24.56	.77001
1033		249.31	13.782	178.50	27.77	.75841	1083		261.83	16.377	187.58	24.50	.77024
1034		249.56	13.830	178.68	27.70	.75865	1084		262.08	16.433	187.76	24.44	.77047
1035	575.33	249.81	13.879	178.86	27.63	.75889	1085	625.33	262.33	16.488	187.95	24.38	.77070
1036		250.06	13.928	179.04	27.56	.75913	1086		262.58	16.544	188.13	24.32	.77093
1037		250.31	13.977	179.22	27.49	.75937	1087		262.83	16.600	188.31	24.26	.77117
1038		250.56	14.026	179.40	27.42	.75962	1088		263.08	16.656	188.49	24.20	.77140
1039		250.81	14.075	179.58	27.35	.75986	1089		263.33	16.712	188.68	24.14	.77163
1040	580.33	251.06	14.125	179.76	27.28	.76010	1090	630.33	263.59	16.768	188.86	24.08	.77186
1041		251.31	14.174	179.94	27.21	.76034	1091		263.84	16.825	189.04	24.02	.77209
1042		251.56	14.224	180.13	27.14	.76058	1092		264.09	16.881	189.23	23.96	.77232
1043		251.81	14.274	180.31	27.07	.76082	1093		264.34	16.938	189.41	23.91	.77255
1044		252.06	14.324	180.49	27.00	.76105	1094		264.59	16.995	189.59	23.85	.77278
1045	585.33	252.31	14.374	180.67	26.93	.76129	1095	635.33	264.84	17.052	189.77	23.79	.77301
1046		252.56	14.424	180.85	26.87	.76153	1096		265.09	17.109	189.96	23.73	.77324
1047		252.81	14.474	181.03	26.80	.76177	1097		265.35	17.166	190.14	23.67	.77347
1048		253.06	14.525	181.21	26.73	.76201	1098		265.60	17.224	190.32	23.62	.77370
1049		253.31	14.575	181.40	26.66	.76225	1099		265.85	17.282	190.51	23.56	.77393

Table 1 Air at Low Pressures (for One Pound)

T	t	h	p_r	u	v_r	ϕ	T	t	h	p_r	u	v_r	ϕ
1100	640.33	266.10	17.339	190.69	23.50	.77415	1150	690.33	278.72	20.42	199.88	20.863	.78537
1101		266.35	17.397	190.87	23.45	.77438	1151		278.97	20.49	200.06	20.814	.78559
1102		266.60	17.455	191.05	23.39	.77461	1152		279.22	20.55	200.25	20.765	.78581
1103		266.85	17.514	191.24	23.33	.77484	1153		279.48	20.62	200.43	20.717	.78603
1104		267.11	17.572	191.42	23.28	.77507	1154		279.73	20.69	200.62	20.668	.78625
1105	645.33	267.36	17.630	191.60	23.22	.77530	1155	695.33	279.98	20.75	200.80	20.620	.78647
1106		267.61	17.689	191.79	23.16	.77552	1156		280.24	20.82	200.99	20.572	.78669
1107		267.86	17.748	191.97	23.11	.77575	1157		280.49	20.88	201.17	20.524	.78691
1108		268.11	17.807	192.15	23.05	.77598	1158		280.74	20.95	201.36	20.477	.78713
1109		268.37	17.866	192.34	23.00	.77620	1159		281.00	21.02	201.54	20.429	.78735
1110	650.33	268.62	17.925	192.52	22.94	.77643	1160	700.33	281.25	21.09	201.72	20.381	.78756
1111		268.87	17.985	192.70	22.89	.77666	1161		281.50	21.15	201.91	20.334	.78778
1112		269.12	18.044	192.89	22.83	.77689	1162		281.76	21.22	202.09	20.287	.78800
1113		269.37	18.104	193.07	22.78	.77711	1163		282.01	21.29	202.28	20.240	.78822
1114		269.63	18.164	193.25	22.72	.77734	1164		282.26	21.36	202.46	20.193	.78844
1115	655.33	269.88	18.224	193.44	22.67	.77756	1165	705.33	282.52	21.42	202.65	20.146	.78865
1116		270.13	18.284	193.62	22.61	.77779	1166		282.77	21.49	202.83	20.100	.78887
1117		270.38	18.344	193.80	22.56	.77802	1167		283.03	21.56	203.02	20.053	.78909
1118		270.63	18.405	193.99	22.50	.77824	1168		283.28	21.63	203.21	20.007	.78931
1119		270.89	18.465	194.17	22.45	.77847	1169		283.53	21.70	203.39	19.961	.78952
1120	660.33	271.14	18.526	194.35	22.40	.77869	1170	710.33	283.79	21.77	203.58	19.915	.78974
1121		271.39	18.587	194.54	22.34	.77892	1171		284.04	21.83	203.76	19.869	.78996
1122		271.64	18.648	194.72	22.29	.77914	1172		284.29	21.90	203.95	19.823	.79017
1123		271.89	18.709	194.91	22.24	.77937	1173		284.55	21.97	204.13	19.777	.79039
1124		272.15	18.771	195.09	22.18	.77959	1174		284.80	22.04	204.32	19.732	.79061
1125	665.33	272.40	18.832	195.27	22.13	.77982	1175	715.33	285.06	22.11	204.50	19.686	.79082
1126		272.65	18.894	195.46	22.08	.78004	1176		285.31	22.18	204.69	19.641	.79104
1127		272.90	18.956	195.64	22.03	.78026	1177		285.56	22.25	204.87	19.596	.79125
1128		273.16	19.018	195.82	21.97	.78049	1178		285.82	22.32	205.06	19.551	.79147
1129		273.41	19.080	196.01	21.92	.78071	1179		286.07	22.39	205.24	19.506	.79169
1130	670.33	273.66	19.142	196.19	21.87	.78094	1180	720.33	286.33	22.46	205.43	19.462	.79190
1131		273.91	19.205	196.38	21.82	.78116	1181		286.58	22.53	205.61	19.417	.79212
1132		274.17	19.267	196.56	21.77	.78138	1182		286.83	22.60	205.80	19.373	.79233
1133		274.42	19.330	196.74	21.71	.78160	1183		287.09	22.68	205.99	19.328	.79255
1134		274.67	19.393	196.93	21.66	.78183	1184		287.34	22.75	206.17	19.284	.79276
1135	675.33	274.92	19.456	197.11	21.61	.78205	1185	725.33	287.60	22.82	206.36	19.240	.79298
1136		275.18	19.519	197.30	21.56	.78227	1186		287.85	22.89	206.54	19.196	.79319
1137		275.43	19.583	197.48	21.51	.78250	1187		288.10	22.96	206.73	19.152	.79341
1138		275.68	19.646	197.67	21.46	.78272	1188		288.36	23.03	206.91	19.109	.79362
1139		275.93	19.710	197.85	21.41	.78294	1189		288.61	23.10	207.10	19.065	.79383
1140	680.33	276.19	19.774	198.03	21.36	.78316	1190	730.33	288.87	23.18	207.29	19.022	.79405
1141		276.44	19.838	198.22	21.31	.78338	1191		289.12	23.25	207.47	18.979	.79426
1142		276.69	19.902	198.40	21.26	.78360	1192		289.38	23.32	207.66	18.935	.79447
1143		276.95	19.967	198.59	21.21	.78383	1193		289.63	23.39	207.84	18.892	.79469
1144		277.20	20.031	198.77	21.16	.78405	1194		289.89	23.47	208.03	18.850	.79490
1145	685.33	277.45	20.096	198.96	21.11	.78427	1195	735.33	290.14	23.54	208.22	18.807	.79511
1146		277.71	20.161	199.14	21.06	.78449	1196		290.39	23.61	208.40	18.764	.79533
1147		277.96	20.226	199.32	21.01	.78471	1197		290.65	23.69	208.59	18.722	.79554
1148		278.21	20.291	199.51	20.96	.78493	1198		290.90	23.76	208.77	18.679	.79575
1149		278.46	20.356	199.69	20.91	.78515	1199		291.16	23.83	208.96	18.637	.79596

Table 1   Air at Low Pressures (for One Pound)

T	t	h	p_r	u	v_r	ϕ	T	t	h	p_r	u	v_r	ϕ
1200	740.33	291.41	23.91	209.15	18.595	.79618	1250	790.33	304.19	27.84	218.49	16.636	.80661
1201		291.67	23.98	209.33	18.553	.79639	1251		304.45	27.92	218.68	16.599	.80681
1202		291.92	24.06	209.52	18.511	.79660	1252		304.70	28.00	218.87	16.563	.80702
1203		292.18	24.13	209.70	18.469	.79681	1253		304.96	28.09	219.06	16.527	.80722
1204		292.43	24.21	209.89	18.428	.79703	1254		305.21	28.17	219.25	16.491	.80743
1205	745.33	292.69	24.28	210.08	18.386	.79724	1255	795.33	305.47	28.26	219.43	16.455	.80763
1206		292.94	24.36	210.26	18.345	.79745	1256		305.73	28.34	219.62	16.419	.80784
1207		293.20	24.43	210.45	18.303	.79766	1257		305.98	28.42	219.81	16.383	.80804
1208		293.45	24.51	210.64	18.262	.79787	1258		306.24	28.51	220.00	16.348	.80824
1209		293.71	24.58	210.82	18.221	.79808	1259		306.50	28.59	220.19	16.312	.80845
1210	750.33	293.96	24.66	211.01	18.180	.79829	1260	800.33	306.75	28.68	220.37	16.276	.80865
1211		294.22	24.73	211.20	18.139	.79850	1261		307.01	28.76	220.56	16.241	.80885
1212		294.47	24.81	211.38	18.099	.79871	1262		307.27	28.85	220.75	16.206	.80906
1213		294.73	24.89	211.57	18.058	.79892	1263		307.52	28.94	220.94	16.171	.80926
1214		294.98	24.96	211.76	18.018	.79913	1264		307.78	29.02	221.13	16.135	.80946
1215	755.33	295.24	25.04	211.94	17.977	.79934	1265	805.33	308.04	29.11	221.31	16.100	.80967
1216		295.49	25.12	212.13	17.937	.79955	1266		308.29	29.19	221.50	16.066	.80987
1217		295.75	25.19	212.32	17.897	.79976	1267		308.55	29.28	221.69	16.031	.81007
1218		296.00	25.27	212.50	17.857	.79997	1268		308.81	29.37	221.88	15.996	.81028
1219		296.26	25.35	212.69	17.817	.80018	1269		309.06	29.45	222.07	15.961	.81048
1220	760.33	296.51	25.42	212.88	17.777	.80039	1270	810.33	309.32	29.54	222.26	15.927	.81068
1221		296.77	25.50	213.06	17.738	.80060	1271		309.58	29.63	222.44	15.893	.81088
1222		297.02	25.58	213.25	17.698	.80081	1272		309.84	29.72	222.63	15.858	.81109
1223		297.28	25.66	213.44	17.659	.80102	1273		310.09	29.80	222.82	15.824	.81129
1224		297.54	25.74	213.62	17.619	.80123	1274		310.35	29.89	223.01	15.790	.81149
1225	765.33	297.79	25.81	213.81	17.580	.80144	1275	815.33	310.61	29.98	223.20	15.756	.81169
1226		298.05	25.89	214.00	17.541	.80165	1276		310.86	30.07	223.39	15.722	.81189
1227		298.30	25.97	214.18	17.502	.80185	1277		311.12	30.16	223.58	15.688	.81209
1228		298.56	26.05	214.37	17.463	.80206	1278		311.38	30.25	223.76	15.654	.81230
1229		298.81	26.13	214.56	17.425	.80227	1279		311.64	30.33	223.95	15.621	.81250
1230	770.33	299.07	26.21	214.75	17.386	.80248	1280	820.33	311.89	30.42	224.14	15.587	.81270
1231		299.32	26.29	214.93	17.347	.80269	1281		312.15	30.51	224.33	15.554	.81290
1232		299.58	26.37	215.12	17.309	.80289	1282		312.41	30.60	224.52	15.520	.81310
1233		299.84	26.45	215.31	17.271	.80310	1283		312.66	30.69	224.71	15.487	.81330
1234		300.09	26.53	215.49	17.232	.80331	1284		312.92	30.78	224.90	15.454	.81350
1235	775.33	300.35	26.61	215.68	17.194	.80352	1285	825.33	313.18	30.87	225.09	15.421	.81370
1236		300.60	26.69	215.87	17.156	.80372	1286		313.44	30.96	225.27	15.388	.81390
1237		300.86	26.77	216.06	17.119	.80393	1287		313.69	31.05	225.46	15.355	.81410
1238		301.12	26.85	216.24	17.081	.80414	1288		313.95	31.14	225.65	15.322	.81430
1239		301.37	26.93	216.43	17.043	.80434	1289		314.21	31.23	225.84	15.289	.81450
1240	780.33	301.63	27.01	216.62	17.006	.80455	1290	830.33	314.47	31.33	226.03	15.256	.81470
1241		301.88	27.10	216.81	16.968	.80476	1291		314.72	31.42	226.22	15.224	.81490
1242		302.14	27.18	216.99	16.931	.80496	1292		314.98	31.51	226.41	15.191	.81510
1243		302.40	27.26	217.18	16.894	.80517	1293		315.24	31.60	226.60	15.159	.81530
1244		302.65	27.34	217.37	16.856	.80537	1294		315.50	31.69	226.79	15.127	.81550
1245	785.33	302.91	27.42	217.56	16.819	.80558	1295	835.33	315.76	31.78	226.98	15.095	.81570
1246		303.16	27.51	217.74	16.782	.80579	1296		316.01	31.88	227.16	15.062	.81590
1247		303.42	27.59	217.93	16.746	.80599	1297		316.27	31.97	227.35	15.030	.81610
1248		303.68	27.67	218.12	16.709	.80620	1298		316.53	32.06	227.54	14.998	.81629
1249		303.93	27.75	218.31	16.672	.80640	1299		316.79	32.15	227.73	14.967	.81649

Table 1   Air at Low Pressures (for One Pound)

T	t	h	p_r	u	v_r	ϕ	T	t	h	p_r	u	v_r	ϕ
1300	840.33	317.04	32.25	227.92	14.935	.81669	1350	890.33	329.98	37.18	237.43	13.451	.82646
1301		317.30	32.34	228.11	14.903	.81689	1351		330.24	37.29	237.62	13.423	.82665
1302		317.56	32.44	228.30	14.871	.81709	1352		330.50	37.39	237.81	13.395	.82684
1303		317.82	32.53	228.49	14.840	.81729	1353		330.76	37.50	238.00	13.368	.82703
1304		318.08	32.62	228.68	14.809	.81748	1354		331.02	37.60	238.19	13.340	.82722
1305	845.33	318.33	32.72	228.87	14.777	.81768	1355	895.33	331.28	37.71	238.38	13.313	.82741
1306		318.59	32.81	229.06	14.746	.81788	1356		331.54	37.81	238.58	13.285	.82761
1307		318.85	32.91	229.25	14.715	.81808	1357		331.80	37.92	238.77	13.258	.82780
1308		319.11	33.00	229.44	14.684	.81827	1358		332.06	38.03	238.96	13.231	.82799
1309		319.37	33.10	229.63	14.653	.81847	1359		332.32	38.13	239.15	13.204	.82818
1310	850.33	319.63	33.19	229.82	14.622	.81867	1360	900.33	332.58	38.24	239.34	13.177	.82837
1311		319.88	33.29	230.01	14.591	.81887	1361		332.84	38.34	239.53	13.150	.82856
1312		320.14	33.38	230.20	14.560	.81906	1362		333.10	38.45	239.72	13.123	.82875
1313		320.40	33.48	230.39	14.529	.81926	1363		333.36	38.56	239.91	13.096	.82894
1314		320.66	33.58	230.58	14.499	.81946	1364		333.62	38.67	240.11	13.069	.82913
1315	855.33	320.92	33.67	230.77	14.468	.81965	1365	905.33	333.88	38.77	240.30	13.042	.82933
1316		321.18	33.77	230.96	14.438	.81985	1366		334.14	38.88	240.49	13.016	.82952
1317		321.43	33.87	231.15	14.408	.82005	1367		334.40	38.99	240.68	12.989	.82971
1318		321.69	33.96	231.34	14.377	.82024	1368		334.66	39.10	240.87	12.963	.82990
1319		321.95	34.06	231.53	14.347	.82044	1369		334.92	39.21	241.06	12.936	.83009
1320	860.33	322.21	34.16	231.72	14.317	.82063	1370	910.33	335.18	39.31	241.25	12.910	.83028
1321		322.47	34.25	231.91	14.287	.82083	1371		335.44	39.42	241.45	12.884	.83047
1322		322.73	34.35	232.10	14.257	.82103	1372		335.70	39.53	241.64	12.857	.83066
1323		322.99	34.45	232.29	14.227	.82122	1373		335.96	39.64	241.83	12.831	.83085
1324		323.24	34.55	232.48	14.197	.82142	1374		336.22	39.75	242.02	12.805	.83103
1325	865.33	323.50	34.65	232.67	14.168	.82161	1375	915.33	336.48	39.86	242.21	12.779	.83122
1326		323.76	34.75	232.86	14.138	.82181	1376		336.74	39.97	242.40	12.753	.83141
1327		324.02	34.85	233.05	14.109	.82200	1377		337.00	40.08	242.60	12.727	.83160
1328		324.28	34.94	233.24	14.079	.82220	1378		337.26	40.19	242.79	12.701	.83179
1329		324.54	35.04	233.43	14.050	.82239	1379		337.52	40.30	242.98	12.676	.83198
1330	870.33	324.80	35.14	233.62	14.020	.82259	1380	920.33	337.78	40.42	243.17	12.650	.83217
1331		325.06	35.24	233.81	13.991	.82278	1381		338.04	40.53	243.36	12.624	.83236
1332		325.31	35.34	234.00	13.962	.82298	1382		338.30	40.64	243.56	12.599	.83255
1333		325.57	35.44	234.19	13.933	.82317	1383		338.56	40.75	243.75	12.573	.83273
1334		325.83	35.54	234.38	13.904	.82336	1384		338.82	40.86	243.94	12.548	.83292
1335	875.33	326.09	35.65	234.57	13.875	.82356	1385	925.33	339.08	40.97	244.13	12.522	.83311
1336		326.35	35.75	234.76	13.846	.82375	1386		339.34	41.09	244.32	12.497	.83330
1337		326.61	35.85	234.95	13.818	.82395	1387		339.60	41.20	244.52	12.472	.83349
1338		326.87	35.95	235.14	13.789	.82414	1388		339.86	41.31	244.71	12.447	.83368
1339		327.13	36.05	235.33	13.760	.82433	1389		340.12	41.43	244.90	12.422	.83386
1340	880.33	327.39	36.15	235.52	13.732	.82453	1390	930.33	340.39	41.54	245.09	12.397	.83405
1341		327.65	36.25	235.71	13.703	.82472	1391		340.65	41.65	245.28	12.372	.83424
1342		327.91	36.36	235.90	13.675	.82491	1392		340.91	41.77	245.48	12.347	.83443
1343		328.16	36.46	236.09	13.647	.82511	1393		341.17	41.88	245.67	12.322	.83461
1344		328.42	36.56	236.28	13.618	.82530	1394		341.43	42.00	245.86	12.297	.83480
1345	885.33	328.68	36.67	236.47	13.590	.82549	1395	935.33	341.69	42.11	246.05	12.272	.83499
1346		328.94	36.77	236.67	13.562	.82569	1396		341.95	42.23	246.25	12.248	.83517
1347		329.20	36.87	236.86	13.534	.82588	1397		342.21	42.34	246.44	12.223	.83536
1348		329.46	36.98	237.05	13.506	.82607	1398		342.47	42.46	246.63	12.199	.83555
1349		329.72	37.08	237.24	13.478	.82626	1399		342.73	42.57	246.82	12.174	.83573

Table 1 Air at Low Pressures (for One Pound)

T	t	h	p_r	u	v_r	ϕ	T	t	h	p_r	u	v_r	ϕ
1400	940.33	342.99	42.69	247.02	12.150	.83592	1450	990.33	356.09	48.81	256.68	11.005	.84511
1401		343.26	42.81	247.21	12.125	.83611	1451		356.35	48.94	256.87	10.984	.84529
1402		343.52	42.92	247.40	12.101	.83629	1452		356.61	49.07	257.07	10.962	.84547
1403		343.78	43.04	247.59	12.077	.83648	1453		356.87	49.20	257.26	10.941	.84565
1404		344.04	43.16	247.79	12.053	.83667	1454		357.14	49.33	257.46	10.920	.84583
1405	945.33	344.30	43.27	247.98	12.029	.83685	1455	995.33	357.40	49.46	257.65	10.899	.84601
1406		344.56	43.39	248.17	12.005	.83704	1456		357.66	49.59	257.84	10.877	.84619
1407		344.82	43.51	248.36	11.981	.83722	1457		357.93	49.72	258.04	10.856	.84637
1408		345.08	43.63	248.56	11.957	.83741	1458		358.19	49.85	258.23	10.835	.84655
1409		345.35	43.74	248.75	11.933	.83759	1459		358.45	49.98	258.43	10.814	.84673
1410	950.33	345.61	43.86	248.94	11.909	.83778	1460	1000.33	358.71	50.11	258.62	10.793	.84691
1411		345.87	43.98	249.14	11.885	.83797	1461		358.98	50.25	258.82	10.772	.84709
1412		346.13	44.10	249.33	11.862	.83815	1462		359.24	50.38	259.01	10.751	.84727
1413		346.39	44.22	249.52	11.838	.83834	1463		359.50	50.51	259.20	10.731	.84745
1414		346.65	44.34	249.71	11.815	.83852	1464		359.77	50.64	259.40	10.710	.84763
1415	955.33	346.91	44.46	249.91	11.791	.83871	1465	1005.33	360.03	50.78	259.59	10.689	.84781
1416		347.18	44.58	250.10	11.768	.83889	1466		360.29	50.91	259.79	10.668	.84799
1417		347.44	44.70	250.29	11.744	.83907	1467		360.55	51.04	259.98	10.648	.84817
1418		347.70	44.82	250.49	11.721	.83926	1468		360.82	51.18	260.18	10.627	.84835
1419		347.96	44.94	250.68	11.698	.83944	1469		361.08	51.31	260.37	10.607	.84853
1420	960.33	348.22	45.06	250.87	11.675	.83963	1470	1010.33	361.34	51.44	260.57	10.586	.84871
1421		348.48	45.18	251.07	11.651	.83981	1471		361.61	51.58	260.76	10.566	.84889
1422		348.75	45.30	251.26	11.628	.84000	1472		361.87	51.71	260.96	10.545	.84907
1423		349.01	45.43	251.45	11.605	.84018	1473		362.13	51.85	261.15	10.525	.84925
1424		349.27	45.55	251.64	11.582	.84036	1474		362.40	51.98	261.35	10.505	.84943
1425	965.33	349.53	45.67	251.84	11.560	.84055	1475	1015.33	362.66	52.12	261.54	10.485	.84960
1426		349.79	45.79	252.03	11.537	.84073	1476		362.92	52.26	261.74	10.464	.84978
1427		350.05	45.92	252.22	11.514	.84092	1477		363.19	52.39	261.93	10.444	.84996
1428		350.32	46.04	252.42	11.491	.84110	1478		363.45	52.53	262.12	10.424	.85014
1429		350.58	46.16	252.61	11.469	.84128	1479		363.71	52.66	262.32	10.404	.85032
1430	970.33	350.84	46.29	252.80	11.446	.84147	1480	1020.33	363.98	52.80	262.51	10.384	.85050
1431		351.10	46.41	253.00	11.423	.84165	1481		364.24	52.94	262.71	10.364	.85067
1432		351.36	46.53	253.19	11.401	.84183	1482		364.51	53.08	262.90	10.344	.85085
1433		351.63	46.66	253.39	11.378	.84201	1483		364.77	53.21	263.10	10.325	.85103
1434		351.89	46.78	253.58	11.356	.84220	1484		365.03	53.35	263.29	10.305	.85121
1435	975.33	352.15	46.91	253.77	11.334	.84238	1485	1025.33	365.30	53.49	263.49	10.285	.85139
1436		352.41	47.03	253.97	11.311	.84256	1486		365.56	53.63	263.69	10.265	.85156
1437		352.67	47.16	254.16	11.289	.84275	1487		365.82	53.77	263.88	10.246	.85174
1438		352.94	47.28	254.35	11.267	.84293	1488		366.09	53.91	264.08	10.226	.85192
1439		353.20	47.41	254.55	11.245	.84311	1489		366.35	54.05	264.27	10.207	.85209
1440	980.33	353.46	47.54	254.74	11.223	.84329	1490	1030.33	366.61	54.19	264.47	10.187	.85227
1441		353.72	47.66	254.93	11.201	.84347	1491		366.88	54.33	264.66	10.168	.85245
1442		353.99	47.79	255.13	11.179	.84366	1492		367.14	54.47	264.86	10.148	.85263
1443		354.25	47.92	255.32	11.157	.84384	1493		367.41	54.61	265.05	10.129	.85280
1444		354.51	48.04	255.52	11.135	.84402	1494		367.67	54.75	265.25	10.110	.85298
1445	985.33	354.77	48.17	255.71	11.113	.84420	1495	1035.33	367.93	54.89	265.44	10.090	.85316
1446		355.04	48.30	255.90	11.092	.84438	1496		368.20	55.03	265.64	10.071	.85333
1447		355.30	48.43	256.10	11.070	.84457	1497		368.46	55.17	265.83	10.052	.85351
1448		355.56	48.55	256.29	11.048	.84475	1498		368.73	55.32	266.03	10.033	.85368
1449		355.82	48.68	256.49	11.027	.84493	1499		368.99	55.46	266.22	10.014	.85386

Table 1 Air at Low Pressures (for One Pound)

T	t	h	p_r	u	v_r	φ	T	t	h	p_r	u	v_r	φ
1500	1040.33	369.25	55.60	266.42	9.995	.85404	1550	1090.33	382.50	63.11	276.24	9.099	.86272
1501		369.52	55.74	266.62	9.976	.85421	1551		382.76	63.27	276.43	9.082	.86289
1502		369.78	55.89	266.81	9.957	.85439	1552		383.03	63.43	276.63	9.065	.86306
1503		370.05	56.03	267.01	9.938	.85457	1553		383.29	63.58	276.83	9.049	.86324
1504		370.31	56.17	267.20	9.919	.85474	1554		383.56	63.74	277.02	9.032	.86341
1505	1045.33	370.58	56.32	267.40	9.900	.85492	1555	1095.33	383.83	63.90	277.22	9.015	.86358
1506		370.84	56.46	267.59	9.881	.85509	1556		384.09	64.06	277.42	8.998	.86375
1507		371.10	56.61	267.79	9.863	.85527	1557		384.36	64.22	277.62	8.982	.86392
1508		371.37	56.75	267.99	9.844	.85544	1558		384.62	64.38	277.81	8.965	.86409
1509		371.63	56.90	268.18	9.825	.85562	1559		384.89	64.54	278.01	8.949	.86426
1510	1050.33	371.90	57.04	268.38	9.807	.85579	1560	1100.33	385.15	64.70	278.21	8.932	.86443
1511		372.16	57.19	268.57	9.788	.85597	1561		385.42	64.86	278.40	8.916	.86460
1512		372.43	57.34	268.77	9.770	.85614	1562		385.69	65.03	278.60	8.899	.86477
1513		372.69	57.48	268.96	9.751	.85632	1563		385.95	65.19	278.80	8.883	.86494
1514		372.95	57.63	269.16	9.733	.85649	1564		386.22	65.35	279.00	8.867	.86511
1515	1055.33	373.22	57.78	269.36	9.715	.85667	1565	1105.33	386.48	65.51	279.19	8.850	.86528
1516		373.48	57.92	269.55	9.696	.85684	1566		386.75	65.67	279.39	8.834	.86545
1517		373.75	58.07	269.75	9.678	.85702	1567		387.02	65.84	279.59	8.818	.86562
1518		374.01	58.22	269.95	9.660	.85719	1568		387.28	66.00	279.79	8.802	.86579
1519		374.28	58.37	270.14	9.642	.85737	1569		387.55	66.16	279.98	8.785	.86596
1520	1060.33	374.54	58.52	270.34	9.623	.85754	1570	1110.33	387.82	66.33	280.18	8.769	.86613
1521		374.81	58.66	270.53	9.605	.85771	1571		388.08	66.49	280.38	8.753	.86630
1522		375.07	58.81	270.73	9.587	.85789	1572		388.35	66.66	280.58	8.737	.86647
1523		375.34	58.96	270.93	9.569	.85806	1573		388.61	66.82	280.77	8.721	.86664
1524		375.60	59.11	271.12	9.551	.85824	1574		388.88	66.99	280.97	8.705	.86681
1525	1065.33	375.87	59.26	271.32	9.533	.85841	1575	1115.33	389.15	67.15	281.17	8.689	.86698
1526		376.13	59.41	271.51	9.516	.85858	1576		389.41	67.32	281.37	8.673	.86715
1527		376.40	59.56	271.71	9.498	.85876	1577		389.68	67.48	281.57	8.658	.86732
1528		376.66	59.71	271.91	9.480	.85893	1578		389.95	67.65	281.76	8.642	.86748
1529		376.93	59.87	272.10	9.462	.85910	1579		390.21	67.82	281.96	8.626	.86765
1530	1070.33	377.19	60.02	272.30	9.444	.85928	1580	1120.33	390.48	67.98	282.16	8.610	.86782
1531		377.46	60.17	272.50	9.427	.85945	1581		390.74	68.15	282.36	8.594	.86799
1532		377.72	60.32	272.69	9.409	.85962	1582		391.01	68.32	282.56	8.579	.86816
1533		377.99	60.47	272.89	9.392	.85980	1583		391.28	68.49	282.75	8.563	.86833
1534		378.25	60.63	273.09	9.374	.85997	1584		391.54	68.66	282.95	8.548	.86850
1535	1075.33	378.52	60.78	273.28	9.357	.86014	1585	1125.33	391.81	68.82	283.15	8.532	.86866
1536		378.78	60.93	273.48	9.339	.86031	1586		392.08	68.99	283.35	8.516	.86883
1537		379.05	61.09	273.68	9.322	.86049	1587		392.34	69.16	283.55	8.501	.86900
1538		379.31	61.24	273.87	9.304	.86066	1588		392.61	69.33	283.74	8.485	.86917
1539		379.58	61.39	274.07	9.287	.86083	1589		392.88	69.50	283.94	8.470	.86934
1540	1080.33	379.84	61.55	274.27	9.270	.86100	1590	1130.33	393.14	69.67	284.14	8.455	.86950
1541		380.11	61.70	274.46	9.252	.86118	1591		393.41	69.84	284.34	8.439	.86967
1542		380.37	61.86	274.66	9.235	.86135	1592		393.68	70.01	284.54	8.424	.86984
1543		380.64	62.01	274.86	9.218	.86152	1593		393.94	70.19	284.73	8.409	.87001
1544		380.90	62.17	275.05	9.201	.86169	1594		394.21	70.36	284.93	8.393	.87017
1545	1085.33	381.17	62.33	275.25	9.184	.86186	1595	1135.33	394.48	70.53	285.13	8.378	.87034
1546		381.44	62.48	275.45	9.167	.86204	1596		394.75	70.70	285.33	8.363	.87051
1547		381.70	62.64	275.64	9.150	.86221	1597		395.01	70.87	285.53	8.348	.87068
1548		381.97	62.80	275.84	9.133	.86238	1598		395.28	71.05	285.73	8.333	.87084
1549		382.23	62.95	276.04	9.116	.86255	1599		395.55	71.22	285.92	8.318	.87101

Table 1 Air at Low Pressures (for One Pound)

T	t	h	p_r	u	v_r	ϕ	T	t	h	p_r	u	v_r	ϕ
1600	1140.33	395.81	71.39	286.12	8.303	.87118	1650	1190.33	409.20	80.51	296.08	7.593	.87942
1601		396.08	71.57	286.32	8.288	.87134	1651		409.47	80.70	296.28	7.579	.87958
1602		396.35	71.74	286.52	8.273	.87151	1652		409.74	80.89	296.48	7.566	.87974
1603		396.61	71.92	286.72	8.258	.87168	1653		410.01	81.09	296.68	7.552	.87990
1604		396.88	72.09	286.92	8.243	.87184	1654		410.27	81.28	296.88	7.539	.88007
1605	1145.33	397.15	72.27	287.12	8.228	.87201	1655	1195.33	410.54	81.47	297.08	7.526	.88023
1606		397.42	72.44	287.31	8.213	.87218	1656		410.81	81.66	297.28	7.513	.88039
1607		397.68	72.62	287.51	8.198	.87234	1657		411.08	81.86	297.48	7.499	.88055
1608		397.95	72.79	287.71	8.184	.87251	1658		411.35	82.05	297.68	7.486	.88071
1609		398.22	72.97	287.91	8.169	.87268	1659		411.62	82.24	297.88	7.473	.88088
1610	1150.33	398.49	73.15	288.11	8.154	.87284	1660	1200.33	411.89	82.44	298.08	7.460	.88104
1611		398.75	73.33	288.31	8.139	.87301	1661		412.15	82.63	298.28	7.447	.88120
1612		399.02	73.50	288.51	8.125	.87317	1662		412.42	82.83	298.48	7.434	.88136
1613		399.29	73.68	288.71	8.110	.87334	1663		412.69	83.02	298.68	7.421	.88152
1614		399.55	73.86	288.90	8.096	.87351	1664		412.96	83.22	298.88	7.408	.88169
1615	1155.33	399.82	74.04	289.10	8.081	.87367	1665	1205.33	413.23	83.42	299.08	7.395	.88185
1616		400.09	74.22	289.30	8.067	.87384	1666		413.50	83.61	299.28	7.382	.88201
1617		400.36	74.40	289.50	8.052	.87400	1667		413.77	83.81	299.48	7.369	.88217
1618		400.62	74.58	289.70	8.038	.87417	1668		414.04	84.01	299.68	7.356	.88233
1619		400.89	74.76	289.90	8.023	.87433	1669		414.31	84.21	299.89	7.343	.88249
1620	1160.33	401.16	74.94	290.10	8.009	.87450	1670	1210.33	414.57	84.40	300.09	7.330	.88265
1621		401.43	75.12	290.30	7.995	.87466	1671		414.84	84.60	300.29	7.317	.88281
1622		401.69	75.30	290.50	7.980	.87483	1672		415.11	84.80	300.49	7.305	.88298
1623		401.96	75.48	290.70	7.966	.87499	1673		415.38	85.00	300.69	7.292	.88314
1624		402.23	75.66	290.90	7.952	.87516	1674		415.65	85.20	300.89	7.279	.88330
1625	1165.33	402.50	75.84	291.09	7.938	.87532	1675	1215.33	415.92	85.40	301.09	7.266	.88346
1626		402.77	76.03	291.29	7.923	.87549	1676		416.19	85.60	301.29	7.254	.88362
1627		403.03	76.21	291.49	7.909	.87565	1677		416.46	85.80	301.49	7.241	.88378
1628		403.30	76.39	291.69	7.895	.87582	1678		416.73	86.00	301.69	7.228	.88394
1629		403.57	76.58	291.89	7.881	.87598	1679		417.00	86.20	301.89	7.216	.88410
1630	1170.33	403.84	76.76	292.09	7.867	.87615	1680	1220.33	417.27	86.41	302.09	7.203	.88426
1631		404.11	76.94	292.29	7.853	.87631	1681		417.54	86.61	302.29	7.191	.88442
1632		404.37	77.13	292.49	7.839	.87647	1682		417.80	86.81	302.49	7.178	.88458
1633		404.64	77.31	292.69	7.825	.87664	1683		418.07	87.01	302.69	7.166	.88474
1634		404.91	77.50	292.89	7.811	.87680	1684		418.34	87.22	302.89	7.153	.88490
1635	1175.33	405.18	77.68	293.09	7.797	.87697	1685	1225.33	418.61	87.42	303.10	7.141	.88506
1636		405.44	77.87	293.29	7.783	.87713	1686		418.88	87.62	303.30	7.128	.88522
1637		405.71	78.06	293.49	7.770	.87729	1687		419.15	87.83	303.50	7.116	.88538
1638		405.98	78.24	293.69	7.756	.87746	1688		419.42	88.03	303.70	7.104	.88554
1639		406.25	78.43	293.89	7.742	.87762	1689		419.69	88.24	303.90	7.091	.88570
1640	1180.33	406.52	78.62	294.09	7.728	.87778	1690	1230.33	419.96	88.44	304.10	7.079	.88586
1641		406.79	78.81	294.28	7.715	.87795	1691		420.23	88.65	304.30	7.067	.88602
1642		407.05	78.99	294.48	7.701	.87811	1692		420.50	88.86	304.50	7.055	.88618
1643		407.32	79.18	294.68	7.687	.87827	1693		420.77	89.06	304.70	7.042	.88634
1644		407.59	79.37	294.88	7.674	.87844	1694		421.04	89.27	304.90	7.030	.88650
1645	1185.33	407.86	79.56	295.08	7.660	.87860	1695	1235.33	421.31	89.48	305.10	7.018	.88666
1646		408.13	79.75	295.28	7.647	.87876	1696		421.58	89.69	305.31	7.006	.88681
1647		408.40	79.94	295.48	7.633	.87893	1697		421.85	89.89	305.51	6.994	.88697
1648		408.66	80.13	295.68	7.619	.87909	1698		422.12	90.10	305.71	6.982	.88713
1649		408.93	80.32	295.88	7.606	.87925	1699		422.39	90.31	305.91	6.970	.88729

Table 1 Air at Low Pressures (for One Pound)

T	t	h	p_r	u	v_r	ϕ	T	t	h	p_r	u	v_r	ϕ
1700	1240.33	422.66	90.52	306.11	6.958	.88745	1750	1290.33	436.18	101.49	316.21	6.388	.89529
1701		422.93	90.73	306.31	6.946	.88761	1751		436.45	101.72	316.41	6.378	.89544
1702		423.20	90.94	306.51	6.934	.88777	1752		436.72	101.95	316.61	6.367	.89560
1703		423.47	91.15	306.71	6.922	.88793	1753		436.99	102.18	316.81	6.356	.89575
1704		423.74	91.36	306.92	6.910	.88808	1754		437.26	102.41	317.02	6.345	.89591
1705	1245.33	424.01	91.57	307.12	6.898	.88824	1755	1295.33	437.54	102.64	317.22	6.335	.89606
1706		424.28	91.78	307.32	6.886	.88840	1756		437.81	102.87	317.42	6.324	.89622
1707		424.55	92.00	307.52	6.874	.88856	1757		438.08	103.10	317.62	6.313	.89637
1708		424.82	92.21	307.72	6.862	.88872	1758		438.35	103.33	317.83	6.303	.89653
1709		425.09	92.42	307.92	6.851	.88887	1759		438.62	103.57	318.03	6.292	.89668
1710	1250.33	425.36	92.63	308.12	6.839	.88903	1760	1300.33	438.89	103.80	318.23	6.282	.89684
1711		425.63	92.85	308.33	6.827	.88919	1761		439.16	104.03	318.44	6.271	.89699
1712		425.90	93.06	308.53	6.815	.88935	1762		439.43	104.27	318.64	6.261	.89714
1713		426.17	93.28	308.73	6.804	.88951	1763		439.71	104.50	318.84	6.250	.89730
1714		426.44	93.49	308.93	6.792	.88966	1764		439.98	104.74	319.04	6.240	.89745
1715	1255.33	426.71	93.71	309.13	6.780	.88982	1765	1305.33	440.25	104.97	319.25	6.229	.89761
1716		426.98	93.92	309.33	6.769	.88998	1766		440.52	105.21	319.45	6.219	.89776
1717		427.25	94.14	309.54	6.757	.89014	1767		440.79	105.45	319.65	6.208	.89791
1718		427.52	94.35	309.74	6.746	.89029	1768		441.06	105.68	319.86	6.198	.89807
1719		427.79	94.57	309.94	6.734	.89045	1769		441.34	105.92	320.06	6.187	.89822
1720	1260.33	428.06	94.79	310.14	6.723	.89061	1770	1310.33	441.61	106.16	320.26	6.177	.89837
1721		428.33	95.01	310.34	6.711	.89077	1771		441.88	106.39	320.47	6.167	.89853
1722		428.60	95.22	310.54	6.700	.89092	1772		442.15	106.63	320.67	6.156	.89868
1723		428.87	95.44	310.75	6.688	.89108	1773		442.42	106.87	320.87	6.146	.89883
1724		429.14	95.66	310.95	6.677	.89124	1774		442.69	107.11	321.07	6.136	.89899
1725	1265.33	429.41	95.88	311.15	6.665	.89139	1775	1315.33	442.97	107.35	321.28	6.126	.89914
1726		429.68	96.10	311.35	6.654	.89155	1776		443.24	107.59	321.48	6.115	.89929
1727		429.95	96.32	311.55	6.643	.89171	1777		443.51	107.83	321.68	6.105	.89945
1728		430.22	96.54	311.76	6.631	.89186	1778		443.78	108.07	321.89	6.095	.89960
1729		430.49	96.76	311.96	6.620	.89202	1779		444.05	108.31	322.09	6.085	.89975
1730	1270.33	430.76	96.98	312.16	6.609	.89218	1780	1320.33	444.32	108.55	322.29	6.075	.89990
1731		431.03	97.20	312.36	6.597	.89233	1781		444.60	108.80	322.50	6.065	.90006
1732		431.30	97.42	312.56	6.586	.89249	1782		444.87	109.04	322.70	6.055	.90021
1733		431.57	97.65	312.77	6.575	.89265	1783		445.14	109.28	322.90	6.045	.90036
1734		431.84	97.87	312.97	6.564	.89280	1784		445.41	109.52	323.11	6.035	.90051
1735	1275.33	432.12	98.09	313.17	6.553	.89296	1785	1325.33	445.68	109.77	323.31	6.024	.90067
1736		432.39	98.32	313.37	6.542	.89311	1786		445.96	110.01	323.51	6.014	.90082
1737		432.66	98.54	313.57	6.530	.89327	1787		446.23	110.26	323.72	6.004	.90097
1738		432.93	98.76	313.78	6.519	.89343	1788		446.50	110.50	323.92	5.995	.90112
1739		433.20	98.99	313.98	6.508	.89358	1789		446.77	110.75	324.13	5.985	.90128
1740	1280.33	433.47	99.21	314.18	6.497	.89374	1790	1330.33	447.04	110.99	324.33	5.975	.90143
1741		433.74	99.44	314.38	6.486	.89389	1791		447.32	111.24	324.53	5.965	.90158
1742		434.01	99.67	314.59	6.475	.89405	1792		447.59	111.49	324.74	5.955	.90173
1743		434.28	99.89	314.79	6.464	.89420	1793		447.86	111.73	324.94	5.945	.90188
1744		434.55	100.12	314.99	6.453	.89436	1794		448.13	111.98	325.14	5.935	.90204
1745	1285.33	434.82	100.35	315.19	6.443	.89451	1795	1335.33	448.41	112.23	325.35	5.925	.90219
1746		435.10	100.57	315.40	6.432	.89467	1796		448.68	112.48	325.55	5.916	.90234
1747		435.37	100.80	315.60	6.421	.89482	1797		448.95	112.73	325.75	5.906	.90249
1748		435.64	101.03	315.80	6.410	.89498	1798		449.22	112.98	325.96	5.896	.90264
1749		435.91	101.26	316.00	6.399	.89513	1799		449.49	113.23	326.16	5.886	.90279

Table 1 Air at Low Pressures (for One Pound)

T	t	h	p_r	u	v_r	ϕ	T	t	h	p_r	u	v_r	ϕ
1800	1340.33	449.77	113.48	326.37	5.877	.90294	**1850**	1390.33	463.42	126.56	336.59	5.415	.91043
1801		450.04	113.73	326.57	5.867	.90310	**1851**		463.69	126.83	336.79	5.407	.91057
1802		450.31	113.98	326.77	5.857	.90325	**1852**		463.96	127.11	337.00	5.398	.91072
1803		450.58	114.23	326.98	5.848	.90340	**1853**		464.24	127.38	337.20	5.389	.91087
1804		450.86	114.48	327.18	5.838	.90355	**1854**		464.51	127.65	337.41	5.381	.91102
1805	1345.33	451.13	114.73	327.39	5.828	.90370	**1855**	1395.33	464.79	127.93	337.61	5.372	.91116
1806		451.40	114.99	327.59	5.819	.90385	**1856**		465.06	128.21	337.82	5.363	.91131
1807		451.67	115.24	327.79	5.809	.90400	**1857**		465.33	128.48	338.02	5.355	.91146
1808		451.95	115.49	328.00	5.800	.90415	**1858**		465.61	128.76	338.23	5.346	.91161
1809		452.22	115.75	328.20	5.790	.90430	**1859**		465.88	129.04	338.44	5.337	.91175
1810	1350.33	452.49	116.00	328.41	5.781	.90445	**1860**	1400.33	466.16	129.31	338.64	5.329	.91190
1811		452.76	116.26	328.61	5.771	.90461	**1861**		466.43	129.59	338.85	5.320	.91205
1812		453.04	116.51	328.81	5.762	.90476	**1862**		466.70	129.87	339.05	5.312	.91220
1813		453.31	116.77	329.02	5.752	.90491	**1863**		466.98	130.15	339.26	5.303	.91234
1814		453.58	117.03	329.22	5.743	.90506	**1864**		467.25	130.43	339.46	5.295	.91249
1815	1355.33	453.86	117.28	329.43	5.733	.90521	**1865**	1405.33	467.52	130.71	339.67	5.286	.91264
1816		454.13	117.54	329.63	5.724	.90536	**1866**		467.80	130.99	339.87	5.278	.91278
1817		454.40	117.80	329.83	5.714	.90551	**1867**		468.07	131.27	340.08	5.269	.91293
1818		454.67	118.06	330.04	5.705	.90566	**1868**		468.35	131.55	340.28	5.261	.91308
1819		454.95	118.32	330.24	5.696	.90581	**1869**		468.62	131.83	340.49	5.252	.91322
1820	1360.33	455.22	118.57	330.45	5.686	.90596	**1870**	1410.33	468.89	132.11	340.69	5.244	.91337
1821		455.49	118.83	330.65	5.677	.90611	**1871**		469.17	132.40	340.90	5.235	.91352
1822		455.77	119.09	330.86	5.668	.90626	**1872**		469.44	132.68	341.11	5.227	.91366
1823		456.04	119.35	331.06	5.659	.90641	**1873**		469.72	132.96	341.31	5.219	.91381
1824		456.31	119.62	331.27	5.649	.90656	**1874**		469.99	133.25	341.52	5.210	.91396
1825	1365.33	456.58	119.88	331.47	5.640	.90671	**1875**	1415.33	470.27	133.53	341.72	5.202	.91410
1826		456.86	120.14	331.67	5.631	.90686	**1876**		470.54	133.82	341.93	5.194	.91425
1827		457.13	120.40	331.88	5.622	.90701	**1877**		470.81	134.10	342.13	5.185	.91439
1828		457.40	120.66	332.08	5.613	.90715	**1878**		471.09	134.39	342.34	5.177	.91454
1829		457.68	120.93	332.29	5.603	.90730	**1879**		471.36	134.68	342.55	5.169	.91469
1830	1370.33	457.95	121.19	332.49	5.594	.90745	**1880**	1420.33	471.64	134.96	342.75	5.161	.91483
1831		458.22	121.45	332.70	5.585	.90760	**1881**		471.91	135.25	342.96	5.152	.91498
1832		458.50	121.72	332.90	5.576	.90775	**1882**		472.19	135.54	343.16	5.144	.91512
1833		458.77	121.98	333.11	5.567	.90790	**1883**		472.46	135.83	343.37	5.136	.91527
1834		459.04	122.25	333.31	5.558	.90805	**1884**		472.73	136.12	343.57	5.128	.91542
1835	1375.33	459.32	122.52	333.52	5.549	.90820	**1885**	1425.33	473.01	136.40	343.78	5.120	.91556
1836		459.59	122.78	333.72	5.540	.90835	**1886**		473.28	136.69	343.99	5.112	.91571
1837		459.86	123.05	333.92	5.531	.90850	**1887**		473.56	136.99	344.19	5.103	.91585
1838		460.14	123.32	334.13	5.522	.90865	**1888**		473.83	137.28	344.40	5.095	.91600
1839		460.41	123.58	334.33	5.513	.90879	**1889**		474.11	137.57	344.60	5.087	.91614
1840	1380.33	460.68	123.85	334.54	5.504	.90894	**1890**	1430.33	474.38	137.86	344.81	5.079	.91629
1841		460.96	124.12	334.74	5.495	.90909	**1891**		474.66	138.15	345.02	5.071	.91643
1842		461.23	124.39	334.95	5.486	.90924	**1892**		474.93	138.44	345.22	5.063	.91658
1843		461.50	124.66	335.15	5.477	.90939	**1893**		475.21	138.74	345.43	5.055	.91672
1844		461.78	124.93	335.36	5.468	.90954	**1894**		475.48	139.03	345.63	5.047	.91687
1845	1385.33	462.05	125.20	335.56	5.459	.90968	**1895**	1435.33	475.75	139.33	345.84	5.039	.91701
1846		462.32	125.47	335.77	5.451	.90983	**1896**		476.03	139.62	346.05	5.031	.91716
1847		462.60	125.74	335.97	5.442	.90998	**1897**		476.30	139.92	346.25	5.023	.91730
1848		462.87	126.01	336.18	5.433	.91013	**1898**		476.58	140.21	346.46	5.015	.91745
1849		463.14	126.29	336.38	5.424	.91028	**1899**		476.85	140.51	346.67	5.007	.91759

Table 1   Air at Low Pressures (for One Pound)

T	t	h	p_r	u	v_r	φ	T	t	h	p_r	u	v_r	φ
1900	1440.33	477.13	140.81	346.87	4.999	.91774	1950	1490.33	490.90	156.29	357.21	4.622	.92489
1901		477.40	141.10	347.08	4.991	.91788	1951		491.17	156.61	357.42	4.615	.92503
1902		477.68	141.40	347.28	4.983	.91803	1952		491.45	156.94	357.63	4.608	.92517
1903		477.95	141.70	347.49	4.975	.91817	1953		491.73	157.26	357.84	4.601	.92532
1904		478.23	142.00	347.70	4.968	.91832	1954		492.00	157.59	358.04	4.594	.92546
1905	1445.33	478.50	142.30	347.90	4.960	.91846	1955	1495.33	492.28	157.91	358.25	4.587	.92560
1906		478.78	142.60	348.11	4.952	.91860	1956		492.55	158.24	358.46	4.580	.92574
1907		479.05	142.90	348.32	4.944	.91875	1957		492.83	158.56	358.67	4.572	.92588
1908		479.33	143.20	348.52	4.936	.91889	1958		493.11	158.89	358.87	4.565	.92602
1909		479.60	143.50	348.73	4.929	.91904	1959		493.38	159.22	359.08	4.558	.92616
1910	1450.33	479.88	143.80	348.94	4.921	.91918	1960	1500.33	493.66	159.54	359.29	4.551	.92630
1911		480.15	144.10	349.14	4.913	.91933	1961		493.94	159.87	359.50	4.544	.92644
1912		480.43	144.41	349.35	4.905	.91947	1962		494.21	160.20	359.70	4.537	.92659
1913		480.70	144.71	349.56	4.898	.91961	1963		494.49	160.53	359.91	4.530	.92673
1914		480.98	145.01	349.76	4.890	.91976	1964		494.76	160.86	360.12	4.523	.92687
1915	1455.33	481.25	145.32	349.97	4.882	.91990	1965	1505.33	495.04	161.19	360.33	4.516	.92701
1916		481.53	145.62	350.18	4.874	.92004	1966		495.32	161.52	360.54	4.509	.92715
1917		481.80	145.93	350.38	4.867	.92019	1967		495.59	161.85	360.74	4.502	.92729
1918		482.08	146.23	350.59	4.859	.92033	1968		495.87	162.18	360.95	4.495	.92743
1919		482.35	146.54	350.80	4.852	.92047	1969		496.15	162.52	361.16	4.489	.92757
1920	1460.33	482.63	146.85	351.00	4.844	.92062	1970	1510.33	496.42	162.85	361.37	4.482	.92771
1921		482.90	147.15	351.21	4.836	.92076	1971		496.70	163.18	361.57	4.475	.92785
1922		483.18	147.46	351.42	4.829	.92090	1972		496.98	163.52	361.78	4.468	.92799
1923		483.46	147.77	351.62	4.821	.92105	1973		497.25	163.85	361.99	4.461	.92813
1924		483.73	148.08	351.83	4.814	.92119	1974		497.53	164.19	362.20	4.454	.92827
1925	1465.33	484.01	148.39	352.04	4.806	.92133	1975	1515.33	497.80	164.52	362.41	4.447	.92841
1926		484.28	148.70	352.24	4.799	.92148	1976		498.08	164.86	362.61	4.440	.92855
1927		484.56	149.01	352.45	4.791	.92162	1977		498.36	165.20	362.82	4.434	.92869
1928		484.83	149.32	352.66	4.784	.92176	1978		498.63	165.53	363.03	4.427	.92883
1929		485.11	149.63	352.86	4.776	.92191	1979		498.91	165.87	363.24	4.420	.92897
1930	1470.33	485.38	149.94	353.07	4.769	.92205	1980	1520.33	499.19	166.21	363.45	4.413	.92911
1931		485.66	150.26	353.28	4.761	.92219	1981		499.46	166.55	363.65	4.407	.92925
1932		485.93	150.57	353.48	4.754	.92233	1982		499.74	166.89	363.86	4.400	.92939
1933		486.21	150.88	353.69	4.746	.92248	1983		500.02	167.23	364.07	4.393	.92953
1934		486.49	151.20	353.90	4.739	.92262	1984		500.29	167.57	364.28	4.386	.92967
1935	1475.33	486.76	151.51	354.11	4.731	.92276	1985	1525.33	500.57	167.91	364.49	4.380	.92981
1936		487.04	151.83	354.31	4.724	.92290	1986		500.85	168.25	364.70	4.373	.92995
1937		487.31	152.14	354.52	4.717	.92305	1987		501.12	168.59	364.90	4.366	.93009
1938		487.59	152.46	354.73	4.709	.92319	1988		501.40	168.94	365.11	4.360	.93023
1939		487.86	152.77	354.93	4.702	.92333	1989		501.68	169.28	365.32	4.353	.93036
1940	1480.33	488.14	153.09	355.14	4.695	.92347	1990	1530.33	501.95	169.63	365.53	4.346	.93050
1941		488.42	153.41	355.35	4.687	.92362	1991		502.23	169.97	365.74	4.340	.93064
1942		488.69	153.73	355.56	4.680	.92376	1992		502.51	170.31	365.94	4.333	.93078
1943		488.97	154.05	355.76	4.673	.92390	1993		502.79	170.66	366.15	4.326	.93092
1944		489.24	154.37	355.97	4.666	.92404	1994		503.06	171.01	366.36	4.320	.93106
1945	1485.33	489.52	154.68	356.18	4.658	.92418	1995	1535.33	503.34	171.35	366.57	4.313	.93120
1946		489.79	155.01	356.38	4.651	.92432	1996		503.62	171.70	366.78	4.307	.93134
1947		490.07	155.33	356.59	4.644	.92447	1997		503.89	172.05	366.99	4.300	.93148
1948		490.35	155.65	356.80	4.637	.92461	1998		504.17	172.40	367.20	4.294	.93162
1949		490.62	155.97	357.01	4.629	.92475	1999		504.45	172.75	367.40	4.287	.93175

Table 1   Air at Low Pressures (for One Pound)

T	t	h	p_r	u	v_r	ϕ	T	t	h	p_r	u	v_r	ϕ
2000	1540.33	504.72	173.10	367.61	4.281	.93189	2050	1590.33	518.61	191.30	378.06	3.970	.93875
2001		505.00	173.45	367.82	4.274	.93203	2051		518.88	191.68	378.27	3.964	.93888
2002		505.28	173.80	368.03	4.268	.93217	2052		519.16	192.06	378.48	3.958	.93902
2003		505.56	174.15	368.24	4.261	.93231	2053		519.44	192.44	378.69	3.952	.93915
2004		505.83	174.50	368.45	4.255	.93245	2054		519.72	192.82	378.90	3.947	.93929
2005	1545.33	506.11	174.85	368.66	4.248	.93258	2055	1595.33	520.00	193.20	379.11	3.941	.93943
2006		506.39	175.20	368.86	4.242	.93272	2056		520.27	193.58	379.32	3.935	.93956
2007		506.66	175.56	369.07	4.235	.93286	2057		520.55	193.96	379.53	3.929	.93970
2008		506.94	175.91	369.28	4.229	.93300	2058		520.83	194.35	379.74	3.923	.93983
2009		507.22	176.27	369.49	4.223	.93314	2059		521.11	194.73	379.95	3.917	.93997
2010	1550.33	507.50	176.62	369.70	4.216	.93327	→2060	1600.33	521.39	195.11	380.16	3.911	.94010
2011		507.77	176.98	369.91	4.210	.93341	2061		521.67	195.50	380.37	3.906	.94024
2012		508.05	177.33	370.12	4.203	.93355	2062		521.94	195.88	380.58	3.900	.94037
2013		508.33	177.69	370.32	4.197	.93369	2063		522.22	196.27	380.79	3.894	.94051
2014		508.61	178.05	370.53	4.191	.93383	2064		522.50	196.66	381.00	3.888	.94064
2015	1555.33	508.88	178.41	370.74	4.184	.93396	2065	1605.33	522.78	197.04	381.21	3.883	.94078
2016		509.16	178.76	370.95	4.178	.93410	2066		523.06	197.43	381.42	3.877	.94091
2017		509.44	179.12	371.16	4.172	.93424	2067		523.34	197.82	381.63	3.871	.94105
2018		509.72	179.48	371.37	4.165	.93438	2068		523.62	198.21	381.84	3.865	.94118
2019		509.99	179.84	371.58	4.159	.93451	2069		523.89	198.60	382.05	3.860	.94132
2020	1560.33	510.27	180.20	371.79	4.153	.93465	2070	1610.33	524.17	198.99	382.26	3.854	.94145
2021		510.55	180.56	372.00	4.147	.93479	2071		524.45	199.38	382.47	3.848	.94158
2022		510.83	180.93	372.20	4.140	.93493	2072		524.73	199.77	382.68	3.843	.94172
2023		511.10	181.29	372.41	4.134	.93506	2073		525.01	200.16	382.89	3.837	.94185
2024		511.38	181.65	372.62	4.128	.93520	2074		525.29	200.56	383.10	3.831	.94199
2025	1565.33	511.66	182.02	372.83	4.122	.93534	2075	1615.33	525.57	200.95	383.31	3.826	.94212
2026		511.94	182.38	373.04	4.115	.93547	2076		525.84	201.34	383.52	3.820	.94226
2027		512.21	182.75	373.25	4.109	.93561	2077		526.12	201.74	383.73	3.814	.94239
2028		512.49	183.11	373.46	4.103	.93575	2078		526.40	202.13	383.94	3.809	.94252
2029		512.77	183.48	373.67	4.097	.93589	2079		526.68	202.53	384.15	3.803	.94266
2030	1570.33	513.05	183.84	373.88	4.091	.93602	2080	1620.33	526.96	202.93	384.36	3.797	.94279
2031		513.32	184.21	374.09	4.085	.93616	2081		527.24	203.32	384.57	3.792	.94293
2032		513.60	184.58	374.30	4.079	.93630	2082		527.52	203.72	384.78	3.786	.94306
2033		513.88	184.95	374.51	4.072	.93643	2083		527.80	204.12	384.99	3.781	.94319
2034		514.16	185.32	374.71	4.066	.93657	2084		528.07	204.52	385.20	3.775	.94333
2035	1575.33	514.44	185.69	374.92	4.060	.93671	2085	1625.33	528.35	204.92	385.41	3.770	.94346
2036		514.71	186.06	375.13	4.054	.93684	2086		528.63	205.32	385.62	3.764	.94360
2037		514.99	186.43	375.34	4.048	.93698	2087		528.91	205.72	385.83	3.758	.94373
2038		515.27	186.80	375.55	4.042	.93712	2088		529.19	206.12	386.04	3.753	.94386
2039		515.55	187.17	375.76	4.036	.93725	2089		529.47	206.52	386.25	3.747	.94400
2040	1580.33	515.82	187.54	375.97	4.030	.93739	2090	1630.33	529.75	206.92	386.47	3.742	.94413
2041		516.10	187.91	376.18	4.024	.93752	2091		530.03	207.33	386.68	3.736	.94426
2042		516.38	188.29	376.39	4.018	.93766	2092		530.31	207.73	386.89	3.731	.94440
2043		516.66	188.66	376.60	4.012	.93780	2093		530.58	208.14	387.10	3.725	.94453
2044		516.94	189.04	376.81	4.006	.93793	2094		530.86	208.54	387.31	3.720	.94466
2045	1585.33	517.21	189.41	377.02	4.000	.93807	2095	1635.33	531.14	208.95	387.52	3.715	.94480
2046		517.49	189.79	377.23	3.994	.93820	2096		531.42	209.35	387.73	3.709	.94493
2047		517.77	190.16	377.44	3.988	.93834	2097		531.70	209.76	387.94	3.704	.94506
2048		518.05	190.54	377.65	3.982	.93848	2098		531.98	210.17	388.15	3.698	.94520
2049		518.33	190.92	377.86	3.976	.93861	2099		532.26	210.57	388.36	3.693	.94533

Table 1   Air at Low Pressures (for One Pound)

T	t	h	p_r	u	v_r	ϕ	T	t	h	p_r	u	v_r	ϕ
2100	1640.33	532.54	211.0	388.57	3.687	.94546	2150	1690.33	546.52	232.2	399.13	3.430	.95204
2101		532.82	211.4	388.78	3.682	.94560	2151		546.80	232.7	399.34	3.425	.95217
2102		533.10	211.8	388.99	3.677	.94573	2152		547.08	233.1	399.55	3.420	.95230
2103		533.38	212.2	389.20	3.671	.94586	2153		547.36	233.6	399.76	3.415	.95243
2104		533.66	212.6	389.41	3.666	.94599	2154		547.64	234.0	399.97	3.410	.95256
2105	1645.33	533.93	213.0	389.62	3.661	.94613	2155	1695.33	547.92	234.5	400.18	3.405	.95269
2106		534.21	213.4	389.83	3.655	.94626	2156		548.20	234.9	400.40	3.400	.95282
2107		534.49	213.9	390.05	3.650	.94639	2157		548.48	235.3	400.61	3.395	.95295
2108		534.77	214.3	390.26	3.645	.94652	2158		548.76	235.8	400.82	3.391	.95308
2109		535.05	214.7	390.47	3.639	.94666	2159		549.04	236.2	401.03	3.386	.95321
2110	1650.33	535.33	215.1	390.68	3.634	.94679	2160	1700.33	549.32	236.7	401.24	3.381	.95334
2111		535.61	215.5	390.89	3.629	.94692	2161		549.61	237.1	401.46	3.376	.95347
2112		535.89	215.9	391.10	3.623	.94705	2162		549.89	237.6	401.67	3.371	.95360
2113		536.17	216.4	391.31	3.618	.94719	2163		550.17	238.0	401.88	3.366	.95373
2114		536.45	216.8	391.52	3.613	.94732	2164		550.45	238.5	402.09	3.362	.95386
2115	1655.33	536.73	217.2	391.73	3.608	.94745	2165	1705.33	550.73	238.9	402.30	3.357	.95399
2116		537.01	217.6	391.94	3.602	.94758	2166		551.01	239.4	402.51	3.352	.95412
2117		537.29	218.0	392.15	3.597	.94771	2167		551.29	239.8	402.73	3.347	.95425
2118		537.57	218.5	392.36	3.592	.94785	2168		551.57	240.3	402.94	3.343	.95438
2119		537.85	218.9	392.58	3.587	.94798	2169		551.85	240.7	403.15	3.338	.95451
2120	1660.33	538.13	219.3	392.79	3.582	.94811	2170	1710.33	552.13	241.2	403.36	3.333	.95464
2121		538.41	219.7	393.00	3.576	.94824	2171		552.41	241.7	403.57	3.328	.95477
2122		538.69	220.1	393.21	3.571	.94837	2172		552.69	242.1	403.79	3.324	.95490
2123		538.96	220.6	393.42	3.566	.94851	2173		552.97	242.6	404.00	3.319	.95503
2124		539.24	221.0	393.63	3.561	.94864	2174		553.25	243.0	404.21	3.314	.95516
2125	1665.33	539.52	221.4	393.84	3.556	.94877	2175	1715.33	553.53	243.5	404.42	3.309	.95529
2126		539.80	221.8	394.05	3.551	.94890	2176		553.81	243.9	404.63	3.305	.95541
2127		540.08	222.3	394.26	3.545	.94903	2177		554.09	244.4	404.85	3.300	.95554
2128		540.36	222.7	394.48	3.540	.94916	2178		554.37	244.9	405.06	3.295	.95567
2129		540.64	223.1	394.69	3.535	.94930	2179		554.66	245.3	405.27	3.291	.95580
2130	1670.33	540.92	223.5	394.90	3.530	.94943	2180	1720.33	554.94	245.8	405.48	3.286	.95593
2131		541.20	224.0	395.11	3.525	.94956	2181		555.22	246.2	405.70	3.281	.95606
2132		541.48	224.4	395.32	3.520	.94969	2182		555.50	246.7	405.91	3.277	.95619
2133		541.76	224.8	395.53	3.515	.94982	2183		555.78	247.2	406.12	3.272	.95632
2134		542.04	225.3	395.74	3.510	.94995	2184		556.06	247.6	406.33	3.267	.95644
2135	1675.33	542.32	225.7	395.95	3.505	.95008	2185	1725.33	556.34	248.1	406.54	3.263	.95657
2136		542.60	226.1	396.17	3.500	.95021	2186		556.62	248.6	406.76	3.258	.95670
2137		542.88	226.6	396.38	3.494	.95034	2187		556.90	249.0	406.97	3.253	.95683
2138		543.16	227.0	396.59	3.489	.95048	2188		557.18	249.5	407.18	3.249	.95696
2139		543.44	227.4	396.80	3.484	.95061	2189		557.46	250.0	407.39	3.244	.95709
2140	1680.33	543.72	227.9	397.01	3.479	.95074	2190	1730.33	557.74	250.4	407.61	3.240	.95721
2141		544.00	228.3	397.22	3.474	.95087	2191		558.03	250.9	407.82	3.235	.95734
2142		544.28	228.7	397.43	3.469	.95100	2192		558.31	251.4	408.03	3.231	.95747
2143		544.56	229.2	397.65	3.464	.95113	2193		558.59	251.8	408.24	3.226	.95760
2144		544.84	229.6	397.86	3.459	.95126	2194		558.87	252.3	408.46	3.221	.95773
2145	1685.33	545.12	230.0	398.07	3.454	.95139	2195	1735.33	559.15	252.8	408.67	3.217	.95786
2146		545.40	230.5	398.28	3.449	.95152	2196		559.43	253.3	408.88	3.212	.95798
2147		545.68	230.9	398.49	3.445	.95165	2197		559.71	253.7	409.09	3.208	.95811
2148		545.96	231.4	398.70	3.440	.95178	2198		559.99	254.2	409.31	3.203	.95824
2149		546.24	231.8	398.91	3.435	.95191	2199		560.27	254.7	409.52	3.199	.95837

Table 1 Air at Low Pressures (for One Pound)

T	t	h	p_r	u	v_r	φ	T	t	h	p_r	u	v_r	φ
2200	1740.33	560.55	255.2	409.73	3.194	.95850	2250	1790.33	574.63	279.8	420.38	2.979	.96482
2201		560.84	255.6	409.94	3.190	.95862	2251		574.92	280.3	420.60	2.975	.96495
2202		561.12	256.1	410.16	3.185	.95875	2252		575.20	280.9	420.81	2.971	.96507
2203		561.40	256.6	410.37	3.181	.95888	2253		575.48	281.4	421.02	2.966	.96520
2204		561.68	257.1	410.58	3.176	.95901	2254		575.76	281.9	421.24	2.962	.96532
2205	1745.33	561.96	257.5	410.79	3.172	.95913	2255	1795.33	576.05	282.4	421.45	2.958	.96545
2206		562.24	258.0	411.01	3.167	.95926	2256		576.33	282.9	421.66	2.954	.96557
2207		562.52	258.5	411.22	3.163	.95939	2257		576.61	283.4	421.88	2.950	.96570
2208		562.80	259.0	411.43	3.158	.95952	2258		576.89	284.0	422.09	2.946	.96582
2209		563.09	259.5	411.65	3.154	.95964	2259		577.17	284.5	422.31	2.942	.96595
2210	1750.33	563.37	260.0	411.86	3.150	.95977	2260	1800.33	577.46	285.0	422.52	2.938	.96607
2211		563.65	260.4	412.07	3.145	.95990	2261		577.74	285.5	422.73	2.934	.96620
2212		563.93	260.9	412.28	3.141	.96003	2262		578.02	286.0	422.95	2.930	.96632
2213		564.21	261.4	412.50	3.136	.96015	2263		578.30	286.5	423.16	2.926	.96645
2214		564.49	261.9	412.71	3.132	.96028	2264		578.59	287.1	423.37	2.922	.96657
2215	1755.33	564.77	262.4	412.92	3.128	.96041	2265	1805.33	578.87	287.6	423.59	2.918	.96670
2216		565.06	262.9	413.13	3.123	.96053	2266		579.15	288.1	423.80	2.914	.96682
2217		565.34	263.3	413.35	3.119	.96066	2267		579.43	288.6	424.02	2.910	.96695
2218		565.62	263.8	413.56	3.114	.96079	2268		579.71	289.2	424.23	2.906	.96707
2219		565.90	264.3	413.77	3.110	.96091	2269		580.00	289.7	424.44	2.902	.96720
2220	1760.33	566.18	264.8	413.99	3.106	.96104	2270	1810.33	580.28	290.2	424.66	2.898	.96732
2221		566.46	265.3	414.20	3.101	.96117	2271		580.56	290.7	424.87	2.894	.96745
2222		566.74	265.8	414.41	3.097	.96129	2272		580.84	291.3	425.08	2.890	.96757
2223		567.03	266.3	414.63	3.093	.96142	2273		581.13	291.8	425.30	2.886	.96769
2224		567.31	266.8	414.84	3.088	.96155	2274		581.41	292.3	425.51	2.882	.96782
2225	1765.33	567.59	267.3	415.05	3.084	.96167	2275	1815.33	581.69	292.9	425.73	2.878	.96794
2226		567.87	267.8	415.26	3.080	.96180	2276		581.97	293.4	425.94	2.874	.96807
2227		568.15	268.3	415.48	3.076	.96193	2277		582.26	293.9	426.15	2.870	.96819
2228		568.43	268.8	415.69	3.071	.96205	2278		582.54	294.5	426.37	2.866	.96832
2229		568.72	269.3	415.90	3.067	.96218	2279		582.82	295.0	426.58	2.862	.96844
2230	1770.33	569.00	269.7	416.12	3.063	.96231	2280	1820.33	583.10	295.5	426.80	2.858	.96856
2231		569.28	270.2	416.33	3.058	.96243	2281		583.39	296.1	427.01	2.854	.96869
2232		569.56	270.7	416.54	3.054	.96256	2282		583.67	296.6	427.22	2.850	.96881
2233		569.84	271.2	416.76	3.050	.96269	2283		583.95	297.1	427.44	2.847	.96893
2234		570.12	271.7	416.97	3.046	.96281	2284		584.24	297.7	427.65	2.843	.96906
2235	1775.33	570.41	272.2	417.18	3.041	.96294	2285	1825.33	584.52	298.2	427.87	2.839	.96918
2236		570.69	272.7	417.40	3.037	.96306	2286		584.80	298.7	428.08	2.835	.96931
2237		570.97	273.2	417.61	3.033	.96319	2287		585.08	299.3	428.30	2.831	.96943
2238		571.25	273.7	417.82	3.029	.96332	2288		585.37	299.8	428.51	2.827	.96955
2239		571.53	274.2	418.04	3.025	.96344	2289		585.65	300.4	428.72	2.823	.96968
2240	1780.33	571.82	274.8	418.25	3.020	.96357	2290	1830.33	585.93	300.9	428.94	2.819	.96980
2241		572.10	275.3	418.46	3.016	.96369	2291		586.21	301.4	429.15	2.816	.96992
2242		572.38	275.8	418.68	3.012	.96382	2292		586.50	302.0	429.37	2.812	.97005
2243		572.66	276.3	418.89	3.008	.96394	2293		586.78	302.5	429.58	2.808	.97017
2244		572.94	276.8	419.10	3.004	.96407	2294		587.06	303.1	429.79	2.804	.97029
2245	1785.33	573.22	277.3	419.32	3.000	.96420	2295	1835.33	587.35	303.6	430.01	2.800	.97042
2246		573.51	277.8	419.53	2.995	.96432	2296		587.63	304.2	430.22	2.797	.97054
2247		573.79	278.3	419.74	2.991	.96445	2297		587.91	304.7	430.44	2.793	.97066
2248		574.07	278.8	419.96	2.987	.96457	2298		588.19	305.3	430.65	2.789	.97079
2249		574.35	279.3	420.17	2.983	.96470	2299		588.48	305.8	430.87	2.785	.97091

Table 1 Air at Low Pressures (for One Pound)

T	t	h	p_r	u	v_r	ϕ	T	t	h	p_r	u	v_r	ϕ
2300	1840.33	588.76	306.4	431.08	2.781	.97103	2350	1890.33	602.93	334.8	441.82	2.600	.97713
2301		589.04	306.9	431.30	2.778	.97116	2351		603.21	335.4	442.04	2.597	.97725
2302		589.33	307.5	431.51	2.774	.97128	2352		603.50	336.0	442.25	2.593	.97737
2303		589.61	308.0	431.72	2.770	.97140	2353		603.78	336.6	442.47	2.590	.97749
2304		589.89	308.6	431.94	2.766	.97152	2354		604.06	337.2	442.68	2.586	.97761
2305	1845.33	590.17	309.1	432.15	2.763	.97165	2355	1895.33	604.35	337.8	442.90	2.583	.97773
2306		590.46	309.7	432.37	2.759	.97177	2356		604.63	338.4	443.11	2.579	.97785
2307		590.74	310.2	432.58	2.755	.97189	2357		604.92	339.0	443.33	2.576	.97797
2308		591.02	310.8	432.80	2.751	.97202	2358		605.20	339.6	443.54	2.572	.97809
2309		591.31	311.3	433.01	2.748	.97214	2359		605.48	340.2	443.76	2.569	.97821
2310	1850.33	591.59	311.9	433.23	2.744	.97226	2360	1900.33	605.77	340.8	443.98	2.566	.97833
2311		591.87	312.5	433.44	2.740	.97238	2361		606.05	341.4	444.19	2.562	.97845
2312		592.16	313.0	433.65	2.736	.97251	2362		606.34	342.0	444.41	2.559	.97857
2313		592.44	313.6	433.87	2.733	.97263	2363		606.62	342.6	444.62	2.555	.97869
2314		592.72	314.1	434.08	2.729	.97275	2364		606.90	343.2	444.84	2.552	.97881
2315	1855.33	593.01	314.7	434.30	2.725	.97287	2365	1905.33	607.19	343.8	445.05	2.549	.97893
2316		593.29	315.3	434.51	2.722	.97300	2366		607.47	344.4	445.27	2.545	.97905
2317		593.57	315.8	434.73	2.718	.97312	2367		607.76	345.0	445.48	2.542	.97917
2318		593.86	316.4	434.94	2.714	.97324	2368		608.04	345.6	445.70	2.538	.97929
2319		594.14	316.9	435.16	2.711	.97336	2369		608.32	346.2	445.91	2.535	.97941
2320	1860.33	594.42	317.5	435.37	2.707	.97348	2370	1910.33	608.61	346.8	446.13	2.532	.97953
2321		594.71	318.1	435.59	2.703	.97361	2371		608.89	347.4	446.35	2.528	.97965
2322		594.99	318.6	435.80	2.700	.97373	2372		609.18	348.0	446.56	2.525	.97977
2323		595.27	319.2	436.02	2.696	.97385	2373		609.46	348.6	446.78	2.522	.97989
2324		595.56	319.8	436.23	2.692	.97397	2374		609.75	349.2	446.99	2.518	.98001
2325	1865.33	595.84	320.4	436.45	2.689	.97409	2375	1915.33	610.03	349.8	447.21	2.515	.98013
2326		596.12	320.9	436.66	2.685	.97422	2376		610.31	350.5	447.42	2.512	.98025
2327		596.41	321.5	436.88	2.682	.97434	2377		610.60	351.1	447.64	2.508	.98037
2328		596.69	322.1	437.09	2.678	.97446	2378		610.88	351.7	447.86	2.505	.98049
2329		596.97	322.6	437.31	2.674	.97458	2379		611.17	352.3	448.07	2.502	.98061
2330	1870.33	597.26	323.2	437.52	2.671	.97470	2380	1920.33	611.45	352.9	448.29	2.498	.98073
2331		597.54	323.8	437.74	2.667	.97482	2381		611.74	353.5	448.50	2.495	.98085
2332		597.82	324.4	437.95	2.664	.97495	2382		612.02	354.1	448.72	2.492	.98097
2333		598.11	324.9	438.17	2.660	.97507	2383		612.30	354.8	448.93	2.489	.98109
2334		598.39	325.5	438.38	2.656	.97519	2384		612.59	355.4	449.15	2.485	.98121
2335	1875.33	598.67	326.1	438.60	2.653	.97531	2385	1925.33	612.87	356.0	449.37	2.482	.98133
2336		598.96	326.7	438.81	2.649	.97543	2386		613.16	356.6	449.58	2.479	.98145
2337		599.24	327.2	439.03	2.646	.97555	2387		613.44	357.2	449.80	2.475	.98157
2338		599.52	327.8	439.24	2.642	.97567	2388		613.73	357.9	450.01	2.472	.98168
2339		599.81	328.4	439.46	2.639	.97580	2389		614.01	358.5	450.23	2.469	.98180
2340	1880.33	600.09	329.0	439.67	2.635	.97592	2390	1930.33	614.29	359.1	450.45	2.466	.98192
2341		600.38	329.6	439.89	2.632	.97604	2391		614.58	359.7	450.66	2.462	.98204
2342		600.66	330.1	440.10	2.628	.97616	2392		614.86	360.4	450.88	2.459	.98216
2343		600.94	330.7	440.32	2.625	.97628	2393		615.15	361.0	451.09	2.456	.98228
2344		601.23	331.3	440.53	2.621	.97640	2394		615.43	361.6	451.31	2.453	.98240
2345	1885.33	601.51	331.9	440.75	2.618	.97652	2395	1935.33	615.72	362.2	451.53	2.449	.98252
2346		601.79	332.5	440.96	2.614	.97664	2396		616.00	362.9	451.74	2.446	.98264
2347		602.08	333.1	441.18	2.611	.97676	2397		616.29	363.5	451.96	2.443	.98276
2348		602.36	333.7	441.39	2.607	.97689	2398		616.57	364.1	452.17	2.440	.98287
2349		602.65	334.3	441.61	2.604	.97701	2399		616.86	364.8	452.39	2.437	.98299

Table 1 Air at Low Pressures (for One Pound)

T	t	h	p_r	u	v_r	ϕ	T	t	h	p_r	u	v_r	ϕ
2400	1940.33	617.14	365.4	452.61	2.433	.98311	2450	1990.33	631.39	398.1	463.43	2.280	.98899
2401		617.42	366.0	452.82	2.430	.98323	2451		631.68	398.8	463.65	2.277	.98910
2402		617.71	366.6	453.04	2.427	.98335	2452		631.96	399.4	463.86	2.274	.98922
2403		617.99	367.3	453.25	2.424	.98347	2453		632.25	400.1	464.08	2.271	.98934
2404		618.28	367.9	453.47	2.421	.98359	2454		632.53	400.8	464.30	2.268	.98945
2405	1945.33	618.56	368.6	453.69	2.418	.98370	2455	1995.33	632.82	401.5	464.51	2.265	.98957
2406		618.85	369.2	453.90	2.414	.98382	2456		633.11	402.2	464.73	2.262	.98969
2407		619.13	369.8	454.12	2.411	.98394	2457		633.39	402.8	464.95	2.260	.98980
2408		619.42	370.5	454.33	2.408	.98406	2458		633.68	403.5	465.17	2.257	.98992
2409		619.70	371.1	454.55	2.405	.98418	2459		633.96	404.2	465.38	2.254	.99004
2410	1950.33	619.99	371.7	454.77	2.402	.98429	2460	2000.33	634.25	404.9	465.60	2.251	.99015
2411		620.27	372.4	454.98	2.399	.98441	2461		634.53	405.6	465.82	2.248	.99027
2412		620.56	373.0	455.20	2.395	.98453	2462		634.82	406.3	466.03	2.245	.99038
2413		620.84	373.7	455.42	2.392	.98465	2463		635.10	407.0	466.25	2.242	.99050
2414		621.13	374.3	455.63	2.389	.98477	2464		635.39	407.6	466.47	2.239	.99062
2415	1955.33	621.41	375.0	455.85	2.386	.98489	2465	2005.33	635.68	408.3	466.68	2.236	.99073
2416		621.70	375.6	456.06	2.383	.98500	2466		635.96	409.0	466.90	2.234	.99085
2417		621.98	376.3	456.28	2.380	.98512	2467		636.25	409.7	467.12	2.231	.99096
2418		622.27	376.9	456.50	2.377	.98524	2468		636.53	410.4	467.34	2.228	.99108
2419		622.55	377.5	456.71	2.374	.98536	2469		636.82	411.1	467.55	2.225	.99119
2420	1960.33	622.84	378.2	456.93	2.371	.98547	2470	2010.33	637.10	411.8	467.77	2.222	.99131
2421		623.12	378.8	457.15	2.367	.98559	2471		637.39	412.5	467.99	2.219	.99143
2422		623.41	379.5	457.36	2.364	.98571	2472		637.68	413.2	468.20	2.216	.99154
2423		623.69	380.2	457.58	2.361	.98583	2473		637.96	413.9	468.42	2.214	.99166
2424		623.98	380.8	457.80	2.358	.98595	2474		638.25	414.6	468.64	2.211	.99177
2425	1965.33	624.26	381.5	458.01	2.355	.98606	2475	2015.33	638.53	415.3	468.86	2.208	.99189
2426		624.55	382.1	458.23	2.352	.98618	2476		638.82	416.0	469.07	2.205	.99200
2427		624.83	382.8	458.45	2.349	.98630	2477		639.11	416.7	469.29	2.202	.99212
2428		625.12	383.4	458.66	2.346	.98642	2478		639.39	417.4	469.51	2.199	.99223
2429		625.40	384.1	458.88	2.343	.98653	2479		639.68	418.1	469.73	2.197	.99235
2430	1970.33	625.69	384.7	459.10	2.340	.98665	2480	2020.33	639.96	418.8	469.94	2.194	.99247
2431		625.97	385.4	459.31	2.337	.98677	2481		640.25	419.5	470.16	2.191	.99258
2432		626.26	386.1	459.53	2.334	.98688	2482		640.53	420.2	470.38	2.188	.99270
2433		626.54	386.7	459.75	2.331	.98700	2483		640.82	420.9	470.60	2.185	.99281
2434		626.83	387.4	459.96	2.328	.98712	2484		641.11	421.6	470.81	2.183	.99293
2435	1975.33	627.11	388.0	460.18	2.325	.98724	2485	2025.33	641.39	422.3	471.03	2.180	.99304
2436		627.40	388.7	460.39	2.322	.98735	2486		641.68	423.0	471.25	2.177	.99316
2437		627.68	389.4	460.61	2.319	.98747	2487		641.96	423.8	471.47	2.174	.99327
2438		627.97	390.0	460.83	2.316	.98759	2488		642.25	424.5	471.68	2.172	.99339
2439		628.25	390.7	461.04	2.313	.98770	2489		642.54	425.2	471.90	2.169	.99350
2440	1980.33	628.54	391.4	461.26	2.310	.98782	2490	2030.33	642.82	425.9	472.12	2.166	.99362
2441		628.82	392.0	461.48	2.307	.98794	2491		643.11	426.6	472.34	2.163	.99373
2442		629.11	392.7	461.70	2.304	.98805	2492		643.39	427.3	472.55	2.161	.99385
2443		629.39	393.4	461.91	2.301	.98817	2493		643.68	428.0	472.77	2.158	.99396
2444		629.68	394.0	462.13	2.298	.98829	2494		643.97	428.8	472.99	2.155	.99408
2445	1985.33	629.97	394.7	462.35	2.295	.98841	2495	2035.33	644.25	429.5	473.21	2.152	.99419
2446		630.25	395.4	462.56	2.292	.98852	2496		644.54	430.2	473.42	2.150	.99430
2447		630.54	396.1	462.78	2.289	.98864	2497		644.83	430.9	473.64	2.147	.99442
2448		630.82	396.7	463.00	2.286	.98876	2498		645.11	431.6	473.86	2.144	.99453
2449		631.11	397.4	463.21	2.283	.98887	2499		645.40	432.3	474.08	2.141	.99465

Table 1 Air at Low Pressures (for One Pound)

T	t	h	p_r	u	v_r	φ	T	t	h	p_r	u	v_r	φ
2500	2040.33	645.68	433.1	474.29	2.1386	.99476	2750	2290.33	717.69	646.3	529.16	1.5764	1.02221
2505		647.12	436.7	475.38	2.1251	.99533	2755		719.13	651.3	530.26	1.5671	1.02274
2510		648.55	440.4	476.47	2.1117	.99591	2760		720.58	656.3	531.37	1.5580	1.02326
2515		649.98	444.0	477.56	2.0984	.99648	2765		722.03	661.3	532.48	1.5489	1.02379
2520		651.41	447.7	478.65	2.0852	.99704	2770		723.48	666.4	533.58	1.5399	1.02431
2525	2065.33	652.84	451.5	479.74	2.0721	.99761	2775	2315.33	724.93	671.5	534.69	1.5310	1.02483
2530		654.28	455.2	480.83	2.0591	.99818	2780		726.38	676.6	535.80	1.5221	1.02536
2535		655.71	459.0	481.92	2.0462	.99875	2785		727.83	681.8	536.90	1.5133	1.02588
2540		657.14	462.8	483.01	2.0334	.99931	2790		729.28	687.0	538.01	1.5045	1.02640
2545		658.58	466.6	484.10	2.0207	.99987	2795		730.74	692.2	539.12	1.4958	1.02692
2550	2090.33	660.01	470.4	485.20	2.0081	1.00044	2800	2340.33	732.19	697.5	540.23	1.4872	1.02744
2555		661.45	474.3	486.29	1.9956	1.00100	2805		733.64	702.8	541.34	1.4787	1.02795
2560		662.88	478.2	487.38	1.9832	1.00156	2810		735.09	708.1	542.45	1.4702	1.02847
2565		664.32	482.1	488.47	1.9709	1.00212	2815		736.54	713.5	543.56	1.4617	1.02899
2570		665.76	486.1	489.57	1.9587	1.00268	2820		738.00	718.9	544.67	1.4533	1.02950
2575	2115.33	667.19	490.1	490.66	1.9466	1.00324	2825	2365.33	739.45	724.3	545.78	1.4450	1.03002
2580		668.63	494.1	491.75	1.9346	1.00380	2830		740.90	729.7	546.89	1.4368	1.03053
2585		670.07	498.1	492.85	1.9227	1.00435	2835		742.36	735.2	548.00	1.4286	1.03105
2590		671.50	502.2	493.94	1.9108	1.00491	2840		743.81	740.7	549.11	1.4204	1.03156
2595		672.94	506.2	495.04	1.8991	1.00546	2845		745.27	746.3	550.22	1.4124	1.03207
2600	2140.33	674.38	510.3	496.13	1.8875	1.00602	2850	2390.33	746.72	751.8	551.33	1.4043	1.03258
2605		675.82	514.5	497.23	1.8759	1.00657	2855		748.17	757.5	552.45	1.3964	1.03309
2610		677.26	518.6	498.33	1.8644	1.00712	2860		749.63	763.1	553.56	1.3885	1.03360
2615		678.70	522.8	499.42	1.8530	1.00767	2865		751.09	768.8	554.67	1.3806	1.03411
2620		680.14	527.0	500.52	1.8418	1.00822	2870		752.54	774.5	555.78	1.3728	1.03462
2625	2165.33	681.58	531.3	501.62	1.8305	1.00877	2875	2415.33	754.00	780.3	556.90	1.3651	1.03512
2630		683.02	535.5	502.71	1.8194	1.00932	2880		755.45	786.0	558.01	1.3574	1.03563
2635		684.46	539.8	503.81	1.8084	1.00987	2885		756.91	791.9	559.13	1.3498	1.03613
2640		685.90	544.1	504.91	1.7974	1.01041	2890		758.37	797.7	560.24	1.3422	1.03664
2645		687.34	548.5	506.01	1.7866	1.01096	2895		759.82	803.6	561.35	1.3347	1.03714
2650	2190.33	688.78	552.9	507.11	1.7758	1.01150	2900	2440.33	761.28	809.5	562.47	1.3272	1.03765
2655		690.22	557.3	508.21	1.7651	1.01205	2905		762.74	815.5	563.58	1.3198	1.03815
2660		691.67	561.7	509.31	1.7545	1.01259	2910		764.20	821.4	564.70	1.3124	1.03865
2665		693.11	566.1	510.41	1.7439	1.01313	2915		765.66	827.5	565.82	1.3051	1.03915
2670		694.55	570.6	511.51	1.7335	1.01367	2920		767.12	833.5	566.93	1.2978	1.03965
2675	2215.33	695.99	575.1	512.61	1.7231	1.01421	2925	2465.33	768.57	839.6	568.05	1.2906	1.04015
2680		697.44	579.7	513.71	1.7128	1.01475	2930		770.03	845.7	569.16	1.2835	1.04065
2685		698.88	584.3	514.81	1.7026	1.01529	2935		771.49	851.9	570.28	1.2764	1.04115
2690		700.33	588.9	515.91	1.6924	1.01583	2940		772.95	858.1	571.40	1.2693	1.04164
2695		701.77	593.5	517.01	1.6823	1.01636	2945		774.41	864.3	572.52	1.2623	1.04214
2700	2240.33	703.22	598.1	518.12	1.6723	1.01690	2950	2490.33	775.87	870.6	573.63	1.2553	1.04263
2705		704.66	602.8	519.22	1.6624	1.01744	2955		777.33	876.9	574.75	1.2484	1.04313
2710		706.11	607.5	520.32	1.6526	1.01797	2960		778.80	883.3	575.87	1.2416	1.04362
2715		707.55	612.3	521.42	1.6428	1.01850	2965		780.26	889.6	576.99	1.2347	1.04412
2720		709.00	617.0	522.53	1.6331	1.01903	2970		781.72	896.0	578.11	1.2280	1.04461
2725	2265.33	710.45	621.9	523.63	1.6234	1.01957	2975	2515.33	783.18	902.5	579.23	1.2212	1.04510
2730		711.89	626.7	524.74	1.6139	1.02010	2980		784.64	909.0	580.35	1.2146	1.04559
2735		713.34	631.5	525.84	1.6044	1.02063	2985		786.11	915.5	581.46	1.2079	1.04608
2740		714.79	636.4	526.95	1.5950	1.02116	2990		787.57	922.1	582.58	1.2013	1.04657
2745		716.24	641.4	528.05	1.5856	1.02168	2995		789.03	928.7	583.71	1.1948	1.04706

Table 1   Air at Low Pressures (for One Pound)

T	t	h	p_r	u	v_r	φ	T	t	h	p_r	u	v_r	φ
3000	2540.33	790.49	935.3	584.83	1.1883	1.04755	3500	3040.33	938.09	1816.1	698.14	.7140	1.09304
3010		793.42	948.7	587.07	1.1754	1.04852	3510		941.06	1838.8	700.43	.7072	1.09389
3020		796.35	962.2	589.31	1.1628	1.04949	3520		944.04	1861.6	702.72	.7005	1.09474
3030		799.28	975.9	591.55	1.1502	1.05046	3530		947.01	1884.7	705.01	.6939	1.09558
3040		802.21	989.8	593.80	1.1379	1.05143	3540		949.99	1908.0	707.30	.6874	1.09642
3050	2590.33	805.14	1003.8	596.05	1.1257	1.05239	3550	3090.33	952.97	1931.5	709.60	.6809	1.09727
3060		808.07	1017.9	598.29	1.1137	1.05335	3560		955.95	1955.3	711.89	.6745	1.09810
3070		811.01	1032.2	600.54	1.1019	1.05431	3570		958.93	1979.3	714.18	.6682	1.09894
3080		813.94	1046.7	602.79	1.0902	1.05526	3580		961.91	2003.5	716.48	.6620	1.09977
3090		816.88	1061.3	605.04	1.0786	1.05621	3590		964.89	2027.9	718.78	.6558	1.10060
3100	2640.33	819.82	1076.1	607.29	1.0672	1.05716	3600	3140.33	967.88	2052.6	721.07	.6498	1.10143
3110		822.75	1091.1	609.54	1.0560	1.05811	3610		970.86	2077.6	723.37	.6437	1.10226
3120		825.69	1106.2	611.80	1.0449	1.05905	3620		973.84	2102.7	725.67	.6378	1.10309
3130		828.63	1121.5	614.05	1.0340	1.05999	3630		976.83	2128.1	727.97	.6319	1.10391
3140		831.57	1136.9	616.31	1.0232	1.06093	3640		979.81	2153.8	730.27	.6261	1.10473
3150	2690.33	834.52	1152.6	618.56	1.0125	1.06187	3650	3190.33	982.80	2179.7	732.57	.6204	1.10555
3160		837.46	1168.3	620.82	1.0020	1.06280	3660		985.79	2205.8	734.87	.6147	1.10637
3170		840.40	1184.3	623.08	.9916	1.06373	3670		988.78	2232.2	737.17	.6091	1.10718
3180		843.35	1200.4	625.34	.9814	1.06466	3680		991.76	2258.8	739.48	.6036	1.10800
3190		846.29	1216.7	627.60	.9713	1.06558	3690		994.75	2285.7	741.78	.5981	1.10881
3200	2740.33	849.24	1233.2	629.86	.9613	1.06651	3700	3240.33	997.74	2312.9	744.09	.5927	1.10962
3210		852.19	1249.9	632.12	.9515	1.06743	3710		1000.74	2340.3	746.39	.5873	1.11043
3220		855.14	1266.7	634.39	.9417	1.06834	3720		1003.73	2367.9	748.70	.5820	1.11123
3230		858.09	1283.7	636.65	.9322	1.06926	3730		1006.72	2395.8	751.01	.5768	1.11203
3240		861.04	1300.9	638.92	.9227	1.07017	3740		1009.71	2424.0	753.31	.5716	1.11284
3250	2790.33	863.99	1318.3	641.18	.9133	1.07108	3750	3290.33	1012.71	2452.4	755.62	.5665	1.11363
3260		866.94	1335.9	643.45	.9041	1.07199	3760		1015.70	2481.1	757.93	.5614	1.11443
3270		869.90	1353.6	645.72	.8950	1.07289	3770		1018.70	2510.1	760.24	.5564	1.11523
3280		872.85	1371.5	647.99	.8860	1.07379	3780		1021.69	2539.3	762.55	.5515	1.11602
3290		875.81	1389.7	650.26	.8771	1.07469	3790		1024.69	2568.8	764.86	.5466	1.11681
3300	2840.33	878.77	1408.0	652.53	.8683	1.07559	3800	3340.33	1027.69	2598.6	767.18	.5418	1.11760
3310		881.72	1426.5	654.80	.8597	1.07649	3810		1030.69	2628.7	769.49	.5370	1.11839
3320		884.68	1445.2	657.07	.8511	1.07738	3820		1033.69	2659.0	771.80	.5322	1.11918
3330		887.64	1464.0	659.35	.8427	1.07827	3830		1036.69	2689.6	774.12	.5276	1.11996
3340		890.60	1483.1	661.62	.8343	1.07916	3840		1039.69	2720.4	776.43	.5229	1.12074
3350	2890.33	893.56	1502.4	663.90	.8261	1.08004	3850	3390.33	1042.69	2751.6	778.75	.5184	1.12153
3360		896.52	1521.9	666.18	.8179	1.08092	3860		1045.69	2783.0	781.06	.5138	1.12230
3370		899.49	1541.5	668.45	.8099	1.08180	3870		1048.69	2814.7	783.38	.5094	1.12308
3380		902.45	1561.4	670.73	.8020	1.08268	3880		1051.70	2846.7	785.70	.5049	1.12386
3390		905.42	1581.5	673.01	.7941	1.08356	3890		1054.70	2879.0	788.02	.5006	1.12463
3400	2940.33	908.38	1601.8	675.29	.7864	1.08443	3900	3440.33	1057.71	2911.6	790.34	.4962	1.12540
3410		911.35	1622.3	677.57	.7787	1.08530	3910		1060.71	2944.5	792.66	.4920	1.12617
3420		914.32	1643.0	679.85	.7712	1.08617	3920		1063.72	2977.7	794.98	.4877	1.12694
3430		917.28	1663.9	682.14	.7637	1.08704	3930		1066.73	3011.1	797.30	.4835	1.12770
3440		920.25	1685.0	684.42	.7564	1.08790	3940		1069.73	3044.9	799.62	.4794	1.12847
3450	2990.33	923.22	1706.3	686.70	.7491	1.08877	3950	3490.33	1072.74	3078.9	801.94	.4753	1.12923
3460		926.19	1727.8	688.99	.7419	1.08963	3960		1075.75	3113.3	804.27	.4712	1.12999
3470		929.17	1749.6	691.28	.7348	1.09048	3970		1078.76	3148.0	806.59	.4672	1.13075
3480		932.14	1771.6	693.56	.7278	1.09134	3980		1081.77	3182.9	808.92	.4632	1.13151
3490		935.11	1793.7	695.85	.7208	1.09219	3990		1084.78	3218.2	811.24	.4593	1.13226

Table 1 Air at Low Pressures (for One Pound)

T	t	h	p_r	u	v_r	φ	T	t	h	p_r	u	v_r	φ
4000	3540.33	1087.79	3254	813.57	.4554	1.13302	4500	4040.33	1239.18	5474	930.68	.3046	1.16868
4010		1090.81	3290	815.90	.4516	1.13377	4510		1242.22	5528	933.03	.3023	1.16935
4020		1093.82	3326	818.22	.4478	1.13452	4520		1245.27	5582	935.39	.3000	1.17003
4030		1096.83	3362	820.55	.4440	1.13527	4530		1248.31	5638	937.75	.2977	1.17070
4040		1099.85	3399	822.88	.4403	1.13602	4540		1251.35	5693	940.11	.2954	1.17137
4050	3590.33	1102.86	3436	825.21	.4366	1.13676	4550	4090.33	1254.40	5749	942.47	.2932	1.17204
4060		1105.88	3474	827.54	.4330	1.13751	4560		1257.44	5805	944.83	.2910	1.17271
4070		1108.89	3512	829.87	.4294	1.13825	4570		1260.49	5862	947.19	.2888	1.17338
4080		1111.91	3550	832.20	.4258	1.13899	4580		1263.54	5919	949.55	.2867	1.17404
4090		1114.93	3588	834.53	.4223	1.13973	4590		1266.58	5977	951.91	.2845	1.17471
4100	3640.33	1117.95	3627	836.87	.4188	1.14046	4600	4140.33	1269.63	6035	954.27	.2824	1.17537
4110		1120.97	3666	839.20	.4153	1.14120	4610		1272.68	6094	956.63	.2803	1.17603
4120		1123.99	3706	841.54	.4119	1.14193	4620		1275.73	6153	959.00	.2782	1.17669
4130		1127.01	3745	843.87	.4085	1.14267	4630		1278.78	6212	961.36	.2761	1.17735
4140		1130.03	3786	846.21	.4052	1.14340	4640		1281.83	6272	963.73	.2741	1.17801
4150	3690.33	1133.05	3826	848.54	.4018	1.14412	4650	4190.33	1284.88	6332	966.09	.2721	1.17867
4160		1136.07	3867	850.88	.3986	1.14485	4660		1287.93	6393	968.45	.2700	1.17932
4170		1139.09	3908	853.21	.3953	1.14558	4670		1290.98	6454	970.82	.2681	1.17997
4180		1142.12	3950	855.55	.3921	1.14630	4680		1294.03	6516	973.19	.2661	1.18063
4190		1145.14	3991	857.89	.3889	1.14702	4690		1297.08	6578	975.55	.2641	1.18128
4200	3740.33	1148.17	4034	860.23	.3858	1.14775	4700	4240.33	1300.13	6641	977.92	.2622	1.18193
4210		1151.19	4076	862.57	.3826	1.14847	4710		1303.19	6704	980.29	.2603	1.18258
4220		1154.22	4119	864.91	.3796	1.14918	4720		1306.24	6768	982.66	.2584	1.18323
4230		1157.24	4162	867.25	.3765	1.14990	4730		1309.30	6832	985.02	.2565	1.18387
4240		1160.27	4206	869.59	.3735	1.15061	4740		1312.35	6897	987.39	.2546	1.18452
4250	3790.33	1163.30	4250	871.93	.3705	1.15133	4750	4290.33	1315.41	6962	989.76	.2528	1.18516
4260		1166.33	4294	874.28	.3675	1.15204	4760		1318.46	7027	992.13	.2509	1.18580
4270		1169.36	4339	876.62	.3646	1.15275	4770		1321.52	7093	994.50	.2491	1.18644
4280		1172.39	4384	878.96	.3617	1.15346	4780		1324.57	7160	996.88	.2473	1.18709
4290		1175.42	4429	881.31	.3588	1.15417	4790		1327.63	7227	999.25	.2456	1.18772
4300	3840.33	1178.45	4475	883.65	.3560	1.15487	4800	4340.33	1330.69	7294	1001.62	.2438	1.18836
4310		1181.48	4522	886.00	.3531	1.15557	4810		1333.75	7362	1003.99	.2420	1.18900
4320		1184.51	4568	888.35	.3504	1.15628	4820		1336.81	7431	1006.36	.2403	1.18963
4330		1187.54	4615	890.69	.3476	1.15698	4830		1339.86	7500	1008.74	.2386	1.19027
4340		1190.57	4662	893.04	.3449	1.15768	4840		1342.92	7569	1011.11	.2369	1.19090
4350	3890.33	1193.61	4710	895.39	.3421	1.15838	4850	4390.33	1345.98	7640	1013.49	.2352	1.19153
4360		1196.64	4758	897.74	.3395	1.15907	4860		1349.05	7710	1015.86	.2335	1.19216
4370		1199.68	4807	900.09	.3368	1.15977	4870		1352.11	7781	1018.24	.2319	1.19279
4380		1202.71	4856	902.44	.3342	1.16046	4880		1355.17	7853	1020.61	.2302	1.19342
4390		1205.75	4905	904.79	.3316	1.16115	4890		1358.23	7925	1022.99	.2286	1.19405
4400	3940.33	1208.78	4955	907.14	.3290	1.16185	4900	4440.33	1361.29	7998	1025.37	.2270	1.19467
4410		1211.82	5005	909.49	.3265	1.16253	4910		1364.36	8071	1027.74	.2254	1.19530
4420		1214.86	5055	911.84	.3239	1.16322	4920		1367.42	8144	1030.12	.2238	1.19592
4430		1217.90	5106	914.19	.3214	1.16391	4930		1370.48	8219	1032.50	.2222	1.19654
4440		1220.94	5157	916.55	.3189	1.16459	4940		1373.55	8293	1034.88	.2207	1.19716
4450	3990.33	1223.98	5209	918.90	.3165	1.16528	4950	4490.33	1376.61	8369	1037.26	.2191	1.19778
4460		1227.02	5261	921.25	.3141	1.16596	4960		1379.68	8445	1039.64	.2176	1.19840
4470		1230.06	5314	923.61	.3117	1.16664	4970		1382.74	8521	1042.02	.2161	1.19902
4480		1233.10	5367	925.97	.3093	1.16732	4980		1385.81	8598	1044.40	.2146	1.19963
4490		1236.14	5420	928.32	.3069	1.16800	4990		1388.88	8676	1046.78	.2131	1.20025

Table 1   Air at Low Pressures (for One Pound)

T	t	h	p_r	u	v_r	ϕ	T	t	h	p_r	u	v_r	ϕ
5000	4540.33	1391.94	8754	1049.16	.21161	1.20086	5500	5040.33	1545.86	13429	1168.80	.15173	1.23020
5010		1395.01	8832	1051.54	.21015	1.20148	5510		1548.95	13539	1171.20	.15077	1.23076
5020		1398.08	8911	1053.93	.20870	1.20209	5520		1552.04	13650	1173.61	.14982	1.23132
5030		1401.15	8991	1056.31	.20726	1.20270	5530		1555.13	13762	1176.01	.14887	1.23188
5040		1404.22	9071	1058.69	.20583	1.20331	5540		1558.22	13875	1178.41	.14793	1.23244
5050	4590.33	1407.29	9152	1061.08	.20442	1.20392	5550	5090.33	1561.31	13988	1180.82	.14700	1.23300
5060		1410.36	9234	1063.46	.20302	1.20452	5560		1564.40	14102	1183.22	.14607	1.23355
5070		1413.43	9316	1065.85	.20163	1.20513	5570		1567.49	14216	1185.63	.14515	1.23411
5080		1416.50	9398	1068.23	.20025	1.20574	5580		1570.58	14332	1188.04	.14424	1.23466
5090		1419.57	9481	1070.62	.19888	1.20634	5590		1573.67	14448	1190.44	.14334	1.23522
5100	4640.33	1422.64	9565	1073.00	.19753	1.20694	5600	5140.33	1576.76	14565	1192.85	.14244	1.23577
5110		1425.71	9650	1075.39	.19619	1.20754	5610		1579.86	14683	1195.26	.14155	1.23632
5120		1428.79	9734	1077.78	.19486	1.20815	5620		1582.95	14801	1197.66	.14067	1.23687
5130		1431.86	9820	1080.16	.19354	1.20875	5630		1586.04	14920	1200.07	.13979	1.23742
5140		1434.93	9906	1082.55	.19223	1.20934	5640		1589.14	15040	1202.48	.13892	1.23797
5150	4690.33	1438.01	9993	1084.94	.19093	1.20994	5650	5190.33	1592.23	15161	1204.89	.13806	1.23852
5160		1441.08	10080	1087.33	.18965	1.21054	5660		1595.32	15283	1207.30	.13721	1.23907
5170		1444.15	10168	1089.72	.18837	1.21113	5670		1598.42	15405	1209.71	.13636	1.23961
5180		1447.23	10257	1092.11	.18711	1.21173	5680		1601.51	15528	1212.11	.13552	1.24016
5190		1450.31	10346	1094.50	.18585	1.21232	5690		1604.61	15652	1214.52	.13468	1.24070
5200	4740.33	1453.38	10435	1096.89	.18461	1.21291	5700	5240.33	1607.71	15776	1216.94	.13385	1.24125
5210		1456.46	10526	1099.28	.18338	1.21350	5710		1610.80	15902	1219.35	.13303	1.24179
5220		1459.54	10617	1101.67	.18215	1.21409	5720		1613.90	16028	1221.76	.13221	1.24233
5230		1462.61	10708	1104.06	.18094	1.21468	5730		1617.00	16155	1224.17	.13140	1.24287
5240		1465.69	10801	1106.46	.17974	1.21527	5740		1620.09	16283	1226.58	.13060	1.24341
5250	4790.33	1468.77	10893	1108.85	.17855	1.21586	5750	5290.33	1623.19	16411	1228.99	.12980	1.24395
5260		1471.85	10987	1111.24	.17736	1.21644	5760		1626.29	16541	1231.40	.12901	1.24449
5270		1474.93	11081	1113.64	.17619	1.21703	5770		1629.39	16671	1233.82	.12823	1.24503
5280		1478.01	11176	1116.03	.17503	1.21761	5780		1632.49	16802	1236.23	.12745	1.24556
5290		1481.09	11271	1118.42	.17388	1.21819	5790		1635.59	16934	1238.64	.12667	1.24610
5300	4840.33	1484.17	11367	1120.82	.17273	1.21878	5800	5340.33	1638.68	17066	1241.06	.12591	1.24663
5310		1487.25	11464	1123.21	.17160	1.21936	5810		1641.78	17200	1243.47	.12515	1.24717
5320		1490.33	11561	1125.61	.17048	1.21994	5820		1644.88	17334	1245.89	.12439	1.24770
5330		1493.41	11659	1128.00	.16936	1.22052	5830		1647.99	17469	1248.30	.12364	1.24823
5340		1496.49	11758	1130.40	.16825	1.22109	5840		1651.09	17605	1250.72	.12290	1.24877
5350	4890.33	1499.57	11857	1132.80	.16716	1.22167	5850	5390.33	1654.19	17742	1253.13	.12216	1.24930
5360		1502.66	11957	1135.20	.16607	1.22225	5860		1657.29	17879	1255.55	.12142	1.24983
5370		1505.74	12058	1137.59	.16499	1.22282	5870		1660.39	18018	1257.97	.12070	1.25036
5380		1508.82	12159	1139.99	.16392	1.22339	5880		1663.49	18157	1260.38	.11997	1.25088
5390		1511.91	12261	1142.39	.16286	1.22397	5890		1666.60	18297	1262.80	.11926	1.25141
5400	4940.33	1514.99	12364	1144.79	.16181	1.22454	5900	5440.33	1669.70	18438	1265.22	.11855	1.25194
5410		1518.08	12467	1147.19	.16076	1.22511	5910		1672.80	18580	1267.64	.11784	1.25246
5420		1521.16	12571	1149.59	.15973	1.22568	5920		1675.91	18723	1270.05	.11714	1.25299
5430		1524.25	12676	1151.99	.15870	1.22625	5930		1679.01	18867	1272.47	.11644	1.25351
5440		1527.33	12782	1154.39	.15768	1.22682	5940		1682.11	19011	1274.89	.11575	1.25403
5450	4990.33	1530.42	12888	1156.79	.15667	1.22738	5950	5490.33	1685.22	19157	1277.31	.11507	1.25456
5460		1533.51	12994	1159.19	.15567	1.22795	5960		1688.32	19303	1279.73	.11439	1.25508
5470		1536.59	13102	1161.59	.15467	1.22851	5970		1691.43	19450	1282.15	.11371	1.25560
5480		1539.68	13210	1163.99	.15368	1.22908	5980		1694.54	19598	1284.57	.11304	1.25612
5490		1542.77	13319	1166.40	.15271	1.22964	5990		1697.64	19747	1286.99	.11238	1.25664

Table 2 Air at Low Pressures

T	t	c_p	c_v	$k = \dfrac{c_p}{c_v}$	a	$\dfrac{G_{max}}{p_1}$	$\mu \times 10^7$	λ	$Pr = \dfrac{3600\,c_p\,\mu}{\lambda}$
R	F	$\dfrac{Btu}{lb\ R}$	$\dfrac{Btu}{lb\ R}$		$\dfrac{ft}{sec}$	$\dfrac{lbm}{sec\ ft^2}\Big/\dfrac{lbf}{in.^2}$	$\dfrac{lbm}{sec\ ft}$	$\dfrac{Btu}{hr\ ft\ R}$	
100	−359.67	.2393	.1707	1.402	490.5	7.6601			
150	−309.67	.2393	.1707	1.402	600.7	6.2545	39.1	.0044	.758
200	−259.67	.2393	.1708	1.401	693.6	5.4165	52.6	.0059	.765
250	−209.67	.2393	.1708	1.401	775.5	4.8446	65.0	.0074	.759
300	−159.67	.2393	.1708	1.401	849.5	4.4225	76.7	.0088	.752
350	−109.67	.2394	.1708	1.401	917.5	4.0944	87.7	.0102	.742
400	−59.67	.2394	.1709	1.401	980.8	3.8299	98.0	.0116	.732
450	−9.67	.2395	.1710	1.401	1040.2	3.6107	107.9	.0129	.724
500	40.33	.2397	.1711	1.401	1096.3	3.4252	117.3	.0141	.717
550	90.33	.2400	.1714	1.400	1149.6	3.2655	126.3	.0153	.711
600	140.33	.2404	.1719	1.399	1200.3	3.1260	134.9	.0165	.708
650	190.33	.2410	.1724	1.398	1248.7	3.0028	143.1	.0176	.705
700	240.33	.2417	.1731	1.396	1295.1	2.8928	151.0	.0187	.703
750	290.33	.2425	.1739	1.394	1339.7	2.7937	158.7	.0197	.702
800	340.33	.2435	.1749	1.392	1382.5	2.7039	166.0	.0208	.700
900	440.33	.2458	.1773	1.387	1463.6	2.5468	180.2	.0228	.699
1000	540.33	.2486	.1800	1.381	1539.5	2.4132	193.5	.0248	.699
1100	640.33	.2516	.1830	1.375	1611.0	2.2978	206.1	.0268	.698
1200	740.33	.2547	.1862	1.368	1678.7	2.1968	218.1	.0285	.701
1300	840.33	.2579	.1894	1.362	1743.3	2.1074	229.5	.0303	.703
1400	940.33	.2611	.1925	1.356	1805.2	2.0278	240.6	.0322	.703
1500	1040.33	.2641	.1956	1.351	1864.7	1.9563	251.0	.0340	.703
1600	1140.33	.2670	.1985	1.345	1922.2	1.8916	261.3	.0357	.703
1700	1240.33	.2698	.2012	1.341	1977.9	1.8328	271.1	.0373	.706
1800	1340.33	.2724	.2038	1.336	2031.9	1.7790	280.7	.0388	.709
1900	1440.33	.2748	.2063	1.332	2084.5	1.7296	289.6	.0403	.711
2000	1540.33	.2771	.2085	1.329	2135.7	1.6841	299.0	.0417	.715
2100	1640.33	.2792	.2106	1.325	2185.8	1.6420	307.8	.0430	.719
2200	1740.33	.2811	.2126	1.323	2234.7	1.6029	315.8	.0444	.720
2300	1840.33	.2829	.2144	1.320	2282.6	1.5664	324.6	.0456	.725
2400	1940.33	.2846	.2161	1.317	2329.4	1.5323	332.6	.0468	.728
2600	2140.33	.2877	.2191	1.313	2420.5	1.4703	348.1	.0492	.733
2800	2340.33	.2903	.2218	1.309	2508.3	1.4152			
3000	2540.33	.2927	.2241	1.306	2593.1	1.3659			
3200	2740.33	.2948	.2262	1.303	2675.3	1.3214			
3400	2940.33	.2966	.2281	1.301	2755.0	1.2810			
3600	3140.33	.2983	.2297	1.298	2832.5	1.2441			
3800	3340.33	.2998	.2313	1.296	2907.9	1.2103			
4000	3540.33	.3012	.2327	1.295	2981.4	1.1790			
4200	3740.33	.3025	.2339	1.293	3053.1	1.1500			
4400	3940.33	.3037	.2351	1.292	3123.2	1.1231			
4600	4140.33	.3048	.2362	1.290	3191.7	1.0980			
4800	4340.33	.3058	.2372	1.289	3258.8	1.0745			
5000	4540.33	.3067	.2382	1.288	3324.5	1.0524			
5200	4740.33	.3076	.2391	1.287	3388.9	1.0316			
5400	4940.33	.3085	.2399	1.286	3452.1	1.0121			
5600	5140.33	.3092	.2407	1.285	3514.2	.9936			
5800	5340.33	.3100	.2414	1.284	3575.2	.9760			
6000	5540.33	.3107	.2421	1.283	3635.2	.9594			
6200	5740.33	.3113	.2427	1.282	3694.2	.9436			

Table 3   Products—400% Theoretical Air (for One Pound-Mole)

T	t	$\bar{h}$	p_r	$\bar{u}$	v_r	$\bar{\phi}$	T	t	$\bar{h}$	p_r	$\bar{u}$	v_r	$\bar{\phi}$
300	−159.67	2087.1	.17192	1491.3	18726	42.230	350	−109.67	2436.5	.2957	1741.4	12701	43.307
301		2094.1	.17395	1496.3	18570	42.253	351		2443.5	.2987	1746.4	12610	43.327
302		2101.1	.17599	1501.3	18416	42.276	352		2450.4	.3017	1751.4	12520	43.347
303		2108.0	.17804	1506.3	18263	42.299	353		2457.4	.3047	1756.4	12431	43.366
304		2115.0	.18012	1511.3	18112	42.322	354		2464.4	.3078	1761.4	12342	43.386
305	−154.67	2122.0	.18221	1516.3	17963	42.345	355	−104.67	2471.4	.3109	1766.5	12255	43.406
306		2129.0	.18432	1521.3	17816	42.368	356		2478.4	.3140	1771.5	12168	43.426
307		2136.0	.18645	1526.3	17670	42.391	357		2485.4	.3171	1776.5	12083	43.445
308		2143.0	.18859	1531.3	17526	42.413	358		2492.4	.3202	1781.5	11998	43.465
309		2149.9	.19075	1536.3	17384	42.436	359		2499.4	.3234	1786.5	11914	43.484
310	−149.67	2156.9	.19293	1541.3	17243	42.459	360	−99.67	2506.4	.3266	1791.5	11830	43.504
311		2163.9	.19513	1546.3	17104	42.481	361		2513.4	.3298	1796.5	11748	43.523
312		2170.9	.19735	1551.3	16966	42.503	362		2520.4	.3330	1801.5	11666	43.542
313		2177.9	.19958	1556.3	16830	42.526	363		2527.4	.3363	1806.5	11585	43.562
314		2184.9	.20183	1561.3	16696	42.548	364		2534.4	.3395	1811.5	11505	43.581
315	−144.67	2191.9	.20410	1566.3	16562	42.570	365	−94.67	2541.4	.3428	1816.5	11426	43.600
316		2198.8	.20639	1571.3	16431	42.592	366		2548.4	.3461	1821.6	11347	43.619
317		2205.8	.20870	1576.3	16301	42.615	367		2555.4	.3495	1826.6	11269	43.638
318		2212.8	.21102	1581.3	16172	42.637	368		2562.4	.3529	1831.6	11192	43.657
319		2219.8	.21336	1586.3	16045	42.658	369		2569.4	.3562	1836.6	11116	43.676
320	−139.67	2226.8	.21573	1591.3	15919	42.680	370	−89.67	2576.4	.3597	1841.6	11040	43.695
321		2233.8	.21811	1596.3	15794	42.702	371		2583.4	.3631	1846.6	10965	43.714
322		2240.8	.22051	1601.3	15671	42.724	372		2590.4	.3666	1851.6	10891	43.733
323		2247.7	.22293	1606.3	15549	42.746	373		2597.4	.3700	1856.6	10817	43.752
324		2254.7	.22536	1611.3	15428	42.767	374		2604.4	.3735	1861.6	10744	43.771
325	−134.67	2261.7	.22782	1616.3	15309	42.789	375	−84.67	2611.4	.3771	1866.7	10672	43.789
326		2268.7	.23030	1621.3	15191	42.810	376		2618.4	.3806	1871.7	10601	43.808
327		2275.7	.23279	1626.3	15074	42.832	377		2625.4	.3842	1876.7	10530	43.827
328		2282.7	.23531	1631.3	14959	42.853	378		2632.4	.3878	1881.7	10460	43.845
329		2289.7	.23784	1636.3	14845	42.874	379		2639.4	.3914	1886.7	10390	43.864
330	−129.67	2296.7	.24040	1641.3	14732	42.895	380	−79.67	2646.4	.3951	1891.7	10321	43.882
331		2303.6	.24297	1646.3	14620	42.916	381		2653.4	.3988	1896.7	10253	43.900
332		2310.6	.24556	1651.3	14509	42.938	382		2660.4	.4025	1901.8	10185	43.919
333		2317.6	.24818	1656.3	14399	42.959	383		2667.4	.4062	1906.8	10118	43.937
334		2324.6	.25081	1661.3	14291	42.980	384		2674.4	.4100	1911.8	10052	43.955
335	−124.67	2331.6	.25346	1666.3	14184	43.000	385	−74.67	2681.4	.4137	1916.8	9986	43.974
336		2338.6	.25613	1671.3	14078	43.021	386		2688.4	.4175	1921.8	9921	43.992
337		2345.6	.25883	1676.3	13973	43.042	387		2695.8	.4214	1926.8	9856	44.010
338		2352.6	.26154	1681.3	13869	43.063	388		2702.4	.4252	1931.9	9792	44.028
339		2359.6	.26427	1686.4	13766	43.083	389		2709.4	.4291	1936.9	9729	44.046
340	−119.67	2366.5	.26703	1691.4	13664	43.104	390	−69.67	2716.4	.4330	1941.9	9666	44.064
341		2373.5	.26980	1696.4	13563	43.125	391		2723.4	.4369	1946.9	9604	44.082
342		2380.5	.27260	1701.4	13464	43.145	392		2730.4	.4409	1951.9	9542	44.100
343		2387.5	.27542	1706.4	13365	43.165	393		2737.4	.4449	1956.9	9481	44.118
344		2394.5	.27825	1711.4	13267	43.186	394		2744.4	.4489	1962.0	9420	44.135
345	−114.67	2401.5	.28111	1716.4	13170	43.206	395	−64.67	2751.4	.4529	1967.0	9360	44.153
346		2408.5	.28399	1721.4	13075	43.226	396		2758.4	.4570	1972.0	9300	44.171
347		2415.5	.28689	1726.4	12980	43.246	397		2765.4	.4610	1977.0	9241	44.188
348		2422.5	.28981	1731.4	12886	43.267	398		2772.4	.4651	1982.0	9182	44.206
349		2429.5	.29276	1736.4	12793	43.287	399		2779.4	.4693	1987.1	9124	44.224

Table 3 Products—400% Theoretical Air (for One Pound-Mole)

T	t	$\bar{h}$	p_r	$\bar{u}$	v_r	$\bar{\phi}$	T	t	$\bar{h}$	p_r	$\bar{u}$	v_r	$\bar{\phi}$
400	−59.67	2786.4	.4734	1992.1	9067	44.241	450	−9.67	3137.0	.7176	2243.4	6729	45.067
401		2793.4	.4776	1997.1	9010	44.259	451		3144.1	.7233	2248.4	6692	45.083
402		2800.4	.4818	2002.1	8953	44.276	452		3151.1	.7290	2253.5	6654	45.098
403		2807.4	.4861	2007.1	8897	44.294	453		3158.1	.7347	2258.5	6617	45.114
404		2814.4	.4904	2012.2	8842	44.311	454		3165.1	.7404	2263.5	6580	45.129
405	−54.67	2821.4	.4947	2017.2	8787	44.328	455	−4.67	3172.1	.7462	2268.6	6544	45.145
406		2828.5	.4990	2022.2	8732	44.346	456		3179.2	.7520	2273.6	6507	45.160
407		2835.5	.5033	2027.2	8678	44.363	457		3186.2	.7579	2278.6	6471	45.176
408		2842.5	.5077	2032.2	8624	44.380	458		3193.2	.7637	2283.7	6435	45.191
409		2849.5	.5121	2037.3	8571	44.397	459		3200.2	.7697	2288.7	6400	45.206
410	−49.67	2856.5	.5165	2042.3	8518	44.414	460	.33	3207.2	.7756	2293.8	6365	45.221
411		2863.5	.5210	2047.3	8466	44.431	461		3214.3	.7816	2298.8	6330	45.237
412		2870.5	.5255	2052.3	8414	44.448	462		3221.3	.7876	2303.8	6295	45.252
413		2877.5	.5300	2057.3	8362	44.465	463		3228.3	.7936	2308.9	6261	45.267
414		2884.5	.5345	2062.4	8311	44.482	464		3235.3	.7997	2313.9	6226	45.282
415	−44.67	2891.5	.5391	2067.4	8261	44.499	465	5.33	3242.4	.8058	2318.9	6193	45.297
416		2898.5	.5437	2072.4	8211	44.516	466		3249.4	.8120	2324.0	6159	45.312
417		2905.5	.5483	2077.4	8161	44.533	467		3256.4	.8182	2329.0	6126	45.328
418		2912.6	.5530	2082.5	8112	44.550	468		3263.4	.8244	2334.1	6092	45.343
419		2919.6	.5577	2087.5	8063	44.566	469		3270.5	.8306	2339.1	6059	45.358
420	−39.67	2926.6	.5624	2092.5	8014	44.583	470	10.33	3277.5	.8369	2344.1	6027	45.373
421		2933.6	.5671	2097.5	7966	44.600	471		3284.5	.8432	2349.2	5994	45.387
422		2940.6	.5719	2102.6	7919	44.616	472		3291.5	.8496	2354.2	5962	45.402
423		2947.6	.5767	2107.6	7871	44.633	473		3298.6	.8559	2359.3	5930	45.417
424		2954.6	.5815	2112.6	7824	44.650	474		3305.6	.8624	2364.3	5899	45.432
425	−34.67	2961.6	.5864	2117.6	7778	44.666	475	15.33	3312.6	.8688	2369.3	5867	45.447
426		2968.6	.5913	2122.7	7732	44.683	476		3319.6	.8753	2374.4	5836	45.462
427		2975.7	.5962	2127.7	7686	44.699	477		3326.7	.8818	2379.4	5805	45.476
428		2982.7	.6011	2132.7	7641	44.715	478		3333.7	.8884	2384.5	5774	45.491
429		2989.7	.6061	2137.8	7596	44.732	479		3340.7	.8950	2389.5	5743	45.506
430	−29.67	2996.7	.6111	2142.8	7551	44.748	480	20.33	3347.8	.9016	2394.5	5713	45.520
431		3003.7	.6162	2147.8	7507	44.764	481		3354.8	.9083	2399.6	5683	45.535
432		3010.7	.6212	2152.8	7463	44.781	482		3361.8	.9150	2404.6	5653	45.550
433		3017.7	.6263	2157.9	7419	44.797	483		3368.9	.9217	2409.7	5623	45.564
434		3024.8	.6314	2162.9	7376	44.813	484		3375.9	.9285	2414.7	5594	45.579
435	−24.67	3031.8	.6366	2167.9	7333	44.829	485	25.33	3382.9	.9353	2419.8	5565	45.593
436		3038.8	.6418	2173.0	7291	44.845	486		3389.9	.9422	2424.8	5536	45.608
437		3045.8	.6470	2178.0	7248	44.861	487		3397.0	.9491	2429.9	5507	45.622
438		3052.8	.6522	2183.0	7207	44.877	488		3404.0	.9560	2434.9	5478	45.637
439		3059.8	.6575	2188.0	7165	44.893	489		3411.0	.9629	2439.9	5450	45.651
440	−19.67	3066.9	.6628	2193.1	7124	44.909	490	30.33	3418.1	.9699	2445.0	5422	45.665
441		3073.9	.6682	2198.1	7083	44.925	491		3425.1	.9769	2450.0	5394	45.680
442		3080.9	.6735	2203.1	7042	44.941	492		3432.1	.9840	2455.1	5366	45.694
443		3087.9	.6789	2208.2	7002	44.957	493		3439.2	.9911	2460.1	5338	45.708
444		3094.9	.6844	2213.2	6962	44.973	494		3446.2	.9982	2465.2	5311	45.723
445	−14.67	3101.9	.6898	2218.2	6923	44.989	495	35.33	3453.2	1.0054	2470.2	5283	45.737
446		3109.0	.6953	2223.3	6884	45.004	496		3460.3	1.0126	2475.3	5256	45.751
447		3116.0	.7008	2228.3	6845	45.020	497		3467.3	1.0199	2480.3	5230	45.765
448		3123.0	.7064	2233.3	6806	45.036	498		3474.3	1.0272	2485.4	5203	45.779
449		3130.0	.7120	2238.4	6768	45.052	499		3481.4	1.0345	2490.4	5176	45.793

Table 3　Products—400% Theoretical Air (for One Pound-Mole)

T	t	$\bar{h}$	p_r	$\bar{u}$	v_r	$\bar{\phi}$	T	t	$\bar{h}$	p_r	$\bar{u}$	v_r	$\bar{\phi}$
500	40.33	3488.4	1.0419	2495.5	5150	45.808	550	90.33	3840.6	1.4610	2748.4	4040	46.479
501		3495.4	1.0493	2500.5	5124	45.822	551		3847.7	1.4704	2753.5	4021	46.492
502		3502.5	1.0567	2505.6	5098	45.836	552		3854.7	1.4799	2758.6	4003	46.505
503		3509.5	1.0642	2510.6	5072	45.850	553		3861.8	1.4895	2763.6	3984	46.517
504		3516.6	1.0717	2515.7	5047	45.864	554		3868.9	1.4991	2768.7	3966	46.530
505	45.33	3523.6	1.0793	2520.7	5021	45.878	555	95.33	3875.9	1.5087	2773.8	3948	46.543
506		3530.6	1.0868	2525.8	4996	45.892	556		3883.0	1.5184	2778.8	3930	46.556
507		3537.7	1.0945	2530.8	4971	45.905	557		3890.0	1.5281	2783.9	3912	46.568
508		3544.7	1.1021	2535.9	4946	45.919	558		3897.1	1.5379	2789.0	3894	46.581
509		3551.7	1.1099	2540.9	4922	45.933	559		3904.1	1.5477	2794.0	3876	46.593
510	50.33	3558.8	1.1176	2546.0	4897	45.947	560	100.33	3911.2	1.5576	2799.1	3858	46.606
511		3565.8	1.1254	2551.0	4873	45.961	561		3918.3	1.5675	2804.2	3841	46.619
512		3572.9	1.1332	2556.1	4849	45.974	562		3925.3	1.5774	2809.3	3823	46.631
513		3579.9	1.1411	2561.2	4825	45.988	563		3932.4	1.5874	2814.3	3806	46.644
514		3586.9	1.1490	2566.2	4801	46.002	564		3939.4	1.5975	2819.4	3789	46.656
515	55.33	3594.0	1.1569	2571.3	4777	46.016	565	105.33	3946.5	1.6076	2824.5	3772	46.669
516		3601.0	1.1649	2576.3	4754	46.029	566		3953.5	1.6177	2829.6	3755	46.681
517		3608.1	1.1729	2581.4	4730	46.043	567		3960.6	1.6279	2834.6	3738	46.694
518		3615.1	1.1810	2586.4	4707	46.057	568		3967.7	1.6381	2839.7	3721	46.706
519		3622.1	1.1891	2591.5	4684	46.070	569		3974.7	1.6484	2844.8	3704	46.719
520	60.33	3629.2	1.1973	2596.5	4661	46.084	570	110.33	3981.8	1.6587	2849.9	3688	46.731
521		3636.2	1.2054	2601.6	4638	46.097	571		3988.9	1.6691	2854.9	3671	46.743
522		3643.3	1.2137	2606.7	4616	46.111	572		3995.9	1.6795	2860.0	3655	46.756
523		3650.3	1.2219	2611.7	4593	46.124	573		4003.0	1.6900	2865.1	3639	46.768
524		3657.4	1.2302	2616.8	4571	46.138	574		4010.0	1.7005	2870.2	3622	46.780
525	65.33	3664.4	1.2386	2621.8	4549	46.151	575	115.33	4017.1	1.7111	2875.2	3606	46.793
526		3671.4	1.2470	2626.9	4527	46.164	576		4024.2	1.7217	2880.3	3590	46.805
527		3678.5	1.2554	2631.9	4505	46.178	577		4031.2	1.7323	2885.4	3574	46.817
528		3685.5	1.2639	2637.0	4483	46.191	578		4038.3	1.7430	2890.5	3559	46.829
529		3692.6	1.2724	2642.1	4462	46.205	579		4045.4	1.7538	2895.6	3543	46.842
530	70.33	3699.6	1.2809	2647.1	4440	46.218	580	120.33	4052.4	1.7646	2900.6	3527	46.854
531		3706.7	1.2895	2652.2	4419	46.231	581		4059.5	1.7754	2905.7	3512	46.866
532		3713.7	1.2982	2657.2	4398	46.244	582		4066.6	1.7863	2910.8	3496	46.878
533		3720.8	1.3069	2662.3	4377	46.258	583		4073.6	1.7973	2915.9	3481	46.890
534		3727.8	1.3156	2667.4	4356	46.271	584		4080.7	1.8083	2921.0	3466	46.902
535	75.33	3734.9	1.3243	2672.4	4335	46.284	585	125.33	4087.8	1.8193	2926.0	3451	46.915
536		3741.9	1.3332	2677.5	4315	46.297	586		4094.8	1.8304	2931.1	3436	46.927
537		3749.0	1.3420	2682.6	4294	46.310	587		4101.9	1.8415	2936.2	3421	46.939
538		3756.0	1.3509	2687.6	4274	46.323	588		4109.0	1.8527	2941.3	3406	46.951
539		3763.1	1.3598	2692.7	4254	46.336	589		4116.0	1.8640	2946.4	3391	46.963
540	80.33	3770.1	1.3688	2697.8	4234	46.350	590	130.33	4123.1	1.8753	2951.5	3376	46.975
541		3777.2	1.3778	2702.8	4214	46.363	591		4130.2	1.8866	2956.5	3362	46.987
542		3784.2	1.3869	2707.9	4194	46.376	592		4137.3	1.8980	2961.6	3347	46.999
543		3791.3	1.3960	2712.9	4174	46.389	593		4144.3	1.9094	2966.7	3333	47.011
544		3798.3	1.4051	2718.0	4155	46.402	594		4151.4	1.9209	2971.8	3318	47.023
545	85.33	3805.4	1.4143	2723.1	4135	46.415	595	135.33	4158.5	1.9325	2976.9	3304	47.034
546		3812.4	1.4236	2728.1	4116	46.427	596		4165.6	1.9441	2982.0	3290	47.046
547		3819.5	1.4329	2733.2	4097	46.440	597		4172.6	1.9557	2987.1	3276	47.058
548		3826.5	1.4422	2738.3	4078	46.453	598		4179.7	1.9674	2992.2	3262	47.070
549		3833.6	1.4516	2743.3	4059	46.466	599		4186.8	1.9791	2997.2	3248	47.082

Table 3 Products—400% Theoretical Air (for One Pound-Mole)

T	t	$\bar{h}$	p_r	$\bar{u}$	v_r	$\bar{\phi}$	T	t	$\bar{h}$	p_r	$\bar{u}$	v_r	$\bar{\phi}$
600	140.33	4193.8	1.991	3002.3	3234	47.094	650	190.33	4548.2	2.649	3257.4	2633	47.661
601		4200.9	2.003	3007.4	3220	47.105	651		4555.3	2.664	3262.5	2623	47.672
602		4208.0	2.015	3012.5	3207	47.117	652		4562.4	2.678	3267.6	2612	47.683
603		4215.1	2.027	3017.6	3193	47.129	653		4569.5	2.693	3272.7	2602	47.694
604		4222.1	2.039	3022.7	3179	47.141	654		4576.6	2.708	3277.8	2592	47.704
605	145.33	4229.2	2.051	3027.8	3166	47.152	655	195.33	4583.7	2.723	3282.9	2582	47.715
606		4236.3	2.063	3032.9	3153	47.164	656		4590.8	2.738	3288.1	2572	47.726
607		4243.4	2.075	3038.0	3139	47.176	657		4597.9	2.753	3293.2	2561	47.737
608		4250.5	2.087	3043.1	3126	47.187	658		4605.0	2.768	3298.3	2551	47.748
609		4257.5	2.099	3048.1	3113	47.199	659		4612.1	2.783	3303.4	2541	47.758
610	150.33	4264.6	2.112	3053.2	3100	47.211	660	200.33	4619.2	2.798	3308.5	2532	47.769
611		4271.7	2.124	3058.3	3087	47.222	661		4626.3	2.813	3313.6	2522	47.780
612		4278.8	2.137	3063.4	3074	47.234	662		4633.4	2.828	3318.8	2512	47.791
613		4285.9	2.149	3068.5	3061	47.245	663		4640.5	2.844	3323.9	2502	47.801
614		4292.9	2.162	3073.6	3048	47.257	664		4647.6	2.859	3329.0	2492	47.812
615	155.33	4300.0	2.174	3078.7	3036	47.268	665	205.33	4654.7	2.874	3334.1	2483	47.823
616		4307.1	2.187	3083.8	3023	47.280	666		4661.8	2.890	3339.2	2473	47.834
617		4314.2	2.199	3088.9	3010	47.291	667		4668.9	2.905	3344.4	2464	47.844
618		4321.3	2.212	3094.0	2998	47.303	668		4676.0	2.921	3349.5	2454	47.855
619		4328.4	2.225	3099.1	2986	47.314	669		4683.1	2.937	3354.6	2445	47.866
620	160.33	4335.4	2.238	3104.2	2973	47.326	670	210.33	4690.3	2.953	3359.7	2435	47.876
621		4342.5	2.251	3109.3	2961	47.337	671		4697.4	2.968	3364.9	2426	47.887
622		4349.6	2.264	3114.4	2949	47.349	672		4704.5	2.984	3370.0	2417	47.897
623		4356.7	2.277	3119.5	2937	47.360	673		4711.6	3.000	3375.1	2407	47.908
624		4363.8	2.290	3124.6	2925	47.371	674		4718.7	3.016	3380.2	2398	47.918
625	165.33	4370.9	2.303	3129.7	2913	47.383	675	215.33	4725.8	3.032	3385.4	2389	47.929
626		4377.9	2.316	3134.8	2901	47.394	676		4732.9	3.048	3390.5	2380	47.940
627		4385.0	2.329	3139.9	2889	47.405	677		4740.0	3.064	3395.6	2371	47.950
628		4392.1	2.343	3145.0	2877	47.417	678		4747.2	3.081	3400.7	2362	47.961
629		4399.2	2.356	3150.1	2865	47.428	679		4754.3	3.097	3405.9	2353	47.971
630	170.33	4406.3	2.369	3155.2	2854	47.439	680	220.33	4761.4	3.113	3411.0	2344	47.981
631		4413.4	2.383	3160.3	2842	47.450	681		4768.5	3.130	3416.1	2335	47.992
632		4420.5	2.396	3165.4	2830	47.462	682		4775.6	3.146	3421.3	2326	48.002
633		4427.6	2.410	3170.5	2819	47.473	683		4782.7	3.163	3426.4	2317	48.013
634		4434.7	2.423	3175.6	2807	47.484	684		4789.8	3.180	3431.5	2309	48.023
635	175.33	4441.7	2.437	3180.7	2796	47.495	685	225.33	4797.0	3.196	3436.7	2300	48.034
636		4448.8	2.451	3185.8	2785	47.506	686		4804.1	3.213	3441.8	2291	48.044
637		4455.9	2.465	3190.9	2774	47.517	687		4811.2	3.230	3446.9	2283	48.054
638		4463.0	2.479	3196.0	2762	47.529	688		4818.3	3.247	3452.0	2274	48.065
639		4470.1	2.492	3201.2	2751	47.540	689		4825.4	3.264	3457.2	2266	48.075
640	180.33	4477.2	2.506	3206.3	2740	47.551	690	230.33	4832.6	3.281	3462.3	2257	48.085
641		4484.3	2.520	3211.4	2729	47.562	691		4839.7	3.298	3467.5	2249	48.096
642		4491.4	2.534	3216.5	2718	47.573	692		4846.8	3.315	3472.6	2240	48.106
643		4498.5	2.549	3221.6	2708	47.584	693		4853.9	3.332	3477.7	2232	48.116
644		4505.6	2.563	3226.7	2697	47.595	694		4861.1	3.349	3482.9	2224	48.127
645	185.33	4512.7	2.577	3231.8	2686	47.606	695	235.33	4868.2	3.367	3488.0	2215	48.137
646		4519.8	2.591	3236.9	2675	47.617	696		4875.3	3.384	3493.1	2207	48.147
647		4526.9	2.606	3242.0	2665	47.628	697		4882.4	3.402	3498.3	2199	48.157
648		4534.0	2.620	3247.1	2654	47.639	698		4889.6	3.419	3503.4	2191	48.168
649		4541.1	2.635	3252.3	2644	47.650	699		4896.7	3.437	3508.6	2183	48.178

Table 3 Products—400% Theoretical Air (for One Pound-Mole)

T	t	$\bar{h}$	p_r	$\bar{u}$	v_r	$\bar{\phi}$	T	t	$\bar{h}$	p_r	$\bar{u}$	v_r	$\bar{\phi}$
700	240.33	4903.8	3.454	3513.7	2174.6	48.188	750	290.33	5260.9	4.427	3771.5	1818.0	48.681
701		4910.9	3.472	3518.8	2166.6	48.198	751		5268.0	4.449	3776.7	1811.7	48.690
702		4918.1	3.490	3524.0	2158.6	48.208	752		5275.2	4.470	3781.8	1805.4	48.700
703		4925.2	3.508	3529.1	2150.7	48.218	753		5282.4	4.491	3787.0	1799.2	48.709
704		4932.3	3.526	3534.3	2142.8	48.229	754		5289.5	4.513	3792.2	1793.0	48.719
705	245.33	4939.4	3.544	3539.4	2134.9	48.239	755	295.33	5296.7	4.535	3797.4	1786.8	48.728
706		4946.6	3.562	3544.6	2127.1	48.249	756		5303.9	4.556	3802.5	1780.6	48.738
707		4953.7	3.580	3549.7	2119.3	48.259	757		5311.0	4.578	3807.7	1774.5	48.747
708		4960.8	3.598	3554.8	2111.5	48.269	758		5318.2	4.600	3812.9	1768.4	48.757
709		4968.0	3.617	3560.0	2103.8	48.279	759		5325.3	4.622	3818.1	1762.3	48.766
710	250.33	4975.1	3.635	3565.1	2096.2	48.289	760	300.33	5332.5	4.644	3823.2	1756.3	48.775
711		4982.2	3.653	3570.3	2088.5	48.299	761		5339.7	4.666	3828.4	1750.3	48.785
712		4989.4	3.672	3575.4	2080.9	48.309	762		5346.8	4.688	3833.6	1744.3	48.794
713		4996.5	3.690	3580.6	2073.4	48.319	763		5354.0	4.710	3838.8	1738.3	48.804
714		5003.6	3.709	3585.7	2065.9	48.329	764		5361.2	4.733	3844.0	1732.4	48.813
715	255.33	5010.8	3.728	3590.9	2058.4	48.339	765	305.33	5368.3	4.755	3849.2	1726.5	48.822
716		5017.9	3.746	3596.0	2050.9	48.349	766		5375.5	4.777	3854.3	1720.6	48.832
717		5025.0	3.765	3601.2	2043.5	48.359	767		5382.7	4.800	3859.5	1714.8	48.841
718		5032.2	3.784	3606.3	2036.1	48.369	768		5389.8	4.823	3864.7	1709.0	48.851
719		5039.3	3.803	3611.5	2028.8	48.379	769		5397.0	4.845	3869.9	1703.2	48.860
720	260.33	5046.5	3.822	3616.6	2021.5	48.389	770	310.33	5404.2	4.868	3875.1	1697.4	48.869
721		5053.6	3.841	3621.8	2014.2	48.399	771		5411.4	4.891	3880.3	1691.7	48.878
722		5060.7	3.861	3626.9	2007.0	48.409	772		5418.5	4.914	3885.4	1685.9	48.888
723		5067.9	3.880	3632.1	1999.8	48.419	773		5425.7	4.937	3890.6	1680.3	48.897
724		5075.0	3.899	3637.2	1992.6	48.428	774		5432.9	4.960	3895.8	1674.6	48.906
725	265.33	5082.2	3.919	3642.4	1985.5	48.438	775	315.33	5440.0	4.983	3901.0	1668.9	48.916
726		5089.3	3.938	3647.6	1978.4	48.448	776		5447.2	5.007	3906.2	1663.3	48.925
727		5096.4	3.958	3652.7	1971.4	48.458	777		5454.4	5.030	3911.4	1657.7	48.934
728		5103.6	3.977	3657.9	1964.3	48.468	778		5461.6	5.053	3916.6	1652.2	48.943
729		5110.7	3.997	3663.0	1957.3	48.478	779		5468.8	5.077	3921.8	1646.6	48.953
730	270.33	5117.9	4.017	3668.2	1950.4	48.487	780	320.33	5475.9	5.100	3927.0	1641.1	48.962
731		5125.0	4.036	3673.3	1943.5	48.497	781		5483.1	5.124	3932.1	1635.6	48.971
732		5132.2	4.056	3678.5	1936.6	48.507	782		5490.3	5.148	3937.3	1630.2	48.980
733		5139.3	4.076	3683.7	1929.7	48.517	783		5497.5	5.172	3942.5	1624.7	48.989
734		5146.5	4.096	3688.8	1922.9	48.526	784		5504.6	5.196	3947.7	1619.3	48.998
735	275.33	5153.6	4.117	3694.0	1916.1	48.536	785	325.33	5511.8	5.220	3952.9	1613.9	49.008
736		5160.7	4.137	3699.2	1909.3	48.546	786		5519.0	5.244	3958.1	1608.6	49.017
737		5167.9	4.157	3704.3	1902.6	48.556	787		5526.2	5.268	3963.3	1603.2	49.026
738		5175.0	4.177	3709.5	1895.9	48.565	788		5533.4	5.292	3968.5	1597.9	49.035
739		5182.2	4.198	3714.6	1889.2	48.575	789		5540.6	5.317	3973.7	1592.6	49.044
740	280.33	5189.3	4.218	3719.8	1882.6	48.585	790	330.33	5547.7	5.341	3978.9	1587.3	49.053
741		5196.5	4.239	3725.0	1876.0	48.594	791		5554.9	5.365	3984.1	1582.1	49.062
742		5203.7	4.259	3730.1	1869.4	48.604	792		5562.1	5.390	3989.3	1576.9	49.071
743		5210.8	4.280	3735.3	1862.9	48.614	793		5569.3	5.415	3994.5	1571.7	49.080
744		5218.0	4.301	3740.5	1856.4	48.623	794		5576.5	5.439	3999.7	1566.5	49.090
745	285.33	5225.1	4.322	3745.6	1849.9	48.633	795	335.33	5583.7	5.464	4004.9	1561.3	49.099
746		5232.3	4.343	3750.8	1843.5	48.642	796		5590.9	5.489	4010.1	1556.2	49.108
747		5239.4	4.364	3756.0	1837.0	48.652	797		5598.0	5.514	4015.3	1551.1	49.117
748		5246.6	4.385	3761.2	1830.7	48.662	798		5605.2	5.539	4020.5	1546.0	49.126
749		5253.7	4.406	3766.3	1824.3	48.671	799		5612.4	5.565	4025.7	1540.9	49.135

Table 3 Products—400% Theoretical Air (for One Pound-Mole)

T	t	$\bar{h}$	p_r	$\bar{u}$	v_r	$\bar{\phi}$	T	t	$\bar{h}$	p_r	$\bar{u}$	v_r	$\bar{\phi}$
800	340.33	5619.6	5.590	4030.9	1535.9	49.144	850	390.33	5980.2	6.966	4292.2	1309.4	49.581
801		5626.8	5.615	4036.1	1530.9	49.153	851		5987.4	6.996	4297.4	1305.4	49.589
802		5634.0	5.641	4041.4	1525.9	49.162	852		5994.6	7.026	4302.7	1301.3	49.598
803		5641.2	5.666	4046.6	1520.9	49.171	853		6001.9	7.056	4307.9	1297.3	49.606
804		5648.4	5.692	4051.8	1515.9	49.180	854		6009.1	7.086	4313.2	1293.3	49.615
805	345.33	5655.6	5.717	4057.0	1511.0	49.188	855	395.33	6016.3	7.117	4318.4	1289.3	49.623
806		5662.8	5.743	4062.2	1506.1	49.197	856		6023.6	7.147	4323.7	1285.3	49.632
807		5670.0	5.769	4067.4	1501.2	49.206	857		6030.8	7.177	4328.9	1281.4	49.640
808		5677.2	5.795	4072.6	1496.3	49.215	858		6038.0	7.208	4334.2	1277.4	49.649
809		5684.4	5.821	4077.8	1491.5	49.224	859		6045.3	7.239	4339.4	1273.5	49.657
810	350.33	5691.6	5.847	4083.0	1486.6	49.233	860	400.33	6052.5	7.269	4344.7	1269.6	49.665
811		5698.8	5.873	4088.2	1481.8	49.242	861		6059.7	7.300	4349.9	1265.7	49.674
812		5706.0	5.900	4093.5	1477.0	49.251	862		6067.0	7.331	4355.2	1261.8	49.682
813		5713.2	5.926	4098.7	1472.3	49.260	863		6074.2	7.362	4360.4	1257.9	49.691
814		5720.4	5.953	4103.9	1467.5	49.269	864		6081.5	7.393	4365.7	1254.1	49.699
815	355.33	5727.6	5.979	4109.1	1462.8	49.277	865	405.33	6088.7	7.425	4370.9	1250.3	49.707
816		5734.8	6.006	4114.3	1458.1	49.286	866		6095.9	7.456	4376.2	1246.4	49.716
817		5742.0	6.032	4119.5	1453.4	49.295	867		6103.2	7.487	4381.4	1242.6	49.724
818		5749.2	6.059	4124.8	1448.7	49.304	868		6110.4	7.519	4386.7	1238.9	49.732
819		5756.4	6.086	4130.0	1444.1	49.313	869		6117.7	7.551	4392.0	1235.1	49.741
820	360.33	5763.6	6.113	4135.2	1439.5	49.321	870	410.33	6124.9	7.582	4397.2	1231.3	49.749
821		5770.8	6.140	4140.4	1434.9	49.330	871		6132.2	7.614	4402.5	1227.6	49.757
822		5778.0	6.168	4145.7	1430.3	49.339	872		6139.4	7.646	4407.7	1223.9	49.766
823		5785.2	6.195	4150.9	1425.7	49.348	873		6146.7	7.678	4413.0	1220.1	49.774
824		5792.4	6.222	4156.1	1421.2	49.357	874		6153.9	7.710	4418.3	1216.4	49.782
825	365.33	5799.7	6.250	4161.3	1416.6	49.365	875	415.33	6161.2	7.743	4423.5	1212.8	49.791
826		5806.9	6.277	4166.6	1412.1	49.374	876		6168.4	7.775	4428.8	1209.1	49.799
827		5814.1	6.305	4171.8	1407.6	49.383	877		6175.7	7.807	4434.1	1205.4	49.807
828		5821.3	6.333	4177.0	1403.2	49.391	878		6182.9	7.840	4439.3	1201.8	49.815
829		5828.5	6.360	4182.2	1398.7	49.400	879		6190.2	7.873	4444.6	1198.2	49.824
830	370.33	5835.7	6.388	4187.5	1394.3	49.409	880	420.33	6197.4	7.905	4449.9	1194.6	49.832
831		5842.9	6.416	4192.7	1389.9	49.418	881		6204.7	7.938	4455.1	1191.0	49.840
832		5850.2	6.444	4197.9	1385.5	49.426	882		6211.9	7.971	4460.4	1187.4	49.848
833		5857.4	6.473	4203.1	1381.1	49.435	883		6219.2	8.004	4465.7	1183.8	49.857
834		5864.6	6.501	4208.4	1376.7	49.444	884		6226.4	8.038	4470.9	1180.3	49.865
835	375.33	5871.8	6.529	4213.6	1372.4	49.452	885	425.33	6233.7	8.071	4476.2	1176.8	49.873
836		5879.0	6.558	4218.8	1368.1	49.461	886		6241.0	8.104	4481.5	1173.2	49.881
837		5886.2	6.586	4224.1	1363.8	49.469	887		6248.2	8.138	4486.8	1169.7	49.889
838		5893.5	6.615	4229.3	1359.5	49.478	888		6255.5	8.171	4492.0	1166.2	49.898
839		5900.7	6.644	4234.5	1355.2	49.487	889		6262.7	8.205	4497.3	1162.7	49.906
840	380.33	5907.9	6.673	4239.8	1351.0	49.495	890	430.33	6270.0	8.239	4502.6	1159.3	49.914
841		5915.1	6.701	4245.0	1346.7	49.504	891		6277.3	8.273	4507.9	1155.8	49.922
842		5922.4	6.731	4250.3	1342.5	49.512	892		6284.5	8.307	4513.1	1152.4	49.930
843		5929.6	6.760	4255.5	1338.3	49.521	893		6291.8	8.341	4518.4	1149.0	49.938
844		5936.8	6.789	4260.7	1334.2	49.530	894		6299.1	8.375	4523.7	1145.5	49.947
845	385.33	5944.0	6.818	4266.0	1330.0	49.538	895	435.33	6306.3	8.409	4529.0	1142.1	49.955
846		5951.3	6.848	4271.2	1325.8	49.547	896		6313.6	8.444	4534.3	1138.8	49.963
847		5958.5	6.877	4276.5	1321.7	49.555	897		6320.9	8.478	4539.5	1135.4	49.971
848		5965.7	6.907	4281.7	1317.6	49.564	898		6328.1	8.513	4544.8	1132.0	49.979
849		5972.9	6.936	4286.9	1313.5	49.572	899		6335.4	8.548	4550.1	1128.7	49.987

Table 3   Products—400% Theoretical Air (for One Pound-Mole)

T	t	$\bar{h}$	p_r	$\bar{u}$	v_r	$\bar{\phi}$	T	t	$\bar{h}$	p_r	$\bar{u}$	v_r	$\bar{\phi}$
900	440.33	6342.7	8.583	4555.4	1125.3	49.995	950	490.33	6707.3	10.467	4820.7	974.0	50.389
901		6349.9	8.618	4560.7	1122.0	50.003	951		6714.6	10.508	4826.0	971.2	50.397
902		6357.2	8.653	4566.0	1118.7	50.011	952		6721.9	10.549	4831.4	968.5	50.405
903		6364.5	8.688	4571.3	1115.4	50.019	953		6729.2	10.590	4836.7	965.8	50.413
904		6371.8	8.723	4576.5	1112.1	50.027	954		6736.5	10.631	4842.0	963.0	50.420
905	445.33	6379.0	8.758	4581.8	1108.9	50.035	955	495.33	6743.9	10.672	4847.4	960.3	50.428
906		6386.3	8.794	4587.1	1105.6	50.044	956		6751.2	10.713	4852.7	957.6	50.436
907		6393.6	8.830	4592.4	1102.4	50.052	957		6758.5	10.754	4858.0	955.0	50.443
908		6400.9	8.865	4597.7	1099.1	50.060	958		6765.8	10.796	4863.4	952.3	50.451
909		6408.1	8.901	4603.0	1095.9	50.068	959		6773.1	10.837	4868.7	949.6	50.458
910	450.33	6415.4	8.937	4608.3	1092.7	50.076	960	500.33	6780.5	10.879	4874.0	947.0	50.466
911		6422.7	8.973	4613.6	1089.5	50.084	961		6787.8	10.921	4879.4	944.3	50.474
912		6430.0	9.009	4618.9	1086.3	50.092	962		6795.1	10.963	4884.7	941.7	50.481
913		6437.3	9.046	4624.2	1083.2	50.100	963		6802.5	11.005	4890.1	939.1	50.489
914		6444.5	9.082	4629.5	1080.0	50.108	964		6809.8	11.047	4895.4	936.4	50.497
915	455.33	6451.8	9.118	4634.8	1076.9	50.115	965	505.33	6817.1	11.090	4900.7	933.8	50.504
916		6459.1	9.155	4640.1	1073.7	50.123	966		6824.4	11.132	4906.1	931.2	50.512
917		6466.4	9.192	4645.4	1070.6	50.131	967		6831.7	11.175	4911.4	928.6	50.519
918		6473.7	9.229	4650.7	1067.5	50.139	968		6839.1	11.217	4916.8	926.1	50.527
919		6481.0	9.266	4656.0	1064.4	50.147	969		6846.4	11.260	4922.1	923.5	50.534
920	460.33	6488.3	9.303	4661.3	1061.3	50.155	970	510.33	6853.7	11.303	4927.5	920.9	50.542
921		6495.5	9.340	4666.6	1058.2	50.163	971		6861.1	11.346	4932.8	918.4	50.550
922		6502.8	9.377	4671.9	1055.2	50.171	972		6868.4	11.390	4938.1	915.8	50.557
923		6510.1	9.414	4677.2	1052.1	50.179	973		6875.7	11.433	4943.5	913.3	50.565
924		6517.4	9.452	4682.5	1049.1	50.187	974		6883.1	11.476	4948.8	910.8	50.572
925	465.33	6524.7	9.490	4687.8	1046.1	50.195	975	515.33	6890.4	11.520	4954.2	908.3	50.580
926		6532.0	9.527	4693.1	1043.0	50.203	976		6897.7	11.564	4959.5	905.8	50.587
927		6539.3	9.565	4698.4	1040.0	50.210	977		6905.1	11.608	4964.9	903.3	50.595
928		6546.6	9.603	4703.7	1037.0	50.218	978		6912.4	11.651	4970.2	900.8	50.602
929		6553.9	9.641	4709.0	1034.1	50.226	979		6919.8	11.696	4975.6	898.3	50.610
930	470.33	6561.2	9.679	4714.3	1031.1	50.234	980	520.33	6927.1	11.740	4981.0	895.8	50.617
931		6568.5	9.718	4719.6	1028.1	50.242	981		6934.4	11.784	4986.3	893.4	50.625
932		6575.8	9.756	4724.9	1025.2	50.250	982		6941.8	11.829	4991.7	890.9	50.632
933		6583.1	9.795	4730.3	1022.2	50.258	983		6949.1	11.873	4997.0	888.5	50.640
934		6590.4	9.833	4735.6	1019.3	50.265	984		6956.5	11.918	5002.4	886.0	50.647
935	475.33	6597.7	9.872	4740.9	1016.4	50.273	985	525.33	6963.8	11.963	5007.7	883.6	50.655
936		6605.0	9.911	4746.2	1013.5	50.281	986		6971.2	12.008	5013.1	881.2	50.662
937		6612.3	9.950	4751.5	1010.6	50.289	987		6978.5	12.053	5018.5	878.8	50.670
938		6619.6	9.989	4756.8	1007.7	50.297	988		6985.9	12.098	5023.8	876.4	50.677
939		6626.9	10.028	4762.2	1004.9	50.304	989		6993.2	12.144	5029.2	874.0	50.684
940	480.33	6634.2	10.068	4767.5	1002.0	50.312	990	530.33	7000.6	12.189	5034.6	871.6	50.692
941		6641.5	10.107	4772.8	999.1	50.320	991		7007.9	12.235	5039.9	869.2	50.699
942		6648.8	10.147	4778.1	996.3	50.328	992		7015.3	12.280	5045.3	866.9	50.707
943		6656.1	10.186	4783.4	993.5	50.335	993		7022.6	12.326	5050.7	864.5	50.714
944		6663.4	10.226	4788.8	990.7	50.343	994		7030.0	12.372	5056.0	862.2	50.721
945	485.33	6670.7	10.266	4794.1	987.8	50.351	995	535.33	7037.3	12.419	5061.4	859.8	50.729
946		6678.0	10.306	4799.4	985.1	50.359	996		7044.7	12.465	5066.8	857.5	50.736
947		6685.3	10.346	4804.7	982.3	50.366	997		7052.0	12.511	5072.1	855.2	50.744
948		6692.6	10.386	4810.1	979.5	50.374	998		7059.4	12.558	5077.5	852.9	50.751
949		6700.0	10.427	4815.4	976.7	50.382	999		7066.7	12.605	5082.9	850.6	50.758

Table 3   Products—400% Theoretical Air (for One Pound-Mole)

T	t	$\bar{h}$	p_r	$\bar{u}$	v_r	$\bar{\phi}$	T	t	$\bar{h}$	p_r	$\bar{u}$	v_r	$\bar{\phi}$
1000	540.33	7074.1	12.651	5088.2	848.2	50.766	1050	590.33	7443.3	15.167	5358.1	742.9	51.126
1001		7081.5	12.698	5093.6	846.0	50.773	1051		7450.7	15.221	5363.5	741.0	51.133
1002		7088.8	12.745	5099.0	843.7	50.780	1052		7458.1	15.275	5368.9	739.1	51.140
1003		7096.2	12.793	5104.4	841.4	50.788	1053		7465.5	15.330	5374.4	737.1	51.147
1004		7103.6	12.840	5109.8	839.1	50.795	1054		7472.9	15.384	5379.8	735.2	51.154
1005	545.33	7110.9	12.887	5115.1	836.9	50.802	1055	595.33	7480.3	15.439	5385.2	733.3	51.161
1006		7118.3	12.935	5120.5	834.6	50.810	1056		7487.7	15.493	5390.6	731.4	51.168
1007		7125.6	12.983	5125.9	832.4	50.817	1057		7495.1	15.548	5396.1	729.6	51.175
1008		7133.0	13.031	5131.3	830.1	50.824	1058		7502.5	15.603	5401.5	727.7	51.182
1009		7140.4	13.079	5136.6	827.9	50.832	1059		7510.0	15.658	5406.9	725.8	51.189
1010	550.33	7147.8	13.127	5142.0	825.7	50.839	1060	600.33	7517.4	15.714	5412.4	723.9	51.196
1011		7155.1	13.175	5147.4	823.5	50.846	1061		7524.8	15.769	5417.8	722.1	51.203
1012		7162.5	13.224	5152.8	821.3	50.854	1062		7532.2	15.825	5423.2	720.2	51.210
1013		7169.9	13.272	5158.2	819.1	50.861	1063		7539.6	15.880	5428.7	718.3	51.217
1014		7177.2	13.321	5163.6	816.9	50.868	1064		7547.0	15.936	5434.1	716.5	51.224
1015	555.33	7184.6	13.370	5169.0	814.7	50.875	1065	605.33	7554.5	15.992	5439.5	714.7	51.231
1016		7192.0	13.419	5174.3	812.5	50.883	1066		7561.9	16.048	5445.0	712.8	51.238
1017		7199.4	13.468	5179.7	810.4	50.890	1067		7569.3	16.105	5450.4	711.0	51.245
1018		7206.7	13.517	5185.1	808.2	50.897	1068		7576.7	16.161	5455.8	709.2	51.252
1019		7214.1	13.566	5190.5	806.1	50.904	1069		7584.2	16.218	5461.3	707.4	51.259
1020	560.33	7221.5	13.616	5195.9	803.9	50.912	1070	610.33	7591.6	16.275	5466.7	705.6	51.266
1021		7228.9	13.666	5201.3	801.8	50.919	1071		7599.0	16.332	5472.2	703.7	51.273
1022		7236.2	13.715	5206.7	799.7	50.926	1072		7606.4	16.389	5477.6	702.0	51.280
1023		7243.6	13.765	5212.1	797.5	50.933	1073		7613.9	16.446	5483.0	700.2	51.287
1024		7251.0	13.815	5217.5	795.4	50.941	1074		7621.3	16.503	5488.5	698.4	51.294
1025	565.33	7258.4	13.866	5222.9	793.3	50.948	1075	615.33	7628.7	16.561	5493.9	696.6	51.301
1026		7265.8	13.916	5228.3	791.2	50.955	1076		7636.2	16.619	5499.4	694.8	51.307
1027		7273.2	13.967	5233.7	789.1	50.962	1077		7643.6	16.677	5504.8	693.1	51.314
1028		7280.5	14.017	5239.1	787.0	50.969	1078		7651.0	16.735	5510.3	691.3	51.321
1029		7287.9	14.068	5244.5	785.0	50.977	1079		7658.5	16.793	5515.7	689.5	51.328
1030	570.33	7295.3	14.119	5249.9	782.9	50.984	1080	620.33	7665.9	16.851	5521.2	687.8	51.335
1031		7302.7	14.170	5255.3	780.8	50.991	1081		7673.3	16.910	5526.6	686.0	51.342
1032		7310.1	14.221	5260.7	778.8	50.998	1082		7680.8	16.968	5532.1	684.3	51.349
1033		7317.5	14.272	5266.1	776.7	51.005	1083		7688.2	17.027	5537.5	682.6	51.356
1034		7324.9	14.324	5271.5	774.7	51.012	1084		7695.6	17.086	5543.0	680.8	51.363
1035	575.33	7332.3	14.376	5276.9	772.6	51.019	1085	625.33	7703.1	17.145	5548.4	679.1	51.369
1036		7339.7	14.427	5282.3	770.6	51.027	1086		7710.5	17.205	5553.9	677.4	51.376
1037		7347.1	14.479	5287.7	768.6	51.034	1087		7718.0	17.264	5559.3	675.7	51.383
1038		7354.4	14.531	5293.1	766.6	51.041	1088		7725.4	17.324	5564.8	674.0	51.390
1039		7361.8	14.584	5298.5	764.6	51.048	1089		7732.9	17.383	5570.3	672.3	51.397
1040	580.33	7369.2	14.636	5303.9	762.6	51.055	1090	630.33	7740.3	17.443	5575.7	670.6	51.404
1041		7376.6	14.688	5309.4	760.6	51.062	1091		7747.7	17.503	5581.2	668.9	51.410
1042		7384.0	14.741	5314.8	758.6	51.069	1092		7755.2	17.564	5586.6	667.2	51.417
1043		7391.4	14.794	5320.2	756.6	51.076	1093		7762.6	17.624	5592.1	665.5	51.424
1044		7398.8	14.847	5325.6	754.6	51.084	1094		7770.1	17.685	5597.6	663.9	51.431
1045	585.33	7406.2	14.900	5331.0	752.7	51.091	1095	635.33	7777.5	17.745	5603.0	662.2	51.438
1046		7413.6	14.953	5336.4	750.7	51.098	1096		7785.0	17.806	5608.5	660.5	51.444
1047		7421.0	15.006	5341.8	748.7	51.105	1097		7792.4	17.867	5614.0	658.9	51.451
1048		7428.4	15.060	5347.3	746.8	51.112	1098		7799.9	17.928	5619.4	657.2	51.458
1049		7435.9	15.114	5352.7	744.8	51.119	1099		7807.3	17.990	5624.9	655.6	51.465

Table 3 Products—400% Theoretical Air (for One Pound-Mole)

T	t	$\bar{h}$	p_r	$\bar{u}$	v_r	$\bar{\phi}$	T	t	$\bar{h}$	p_r	$\bar{u}$	v_r	$\bar{\phi}$
1100	640.33	7814.8	18.051	5630.4	653.9	51.472	1150	690.33	8188.8	21.34	5905.1	578.3	51.804
1101		7822.3	18.113	5635.8	652.3	51.478	1151		8196.3	21.41	5910.6	576.9	51.811
1102		7829.7	18.175	5641.3	650.7	51.485	1152		8203.8	21.48	5916.1	575.5	51.817
1103		7837.2	18.237	5646.8	649.1	51.492	1153		8211.3	21.55	5921.6	574.1	51.824
1104		7844.6	18.299	5652.2	647.4	51.499	1154		8218.8	21.62	5927.1	572.7	51.830
1105	645.33	7852.1	18.361	5657.7	645.8	51.505	1155	695.33	8226.3	21.69	5932.7	571.3	51.837
1106		7859.6	18.424	5663.2	644.2	51.512	1156		8233.8	21.77	5938.2	570.0	51.843
1107		7867.0	18.487	5668.7	642.6	51.519	1157		8241.4	21.84	5943.7	568.6	51.850
1108		7874.5	18.549	5674.1	641.0	51.526	1158		8248.9	21.91	5949.2	567.2	51.856
1109		7881.9	18.612	5679.6	639.4	51.532	1159		8256.4	21.98	5954.8	565.9	51.863
1110	650.33	7889.4	18.676	5685.1	637.8	51.539	1160	700.33	8263.9	22.05	5960.3	564.5	51.869
1111		7896.9	18.739	5690.6	636.3	51.546	1161		8271.4	22.12	5965.8	563.2	51.876
1112		7904.3	18.802	5696.1	634.7	51.553	1162		8278.9	22.20	5971.4	561.8	51.882
1113		7911.8	18.866	5701.5	633.1	51.559	1163		8286.4	22.27	5976.9	560.5	51.889
1114		7919.3	18.930	5707.0	631.5	51.566	1164		8294.0	22.34	5982.4	559.1	51.895
1115	655.33	7926.7	18.994	5712.5	630.0	51.573	1165	705.33	8301.5	22.41	5988.0	557.8	51.901
1116		7934.2	19.058	5718.0	628.4	51.579	1166		8309.0	22.49	5993.5	556.5	51.908
1117		7941.7	19.122	5723.5	626.9	51.586	1167		8316.5	22.56	5999.0	555.1	51.914
1118		7949.2	19.187	5729.0	625.3	51.593	1168		8324.0	22.63	6004.6	553.8	51.921
1119		7956.6	19.251	5734.5	623.8	51.599	1169		8331.6	22.71	6010.1	552.5	51.927
1120	660.33	7964.1	19.316	5739.9	622.2	51.606	1170	710.33	8339.1	22.78	6015.6	551.2	51.934
1121		7971.6	19.381	5745.4	620.7	51.613	1171		8346.6	22.85	6021.2	549.9	51.940
1122		7979.1	19.446	5750.9	619.2	51.619	1172		8354.1	22.93	6026.7	548.6	51.947
1123		7986.5	19.512	5756.4	617.6	51.626	1173		8361.7	23.00	6032.3	547.3	51.953
1124		7994.0	19.577	5761.9	616.1	51.633	1174		8369.2	23.08	6037.8	546.0	51.959
1125	665.33	8001.5	19.643	5767.4	614.6	51.639	1175	715.33	8376.7	23.15	6043.3	544.7	51.966
1126		8009.0	19.709	5772.9	613.1	51.646	1176		8384.3	23.23	6048.9	543.4	51.972
1127		8016.5	19.775	5778.4	611.6	51.653	1177		8391.8	23.30	6054.4	542.1	51.979
1128		8023.9	19.841	5783.9	610.1	51.659	1178		8399.3	23.38	6060.0	540.8	51.985
1129		8031.4	19.907	5789.4	608.6	51.666	1179		8406.9	23.45	6065.5	539.5	51.991
1130	670.33	8038.9	19.974	5794.9	607.1	51.673	1180	720.33	8414.4	23.53	6071.1	538.2	51.998
1131		8046.4	20.041	5800.4	605.6	51.679	1181		8421.9	23.60	6076.6	537.0	52.004
1132		8053.9	20.108	5805.9	604.2	51.686	1182		8429.5	23.68	6082.2	535.7	52.011
1133		8061.4	20.175	5811.4	602.7	51.692	1183		8437.0	23.76	6087.7	534.4	52.017
1134		8068.9	20.242	5816.9	601.2	51.699	1184		8444.5	23.83	6093.3	533.2	52.023
1135	675.33	8076.3	20.309	5822.4	599.7	51.706	1185	725.33	8452.1	23.91	6098.8	531.9	52.030
1136		8083.8	20.377	5827.9	598.3	51.712	1186		8459.6	23.98	6104.4	530.7	52.036
1137		8091.3	20.445	5833.4	596.8	51.719	1187		8467.2	24.06	6109.9	529.4	52.042
1138		8098.8	20.513	5838.9	595.4	51.725	1188		8474.7	24.14	6115.5	528.2	52.049
1139		8106.3	20.581	5844.4	593.9	51.732	1189		8482.2	24.22	6121.1	526.9	52.055
1140	680.33	8113.8	20.649	5849.9	592.5	51.739	1190	730.33	8489.8	24.29	6126.6	525.7	52.061
1141		8121.3	20.717	5855.4	591.0	51.745	1191		8497.3	24.37	6132.2	524.4	52.068
1142		8128.8	20.786	5860.9	589.6	51.752	1192		8504.9	24.45	6137.7	523.2	52.074
1143		8136.3	20.855	5866.5	588.2	51.758	1193		8512.4	24.53	6143.3	522.0	52.080
1144		8143.8	20.924	5872.0	586.7	51.765	1194		8520.0	24.60	6148.9	520.8	52.087
1145	685.33	8151.3	20.993	5877.5	585.3	51.771	1195	735.33	8527.5	24.68	6154.4	519.5	52.093
1146		8158.8	21.062	5883.0	583.9	51.778	1196		8535.1	24.76	6160.0	518.3	52.099
1147		8166.3	21.132	5888.5	582.5	51.785	1197		8542.6	24.84	6165.5	517.1	52.106
1148		8173.8	21.201	5894.0	581.1	51.791	1198		8550.2	24.92	6171.1	515.9	52.112
1149		8181.3	21.271	5899.5	579.7	51.798	1199		8557.7	25.00	6176.7	514.7	52.118

Table 3 Products—400% Theoretical Air (for One Pound-Mole)

T	t	$\bar{h}$	p_r	$\bar{u}$	v_r	$\bar{\phi}$	T	t	$\bar{h}$	p_r	$\bar{u}$	v_r	$\bar{\phi}$
1200	740.33	8565.3	25.08	6182.2	513.5	52.125	1250	790.33	8944.3	29.31	6461.9	457.7	52.434
1201		8572.8	25.16	6187.8	512.3	52.131	1251		8951.9	29.40	6467.6	456.7	52.440
1202		8580.4	25.24	6193.4	511.1	52.137	1252		8959.5	29.49	6473.2	455.7	52.446
1203		8587.9	25.32	6199.0	509.9	52.143	1253		8967.1	29.58	6478.8	454.6	52.452
1204		8595.5	25.40	6204.5	508.7	52.150	1254		8974.7	29.67	6484.4	453.6	52.458
1205	745.33	8603.1	25.48	6210.1	507.5	52.156	1255	795.33	8982.3	29.76	6490.0	452.6	52.464
1206		8610.6	25.56	6215.7	506.4	52.162	1256		8989.9	29.85	6495.7	451.6	52.470
1207		8618.2	25.64	6221.3	505.2	52.169	1257		8997.5	29.94	6501.3	450.5	52.476
1208		8625.7	25.72	6226.8	504.0	52.175	1258		9005.1	30.03	6506.9	449.5	52.483
1209		8633.3	25.80	6232.4	502.8	52.181	1259		9012.7	30.12	6512.6	448.5	52.489
1210	750.33	8640.9	25.88	6238.0	501.7	52.187	1260	800.33	9020.4	30.22	6518.2	447.5	52.495
1211		8648.4	25.97	6243.6	500.5	52.194	1261		9028.0	30.31	6523.8	446.5	52.501
1212		8656.0	26.05	6249.1	499.4	52.200	1262		9035.6	30.40	6529.4	445.5	52.507
1213		8663.6	26.13	6254.7	498.2	52.206	1263		9043.2	30.49	6535.1	444.5	52.513
1214		8671.1	26.21	6260.3	497.0	52.212	1264		9050.8	30.59	6540.7	443.5	52.519
1215	755.33	8678.7	26.29	6265.9	495.9	52.219	1265	805.33	9058.4	30.68	6546.3	442.5	52.525
1216		8686.3	26.38	6271.5	494.8	52.225	1266		9066.1	30.77	6552.0	441.5	52.531
1217		8693.9	26.46	6277.1	493.6	52.231	1267		9073.7	30.86	6557.6	440.5	52.537
1218		8701.4	26.54	6282.6	492.5	52.237	1268		9081.3	30.96	6563.2	439.5	52.543
1219		8709.0	26.62	6288.2	491.3	52.243	1269		9088.9	31.05	6568.9	438.6	52.549
1220	760.33	8716.6	26.71	6293.8	490.2	52.250	1270	810.33	9096.6	31.15	6574.5	437.6	52.555
1221		8724.1	26.79	6299.4	489.1	52.256	1271		9104.2	31.24	6580.2	436.6	52.561
1222		8731.7	26.88	6305.0	487.9	52.262	1272		9111.8	31.33	6585.8	435.6	52.567
1223		8739.3	26.96	6310.6	486.8	52.268	1273		9119.4	31.43	6591.4	434.7	52.573
1224		8746.9	27.04	6316.2	485.7	52.274	1274		9127.1	31.52	6597.1	433.7	52.579
1225	765.33	8754.5	27.13	6321.8	484.6	52.281	1275	815.33	9134.7	31.62	6602.7	432.7	52.585
1226		8762.0	27.21	6327.4	483.5	52.287	1276		9142.3	31.71	6608.4	431.8	52.591
1227		8769.6	27.30	6333.0	482.4	52.293	1277		9150.0	31.81	6614.0	430.8	52.597
1228		8777.2	27.38	6338.6	481.3	52.299	1278		9157.6	31.91	6619.7	429.8	52.603
1229		8784.8	27.47	6344.2	480.2	52.305	1279		9165.2	32.00	6625.3	428.9	52.609
1230	770.33	8792.4	27.55	6349.8	479.1	52.311	1280	820.33	9172.9	32.10	6631.0	427.9	52.615
1231		8800.0	27.64	6355.4	478.0	52.318	1281		9180.5	32.20	6636.6	427.0	52.621
1232		8807.5	27.73	6361.0	476.9	52.324	1282		9188.1	32.29	6642.3	426.0	52.627
1233		8815.1	27.81	6366.6	475.8	52.330	1283		9195.8	32.39	6647.9	425.1	52.633
1234		8822.7	27.90	6372.2	474.7	52.336	1284		9203.4	32.49	6653.6	424.2	52.639
1235	775.33	8830.3	27.98	6377.8	473.6	52.342	1285	825.33	9211.0	32.58	6659.2	423.2	52.644
1236		8837.9	28.07	6383.4	472.5	52.348	1286		9218.7	32.68	6664.9	422.3	52.650
1237		8845.5	28.16	6389.0	471.4	52.355	1287		9226.3	32.78	6670.5	421.3	52.656
1238		8853.1	28.24	6394.6	470.4	52.361	1288		9234.0	32.88	6676.2	420.4	52.662
1239		8860.7	28.33	6400.2	469.3	52.367	1289		9241.6	32.98	6681.8	419.5	52.668
1240	780.33	8868.3	28.42	6405.8	468.2	52.373	1290	830.33	9249.3	33.07	6687.5	418.6	52.674
1241		8875.9	28.51	6411.4	467.2	52.379	1291		9256.9	33.17	6693.2	417.6	52.680
1242		8883.5	28.60	6417.0	466.1	52.385	1292		9264.5	33.27	6698.8	416.7	52.686
1243		8891.1	28.68	6422.6	465.0	52.391	1293		9272.2	33.37	6704.5	415.8	52.692
1244		8898.7	28.77	6428.2	464.0	52.397	1294		9279.8	33.47	6710.1	414.9	52.698
1245	785.33	8906.3	28.86	6433.9	462.9	52.404	1295	835.33	9287.5	33.57	6715.8	414.0	52.704
1246		8913.9	28.95	6439.5	461.9	52.410	1296		9295.1	33.67	6721.5	413.1	52.710
1247		8921.5	29.04	6445.1	460.8	52.416	1297		9302.8	33.77	6727.1	412.2	52.716
1248		8929.1	29.13	6450.7	459.8	52.422	1298		9310.4	33.87	6732.8	411.2	52.721
1249		8936.7	29.22	6456.3	458.8	52.428	1299		9318.1	33.97	6738.5	410.3	52.727

Table 3   Products—400% Theoretical Air (for One Pound-Mole)

T	t	$\bar{h}$	p_r	$\bar{u}$	v_r	$\bar{\phi}$	T	t	$\bar{h}$	p_r	$\bar{u}$	v_r	$\bar{\phi}$
1300	840.33	9325.8	34.07	6744.1	409.4	52.733	1350	890.33	9709.7	39.43	7028.8	367.5	53.023
1301		9333.4	34.17	6749.8	408.5	52.739	1351		9717.4	39.54	7034.5	366.7	53.029
1302		9341.1	34.28	6755.5	407.6	52.745	1352		9725.1	39.65	7040.3	365.9	53.034
1303		9348.7	34.38	6761.1	406.8	52.751	1353		9732.9	39.77	7046.0	365.1	53.040
1304		9356.4	34.48	6766.8	405.9	52.757	1354		9740.6	39.88	7051.7	364.3	53.046
1305	845.33	9364.0	34.58	6772.5	405.0	52.763	1355	895.33	9748.3	40.00	7057.4	363.6	53.052
1306		9371.7	34.68	6778.2	404.1	52.768	1356		9756.0	40.11	7063.2	362.8	53.057
1307		9379.4	34.79	6783.8	403.2	52.774	1357		9763.7	40.23	7068.9	362.0	53.063
1308		9387.0	34.89	6789.5	402.3	52.780	1358		9771.4	40.34	7074.6	361.2	53.069
1309		9394.7	34.99	6795.2	401.5	52.786	1359		9779.1	40.46	7080.3	360.5	53.074
1310	850.33	9402.4	35.10	6800.9	400.6	52.792	1360	900.33	9786.8	40.57	7086.1	359.7	53.080
1311		9410.0	35.20	6806.6	399.7	52.798	1361		9794.5	40.69	7091.8	359.0	53.086
1312		9417.7	35.30	6812.2	398.8	52.804	1362		9802.3	40.81	7097.5	358.2	53.091
1313		9425.4	35.41	6817.9	398.0	52.809	1363		9810.0	40.92	7103.3	357.4	53.097
1314		9433.0	35.51	6823.6	397.1	52.815	1364		9817.7	41.04	7109.0	356.7	53.103
1315	855.33	9440.7	35.62	6829.3	396.2	52.821	1365	905.33	9825.4	41.16	7114.7	355.9	53.108
1316		9448.4	35.72	6835.0	395.4	52.827	1366		9833.1	41.27	7120.4	355.2	53.114
1317		9456.0	35.82	6840.7	394.5	52.833	1367		9840.9	41.39	7126.2	354.4	53.120
1318		9463.7	35.93	6846.3	393.7	52.839	1368		9848.6	41.51	7131.9	353.7	53.125
1319		9471.4	36.04	6852.0	392.8	52.844	1369		9856.3	41.63	7137.7	352.9	53.131
1320	860.33	9479.1	36.14	6857.7	392.0	52.850	1370	910.33	9864.0	41.74	7143.4	352.2	53.136
1321		9486.7	36.25	6863.4	391.1	52.856	1371		9871.7	41.86	7149.1	351.4	53.142
1322		9494.4	36.35	6869.1	390.3	52.862	1372		9879.5	41.98	7154.9	350.7	53.148
1323		9502.1	36.46	6874.8	389.4	52.868	1373		9887.2	42.10	7160.6	350.0	53.153
1324		9509.8	36.57	6880.5	388.6	52.873	1374		9894.9	42.22	7166.4	349.2	53.159
1325	865.33	9517.4	36.67	6886.2	387.7	52.879	1375	915.33	9902.7	42.34	7172.1	348.5	53.165
1326		9525.1	36.78	6891.9	386.9	52.885	1376		9910.4	42.46	7177.8	347.8	53.170
1327		9532.8	36.89	6897.6	386.1	52.891	1377		9918.1	42.58	7183.6	347.0	53.176
1328		9540.5	37.00	6903.3	385.2	52.897	1378		9925.9	42.70	7189.3	346.3	53.182
1329		9548.2	37.10	6909.0	384.4	52.902	1379		9933.6	42.82	7195.1	345.6	53.187
1330	870.33	9555.8	37.21	6914.7	383.6	52.908	1380	920.33	9941.3	42.94	7200.8	344.9	53.193
1331		9563.5	37.32	6920.4	382.7	52.914	1381		9949.1	43.06	7206.6	344.1	53.198
1332		9571.2	37.43	6926.1	381.9	52.920	1382		9956.8	43.19	7212.3	343.4	53.204
1333		9578.9	37.54	6931.8	381.1	52.926	1383		9964.5	43.31	7218.1	342.7	53.210
1334		9586.6	37.65	6937.5	380.3	52.931	1384		9972.3	43.43	7223.8	342.0	53.215
1335	875.33	9594.3	37.76	6943.2	379.5	52.937	1385	925.33	9980.0	43.55	7229.6	341.3	53.221
1336		9602.0	37.87	6948.9	378.6	52.943	1386		9987.7	43.68	7235.3	340.6	53.226
1337		9609.7	37.98	6954.6	377.8	52.949	1387		9995.5	43.80	7241.1	339.8	53.232
1338		9617.4	38.09	6960.3	377.0	52.954	1388		10003.2	43.92	7246.8	339.1	53.237
1339		9625.0	38.20	6966.0	376.2	52.960	1389		10011.0	44.05	7252.6	338.4	53.243
1340	880.33	9632.7	38.31	6971.7	375.4	52.966	1390	930.33	10018.7	44.17	7258.4	337.7	53.249
1341		9640.4	38.42	6977.4	374.6	52.972	1391		10026.4	44.29	7264.1	337.0	53.254
1342		9648.1	38.53	6983.1	373.8	52.977	1392		10034.2	44.42	7269.9	336.3	53.260
1343		9655.8	38.64	6988.8	373.0	52.983	1393		10041.9	44.54	7275.6	335.6	53.265
1344		9663.5	38.75	6994.5	372.2	52.989	1394		10049.7	44.67	7281.4	334.9	53.271
1345	885.33	9671.2	38.86	7000.2	371.4	52.994	1395	935.33	10057.4	44.79	7287.2	334.2	53.276
1346		9678.9	38.98	7006.0	370.6	53.000	1396		10065.2	44.92	7292.9	333.5	53.282
1347		9686.6	39.09	7011.7	369.8	53.006	1397		10072.9	45.04	7298.7	332.8	53.288
1348		9694.3	39.20	7017.4	369.0	53.012	1398		10080.7	45.17	7304.5	332.1	53.293
1349		9702.0	39.31	7023.1	368.2	53.017	1399		10088.4	45.30	7310.2	331.5	53.299

Table 3   Products—400% Theoretical Air (for One Pound-Mole)

T	t	$\bar{h}$	p_r	$\bar{u}$	v_r	$\bar{\phi}$	T	t	$\bar{h}$	p_r	$\bar{u}$	v_r	$\bar{\phi}$
1400	940.33	10096.2	45.42	7316.0	330.8	53.304	1450	990.33	10485.1	52.11	7605.6	298.6	53.577
1401		10103.9	45.55	7321.8	330.1	53.310	1451		10492.9	52.26	7611.4	298.0	53.582
1402		10111.7	45.68	7327.5	329.4	53.315	1452		10500.7	52.40	7617.2	297.4	53.588
1403		10119.5	45.80	7333.3	328.7	53.321	1453		10508.5	52.54	7623.0	296.8	53.593
1404		10127.2	45.93	7339.1	328.0	53.326	1454		10516.3	52.68	7628.8	296.2	53.599
1405	945.33	10135.0	46.06	7344.8	327.4	53.332	1455	995.33	10524.1	52.82	7634.7	295.6	53.604
1406		10142.7	46.19	7350.6	326.7	53.337	1456		10531.9	52.97	7640.5	295.0	53.609
1407		10150.5	46.32	7356.4	326.0	53.343	1457		10539.7	53.11	7646.3	294.4	53.615
1408		10158.2	46.44	7362.2	325.3	53.348	1458		10547.5	53.25	7652.1	293.8	53.620
1409		10166.0	46.57	7367.9	324.7	53.354	1459		10555.3	53.40	7658.0	293.2	53.625
1410	950.33	10173.8	46.70	7373.7	324.0	53.359	1460	1000.33	10563.1	53.54	7663.8	292.6	53.631
1411		10181.5	46.83	7379.5	323.3	53.365	1461		10571.0	53.69	7669.6	292.0	53.636
1412		10189.3	46.96	7385.3	322.7	53.370	1462		10578.8	53.83	7675.4	291.5	53.641
1413		10197.1	47.09	7391.0	322.0	53.376	1463		10586.6	53.97	7681.3	290.9	53.647
1414		10204.8	47.22	7396.8	321.3	53.381	1464		10594.4	54.12	7687.1	290.3	53.652
1415	955.33	10212.6	47.35	7402.6	320.7	53.387	1465	1005.33	10602.2	54.27	7692.9	289.7	53.657
1416		10220.4	47.48	7408.4	320.0	53.392	1466		10610.0	54.41	7698.8	289.1	53.663
1417		10228.1	47.62	7414.2	319.4	53.398	1467		10617.8	54.56	7704.6	288.6	53.668
1418		10235.9	47.75	7420.0	318.7	53.403	1468		10625.7	54.70	7710.4	288.0	53.673
1419		10243.7	47.88	7425.7	318.0	53.409	1469		10633.5	54.85	7716.3	287.4	53.679
1420	960.33	10251.5	48.01	7431.5	317.4	53.414	1470	1010.33	10641.3	55.00	7722.1	286.8	53.684
1421		10259.2	48.14	7437.3	316.7	53.420	1471		10649.1	55.15	7727.9	286.3	53.689
1422		10267.0	48.28	7443.1	316.1	53.425	1472		10656.9	55.29	7733.8	285.7	53.695
1423		10274.8	48.41	7448.9	315.5	53.431	1473		10664.8	55.44	7739.6	285.1	53.700
1424		10282.6	48.54	7454.7	314.8	53.436	1474		10672.6	55.59	7745.4	284.5	53.705
1425	965.33	10290.3	48.68	7460.5	314.2	53.442	1475	1015.33	10680.4	55.74	7751.3	284.0	53.711
1426		10298.1	48.81	7466.3	313.5	53.447	1476		10688.2	55.89	7757.1	283.4	53.716
1427		10305.9	48.95	7472.1	312.9	53.452	1477		10696.1	56.04	7763.0	282.9	53.721
1428		10313.7	49.08	7477.9	312.2	53.458	1478		10703.9	56.19	7768.8	282.3	53.727
1429		10321.4	49.21	7483.7	311.6	53.463	1479		10711.7	56.34	7774.6	281.7	53.732
1430	970.33	10329.2	49.35	7489.5	311.0	53.469	1480	1020.33	10719.6	56.49	7780.5	281.2	53.737
1431		10337.0	49.48	7495.2	310.3	53.474	1481		10727.4	56.64	7786.3	280.6	53.742
1432		10344.8	49.62	7501.0	309.7	53.480	1482		10735.2	56.79	7792.2	280.1	53.748
1433		10352.6	49.76	7506.8	309.1	53.485	1483		10743.1	56.94	7798.0	279.5	53.753
1434		10360.4	49.89	7512.6	308.4	53.491	1484		10750.9	57.09	7803.9	278.9	53.758
1435	975.33	10368.2	50.03	7518.4	307.8	53.496	1485	1025.33	10758.7	57.24	7809.7	278.4	53.764
1436		10375.9	50.17	7524.2	307.2	53.501	1486		10766.6	57.40	7815.6	277.8	53.769
1437		10383.7	50.30	7530.1	306.6	53.507	1487		10774.4	57.55	7821.4	277.3	53.774
1438		10391.5	50.44	7535.9	305.9	53.512	1488		10782.2	57.70	7827.3	276.7	53.779
1439		10399.3	50.58	7541.7	305.3	53.518	1489		10790.1	57.86	7833.1	276.2	53.785
1440	980.33	10407.1	50.72	7547.5	304.7	53.523	1490	1030.33	10797.9	58.01	7839.0	275.6	53.790
1441		10414.9	50.85	7553.3	304.1	53.529	1491		10805.7	58.16	7844.8	275.1	53.795
1442		10422.7	50.99	7559.1	303.5	53.534	1492		10813.6	58.32	7850.7	274.6	53.800
1443		10430.5	51.13	7564.9	302.9	53.539	1493		10821.4	58.47	7856.5	274.0	53.806
1444		10438.3	51.27	7570.7	302.2	53.545	1494		10829.3	58.63	7862.4	273.5	53.811
1445	985.33	10446.1	51.41	7576.5	301.6	53.550	1495	1035.33	10837.1	58.78	7868.3	272.9	53.816
1446		10453.9	51.55	7582.3	301.0	53.556	1496		10845.0	58.94	7874.1	272.4	53.821
1447		10461.7	51.69	7588.1	300.4	53.561	1497		10852.8	59.09	7880.0	271.9	53.827
1448		10469.5	51.83	7593.9	299.8	53.566	1498		10860.7	59.25	7885.8	271.3	53.832
1449		10477.3	51.97	7599.8	299.2	53.572	1499		10868.5	59.41	7891.7	270.8	53.837

Table 3   Products—400% Theoretical Air (for One Pound-Mole)

T	t	$\bar{h}$	p_r	$\bar{u}$	v_r	$\bar{\phi}$	T	t	$\bar{h}$	p_r	$\bar{u}$	v_r	$\bar{\phi}$
1500	1040.33	10876.4	59.56	7897.6	270.3	53.842	1550	1090.33	11270.0	67.83	8191.9	245.2	54.100
1501		10884.2	59.72	7903.4	269.7	53.848	1551		11277.9	68.00	8197.8	244.8	54.106
1502		10892.1	59.88	7909.3	269.2	53.853	1552		11285.8	68.18	8203.7	244.3	54.111
1503		10899.9	60.03	7915.2	268.7	53.858	1553		11293.7	68.35	8209.6	243.8	54.116
1504		10907.8	60.19	7921.0	268.1	53.863	1554		11301.6	68.53	8215.5	243.4	54.121
1505	1045.33	10915.6	60.35	7926.9	267.6	53.868	1555	1095.33	11309.5	68.70	8221.5	242.9	54.126
1506		10923.5	60.51	7932.8	267.1	53.874	1556		11317.4	68.88	8227.4	242.4	54.131
1507		10931.3	60.67	7938.6	266.6	53.879	1557		11325.3	69.06	8233.3	242.0	54.136
1508		10939.2	60.83	7944.5	266.0	53.884	1558		11333.2	69.23	8239.2	241.5	54.141
1509		10947.0	60.99	7950.4	265.5	53.889	1559		11341.1	69.41	8245.1	241.0	54.146
1510	1050.33	10954.9	61.15	7956.2	265.0	53.895	1560	1100.33	11349.0	69.59	8251.0	240.6	54.151
1511		10962.8	61.31	7962.1	264.5	53.900	1561		11356.9	69.77	8257.0	240.1	54.156
1512		10970.6	61.47	7968.0	264.0	53.905	1562		11364.8	69.94	8262.9	239.7	54.161
1513		10978.5	61.63	7973.9	263.5	53.910	1563		11372.7	70.12	8268.8	239.2	54.166
1514		10986.3	61.79	7979.7	262.9	53.915	1564		11380.6	70.30	8274.7	238.7	54.172
1515	1055.33	10994.2	61.95	7985.6	262.4	53.921	1565	1105.33	11388.5	70.48	8280.6	238.3	54.177
1516		11002.1	62.12	7991.5	261.9	53.926	1566		11396.4	70.66	8286.6	237.8	54.182
1517		11009.9	62.28	7997.4	261.4	53.931	1567		11404.3	70.84	8292.5	237.4	54.187
1518		11017.8	62.44	8003.3	260.9	53.936	1568		11412.3	71.02	8298.4	236.9	54.192
1519		11025.7	62.60	8009.1	260.4	53.941	1569		11420.2	71.20	8304.4	236.5	54.197
1520	1060.33	11033.5	62.77	8015.0	259.9	53.946	1570	1110.33	11428.1	71.38	8310.3	236.0	54.202
1521		11041.4	62.93	8020.9	259.4	53.952	1571		11436.0	71.56	8316.2	235.6	54.207
1522		11049.3	63.10	8026.8	258.9	53.957	1572		11443.9	71.74	8322.1	235.1	54.212
1523		11057.1	63.26	8032.7	258.4	53.962	1573		11451.8	71.93	8328.1	234.7	54.217
1524		11065.0	63.42	8038.6	257.9	53.967	1574		11459.7	72.11	8334.0	234.2	54.222
1525	1065.33	11072.9	63.59	8044.4	257.4	53.972	1575	1115.33	11467.7	72.29	8339.9	233.8	54.227
1526		11080.8	63.75	8050.3	256.9	53.977	1576		11475.6	72.47	8345.9	233.4	54.232
1527		11088.6	63.92	8056.2	256.4	53.983	1577		11483.5	72.66	8351.8	232.9	54.237
1528		11096.5	64.09	8062.1	255.9	53.988	1578		11491.4	72.84	8357.7	232.5	54.242
1529		11104.4	64.25	8068.0	255.4	53.993	1579		11499.3	73.03	8363.7	232.0	54.247
1530	1070.33	11112.3	64.42	8073.9	254.9	53.998	1580	1120.33	11507.3	73.21	8369.6	231.6	54.252
1531		11120.1	64.59	8079.8	254.4	54.003	1581		11515.2	73.40	8375.5	231.2	54.257
1532		11128.0	64.75	8085.7	253.9	54.008	1582		11523.1	73.58	8381.5	230.7	54.262
1533		11135.9	64.92	8091.6	253.4	54.014	1583		11531.0	73.77	8387.4	230.3	54.267
1534		11143.8	65.09	8097.5	252.9	54.019	1584		11539.0	73.95	8393.4	229.9	54.272
1535	1075.33	11151.6	65.26	8103.4	252.4	54.024	1585	1125.33	11546.9	74.14	8399.3	229.4	54.277
1536		11159.5	65.43	8109.2	251.9	54.029	1586		11554.8	74.33	8405.2	229.0	54.282
1537		11167.4	65.60	8115.1	251.4	54.034	1587		11562.7	74.51	8411.2	228.6	54.287
1538		11175.3	65.77	8121.0	251.0	54.039	1588		11570.7	74.70	8417.1	228.1	54.292
1539		11183.2	65.94	8126.9	250.5	54.044	1589		11578.6	74.89	8423.1	227.7	54.297
1540	1080.33	11191.1	66.11	8132.8	250.0	54.049	1590	1130.33	11586.5	75.08	8429.0	227.3	54.302
1541		11199.0	66.28	8138.7	249.5	54.055	1591		11594.5	75.27	8435.0	226.8	54.307
1542		11206.8	66.45	8144.6	249.0	54.060	1592		11602.4	75.46	8440.9	226.4	54.312
1543		11214.7	66.62	8150.6	248.6	54.065	1593		11610.3	75.65	8446.9	226.0	54.317
1544		11222.6	66.79	8156.5	248.1	54.070	1594		11618.3	75.84	8452.8	225.6	54.322
1545	1085.33	11230.5	66.96	8162.4	247.6	54.075	1595	1135.33	11626.2	76.03	8458.8	225.1	54.327
1546		11238.4	67.14	8168.3	247.1	54.080	1596		11634.1	76.22	8464.7	224.7	54.332
1547		11246.3	67.31	8174.2	246.6	54.085	1597		11642.1	76.41	8470.7	224.3	54.337
1548		11254.2	67.48	8180.1	246.2	54.090	1598		11650.0	76.60	8476.6	223.9	54.342
1549		11262.1	67.66	8186.0	245.7	54.095	1599		11658.0	76.79	8482.6	223.5	54.347

Table 3   Products—400% Theoretical Air (for One Pound-Mole)

T	t	$\bar{h}$	p_r	$\bar{u}$	v_r	$\bar{\phi}$	T	t	$\bar{h}$	p_r	$\bar{u}$	v_r	$\bar{\phi}$
1600	1140.33	11665.9	76.98	8488.5	223.0	54.352	1650	1190.33	12064.1	87.09	8787.4	203.31	54.597
1601		11673.8	77.18	8494.5	222.6	54.357	1651		12072.0	87.31	8793.4	202.94	54.602
1602		11681.8	77.37	8500.4	222.2	54.362	1652		12080.0	87.52	8799.4	202.57	54.607
1603		11689.7	77.56	8506.4	221.8	54.367	1653		12088.0	87.73	8805.4	202.20	54.611
1604		11697.7	77.76	8512.4	221.4	54.372	1654		12096.0	87.95	8811.4	201.83	54.616
1605	1145.33	11705.6	77.95	8518.3	221.0	54.377	1655	1195.33	12104.0	88.16	8817.4	201.46	54.621
1606		11713.6	78.15	8524.3	220.5	54.382	1656		12112.0	88.37	8823.4	201.09	54.626
1607		11721.5	78.34	8530.2	220.1	54.387	1657		12120.0	88.59	8829.4	200.72	54.631
1608		11729.5	78.54	8536.2	219.7	54.392	1658		12128.0	88.80	8835.4	200.36	54.636
1609		11737.4	78.73	8542.2	219.3	54.396	1659		12136.0	89.02	8841.4	199.99	54.640
1610	1150.33	11745.4	78.93	8548.1	218.9	54.401	1660	1200.33	12143.9	89.24	8847.4	199.63	54.645
1611		11753.3	79.12	8554.1	218.5	54.406	1661		12151.9	89.45	8853.4	199.27	54.650
1612		11761.3	79.32	8560.1	218.1	54.411	1662		12159.9	89.67	8859.4	198.90	54.655
1613		11769.2	79.52	8566.0	217.7	54.416	1663		12167.9	89.89	8865.5	198.54	54.660
1614		11777.2	79.72	8572.0	217.3	54.421	1664		12175.9	90.11	8871.5	198.18	54.664
1615	1155.33	11785.1	79.91	8578.0	216.9	54.426	1665	1205.33	12183.9	90.32	8877.5	197.82	54.669
1616		11793.1	80.11	8583.9	216.5	54.431	1666		12191.9	90.54	8883.5	197.46	54.674
1617		11801.0	80.31	8589.9	216.1	54.436	1667		12199.9	90.76	8889.5	197.10	54.679
1618		11809.0	80.51	8595.9	215.7	54.441	1668		12207.9	90.98	8895.5	196.74	54.684
1619		11816.9	80.71	8601.8	215.3	54.446	1669		12215.9	91.20	8901.5	196.39	54.688
1620	1160.33	11824.9	80.91	8607.8	214.9	54.451	1670	1210.33	12223.9	91.42	8907.5	196.03	54.693
1621		11832.9	81.11	8613.8	214.5	54.456	1671		12231.9	91.64	8913.6	195.68	54.698
1622		11840.8	81.31	8619.8	214.1	54.460	1672		12239.9	91.86	8919.6	195.32	54.703
1623		11848.8	81.51	8625.7	213.7	54.465	1673		12247.9	92.09	8925.6	194.97	54.708
1624		11856.7	81.71	8631.7	213.3	54.470	1674		12255.9	92.31	8931.6	194.62	54.712
1625	1165.33	11864.7	81.91	8637.7	212.9	54.475	1675	1215.33	12264.0	92.53	8937.6	194.26	54.717
1626		11872.7	82.12	8643.7	212.5	54.480	1676		12272.0	92.75	8943.7	193.91	54.722
1627		11880.6	82.32	8649.6	212.1	54.485	1677		12280.0	92.98	8949.7	193.56	54.727
1628		11888.6	82.52	8655.6	211.7	54.490	1678		12288.0	93.20	8955.7	193.21	54.731
1629		11896.6	82.73	8661.6	211.3	54.495	1679		12296.0	93.42	8961.7	192.86	54.736
1630	1170.33	11904.5	82.93	8667.6	210.9	54.500	1680	1220.33	12304.0	93.65	8967.8	192.52	54.741
1631		11912.5	83.13	8673.6	210.5	54.505	1681		12312.0	93.87	8973.8	192.17	54.746
1632		11920.5	83.34	8679.5	210.1	54.509	1682		12320.0	94.10	8979.8	191.82	54.751
1633		11928.4	83.54	8685.5	209.8	54.514	1683		12328.0	94.33	8985.8	191.48	54.755
1634		11936.4	83.75	8691.5	209.4	54.519	1684		12336.0	94.55	8991.9	191.13	54.760
1635	1175.33	11944.4	83.96	8697.5	209.0	54.524	1685	1225.33	12344.1	94.78	8997.9	190.79	54.765
1636		11952.4	84.16	8703.5	208.6	54.529	1686		12352.1	95.01	9003.9	190.44	54.770
1637		11960.3	84.37	8709.5	208.2	54.534	1687		12360.1	95.23	9009.9	190.10	54.774
1638		11968.3	84.58	8715.5	207.8	54.539	1688		12368.1	95.46	9016.0	189.76	54.779
1639		11976.3	84.78	8721.5	207.5	54.544	1689		12376.1	95.69	9022.0	189.42	54.784
1640	1180.33	11984.3	84.99	8727.4	207.1	54.548	1690	1230.33	12384.1	95.92	9028.0	189.08	54.789
1641		11992.2	85.20	8733.4	206.7	54.553	1691		12392.2	96.15	9034.1	188.74	54.793
1642		12000.2	85.41	8739.4	206.3	54.558	1692		12400.2	96.38	9040.1	188.40	54.798
1643		12008.2	85.62	8745.4	205.9	54.563	1693		12408.2	96.61	9046.2	188.06	54.803
1644		12016.2	85.83	8751.4	205.6	54.568	1694		12416.2	96.84	9052.2	187.72	54.808
1645	1185.33	12024.1	86.04	8757.4	205.2	54.573	1695	1235.33	12424.3	97.07	9058.2	187.39	54.812
1646		12032.1	86.25	8763.4	204.8	54.578	1696		12432.3	97.30	9064.3	187.05	54.817
1647		12040.1	86.46	8769.4	204.4	54.582	1697		12440.3	97.53	9070.3	186.72	54.822
1648		12048.1	86.67	8775.4	204.1	54.587	1698		12448.3	97.77	9076.3	186.38	54.826
1649		12056.1	86.88	8781.4	203.7	54.592	1699		12456.4	98.00	9082.4	186.05	54.831

Table 3   Products—400% Theoretical Air (for One Pound-Mole)

T	t	$\bar{h}$	p_r	$\bar{u}$	v_r	$\bar{\phi}$	T	t	$\bar{h}$	p_r	$\bar{u}$	v_r	$\bar{\phi}$
1700	1240.33	12464.4	98.23	9088.4	185.72	54.836	1750	1290.33	12866.8	110.48	9391.6	169.99	55.069
1701		12472.4	98.47	9094.5	185.38	54.841	1751		12874.9	110.74	9397.7	169.69	55.074
1702		12480.4	98.70	9100.5	185.05	54.845	1752		12883.0	110.99	9403.7	169.39	55.078
1703		12488.5	98.94	9106.6	184.72	54.850	1753		12891.0	111.25	9409.8	169.10	55.083
1704		12496.5	99.17	9112.6	184.39	54.855	1754		12899.1	111.51	9415.9	168.80	55.088
1705	1245.33	12504.5	99.41	9118.6	184.06	54.860	1755	1295.33	12907.2	111.77	9422.0	168.51	55.092
1706		12512.6	99.64	9124.7	183.74	54.864	1756		12915.3	112.03	9428.1	168.21	55.097
1707		12520.6	99.88	9130.7	183.41	54.869	1757		12923.3	112.29	9434.2	167.92	55.101
1708		12528.6	100.12	9136.8	183.08	54.874	1758		12931.4	112.55	9440.3	167.63	55.106
1709		12536.7	100.35	9142.8	182.76	54.878	1759		12939.5	112.81	9446.4	167.33	55.111
1710	1250.33	12544.7	100.59	9148.9	182.43	54.883	1760	1300.33	12947.6	113.07	9452.5	167.04	55.115
1711		12552.7	100.83	9154.9	182.11	54.888	1761		12955.6	113.33	9458.5	166.75	55.120
1712		12560.8	101.07	9161.0	181.78	54.892	1762		12963.7	113.59	9464.6	166.46	55.124
1713		12568.8	101.31	9167.0	181.46	54.897	1763		12971.8	113.86	9470.7	166.17	55.129
1714		12576.9	101.55	9173.1	181.14	54.902	1764		12979.9	114.12	9476.8	165.88	55.134
1715	1255.33	12584.9	101.79	9179.2	180.81	54.907	1765	1305.33	12988.0	114.38	9482.9	165.60	55.138
1716		12592.9	102.03	9185.2	180.49	54.911	1766		12996.0	114.65	9489.0	165.31	55.143
1717		12601.0	102.27	9191.3	180.17	54.916	1767		13004.1	114.91	9495.1	165.02	55.147
1718		12609.0	102.51	9197.3	179.85	54.921	1768		13012.2	115.17	9501.2	164.73	55.152
1719		12617.1	102.75	9203.4	179.53	54.925	1769		13020.3	115.44	9507.3	164.45	55.156
1720	1260.33	12625.1	102.99	9209.4	179.22	54.930	1770	1310.33	13028.4	115.71	9513.4	164.16	55.161
1721		12633.2	103.24	9215.5	178.90	54.935	1771		13036.5	115.97	9519.5	163.88	55.166
1722		12641.2	103.48	9221.6	178.58	54.939	1772		13044.6	116.24	9525.6	163.59	55.170
1723		12649.2	103.72	9227.6	178.27	54.944	1773		13052.6	116.51	9531.7	163.31	55.175
1724		12657.3	103.97	9233.7	177.95	54.949	1774		13060.7	116.78	9537.8	163.03	55.179
1725	1265.33	12665.3	104.21	9239.7	177.63	54.953	1775	1315.33	13068.8	117.04	9543.9	162.75	55.184
1726		12673.4	104.46	9245.8	177.32	54.958	1776		13076.9	117.31	9550.0	162.47	55.188
1727		12681.4	104.70	9251.9	177.01	54.963	1777		13085.0	117.58	9556.1	162.18	55.193
1728		12689.5	104.95	9257.9	176.69	54.967	1778		13093.1	117.85	9562.2	161.90	55.198
1729		12697.6	105.20	9264.0	176.38	54.972	1779		13101.2	118.12	9568.3	161.62	55.202
1730	1270.33	12705.6	105.44	9270.1	176.07	54.977	1780	1320.33	13109.3	118.39	9574.5	161.34	55.207
1731		12713.7	105.69	9276.1	175.76	54.981	1781		13117.4	118.66	9580.6	161.07	55.211
1732		12721.7	105.94	9282.2	175.45	54.986	1782		13125.5	118.94	9586.7	160.79	55.216
1733		12729.8	106.19	9288.3	175.14	54.991	1783		13133.6	119.21	9592.8	160.51	55.220
1734		12737.8	106.44	9294.3	174.83	54.995	1784		13141.7	119.48	9598.9	160.23	55.225
1735	1275.33	12745.9	106.68	9300.4	174.52	55.000	1785	1325.33	13149.8	119.75	9605.0	159.96	55.229
1736		12753.9	106.93	9306.5	174.22	55.004	1786		13157.9	120.03	9611.1	159.68	55.234
1737		12762.0	107.18	9312.6	173.91	55.009	1787		13166.0	120.30	9617.2	159.41	55.238
1738		12770.1	107.44	9318.6	173.61	55.014	1788		13174.1	120.58	9623.3	159.13	55.243
1739		12778.1	107.69	9324.7	173.30	55.018	1789		13182.2	120.85	9629.5	158.86	55.247
1740	1280.33	12786.2	107.94	9330.8	173.00	55.023	1790	1330.33	13190.3	121.13	9635.6	158.59	55.252
1741		12794.2	108.19	9336.9	172.69	55.028	1791		13198.4	121.40	9641.7	158.31	55.257
1742		12802.3	108.44	9342.9	172.39	55.032	1792		13206.5	121.68	9647.8	158.04	55.261
1743		12810.4	108.70	9349.0	172.09	55.037	1793		13214.6	121.96	9653.9	157.77	55.266
1744		12818.4	108.95	9355.1	171.78	55.042	1794		13222.7	122.24	9660.0	157.50	55.270
1745	1285.33	12826.5	109.20	9361.2	171.48	55.046	1795	1335.33	13230.8	122.52	9666.2	157.23	55.275
1746		12834.6	109.46	9367.2	171.18	55.051	1796		13238.9	122.79	9672.3	156.96	55.279
1747		12842.6	109.71	9373.3	170.88	55.055	1797		13247.0	123.07	9678.4	156.69	55.284
1748		12850.7	109.97	9379.4	170.58	55.060	1798		13255.1	123.35	9684.5	156.42	55.288
1749		12858.8	110.22	9385.5	170.29	55.065	1799		13263.2	123.63	9690.7	156.15	55.293

Table 3   Products—400% Theoretical Air (for One Pound-Mole)

T	t	$\bar{h}$	p_r	$\bar{u}$	v_r	$\bar{\phi}$	T	t	$\bar{h}$	p_r	$\bar{u}$	v_r	$\bar{\phi}$
1800	1340.33	13271.3	123.91	9696.8	155.89	55.297	1850	1390.33	13677.8	138.62	10003.9	143.22	55.520
1801		13279.4	124.20	9702.9	155.62	55.302	1851		13686.0	138.93	10010.1	142.98	55.524
1802		13287.6	124.48	9709.0	155.36	55.306	1852		13694.1	139.24	10016.3	142.74	55.529
1803		13295.7	124.76	9715.1	155.09	55.311	1853		13702.2	139.55	10022.4	142.50	55.533
1804		13303.8	125.04	9721.3	154.83	55.315	1854		13710.4	139.85	10028.6	142.26	55.537
1805	1345.33	13311.9	125.33	9727.4	154.56	55.320	1855	1395.33	13718.6	140.17	10034.8	142.02	55.542
1806		13320.0	125.61	9733.5	154.29	55.324	1856		13726.7	140.47	10041.0	141.79	55.546
1807		13328.1	125.89	9739.7	154.03	55.329	1857		13734.9	140.79	10047.1	141.55	55.551
1808		13336.2	126.18	9745.8	153.77	55.333	1858		13743.0	141.10	10053.3	141.32	55.555
1809		13344.3	126.46	9751.9	153.51	55.338	1859		13751.2	141.41	10059.5	141.08	55.559
1810	1350.33	13352.5	126.75	9758.0	153.25	55.342	1860	1400.33	13759.3	141.73	10065.6	140.84	55.564
1811		13360.6	127.04	9764.2	152.98	55.347	1861		13767.5	142.04	10071.8	140.61	55.568
1812		13368.7	127.32	9770.3	152.73	55.351	1862		13775.7	142.35	10078.0	140.37	55.573
1813		13376.8	127.61	9776.5	152.46	55.356	1863		13783.8	142.66	10084.1	140.14	55.577
1814		13385.0	127.90	9782.6	152.20	55.360	1864		13792.0	142.98	10090.3	139.90	55.581
1815	1355.33	13393.1	128.19	9788.7	151.95	55.364	1865	1405.33	13800.1	143.29	10096.5	139.67	55.586
1816		13401.2	128.48	9794.9	151.69	55.369	1866		13808.3	143.61	10102.7	139.44	55.590
1817		13409.3	128.77	9801.0	151.43	55.373	1867		13816.5	143.93	10108.9	139.21	55.594
1818		13417.4	129.06	9807.2	151.17	55.378	1868		13824.6	144.25	10115.0	138.97	55.599
1819		13425.6	129.35	9813.3	150.91	55.382	1869		13832.8	144.56	10121.2	138.75	55.603
1820	1360.33	13433.7	129.64	9819.4	150.66	55.387	1870	1410.33	13840.9	144.88	10127.4	138.51	55.608
1821		13441.8	129.93	9825.6	150.40	55.391	1871		13849.1	145.20	10133.5	138.28	55.612
1822		13449.9	130.22	9831.7	150.15	55.396	1872		13857.3	145.52	10139.7	138.05	55.616
1823		13458.1	130.52	9837.8	149.89	55.400	1873		13865.4	145.84	10145.9	137.83	55.621
1824		13466.2	130.81	9844.0	149.64	55.405	1874		13873.6	146.16	10152.1	137.59	55.625
1825	1365.33	13474.3	131.10	9850.1	149.39	55.409	1875	1415.33	13881.8	146.48	10158.3	137.37	55.629
1826		13482.4	131.40	9856.3	149.13	55.414	1876		13889.9	146.80	10164.5	137.14	55.634
1827		13490.6	131.69	9862.4	148.88	55.418	1877		13898.1	147.12	10170.7	136.91	55.638
1828		13498.7	131.99	9868.6	148.63	55.423	1878		13906.3	147.45	10176.8	136.68	55.642
1829		13506.8	132.28	9874.7	148.38	55.427	1879		13914.4	147.77	10183.0	136.46	55.647
1830	1370.33	13515.0	132.58	9880.9	148.13	55.431	1880	1420.33	13922.6	148.09	10189.2	136.23	55.651
1831		13523.1	132.88	9887.0	147.87	55.436	1881		13930.8	148.42	10195.4	136.01	55.655
1832		13531.2	133.17	9893.1	147.63	55.440	1882		13939.0	148.74	10201.6	135.78	55.660
1833		13539.4	133.47	9899.3	147.38	55.445	1883		13947.1	149.07	10207.8	135.56	55.664
1834		13547.5	133.77	9905.5	147.13	55.449	1884		13955.3	149.40	10213.9	135.33	55.669
1835	1375.33	13555.6	134.07	9911.6	146.88	55.454	1885	1425.33	13963.5	149.72	10220.1	135.11	55.673
1836		13563.8	134.37	9917.8	146.63	55.458	1886		13971.6	150.05	10226.3	134.89	55.677
1837		13571.9	134.67	9923.9	146.39	55.462	1887		13979.8	150.38	10232.5	134.66	55.682
1838		13580.1	134.97	9930.1	146.14	55.467	1888		13988.0	150.71	10238.7	134.44	55.686
1839		13588.2	135.27	9936.2	145.89	55.471	1889		13996.2	151.04	10244.9	134.22	55.690
1840	1380.33	13596.4	135.58	9942.4	145.65	55.476	1890	1430.33	14004.4	151.36	10251.1	134.00	55.695
1841		13604.5	135.88	9948.5	145.40	55.480	1891		14012.5	151.69	10257.3	133.78	55.699
1842		13612.6	136.18	9954.7	145.16	55.485	1892		14020.7	152.03	10263.5	133.56	55.703
1843		13620.8	136.48	9960.8	144.91	55.489	1893		14028.9	152.35	10269.7	133.34	55.707
1844		13628.9	136.79	9967.0	144.67	55.493	1894		14037.1	152.69	10275.9	133.12	55.712
1845	1385.33	13637.1	137.09	9973.2	144.43	55.498	1895	1435.33	14045.3	153.02	10282.1	132.90	55.716
1846		13645.2	137.40	9979.3	144.18	55.502	1896		14053.5	153.35	10288.3	132.68	55.720
1847		13653.3	137.70	9985.5	143.94	55.507	1897		14061.6	153.69	10294.5	132.46	55.725
1848		13661.5	138.01	9991.6	143.70	55.511	1898		14069.8	154.02	10300.7	132.24	55.729
1849		13669.7	138.32	9997.8	143.46	55.515	1899		14078.0	154.35	10306.9	132.03	55.733

Table 3 Products—400% Theoretical Air (for One Pound-Mole)

T	t	$\bar{h}$	p_r	$\bar{u}$	v_r	$\bar{\phi}$	T	t	$\bar{h}$	p_r	$\bar{u}$	v_r	$\bar{\phi}$
1900	1440.33	14086.2	154.69	10313.1	131.81	55.738	1950	1490.33	14496.5	172.22	10624.0	121.51	55.951
1901		14094.4	155.03	10319.3	131.59	55.742	1951		14504.7	172.59	10630.3	121.31	55.955
1902		14102.6	155.37	10325.5	131.38	55.746	1952		14512.9	172.95	10636.5	121.12	55.959
1903		14110.8	155.70	10331.7	131.16	55.751	1953		14521.2	173.32	10642.8	120.92	55.964
1904		14119.0	156.04	10337.9	130.95	55.755	1954		14529.4	173.69	10649.0	120.73	55.968
1905	1445.33	14127.2	156.38	10344.1	130.73	55.759	1955	1495.33	14537.6	174.05	10655.3	120.54	55.972
1906		14135.3	156.72	10350.3	130.52	55.764	1956		14545.8	174.43	10661.5	120.34	55.976
1907		14143.5	157.05	10356.5	130.30	55.768	1957		14554.1	174.79	10667.7	120.15	55.980
1908		14151.7	157.40	10362.7	130.09	55.772	1958		14562.3	175.17	10674.0	119.95	55.985
1909		14159.9	157.74	10368.9	129.88	55.776	1959		14570.5	175.54	10680.2	119.76	55.989
1910	1450.33	14168.1	158.08	10375.1	129.66	55.781	1960	1500.33	14578.7	175.91	10686.5	119.57	55.993
1911		14176.3	158.42	10381.3	129.45	55.785	1961		14587.0	176.28	10692.7	119.38	55.997
1912		14184.5	158.76	10387.5	129.24	55.789	1962		14595.2	176.65	10698.9	119.19	56.001
1913		14192.7	159.10	10393.7	129.03	55.794	1963		14603.4	177.03	10705.2	119.00	56.006
1914		14200.9	159.45	10400.0	128.82	55.798	1964		14611.7	177.40	10711.4	118.81	56.010
1915	1455.33	14209.1	159.79	10406.2	128.61	55.802	1965	1505.33	14619.9	177.77	10717.7	118.62	56.014
1916		14217.3	160.14	10412.4	128.40	55.806	1966		14628.1	178.15	10723.9	118.43	56.018
1917		14225.5	160.48	10418.6	128.19	55.811	1967		14636.4	178.53	10730.2	118.24	56.022
1918		14233.7	160.83	10424.8	127.98	55.815	1968		14644.6	178.90	10736.4	118.05	56.026
1919		14241.9	161.18	10431.0	127.77	55.819	1969		14652.8	179.28	10742.7	117.86	56.031
1920	1460.33	14250.1	161.52	10437.2	127.56	55.823	1970	1510.33	14661.1	179.66	10749.0	117.67	56.035
1921		14258.3	161.87	10443.5	127.35	55.828	1971		14669.3	180.04	10755.2	117.48	56.039
1922		14266.5	162.22	10449.7	127.15	55.832	1972		14677.6	180.42	10761.5	117.30	56.043
1923		14274.7	162.57	10455.9	126.94	55.836	1973		14685.8	180.80	10767.7	117.11	56.047
1924		14282.9	162.92	10462.1	126.74	55.841	1974		14694.1	181.18	10774.0	116.92	56.052
1925	1465.33	14291.1	163.26	10468.3	126.53	55.845	1975	1515.33	14702.3	181.56	10780.2	116.74	56.056
1926		14299.3	163.62	10474.5	126.32	55.849	1976		14710.5	181.95	10786.5	116.55	56.060
1927		14307.5	163.97	10480.8	126.12	55.853	1977		14718.8	182.33	10792.7	116.36	56.064
1928		14315.7	164.32	10487.0	125.91	55.858	1978		14727.0	182.71	10799.0	116.18	56.068
1929		14324.0	164.67	10493.2	125.71	55.862	1979		14735.2	183.09	10805.2	115.99	56.072
1930	1470.33	14332.1	165.03	10499.4	125.50	55.866	1980	1520.33	14743.5	183.48	10811.5	115.81	56.077
1931		14340.4	165.38	10505.7	125.30	55.870	1981		14751.8	183.86	10817.8	115.63	56.081
1932		14348.6	165.74	10511.9	125.10	55.875	1982		14760.0	184.25	10824.0	115.44	56.085
1933		14356.8	166.09	10518.1	124.90	55.879	1983		14768.3	184.63	10830.3	115.26	56.089
1934		14365.0	166.44	10524.3	124.70	55.883	1984		14776.5	185.02	10836.5	115.08	56.093
1935	1475.33	14373.2	166.80	10530.6	124.49	55.887	1985	1525.33	14784.7	185.41	10842.8	114.89	56.097
1936		14381.4	167.16	10536.8	124.29	55.892	1986		14793.0	185.79	10849.1	114.71	56.102
1937		14389.6	167.52	10543.0	124.09	55.896	1987		14801.2	186.19	10855.3	114.53	56.106
1938		14397.8	167.87	10549.2	123.89	55.900	1988		14809.5	186.58	10861.6	114.35	56.110
1939		14406.1	168.24	10555.5	123.69	55.904	1989		14817.7	186.96	10867.9	114.17	56.114
1940	1480.33	14414.3	168.59	10561.7	123.49	55.909	1990	1530.33	14826.0	187.36	10874.1	113.98	56.118
1941		14422.5	168.96	10568.0	123.29	55.913	1991		14834.2	187.75	10880.4	113.80	56.122
1942		14430.7	169.31	10574.2	123.09	55.917	1992		14842.5	188.14	10886.7	113.63	56.126
1943		14438.9	169.67	10580.4	122.89	55.921	1993		14850.7	188.53	10892.9	113.44	56.131
1944		14447.1	170.04	10586.6	122.69	55.926	1994		14859.0	188.93	10899.2	113.26	56.135
1945	1485.33	14455.4	170.40	10592.9	122.49	55.930	1995	1535.33	14867.2	189.32	10905.5	113.09	56.139
1946		14463.6	170.76	10599.1	122.29	55.934	1996		14875.5	189.72	10911.7	112.91	56.143
1947		14471.8	171.13	10605.4	122.10	55.938	1997		14883.8	190.11	10918.0	112.73	56.147
1948		14480.0	171.49	10611.6	121.90	55.942	1998		14892.0	190.50	10924.3	112.55	56.151
1949		14488.3	171.86	10617.8	121.71	55.947	1999		14900.3	190.91	10930.6	112.37	56.155

Table 3   Products—400% Theoretical Air (for One Pound-Mole)

T	t	$\bar{h}$	p_r	$\bar{u}$	v_r	$\bar{\phi}$	T	t	$\bar{h}$	p_r	$\bar{u}$	v_r	$\bar{\phi}$
2000	1540.33	14908.5	191.30	10936.8	112.20	56.160	2250	1790.33	16993.9	313.7	12525.7	76.97	57.142
2005		14949.8	193.30	10968.2	111.31	56.180	2255		17036.0	316.7	12557.8	76.42	57.161
2010		14991.2	195.31	10999.6	110.44	56.201	2260		17078.1	319.7	12590.0	75.87	57.179
2015		15032.5	197.34	11031.0	109.58	56.221	2265		17120.2	322.7	12622.2	75.33	57.198
2020		15073.9	199.39	11062.4	108.72	56.242	2270		17162.4	325.7	12654.5	74.79	57.216
2025	1565.33	15115.2	201.45	11093.9	107.87	56.262	2275	1815.33	17204.5	328.8	12686.7	74.26	57.235
2030		15156.6	203.53	11125.3	107.03	56.283	2280		17246.7	331.9	12718.9	73.73	57.253
2035		15198.0	205.63	11156.8	106.20	56.303	2285		17288.9	335.0	12751.2	73.21	57.272
2040		15239.4	207.75	11188.3	105.38	56.323	2290		17331.1	338.1	12783.5	72.69	57.290
2045		15280.9	209.88	11219.8	104.56	56.344	2295		17373.3	341.2	12815.8	72.18	57.309
2050	1590.33	15322.4	212.03	11251.3	103.76	56.364	2300	1840.33	17415.6	344.4	12848.1	71.67	57.327
2055		15363.8	214.20	11282.9	102.96	56.384	2305		17457.8	347.6	12880.4	71.16	57.346
2060		15405.3	216.38	11314.4	102.17	56.404	2310		17500.0	350.8	12912.7	70.66	57.364
2065		15446.8	218.59	11346.0	101.38	56.424	2315		17542.3	354.1	12945.1	70.17	57.382
2070		15488.3	220.81	11377.6	100.60	56.444	2320		17584.6	357.3	12977.4	69.67	57.400
2075	1615.33	15529.9	223.05	11409.2	99.83	56.464	2325	1865.33	17626.9	360.6	13009.8	69.19	57.419
2080		15571.5	225.31	11440.9	99.07	56.484	2330		17669.2	363.9	13042.2	68.70	57.437
2085		15613.0	227.59	11472.5	98.31	56.504	2335		17711.6	367.3	13074.6	68.23	57.455
2090		15654.6	229.88	11504.2	97.57	56.524	2340		17753.9	370.6	13107.0	67.75	57.473
2095		15696.2	232.20	11535.8	96.83	56.544	2345		17796.3	374.0	13139.4	67.28	57.491
2100	1640.33	15737.8	234.53	11567.5	96.09	56.564	2350	1890.33	17838.6	377.5	13171.9	66.81	57.509
2105		15779.5	236.88	11599.2	95.36	56.584	2355		17881.0	380.9	13204.3	66.35	57.527
2110		15821.1	239.25	11631.0	94.64	56.604	2360		17923.4	384.4	13236.8	65.89	57.545
2115		15862.8	241.64	11662.7	93.93	56.623	2365		17965.8	387.8	13269.3	65.44	57.563
2120		15904.5	244.04	11694.5	93.23	56.643	2370		18008.3	391.4	13301.8	64.99	57.581
2125	1665.33	15946.2	246.46	11726.3	92.53	56.663	2375	1915.33	18050.7	394.9	13334.3	64.54	57.599
2130		15987.9	248.91	11758.1	91.83	56.682	2380		18093.1	398.5	13366.8	64.10	57.617
2135		16029.7	251.38	11789.9	91.14	56.702	2385		18135.6	402.1	13399.3	63.66	57.635
2140		16071.4	253.86	11821.7	90.46	56.721	2390		18178.1	405.7	13431.9	63.22	57.652
2145		16113.2	256.37	11853.5	89.79	56.741	2395		18220.6	409.3	13464.4	62.79	57.670
2150	1690.33	16155.0	258.90	11885.4	89.12	56.760	2400	1940.33	18263.1	413.0	13497.0	62.36	57.688
2155		16196.8	261.43	11917.2	88.46	56.780	2405		18305.6	416.7	13529.6	61.94	57.706
2160		16238.6	264.00	11949.1	87.80	56.799	2410		18348.1	420.4	13562.2	61.51	57.723
2165		16280.4	266.59	11981.0	87.15	56.819	2415		18390.6	424.2	13594.8	61.10	57.741
2170		16322.3	269.19	12013.0	86.51	56.838	2420		18433.2	428.0	13627.4	60.68	57.758
2175	1715.33	16364.1	271.81	12044.9	85.87	56.857	2425	1965.33	18475.8	431.8	13660.1	60.27	57.776
2180		16406.0	274.46	12076.8	85.24	56.876	2430		18518.3	435.6	13692.7	59.87	57.794
2185		16447.9	277.12	12108.8	84.61	56.896	2435		18560.9	439.4	13725.4	59.46	57.811
2190		16489.8	279.81	12140.8	83.99	56.915	2440		18603.6	443.3	13758.1	59.06	57.829
2195		16531.7	282.52	12172.8	83.38	56.934	2445		18646.2	447.3	13790.7	58.67	57.846
2200	1740.33	16573.7	285.25	12204.8	82.77	56.953	2450	1990.33	18688.8	451.2	13823.4	58.27	57.863
2205		16615.6	287.99	12236.8	82.16	56.972	2455		18731.5	455.2	13856.2	57.88	57.881
2210		16657.6	290.77	12268.8	81.56	56.991	2460		18774.1	459.2	13888.9	57.50	57.898
2215		16699.6	293.57	12300.9	80.97	57.010	2465		18816.8	463.2	13921.6	57.11	57.916
2220		16741.6	296.38	12333.0	80.38	57.029	2470		18859.5	467.2	13954.4	56.73	57.933
2225	1765.33	16783.6	299.22	12365.0	79.80	57.048	2475	2015.33	18902.2	471.3	13987.2	56.35	57.950
2230		16825.6	302.07	12397.1	79.22	57.067	2480		18944.9	475.4	14019.9	55.98	57.967
2235		16867.7	304.95	12429.3	78.65	57.086	2485		18987.6	479.6	14052.7	55.61	57.985
2240		16909.7	307.86	12461.4	78.08	57.104	2490		19030.3	483.7	14085.5	55.24	58.002
2245		16951.8	310.78	12493.5	77.52	57.123	2495		19073.1	487.9	14118.4	54.88	58.019

Table 3   Products—400% Theoretical Air (for One Pound-Mole)

T	t	$\bar{h}$	p_r	$\bar{u}$	v_r	$\bar{\phi}$	T	t	$\bar{h}$	p_r	$\bar{u}$	v_r	$\bar{\phi}$
2500	2040.33	19115.8	492.2	14151.2	54.51	58.036	3000	2540.33	23447.6	1089.9	17490.0	29.54	59.615
2510		19201.4	500.7	14216.9	53.80	58.070	3010		23535.2	1106.0	17557.8	29.21	59.644
2520		19287.0	509.3	14282.6	53.10	58.104	3020		23622.9	1122.3	17625.6	28.88	59.673
2530		19372.6	518.1	14348.4	52.40	58.138	3030		23710.7	1138.9	17693.5	28.55	59.702
2540		19458.4	527.0	14414.3	51.72	58.172	3040		23798.4	1155.5	17761.4	28.23	59.731
2550	2090.33	19544.1	536.0	14480.2	51.05	58.206	3050	2590.33	23886.2	1172.5	17829.3	27.92	59.760
2560		19629.9	545.2	14546.1	50.39	58.239	3060		23974.0	1189.5	17897.3	27.61	59.789
2570		19715.7	554.4	14612.1	49.75	58.273	3070		24061.9	1206.8	17965.3	27.30	59.817
2580		19801.6	563.8	14678.1	49.11	58.306	3080		24149.7	1224.3	18033.3	27.00	59.846
2590		19887.6	573.3	14744.2	48.48	58.339	3090		24237.7	1242.0	18101.4	26.70	59.874
2600	2140.33	19973.6	583.0	14810.3	47.86	58.372	3100	2640.33	24325.6	1259.9	18169.5	26.40	59.903
2610		20059.6	592.8	14876.5	47.25	58.405	3110		24413.6	1278.0	18237.6	26.11	59.931
2620		20145.7	602.7	14942.7	46.65	58.438	3120		24501.6	1296.4	18305.8	25.83	59.959
2630		20231.8	612.7	15009.0	46.07	58.471	3130		24589.7	1314.9	18373.9	25.55	59.988
2640		20318.0	622.9	15075.3	45.48	58.504	3140		24677.8	1333.6	18442.2	25.27	60.016
2650	2190.33	20404.2	633.2	15141.6	44.91	58.536	3150	2690.33	24765.9	1352.6	18510.4	24.99	60.044
2660		20490.4	643.6	15208.0	44.35	58.569	3160		24854.1	1371.7	18578.7	24.72	60.072
2670		20576.7	654.2	15274.5	43.80	58.601	3170		24942.2	1391.1	18647.0	24.45	60.099
2680		20663.1	664.9	15341.0	43.25	58.634	3180		25030.5	1410.7	18715.4	24.19	60.127
2690		20749.4	675.8	15407.5	42.72	58.666	3190		25118.7	1430.5	18783.8	23.93	60.155
2700	2240.33	20835.9	686.8	15474.0	42.19	58.698	3200	2740.33	25207.0	1450.6	18852.2	23.67	60.183
2710		20922.3	697.9	15540.7	41.67	58.730	3210		25295.3	1470.8	18920.7	23.42	60.210
2720		21008.9	709.2	15607.3	41.16	58.762	3220		25383.6	1491.3	18989.2	23.17	60.238
2730		21095.4	720.7	15674.0	40.65	58.793	3230		25472.0	1512.1	19057.7	22.92	60.265
2740		21182.0	732.2	15740.8	40.16	58.825	3240		25560.4	1533.0	19126.2	22.68	60.292
2750	2290.33	21268.7	744.0	15807.5	39.67	58.857	3250	2790.33	25648.9	1554.2	19194.8	22.44	60.320
2760		21355.3	755.9	15874.4	39.18	58.888	3260		25737.3	1575.6	19263.4	22.20	60.347
2770		21442.0	767.9	15941.2	38.71	58.919	3270		25825.8	1597.2	19332.1	21.97	60.374
2780		21528.8	780.1	16008.1	38.24	58.951	3280		25914.3	1619.1	19400.7	21.74	60.401
2790		21615.6	792.4	16075.1	37.78	58.982	3290		26002.9	1641.3	19469.5	21.51	60.428
2800	2340.33	21702.5	804.9	16142.1	37.33	59.013	3300	2840.33	26091.5	1663.7	19538.2	21.29	60.455
2810		21789.4	817.5	16209.1	36.89	59.044	3310		26180.1	1686.3	19606.9	21.06	60.482
2820		21876.3	830.4	16276.2	36.45	59.075	3320		26268.8	1709.2	19675.7	20.85	60.508
2830		21963.3	843.3	16343.3	36.01	59.106	3330		26357.5	1732.2	19744.5	20.63	60.535
2840		22050.3	856.5	16410.4	35.59	59.136	3340		26446.1	1755.6	19813.4	20.42	60.562
2850	2390.33	22137.3	869.8	16477.6	35.16	59.167	3350	2890.33	26534.9	1779.3	19882.3	20.21	60.588
2860		22224.4	883.2	16544.9	34.75	59.197	3360		26623.7	1803.1	19951.2	20.00	60.615
2870		22311.5	896.8	16612.1	34.34	59.228	3370		26712.5	1827.2	20020.1	19.79	60.641
2880		22398.7	910.6	16679.4	33.94	59.258	3380		26801.3	1851.6	20089.1	19.59	60.667
2890		22485.9	924.6	16746.8	33.54	59.288	3390		26890.2	1876.2	20158.1	19.39	60.694
2900	2440.33	22573.1	938.8	16814.1	33.15	59.318	3400	2940.33	26979.0	1901.2	20227.1	19.19	60.720
2910		22660.4	953.1	16881.6	32.77	59.348	3410		27067.9	1926.3	20296.2	19.00	60.746
2920		22747.8	967.6	16949.0	32.39	59.378	3420		27156.9	1951.7	20365.2	18.80	60.772
2930		22835.1	982.2	17016.5	32.01	59.408	3430		27245.9	1977.4	20434.4	18.61	60.798
2940		22922.5	997.0	17084.1	31.64	59.438	3440		27334.8	2003.4	20503.5	18.43	60.824
2950	2490.33	23009.9	1012.1	17151.6	31.28	59.468	3450	2990.33	27423.9	2029.6	20572.7	18.24	60.850
2960		23097.4	1027.2	17219.3	30.92	59.497	3460		27512.9	2056.2	20641.8	18.06	60.875
2970		23184.9	1042.6	17286.9	30.57	59.527	3470		27602.0	2083.0	20711.1	17.88	60.901
2980		23272.4	1058.2	17354.6	30.22	59.556	3480		27691.1	2110.0	20780.3	17.70	60.927
2990		23360.0	1074.0	17422.3	29.88	59.586	3490		27780.2	2137.4	20849.6	17.52	60.952

Table 4   Products—400% Theoretical Air (for One Pound-Mole)

T	t	Fuel—(CH₁)ₙ			Fuel—(CH₂)ₙ			Fuel—(CH₃)ₙ		
		$\bar{c}_p$	$\bar{c}_v$	$k = \dfrac{\bar{c}_p}{\bar{c}_v}$	$\bar{c}_p$	$\bar{c}_v$	$k = \dfrac{\bar{c}_p}{\bar{c}_v}$	$\bar{c}_p$	$\bar{c}_v$	$k = \dfrac{\bar{c}_p}{\bar{c}_v}$
		Btu	Btu		Btu	Btu		Btu	Btu	
R	F	lb-mole R	lb-mole R		lb-mole R	lb-mole R		lb-mole R	lb-mole R	
300	−159.67	6.971	4.985	1.398	6.982	4.996	1.397	6.990	5.004	1.397
350	−109.67	6.984	4.998	1.397	6.993	5.007	1.397	6.999	5.013	1.396
400	−59.67	6.999	5.013	1.396	7.005	5.020	1.396	7.010	5.024	1.395
450	−9.67	7.015	5.029	1.395	7.020	5.034	1.395	7.023	5.037	1.394
500	40.33	7.033	5.048	1.393	7.036	5.050	1.393	7.037	5.051	1.393
550	90.33	7.054	5.068	1.392	7.054	5.068	1.392	7.054	5.068	1.392
600	140.33	7.076	5.090	1.390	7.075	5.089	1.390	7.074	5.088	1.390
650	190.33	7.102	5.116	1.388	7.099	5.113	1.388	7.097	5.111	1.389
700	240.33	7.131	5.145	1.386	7.127	5.141	1.386	7.124	5.138	1.387
750	290.33	7.163	5.177	1.384	7.158	5.172	1.384	7.154	5.168	1.384
800	340.33	7.198	5.212	1.381	7.192	5.206	1.381	7.188	5.202	1.382
850	390.33	7.237	5.251	1.378	7.230	5.244	1.379	7.225	5.239	1.379
900	440.33	7.278	5.292	1.375	7.271	5.285	1.376	7.265	5.279	1.376
950	490.33	7.322	5.336	1.372	7.314	5.328	1.373	7.308	5.322	1.373
1000	540.33	7.368	5.382	1.369	7.360	5.374	1.370	7.353	5.367	1.370
1100	640.33	7.465	5.479	1.362	7.455	5.469	1.363	7.448	5.463	1.364
1200	740.33	7.565	5.579	1.356	7.555	5.569	1.357	7.547	5.561	1.357
1300	840.33	7.666	5.680	1.350	7.655	5.669	1.350	7.647	5.661	1.351
1400	940.33	7.764	5.779	1.344	7.754	5.768	1.344	7.746	5.760	1.345
1500	1040.33	7.860	5.874	1.338	7.849	5.863	1.339	7.842	5.856	1.339
1600	1140.33	7.952	5.966	1.333	7.941	5.955	1.333	7.933	5.948	1.334
1700	1240.33	8.038	6.052	1.328	8.028	6.042	1.329	8.021	6.035	1.329
1800	1340.33	8.119	6.134	1.324	8.110	6.124	1.324	8.103	6.117	1.325
1900	1440.33	8.196	6.210	1.320	8.187	6.201	1.320	8.181	6.195	1.321
2000	1540.33	8.267	6.281	1.316	8.259	6.273	1.317	8.254	6.268	1.317
2100	1640.33	8.333	6.347	1.313	8.326	6.340	1.313	8.321	6.336	1.313
2200	1740.33	8.395	6.409	1.310	8.389	6.403	1.310	8.385	6.399	1.310
2300	1840.33	8.453	6.467	1.307	8.448	6.462	1.307	8.444	6.458	1.307
2400	1940.33	8.506	6.520	1.305	8.502	6.516	1.305	8.499	6.514	1.305
2500	2040.33	8.556	6.570	1.302	8.553	6.567	1.302	8.551	6.565	1.302
2600	2140.33	8.602	6.616	1.300	8.601	6.615	1.300	8.599	6.614	1.300
2700	2240.33	8.646	6.660	1.298	8.645	6.659	1.298	8.645	6.659	1.298
2800	2340.33	8.686	6.701	1.296	8.687	6.701	1.296	8.687	6.701	1.296
2900	2440.33	8.725	6.739	1.295	8.726	6.740	1.295	8.727	6.741	1.295
3000	2540.33	8.760	6.774	1.293	8.763	6.777	1.293	8.764	6.778	1.293
3200	2740.33	8.826	6.840	1.290	8.830	6.844	1.290	8.833	6.847	1.290
3400	2940.33	8.884	6.898	1.288	8.890	6.904	1.288	8.894	6.908	1.287
3600	3140.33	8.936	6.950	1.286	8.943	6.958	1.285	8.949	6.963	1.285
3800	3340.33	8.983	6.997	1.284	8.992	7.006	1.283	8.998	7.013	1.283
4000	3540.33	9.026	7.040	1.282	9.036	7.050	1.282	9.043	7.058	1.281

Table 5   Products—200% Theoretical Air (for One Pound-Mole)

T	t	$\bar{h}$	p_r	$\bar{u}$	v_r	$\bar{\phi}$	T	t	$\bar{h}$	p_r	$\bar{u}$	v_r	$\bar{\phi}$
300	−159.67	2097.5	.16707	1501.7	19270	42.173	350	−109.67	2449.5	.2885	1754.4	13017	43.258
301		2104.5	.16905	1506.8	19108	42.196	351		2456.5	.2915	1759.5	12923	43.278
302		2111.6	.17105	1511.8	18947	42.219	352		2463.6	.2944	1764.6	12829	43.298
303		2118.6	.17306	1516.9	18789	42.243	353		2470.6	.2974	1769.6	12737	43.318
304		2125.6	.17509	1521.9	18632	42.266	354		2477.7	.3004	1774.7	12645	43.338
305	−154.67	2132.6	.17714	1527.0	18477	42.289	355	−104.67	2484.7	.3035	1779.8	12555	43.358
306		2139.7	.17920	1532.0	18324	42.312	356		2491.8	.3065	1784.8	12465	43.378
307		2146.7	.18129	1537.0	18173	42.335	357		2498.8	.3096	1789.9	12376	43.398
308		2153.7	.18339	1542.1	18024	42.358	358		2505.9	.3127	1795.0	12288	43.417
309		2160.8	.18550	1547.1	17876	42.381	359		2512.9	.3158	1800.0	12201	43.437
310	−149.67	2167.8	.18764	1552.2	17729	42.403	360	−99.67	2520.0	.3189	1805.1	12114	43.457
311		2174.8	.18979	1557.2	17585	42.426	361		2527.1	.3221	1810.2	12029	43.476
312		2181.9	.19196	1562.3	17442	42.449	362		2534.1	.3252	1815.2	11944	43.496
313		2188.9	.19415	1567.3	17301	42.471	363		2541.2	.3284	1820.3	11860	43.515
314		2195.9	.19636	1572.4	17161	42.494	364		2548.2	.3317	1825.4	11777	43.535
315	−144.67	2203.0	.19858	1577.4	17023	42.516	365	−94.67	2555.3	.3349	1830.4	11695	43.554
316		2210.0	.20082	1582.5	16886	42.538	366		2562.3	.3382	1835.5	11614	43.573
317		2217.1	.20309	1587.5	16751	42.560	367		2569.4	.3415	1840.6	11533	43.592
318		2224.1	.20536	1592.6	16617	42.583	368		2576.5	.3448	1845.7	11453	43.612
319		2231.1	.20766	1597.6	16485	42.605	369		2583.5	.3482	1850.7	11374	43.631
320	−139.67	2238.2	.20998	1602.7	16354	42.627	370	−89.67	2590.6	.3515	1855.8	11296	43.650
321		2245.2	.21231	1607.7	16225	42.649	371		2597.6	.3549	1860.9	11218	43.669
322		2252.2	.21467	1612.8	16097	42.671	372		2604.7	.3583	1865.9	11141	43.688
323		2259.3	.21704	1617.8	15971	42.692	373		2611.7	.3618	1871.0	11065	43.707
324		2266.3	.21943	1622.9	15846	42.714	374		2618.8	.3652	1876.1	10990	43.726
325	−134.67	2273.4	.22184	1628.0	15722	42.736	375	−84.67	2625.9	.3687	1881.2	10915	43.745
326		2280.4	.22427	1633.0	15599	42.757	376		2632.9	.3722	1886.2	10841	43.763
327		2287.4	.22672	1638.1	15478	42.779	377		2640.0	.3757	1891.3	10767	43.782
328		2294.5	.22919	1643.1	15358	42.801	378		2647.1	.3793	1896.4	10695	43.801
329		2301.5	.23167	1648.2	15240	42.822	379		2654.1	.3829	1901.5	10623	43.820
330	−129.67	2308.6	.23418	1653.2	15123	42.843	380	−79.67	2661.2	.3865	1906.6	10551	43.838
331		2315.6	.23670	1658.3	15007	42.865	381		2668.2	.3901	1911.6	10481	43.857
332		2322.6	.23925	1663.3	14892	42.886	382		2675.3	.3938	1916.7	10411	43.875
333		2329.7	.24182	1668.4	14778	42.907	383		2682.4	.3974	1921.8	10341	43.894
334		2336.7	.24440	1673.5	14666	42.928	384		2689.4	.4012	1926.9	10273	43.912
335	−124.67	2343.8	.24701	1678.5	14554	42.949	385	−74.67	2696.5	.4049	1932.0	10204	43.931
336		2350.8	.24963	1683.6	14444	42.970	386		2703.6	.4086	1937.0	10137	43.949
337		2357.9	.25228	1688.6	14335	42.991	387		2710.6	.4124	1942.1	10070	43.967
338		2364.9	.25494	1693.7	14228	43.012	388		2717.7	.4162	1947.2	10004	43.985
339		2372.0	.25763	1698.7	14121	43.033	389		2724.8	.4201	1952.3	9938	44.004
340	−119.67	2379.0	.26034	1703.8	14015	43.054	390	−69.67	2731.9	.4239	1957.4	9873	44.022
341		2386.0	.26306	1708.9	13911	43.074	391		2738.9	.4278	1962.5	9809	44.040
342		2393.1	.26581	1713.9	13807	43.095	392		2746.0	.4317	1967.5	9744	44.058
343		2400.1	.26858	1719.0	13705	43.115	393		2753.1	.4356	1972.6	9681	44.076
344		2407.2	.27137	1724.0	13604	43.136	394		2760.1	.4396	1977.7	9618	44.094
345	−114.67	2414.2	.27418	1729.1	13504	43.156	395	−64.67	2767.2	.4436	1982.8	9556	44.112
346		2421.3	.27701	1734.2	13404	43.177	396		2774.3	.4476	1987.9	9495	44.130
347		2428.3	.27986	1739.2	13306	43.197	397		2781.3	.4516	1993.0	9433	44.148
348		2435.4	.28274	1744.3	13209	43.217	398		2788.4	.4557	1998.0	9373	44.165
349		2442.4	.28563	1749.4	13112	43.238	399		2795.5	.4598	2003.1	9313	44.183

Table 5   Products—200% Theoretical Air (for One Pound-Mole)

T	t	$\bar{h}$	p_r	$\bar{u}$	v_r	$\bar{\phi}$	T	t	$\bar{h}$	p_r	$\bar{u}$	v_r	$\bar{\phi}$
400	−59.67	2802.6	.4639	2008.2	9253	44.201	450	−9.67	3156.9	.7062	2263.2	6838	45.035
401		2809.6	.4680	2013.3	9194	44.218	451		3164.0	.7119	2268.3	6799	45.051
402		2816.7	.4722	2018.4	9136	44.236	452		3171.1	.7175	2273.5	6760	45.067
403		2823.8	.4764	2023.5	9078	44.254	453		3178.2	.7232	2278.6	6722	45.083
404		2830.9	.4806	2028.6	9020	44.271	454		3185.3	.7289	2283.7	6684	45.098
405	−54.67	2837.9	.4849	2033.7	8963	44.289	455	−4.67	3192.4	.7347	2288.8	6646	45.114
406		2845.0	.4892	2038.8	8907	44.306	456		3199.5	.7405	2293.9	6609	45.129
407		2852.1	.4935	2043.8	8851	44.324	457		3206.6	.7463	2299.0	6572	45.145
408		2859.2	.4978	2048.9	8795	44.341	458		3213.7	.7521	2304.1	6535	45.160
409		2866.2	.5022	2054.0	8740	44.358	459		3220.8	.7580	2309.3	6498	45.176
410	−49.67	2873.3	.5066	2059.1	8686	44.376	460	.33	3227.9	.7640	2314.4	6462	45.191
411		2880.4	.5110	2064.2	8632	44.393	461		3235.0	.7699	2319.5	6426	45.207
412		2887.5	.5154	2069.3	8578	44.410	462		3242.1	.7759	2324.6	6390	45.222
413		2894.6	.5199	2074.4	8525	44.427	463		3249.2	.7819	2329.7	6354	45.238
414		2901.6	.5244	2079.5	8472	44.444	464		3256.3	.7880	2334.8	6319	45.253
415	−44.67	2908.7	.5289	2084.6	8420	44.461	465	5.33	3263.4	.7941	2340.0	6284	45.268
416		2915.8	.5335	2089.7	8368	44.478	466		3270.5	.8002	2345.1	6249	45.284
417		2922.9	.5381	2094.8	8317	44.495	467		3277.6	.8064	2350.2	6215	45.299
418		2930.0	.5427	2099.9	8266	44.512	468		3284.7	.8126	2355.3	6181	45.314
419		2937.0	.5473	2105.0	8215	44.529	469		3291.8	.8188	2360.4	6147	45.329
420	−39.67	2944.1	.5520	2110.1	8165	44.546	470	10.33	3298.9	.8251	2365.6	6113	45.344
421		2951.2	.5567	2115.2	8115	44.563	471		3306.0	.8314	2370.7	6080	45.359
422		2958.3	.5615	2120.3	8066	44.580	472		3313.1	.8377	2375.8	6047	45.374
423		2965.4	.5662	2125.4	8017	44.597	473		3320.3	.8441	2380.9	6014	45.389
424		2972.5	.5710	2130.5	7969	44.613	474		3327.4	.8505	2386.1	5981	45.405
425	−34.67	2979.6	.5758	2135.6	7921	44.630	475	15.33	3334.5	.8569	2391.2	5948	45.420
426		2986.6	.5807	2140.7	7873	44.647	476		3341.6	.8634	2396.3	5916	45.434
427		2993.7	.5856	2145.8	7826	44.663	477		3348.7	.8699	2401.4	5884	45.449
428		3000.8	.5905	2150.9	7779	44.680	478		3355.8	.8765	2406.6	5853	45.464
429		3007.9	.5954	2156.0	7732	44.696	479		3362.9	.8831	2411.7	5821	45.479
430	−29.67	3015.0	.6004	2161.1	7686	44.713	480	20.33	3370.0	.8897	2416.8	5790	45.494
431		3022.1	.6054	2166.2	7640	44.729	481		3377.2	.8963	2422.0	5759	45.509
432		3029.2	.6104	2171.3	7595	44.746	482		3384.3	.9030	2427.1	5728	45.524
433		3036.3	.6155	2176.4	7550	44.762	483		3391.4	.9098	2432.2	5697	45.538
434		3043.3	.6205	2181.5	7505	44.779	484		3398.5	.9165	2437.3	5667	45.553
435	−24.67	3050.4	.6257	2186.6	7461	44.795	485	25.33	3405.6	.9233	2442.5	5637	45.568
436		3057.5	.6308	2191.7	7417	44.811	486		3412.7	.9302	2447.6	5607	45.582
437		3064.6	.6360	2196.8	7374	44.827	487		3419.9	.9371	2452.7	5577	45.597
438		3071.7	.6412	2201.9	7330	44.844	488		3427.0	.9440	2457.9	5548	45.612
439		3078.8	.6465	2207.0	7288	44.860	489		3434.1	.9509	2463.0	5519	45.626
440	−19.67	3085.9	.6517	2212.1	7245	44.876	490	30.33	3441.2	.9579	2468.1	5489	45.641
441		3093.0	.6570	2217.2	7203	44.892	491		3448.3	.9649	2473.3	5461	45.655
442		3100.1	.6624	2222.3	7161	44.908	492		3455.5	.9720	2478.4	5432	45.670
443		3107.2	.6677	2227.4	7120	44.924	493		3462.6	.9791	2483.5	5403	45.684
444		3114.3	.6731	2232.6	7078	44.940	494		3469.7	.9863	2488.7	5375	45.699
445	−14.67	3121.4	.6786	2237.7	7037	44.956	495	35.33	3476.8	.9934	2493.8	5347	45.713
446		3128.5	.6840	2242.8	6997	44.972	496		3483.9	1.0007	2498.9	5319	45.727
447		3135.6	.6895	2247.9	6957	44.988	497		3491.1	1.0079	2504.1	5292	45.742
448		3142.7	.6951	2253.0	6917	45.004	498		3498.2	1.0152	2509.2	5264	45.756
449		3149.8	.7006	2258.1	6877	45.020	499		3505.3	1.0225	2514.4	5237	45.770

Table 5   Products—200% Theoretical Air (for One Pound-Mole)

T	t	$\bar{h}$	p_r	$\bar{u}$	v_r	$\bar{\phi}$	T	t	$\bar{h}$	p_r	$\bar{u}$	v_r	$\bar{\phi}$
500	40.33	3512.4	1.0299	2519.5	5210	45.785	550	90.33	3869.4	1.4507	2777.1	4069	46.465
501		3519.6	1.0373	2524.6	5183	45.799	551		3876.5	1.4602	2782.3	4049	46.478
502		3526.7	1.0448	2529.8	5156	45.813	552		3883.7	1.4698	2787.5	4030	46.491
503		3533.8	1.0522	2534.9	5130	45.827	553		3890.8	1.4794	2792.6	4011	46.504
504		3540.9	1.0598	2540.1	5104	45.841	554		3898.0	1.4891	2797.8	3993	46.517
505	45.33	3548.1	1.0673	2545.2	5077	45.856	555	95.33	3905.1	1.4988	2803.0	3974	46.530
506		3555.2	1.0749	2550.3	5052	45.870	556		3912.3	1.5085	2808.1	3955	46.543
507		3562.3	1.0826	2555.5	5026	45.884	557		3919.4	1.5183	2813.3	3937	46.555
508		3569.4	1.0903	2560.6	5000	45.898	558		3926.6	1.5282	2818.5	3919	46.568
509		3576.6	1.0980	2565.8	4975	45.912	559		3933.7	1.5381	2823.7	3900	46.581
510	50.33	3583.7	1.1058	2570.9	4950	45.926	560	100.33	3940.9	1.5480	2828.8	3882	46.594
511		3590.8	1.1136	2576.1	4925	45.940	561		3948.1	1.5580	2834.0	3864	46.607
512		3598.0	1.1214	2581.2	4900	45.954	562		3955.2	1.5680	2839.2	3846	46.619
513		3605.1	1.1293	2586.3	4875	45.968	563		3962.4	1.5781	2844.3	3828	46.632
514		3612.2	1.1372	2591.5	4850	45.981	564		3969.5	1.5883	2849.5	3811	46.645
515	55.33	3619.4	1.1452	2596.6	4826	45.995	565	105.33	3976.7	1.5984	2854.7	3793	46.658
516		3626.5	1.1532	2601.8	4802	46.009	566		3983.9	1.6087	2859.9	3776	46.670
517		3633.6	1.1612	2606.9	4778	46.023	567		3991.0	1.6189	2865.0	3759	46.683
518		3640.8	1.1693	2612.1	4754	46.037	568		3998.2	1.6293	2870.2	3741	46.695
519		3647.9	1.1775	2617.2	4730	46.051	569		4005.3	1.6396	2875.4	3724	46.708
520	60.33	3655.0	1.1856	2622.4	4707	46.064	570	110.33	4012.5	1.6500	2880.6	3707	46.721
521		3662.2	1.1939	2627.5	4683	46.078	571		4019.7	1.6605	2885.8	3690	46.733
522		3669.3	1.2021	2632.7	4660	46.092	572		4026.8	1.6710	2890.9	3673	46.746
523		3676.4	1.2104	2637.8	4637	46.105	573		4034.0	1.6816	2896.1	3657	46.758
524		3683.6	1.2187	2643.0	4614	46.119	574		4041.2	1.6922	2901.3	3640	46.771
525	65.33	3690.7	1.2271	2648.1	4591	46.133	575	115.33	4048.3	1.7029	2906.5	3624	46.783
526		3697.8	1.2355	2653.3	4569	46.146	576		4055.5	1.7136	2911.7	3607	46.796
527		3705.0	1.2440	2658.4	4546	46.160	577		4062.7	1.7243	2916.8	3591	46.808
528		3712.1	1.2525	2663.6	4524	46.173	578		4069.8	1.7352	2922.0	3575	46.821
529		3719.3	1.2611	2668.8	4502	46.187	579		4077.0	1.7460	2927.2	3559	46.833
530	70.33	3726.4	1.2697	2673.9	4480	46.200	580	120.33	4084.2	1.7569	2932.4	3543	46.845
531		3733.6	1.2783	2679.1	4458	46.214	581		4091.4	1.7679	2937.6	3527	46.858
532		3740.7	1.2870	2684.2	4436	46.227	582		4098.5	1.7789	2942.8	3511	46.870
533		3747.8	1.2957	2689.4	4415	46.241	583		4105.7	1.7900	2947.9	3495	46.882
534		3755.0	1.3045	2694.5	4393	46.254	584		4112.9	1.8011	2953.1	3480	46.895
535	75.33	3762.1	1.3133	2699.7	4372	46.267	585	125.33	4120.0	1.8123	2958.3	3464	46.907
536		3769.3	1.3221	2704.8	4351	46.281	586		4127.2	1.8235	2963.5	3449	46.919
537		3776.4	1.3310	2710.0	4330	46.294	587		4134.4	1.8347	2968.7	3433	46.931
538		3783.6	1.3400	2715.2	4309	46.307	588		4141.6	1.8460	2973.9	3418	46.944
539		3790.7	1.3489	2720.3	4288	46.321	589		4148.7	1.8574	2979.1	3403	46.956
540	80.33	3797.9	1.3580	2725.5	4267	46.334	590	130.33	4155.9	1.8688	2984.3	3388	46.968
541		3805.0	1.3670	2730.6	4247	46.347	591		4163.1	1.8803	2989.5	3373	46.980
542		3812.1	1.3762	2735.8	4227	46.360	592		4170.3	1.8918	2994.6	3358	46.992
543		3819.3	1.3853	2741.0	4206	46.373	593		4177.5	1.9034	2999.8	3343	47.004
544		3826.4	1.3945	2746.1	4186	46.387	594		4184.6	1.9150	3005.0	3329	47.016
545	85.33	3833.6	1.4038	2751.3	4166	46.400	595	135.33	4191.8	1.9267	3010.2	3314	47.028
546		3840.7	1.4131	2756.5	4147	46.413	596		4199.0	1.9384	3015.4	3300	47.041
547		3847.9	1.4224	2761.6	4127	46.426	597		4206.2	1.9502	3020.6	3285	47.053
548		3855.0	1.4318	2766.8	4107	46.439	598		4213.4	1.9620	3025.8	3271	47.065
549		3862.2	1.4412	2772.0	4088	46.452	599		4220.5	1.9739	3031.0	3257	47.077

Table 5 Products—200% Theoretical Air (for One Pound-Mole)

T	t	$\bar{h}$	p_r	$\bar{u}$	v_r	$\bar{\phi}$	T	t	$\bar{h}$	p_r	$\bar{u}$	v_r	$\bar{\phi}$
600	140.33	4227.7	1.986	3036.2	3242	47.089	650	190.33	4587.6	2.654	3296.8	2628	47.665
601		4234.9	1.998	3041.4	3228	47.101	651		4594.8	2.669	3302.0	2617	47.676
602		4242.1	2.010	3046.6	3214	47.112	652		4602.0	2.684	3307.3	2607	47.687
603		4249.3	2.022	3051.8	3200	47.124	653		4609.2	2.699	3312.5	2596	47.698
604		4256.4	2.034	3057.0	3186	47.136	654		4616.5	2.714	3317.7	2586	47.709
605	145.33	4263.6	2.046	3062.2	3173	47.148	655	195.33	4623.7	2.729	3322.9	2575	47.720
606		4270.8	2.059	3067.4	3159	47.160	656		4630.9	2.744	3328.2	2565	47.731
607		4278.0	2.071	3072.6	3145	47.172	657		4638.1	2.760	3333.4	2555	47.742
608		4285.2	2.083	3077.8	3132	47.184	658		4645.3	2.775	3338.6	2545	47.753
609		4292.4	2.096	3083.0	3118	47.196	659		4652.5	2.790	3343.9	2534	47.764
610	150.33	4299.6	2.108	3088.2	3105	47.207	660	200.33	4659.8	2.806	3349.1	2524	47.775
611		4306.7	2.121	3093.4	3092	47.219	661		4667.0	2.821	3354.3	2514	47.786
612		4313.9	2.133	3098.6	3078	47.231	662		4674.2	2.837	3359.6	2504	47.797
613		4321.1	2.146	3103.8	3065	47.243	663		4681.4	2.852	3364.8	2494	47.808
614		4328.3	2.159	3109.0	3052	47.254	664		4688.7	2.868	3370.1	2484	47.819
615	155.33	4335.5	2.172	3114.2	3039	47.266	665	205.33	4695.9	2.884	3375.3	2475	47.829
616		4342.7	2.184	3119.4	3026	47.278	666		4703.1	2.900	3380.5	2465	47.840
617		4349.9	2.197	3124.6	3014	47.289	667		4710.3	2.916	3385.8	2455	47.851
618		4357.1	2.210	3129.8	3001	47.301	668		4717.6	2.931	3391.0	2445	47.862
619		4364.3	2.223	3135.0	2988	47.313	669		4724.8	2.947	3396.2	2436	47.873
620	160.33	4371.5	2.236	3140.2	2975	47.324	670	210.33	4732.0	2.964	3401.5	2426	47.884
621		4378.7	2.249	3145.4	2963	47.336	671		4739.2	2.980	3406.7	2417	47.894
622		4385.9	2.262	3150.7	2950	47.347	672		4746.5	2.996	3412.0	2407	47.905
623		4393.1	2.276	3155.9	2938	47.359	673		4753.7	3.012	3417.2	2398	47.916
624		4400.3	2.289	3161.1	2926	47.371	674		4760.9	3.028	3422.5	2388	47.927
625	165.33	4407.5	2.302	3166.3	2913	47.382	675	215.33	4768.2	3.045	3427.7	2379	47.937
626		4414.7	2.316	3171.5	2901	47.394	676		4775.4	3.061	3433.0	2370	47.948
627		4421.9	2.329	3176.7	2889	47.405	677		4782.6	3.078	3438.2	2361	47.959
628		4429.0	2.342	3181.9	2877	47.417	678		4789.9	3.094	3443.4	2351	47.969
629		4436.3	2.356	3187.1	2865	47.428	679		4797.1	3.111	3448.7	2342	47.980
630	170.33	4443.5	2.370	3192.4	2853	47.439	680	220.33	4804.3	3.128	3453.9	2333	47.991
631		4450.7	2.383	3197.6	2841	47.451	681		4811.6	3.145	3459.2	2324	48.001
632		4457.9	2.397	3202.8	2829	47.462	682		4818.8	3.161	3464.4	2315	48.012
633		4465.1	2.411	3208.0	2818	47.474	683		4826.0	3.178	3469.7	2306	48.022
634		4472.3	2.425	3213.2	2806	47.485	684		4833.3	3.195	3474.9	2297	48.033
635	175.33	4479.5	2.439	3218.4	2794	47.496	685	225.33	4840.5	3.212	3480.2	2288	48.044
636		4486.7	2.453	3223.7	2783	47.508	686		4847.7	3.230	3485.4	2280	48.054
637		4493.9	2.467	3228.9	2771	47.519	687		4855.0	3.247	3490.7	2271	48.065
638		4501.1	2.481	3234.1	2760	47.530	688		4862.2	3.264	3496.0	2262	48.075
639		4508.3	2.495	3239.3	2749	47.542	689		4869.5	3.281	3501.2	2253	48.086
640	180.33	4515.5	2.509	3244.5	2737	47.553	690	230.33	4876.7	3.299	3506.5	2245	48.096
641		4522.7	2.523	3249.8	2726	47.564	691		4884.0	3.316	3511.7	2236	48.107
642		4529.9	2.538	3255.0	2715	47.575	692		4891.2	3.334	3517.0	2228	48.117
643		4537.1	2.552	3260.2	2704	47.587	693		4898.4	3.351	3522.2	2219	48.128
644		4544.3	2.566	3265.4	2693	47.598	694		4905.7	3.369	3527.5	2211	48.138
645	185.33	4551.5	2.581	3270.7	2682	47.609	695	235.33	4912.9	3.387	3532.8	2202	48.149
646		4558.7	2.595	3275.9	2671	47.620	696		4920.2	3.405	3538.0	2194	48.159
647		4566.0	2.610	3281.1	2660	47.631	697		4927.4	3.422	3543.3	2186	48.169
648		4573.2	2.625	3286.3	2649	47.642	698		4934.7	3.440	3548.5	2177	48.180
649		4580.4	2.639	3291.6	2639	47.654	699		4941.9	3.458	3553.8	2169	48.190

Table 5 Products—200% Theoretical Air (for One Pound-Mole)

T	t	$\bar{h}$	p_r	$\bar{u}$	v_r	$\bar{\phi}$	T	t	$\bar{h}$	p_r	$\bar{u}$	v_r	$\bar{\phi}$
700	240.33	4949.2	3.477	3559.1	2160.8	48.201	750	290.33	5312.5	4.475	3823.1	1798.6	48.702
701		4956.4	3.495	3564.3	2152.6	48.211	751		5319.8	4.497	3828.4	1792.2	48.712
702		4963.7	3.513	3569.6	2144.5	48.221	752		5327.1	4.519	3833.7	1785.8	48.721
703		4970.9	3.531	3574.9	2136.4	48.232	753		5334.4	4.541	3839.0	1779.5	48.731
704		4978.2	3.550	3580.1	2128.4	48.242	754		5341.7	4.563	3844.3	1773.2	48.741
705	245.33	4985.4	3.568	3585.4	2120.4	48.252	755	295.33	5349.0	4.585	3849.7	1767.0	48.750
706		4992.7	3.587	3590.7	2112.4	48.262	756		5356.3	4.608	3855.0	1760.7	48.760
707		4999.9	3.605	3595.9	2104.5	48.273	757		5363.6	4.630	3860.3	1754.5	48.770
708		5007.2	3.624	3601.2	2096.6	48.283	758		5370.9	4.653	3865.6	1748.3	48.779
709		5014.4	3.643	3606.5	2088.8	48.293	759		5378.2	4.675	3870.9	1742.2	48.789
710	250.33	5021.7	3.661	3611.7	2081.0	48.303	760	300.33	5385.4	4.698	3876.2	1736.1	48.799
711		5029.0	3.680	3617.0	2073.3	48.314	761		5392.7	4.721	3881.5	1730.0	48.808
712		5036.2	3.699	3622.3	2065.5	48.324	762		5400.0	4.744	3886.8	1723.9	48.818
713		5043.5	3.718	3627.5	2057.9	48.334	763		5407.3	4.766	3892.1	1717.9	48.827
714		5050.7	3.737	3632.8	2050.2	48.344	764		5414.6	4.789	3897.4	1711.9	48.837
715	255.33	5058.0	3.756	3638.1	2042.6	48.354	765	305.33	5421.9	4.813	3902.7	1705.9	48.846
716		5065.2	3.776	3643.4	2035.1	48.365	766		5429.2	4.836	3908.1	1699.9	48.856
717		5072.5	3.795	3648.6	2027.5	48.375	767		5436.5	4.859	3913.4	1694.0	48.865
718		5079.8	3.814	3653.9	2020.0	48.385	768		5443.8	4.882	3918.7	1688.1	48.875
719		5087.0	3.834	3659.2	2012.6	48.395	769		5451.1	4.906	3924.0	1682.2	48.884
720	260.33	5094.3	3.853	3664.5	2005.1	48.405	770	310.33	5458.4	4.929	3929.3	1676.4	48.894
721		5101.6	3.873	3669.7	1997.8	48.415	771		5465.7	4.953	3934.6	1670.6	48.903
722		5108.8	3.893	3675.0	1990.4	48.425	772		5473.0	4.976	3940.0	1664.8	48.913
723		5116.1	3.912	3680.3	1983.1	48.435	773		5480.3	5.000	3945.3	1659.0	48.922
724		5123.3	3.932	3685.6	1975.8	48.445	774		5487.6	5.024	3950.6	1653.3	48.932
725	265.33	5130.6	3.952	3690.9	1968.6	48.455	775	315.33	5495.0	5.048	3955.9	1647.6	48.941
726		5137.9	3.972	3696.1	1961.4	48.465	776		5502.3	5.072	3961.2	1641.9	48.951
727		5145.2	3.992	3701.4	1954.2	48.475	777		5509.6	5.096	3966.6	1636.2	48.960
728		5152.4	4.012	3706.7	1947.1	48.485	778		5516.9	5.120	3971.9	1630.6	48.969
729		5159.7	4.033	3712.0	1940.0	48.495	779		5524.2	5.145	3977.2	1625.0	48.979
730	270.33	5167.0	4.053	3717.3	1932.9	48.505	780	320.33	5531.5	5.169	3982.5	1619.4	48.988
731		5174.2	4.073	3722.6	1925.9	48.515	781		5538.8	5.193	3987.9	1613.9	48.998
732		5181.5	4.094	3727.9	1918.9	48.525	782		5546.1	5.218	3993.2	1608.3	49.007
733		5188.8	4.114	3733.1	1911.9	48.535	783		5553.4	5.242	3998.5	1602.8	49.016
734		5196.1	4.135	3738.4	1905.0	48.545	784		5560.7	5.267	4003.8	1597.4	49.026
735	275.33	5203.3	4.156	3743.7	1898.1	48.555	785	325.33	5568.1	5.292	4009.2	1591.9	49.035
736		5210.6	4.176	3749.0	1891.2	48.565	786		5575.4	5.317	4014.5	1586.5	49.044
737		5217.9	4.197	3754.3	1884.4	48.575	787		5582.7	5.342	4019.8	1581.1	49.054
738		5225.2	4.218	3759.6	1877.6	48.585	788		5590.0	5.367	4025.1	1575.7	49.063
739		5232.4	4.239	3764.9	1870.9	48.594	789		5597.3	5.392	4030.5	1570.3	49.072
740	280.33	5239.7	4.260	3770.2	1864.1	48.604	790	330.33	5604.6	5.417	4035.8	1565.0	49.081
741		5247.0	4.281	3775.5	1857.4	48.614	791		5612.0	5.442	4041.1	1559.7	49.091
742		5254.3	4.302	3780.8	1850.8	48.624	792		5619.3	5.468	4046.5	1554.4	49.100
743		5261.6	4.324	3786.1	1844.1	48.634	793		5626.6	5.493	4051.8	1549.1	49.109
744		5268.8	4.345	3791.4	1837.5	48.643	794		5633.9	5.519	4057.1	1543.9	49.118
745	285.33	5276.1	4.367	3796.6	1831.0	48.653	795	335.33	5641.2	5.545	4062.5	1538.7	49.128
746		5283.4	4.388	3801.9	1824.4	48.663	796		5648.6	5.570	4067.8	1533.5	49.137
747		5290.7	4.410	3807.2	1817.9	48.673	797		5655.9	5.596	4073.2	1528.3	49.146
748		5298.0	4.431	3812.5	1811.4	48.682	798		5663.2	5.622	4078.5	1523.2	49.155
749		5305.2	4.453	3817.8	1805.0	48.692	799		5670.5	5.648	4083.8	1518.1	49.164

Table 5 Products—200% Theoretical Air (for One Pound-Mole)

T	t	$\bar{h}$	p_r	$\bar{u}$	v_r	$\bar{\phi}$	T	t	$\bar{h}$	p_r	$\bar{u}$	v_r	$\bar{\phi}$
800	340.33	5677.9	5.674	4089.2	1513.0	49.173	850	390.33	6045.3	7.101	4357.3	1284.5	49.619
801		5685.2	5.701	4094.5	1507.9	49.183	851		6052.7	7.132	4362.7	1280.4	49.628
802		5692.5	5.727	4099.9	1502.9	49.192	852		6060.0	7.164	4368.1	1276.4	49.636
803		5699.8	5.753	4105.2	1497.8	49.201	853		6067.4	7.195	4373.5	1272.3	49.645
804		5707.2	5.780	4110.5	1492.8	49.210	854		6074.8	7.226	4378.8	1268.3	49.654
805	345.33	5714.5	5.806	4115.9	1487.8	49.219	855	395.33	6082.1	7.258	4384.2	1264.3	49.662
806		5721.8	5.833	4121.2	1482.9	49.228	856		6089.5	7.289	4389.6	1260.2	49.671
807		5729.2	5.860	4126.6	1477.9	49.237	857		6096.9	7.321	4395.0	1256.2	49.679
808		5736.5	5.887	4131.9	1473.0	49.246	858		6104.3	7.353	4400.4	1252.3	49.688
809		5743.8	5.914	4137.3	1468.1	49.255	859		6111.6	7.385	4405.8	1248.3	49.697
810	350.33	5751.2	5.941	4142.6	1463.2	49.265	860	400.33	6119.0	7.417	4411.2	1244.4	49.705
811		5758.5	5.968	4148.0	1458.4	49.274	861		6126.4	7.449	4416.6	1240.5	49.714
812		5765.8	5.995	4153.3	1453.5	49.283	862		6133.8	7.481	4422.0	1236.6	49.722
813		5773.2	6.022	4158.7	1448.7	49.292	863		6141.2	7.513	4427.4	1232.7	49.731
814		5780.5	6.050	4164.0	1443.9	49.301	864		6148.6	7.546	4432.8	1228.8	49.739
815	355.33	5787.9	6.077	4169.4	1439.2	49.310	865	405.33	6155.9	7.578	4438.2	1224.9	49.748
816		5795.2	6.105	4174.7	1434.4	49.319	866		6163.3	7.611	4443.6	1221.1	49.757
817		5802.5	6.133	4180.1	1429.7	49.328	867		6170.7	7.643	4449.0	1217.3	49.765
818		5809.9	6.160	4185.5	1425.0	49.337	868		6178.1	7.676	4454.4	1213.5	49.774
819		5817.2	6.188	4190.8	1420.3	49.346	869		6185.5	7.709	4459.8	1209.7	49.782
820	360.33	5824.6	6.216	4196.2	1415.6	49.355	870	410.33	6192.9	7.742	4465.2	1205.9	49.791
821		5831.9	6.244	4201.5	1411.0	49.364	871		6200.3	7.775	4470.6	1202.1	49.799
822		5839.3	6.273	4206.9	1406.3	49.373	872		6207.6	7.809	4476.0	1198.4	49.808
823		5846.6	6.301	4212.2	1401.7	49.381	873		6215.0	7.842	4481.4	1194.6	49.816
824		5854.0	6.329	4217.6	1397.2	49.390	874		6222.4	7.876	4486.8	1190.9	49.824
825	365.33	5861.3	6.358	4223.0	1392.6	49.399	875	415.33	6229.8	7.909	4492.2	1187.2	49.833
826		5868.7	6.386	4228.3	1388.0	49.408	876		6237.2	7.943	4497.6	1183.5	49.841
827		5876.0	6.415	4233.7	1383.5	49.417	877		6244.6	7.977	4503.0	1179.9	49.850
828		5883.3	6.444	4239.1	1379.0	49.426	878		6252.0	8.011	4508.4	1176.2	49.858
829		5890.7	6.472	4244.4	1374.5	49.435	879		6259.4	8.045	4513.8	1172.6	49.867
830	370.33	5898.0	6.501	4249.8	1370.0	49.444	880	420.33	6266.8	8.079	4519.2	1168.9	49.875
831		5905.4	6.530	4255.2	1365.6	49.453	881		6274.2	8.113	4524.7	1165.3	49.883
832		5912.8	6.560	4260.5	1361.1	49.461	882		6281.6	8.147	4530.1	1161.7	49.892
833		5920.1	6.589	4265.9	1356.7	49.470	883		6289.0	8.182	4535.5	1158.2	49.900
834		5927.5	6.618	4271.3	1352.3	49.479	884		6296.4	8.217	4540.9	1154.6	49.909
835	375.33	5934.8	6.648	4276.6	1348.0	49.488	885	425.33	6303.8	8.251	4546.3	1151.0	49.917
836		5942.2	6.677	4282.0	1343.6	49.497	886		6311.2	8.286	4551.7	1147.5	49.925
837		5949.5	6.707	4287.4	1339.3	49.505	887		6318.6	8.321	4557.1	1144.0	49.934
838		5956.9	6.737	4292.8	1334.9	49.514	888		6326.0	8.356	4562.6	1140.5	49.942
839		5964.3	6.766	4298.1	1330.7	49.523	889		6333.4	8.391	4568.0	1137.0	49.950
840	380.33	5971.6	6.796	4303.5	1326.4	49.532	890	430.33	6340.8	8.426	4573.4	1133.5	49.959
841		5979.0	6.826	4308.9	1322.1	49.541	891		6348.2	8.462	4578.8	1130.0	49.967
842		5986.4	6.857	4314.3	1317.9	49.549	892		6355.6	8.497	4584.2	1126.5	49.975
843		5993.7	6.887	4319.6	1313.6	49.558	893		6363.0	8.533	4589.7	1123.1	49.984
844		6001.1	6.917	4325.0	1309.4	49.567	894		6370.4	8.569	4595.1	1119.7	49.992
845	385.33	6008.4	6.948	4330.4	1305.2	49.575	895	435.33	6377.9	8.604	4600.5	1116.3	50.000
846		6015.8	6.978	4335.8	1301.0	49.584	896		6385.3	8.640	4605.9	1112.9	50.008
847		6023.2	7.009	4341.2	1296.9	49.593	897		6392.7	8.676	4611.4	1109.5	50.017
848		6030.5	7.040	4346.5	1292.7	49.602	898		6400.1	8.712	4616.8	1106.1	50.025
849		6037.9	7.070	4351.9	1288.6	49.610	899		6407.5	8.749	4622.2	1102.7	50.033

Table 5 Products—200% Theoretical Air (for One Pound-Mole)

T	t	$\bar{h}$	p_r	$\bar{u}$	v_r	$\bar{\phi}$	T	t	$\bar{h}$	p_r	$\bar{u}$	v_r	$\bar{\phi}$
900	440.33	6414.9	8.785	4627.6	1099.4	50.042	950	490.33	6786.9	10.758	4900.3	947.7	50.444
901		6422.3	8.822	4633.1	1096.1	50.050	951		6794.4	10.800	4905.8	945.0	50.452
902		6429.8	8.858	4638.5	1092.7	50.058	952		6801.8	10.843	4911.3	942.2	50.459
903		6437.2	8.895	4643.9	1089.4	50.066	953		6809.3	10.886	4916.8	939.5	50.467
904		6444.6	8.932	4649.4	1086.1	50.074	954		6816.8	10.929	4922.3	936.8	50.475
905	445.33	6452.0	8.969	4654.8	1082.9	50.083	955	495.33	6824.2	10.972	4927.7	934.1	50.483
906		6459.4	9.006	4660.2	1079.6	50.091	956		6831.7	11.015	4933.2	931.4	50.491
907		6466.9	9.043	4665.7	1076.3	50.099	957		6839.2	11.059	4938.7	928.7	50.499
908		6474.3	9.080	4671.1	1073.1	50.107	958		6846.6	11.102	4944.2	926.0	50.506
909		6481.7	9.118	4676.6	1069.9	50.115	959		6854.1	11.146	4949.7	923.3	50.514
910	450.33	6489.1	9.155	4682.0	1066.6	50.124	960	500.33	6861.6	11.190	4955.2	920.7	50.522
911		6496.6	9.193	4687.4	1063.4	50.132	961		6869.1	11.234	4960.7	918.0	50.530
912		6504.0	9.231	4692.9	1060.3	50.140	962		6876.6	11.278	4966.2	915.4	50.537
913		6511.4	9.269	4698.3	1057.1	50.148	963		6884.0	11.322	4971.6	912.8	50.545
914		6518.8	9.307	4703.8	1053.9	50.156	964		6891.5	11.366	4977.1	910.2	50.553
915	455.33	6526.3	9.345	4709.2	1050.8	50.164	965	505.33	6899.0	11.411	4982.6	907.6	50.561
916		6533.7	9.383	4714.7	1047.6	50.172	966		6906.5	11.455	4988.1	905.0	50.569
917		6541.1	9.422	4720.1	1044.5	50.180	967		6913.9	11.500	4993.6	902.4	50.576
918		6548.6	9.460	4725.5	1041.4	50.189	968		6921.4	11.545	4999.1	899.8	50.584
919		6556.0	9.499	4731.0	1038.3	50.197	969		6928.9	11.590	5004.6	897.2	50.592
920	460.33	6563.4	9.538	4736.4	1035.2	50.205	970	510.33	6936.4	11.635	5010.1	894.7	50.599
921		6570.9	9.576	4741.9	1032.1	50.213	971		6943.9	11.680	5015.6	892.1	50.607
922		6578.3	9.615	4747.3	1029.0	50.221	972		6951.4	11.726	5021.1	889.6	50.615
923		6585.7	9.655	4752.8	1026.0	50.229	973		6958.8	11.771	5026.6	887.1	50.623
924		6593.2	9.694	4758.3	1022.9	50.237	974		6966.3	11.817	5032.1	884.5	50.630
925	465.33	6600.6	9.733	4763.7	1019.9	50.245	975	515.33	6973.8	11.863	5037.6	882.0	50.638
926		6608.1	9.773	4769.2	1016.9	50.253	976		6981.3	11.909	5043.1	879.5	50.646
927		6615.5	9.812	4774.6	1013.8	50.261	977		6988.8	11.955	5048.6	877.0	50.653
928		6622.9	9.852	4780.1	1010.8	50.269	978		6996.3	12.001	5054.1	874.5	50.661
929		6630.4	9.892	4785.5	1007.9	50.277	979		7003.8	12.047	5059.6	872.1	50.669
930	470.33	6637.8	9.932	4791.0	1004.9	50.285	980	520.33	7011.3	12.094	5065.1	869.6	50.676
931		6645.3	9.972	4796.4	1001.9	50.293	981		7018.8	12.140	5070.6	867.1	50.684
932		6652.7	10.012	4801.9	999.0	50.301	982		7026.3	12.187	5076.1	864.7	50.692
933		6660.2	10.053	4807.4	996.0	50.309	983		7033.8	12.234	5081.7	862.3	50.699
934		6667.6	10.093	4812.8	993.1	50.317	984		7041.3	12.281	5087.2	859.8	50.707
935	475.33	6675.1	10.134	4818.3	990.2	50.325	985	525.33	7048.8	12.328	5092.7	857.4	50.714
936		6682.5	10.174	4823.8	987.3	50.333	986		7056.3	12.376	5098.2	855.0	50.722
937		6690.0	10.215	4829.2	984.4	50.341	987		7063.8	12.423	5103.7	852.6	50.730
938		6697.4	10.256	4834.7	981.5	50.349	988		7071.3	12.471	5109.2	850.2	50.737
939		6704.9	10.297	4840.2	978.6	50.357	989		7078.8	12.519	5114.7	847.8	50.745
940	480.33	6712.3	10.338	4845.6	975.7	50.365	990	530.33	7086.3	12.566	5120.3	845.4	50.752
941		6719.8	10.380	4851.1	972.9	50.373	991		7093.8	12.615	5125.8	843.1	50.760
942		6727.2	10.421	4856.6	970.0	50.381	992		7101.3	12.663	5131.3	840.7	50.768
943		6734.7	10.463	4862.0	967.2	50.389	993		7108.8	12.711	5136.8	838.4	50.775
944		6742.2	10.505	4867.5	964.4	50.396	994		7116.3	12.759	5142.3	836.0	50.783
945	485.33	6749.6	10.546	4873.0	961.6	50.404	995	535.33	7123.8	12.808	5147.9	833.7	50.790
946		6757.1	10.588	4878.4	958.8	50.412	996		7131.3	12.857	5153.4	831.4	50.798
947		6764.5	10.630	4883.9	956.0	50.420	997		7138.8	12.906	5158.9	829.0	50.805
948		6772.0	10.673	4889.4	953.2	50.428	998		7146.3	12.955	5164.4	826.7	50.813
949		6779.4	10.715	4894.9	950.5	50.436	999		7153.8	13.004	5170.0	824.4	50.820

Table 5 Products—200% Theoretical Air (for One Pound-Mole)

T	t	$\bar{h}$	p_r	$\bar{u}$	v_r	$\bar{\phi}$	T	t	$\bar{h}$	p_r	$\bar{u}$	v_r	$\bar{\phi}$
1000	540.33	7161.4	13.053	5175.5	822.1	50.828	1050	590.33	7538.3	15.710	5453.2	717.3	51.196
1001		7168.9	13.103	5181.0	819.8	50.835	1051		7545.9	15.767	5458.8	715.4	51.203
1002		7176.4	13.152	5186.5	817.6	50.843	1052		7553.5	15.824	5464.3	713.4	51.210
1003		7183.9	13.202	5192.1	815.3	50.850	1053		7561.0	15.881	5469.9	711.5	51.217
1004		7191.4	13.252	5197.6	813.0	50.858	1054		7568.6	15.939	5475.5	709.7	51.224
1005	545.33	7198.9	13.302	5203.1	810.8	50.865	1055	595.33	7576.2	15.997	5481.1	707.8	51.232
1006		7206.5	13.352	5208.7	808.6	50.873	1056		7583.7	16.054	5486.7	705.9	51.239
1007		7214.0	13.403	5214.2	806.3	50.880	1057		7591.3	16.113	5492.3	704.0	51.246
1008		7221.5	13.453	5219.7	804.1	50.888	1058		7598.9	16.171	5497.8	702.1	51.253
1009		7229.0	13.504	5225.3	801.9	50.895	1059		7606.5	16.229	5503.4	700.3	51.260
1010	550.33	7236.5	13.554	5230.8	799.7	50.903	1060	600.33	7614.0	16.288	5509.0	698.4	51.267
1011		7244.1	13.605	5236.4	797.4	50.910	1061		7621.6	16.346	5514.6	696.6	51.275
1012		7251.6	13.656	5241.9	795.3	50.918	1062		7629.2	16.405	5520.2	694.7	51.282
1013		7259.1	13.708	5247.4	793.1	50.925	1063		7636.8	16.464	5525.8	692.9	51.289
1014		7266.6	13.759	5253.0	790.9	50.932	1064		7644.3	16.523	5531.4	691.0	51.296
1015	555.33	7274.2	13.810	5258.5	788.7	50.940	1065	605.33	7651.9	16.583	5537.0	689.2	51.303
1016		7281.7	13.862	5264.1	786.5	50.947	1066		7659.5	16.642	5542.6	687.4	51.310
1017		7289.2	13.914	5269.6	784.4	50.955	1067		7667.1	16.702	5548.2	685.6	51.317
1018		7296.8	13.966	5275.2	782.2	50.962	1068		7674.7	16.762	5553.8	683.8	51.324
1019		7304.3	14.018	5280.7	780.1	50.969	1069		7682.3	16.822	5559.4	682.0	51.332
1020	560.33	7311.8	14.070	5286.3	778.0	50.977	1070	610.33	7689.8	16.882	5565.0	680.2	51.339
1021		7319.4	14.123	5291.8	775.8	50.984	1071		7697.4	16.942	5570.6	678.4	51.346
1022		7326.9	14.175	5297.4	773.7	50.992	1072		7705.0	17.003	5576.2	676.6	51.353
1023		7334.4	14.228	5302.9	771.6	50.999	1073		7712.6	17.064	5581.8	674.8	51.360
1024		7342.0	14.281	5308.5	769.5	51.006	1074		7720.2	17.124	5587.4	673.1	51.367
1025	565.33	7349.5	14.334	5314.0	767.4	51.014	1075	615.33	7727.8	17.185	5593.0	671.3	51.374
1026		7357.1	14.387	5319.6	765.3	51.021	1076		7735.4	17.247	5598.6	669.5	51.381
1027		7364.6	14.440	5325.1	763.2	51.028	1077		7743.0	17.308	5604.2	667.8	51.388
1028		7372.1	14.494	5330.7	761.2	51.036	1078		7750.6	17.370	5609.8	666.0	51.395
1029		7379.7	14.547	5336.2	759.1	51.043	1079		7758.2	17.431	5615.4	664.3	51.402
1030	570.33	7387.2	14.601	5341.8	757.0	51.050	1080	620.33	7765.8	17.493	5621.0	662.5	51.409
1031		7394.8	14.655	5347.4	755.0	51.058	1081		7773.4	17.555	5626.6	660.8	51.416
1032		7402.3	14.709	5352.9	752.9	51.065	1082		7781.0	17.617	5632.3	659.1	51.423
1033		7409.9	14.763	5358.5	750.9	51.072	1083		7788.6	17.680	5637.9	657.4	51.430
1034		7417.4	14.818	5364.0	748.9	51.080	1084		7796.2	17.742	5643.5	655.7	51.437
1035	575.33	7425.0	14.872	5369.6	746.8	51.087	1085	625.33	7803.8	17.805	5649.1	654.0	51.444
1036		7432.5	14.927	5375.2	744.8	51.094	1086		7811.4	17.868	5654.7	652.3	51.451
1037		7440.1	14.982	5380.7	742.8	51.102	1087		7819.0	17.931	5660.3	650.6	51.458
1038		7447.6	15.037	5386.3	740.8	51.109	1088		7826.6	17.994	5665.9	648.9	51.465
1039		7455.2	15.092	5391.9	738.8	51.116	1089		7834.2	18.058	5671.6	647.2	51.472
1040	580.33	7462.7	15.148	5397.4	736.8	51.123	1090	630.33	7841.8	18.121	5677.2	645.5	51.479
1041		7470.3	15.203	5403.0	734.8	51.131	1091		7849.4	18.185	5682.8	643.8	51.486
1042		7477.8	15.259	5408.6	732.8	51.138	1092		7857.0	18.249	5688.4	642.2	51.493
1043		7485.4	15.314	5414.1	730.9	51.145	1093		7864.6	18.313	5694.1	640.5	51.500
1044		7493.0	15.370	5419.7	728.9	51.152	1094		7872.2	18.377	5699.7	638.8	51.507
1045	585.33	7500.5	15.427	5425.3	727.0	51.160	1095	635.33	7879.8	18.442	5705.3	637.2	51.514
1046		7508.1	15.483	5430.9	725.0	51.167	1096		7887.4	18.507	5710.9	635.5	51.521
1047		7515.6	15.539	5436.4	723.1	51.174	1097		7895.1	18.572	5716.6	633.9	51.528
1048		7523.2	15.596	5442.0	721.1	51.181	1098		7902.7	18.636	5722.2	632.3	51.535
1049		7530.8	15.653	5447.6	719.2	51.189	1099		7910.3	18.702	5727.8	630.6	51.542

Table 5 Products—200% Theoretical Air (for One Pound-Mole)

T	t	$\bar{h}$	p_r	$\bar{u}$	v_r	$\bar{\phi}$	T	t	$\bar{h}$	p_r	$\bar{u}$	v_r	$\bar{\phi}$
1100	640.33	7917.9	18.767	5733.5	629.0	51.549	1150	690.33	8300.1	22.27	6016.4	554.2	51.889
1101		7925.5	18.833	5739.1	627.4	51.556	1151		8307.8	22.34	6022.1	552.8	51.895
1102		7933.1	18.898	5744.7	625.8	51.563	1152		8315.5	22.42	6027.8	551.4	51.902
1103		7940.8	18.964	5750.4	624.2	51.570	1153		8323.1	22.49	6033.5	550.1	51.909
1104		7948.4	19.030	5756.0	622.6	51.577	1154		8330.8	22.57	6039.1	548.7	51.915
1105	645.33	7956.0	19.097	5761.6	621.0	51.583	1155	695.33	8338.5	22.65	6044.8	547.3	51.922
1106		7963.6	19.163	5767.3	619.4	51.590	1156		8346.2	22.72	6050.5	546.0	51.929
1107		7971.3	19.230	5772.9	617.8	51.597	1157		8353.9	22.80	6056.2	544.6	51.935
1108		7978.9	19.296	5778.5	616.2	51.604	1158		8361.5	22.87	6061.9	543.3	51.942
1109		7986.5	19.363	5784.2	614.6	51.611	1159		8369.2	22.95	6067.6	541.9	51.949
1110	650.33	7994.1	19.430	5789.8	613.1	51.618	1160	700.33	8376.9	23.03	6073.3	540.6	51.955
1111		8001.8	19.498	5795.5	611.5	51.625	1161		8384.6	23.10	6079.0	539.3	51.962
1112		8009.4	19.565	5801.1	609.9	51.632	1162		8392.3	23.18	6084.7	537.9	51.968
1113		8017.0	19.633	5806.8	608.4	51.638	1163		8399.9	23.26	6090.4	536.6	51.975
1114		8024.7	19.701	5812.4	606.8	51.645	1164		8407.6	23.34	6096.1	535.3	51.982
1115	655.33	8032.3	19.769	5818.1	605.3	51.652	1165	705.33	8415.3	23.41	6101.8	534.0	51.988
1116		8039.9	19.837	5823.7	603.7	51.659	1166		8423.0	23.49	6107.5	532.7	51.995
1117		8047.6	19.905	5829.4	602.2	51.666	1167		8430.7	23.57	6113.2	531.3	52.001
1118		8055.2	19.974	5835.0	600.7	51.673	1168		8438.4	23.65	6118.9	530.0	52.008
1119		8062.8	20.043	5840.7	599.1	51.679	1169		8446.1	23.73	6124.6	528.7	52.015
1120	660.33	8070.5	20.112	5846.3	597.6	51.686	1170	710.33	8453.8	23.81	6130.3	527.4	52.021
1121		8078.1	20.181	5852.0	596.1	51.693	1171		8461.5	23.88	6136.0	526.1	52.028
1122		8085.8	20.250	5857.6	594.6	51.700	1172		8469.2	23.96	6141.7	524.9	52.034
1123		8093.4	20.320	5863.3	593.1	51.707	1173		8476.8	24.04	6147.4	523.6	52.041
1124		8101.0	20.390	5868.9	591.6	51.714	1174		8484.6	24.12	6153.2	522.3	52.047
1125	665.33	8108.7	20.460	5874.6	590.1	51.720	1175	715.33	8492.2	24.20	6158.9	521.0	52.054
1126		8116.3	20.530	5880.3	588.6	51.727	1176		8499.9	24.28	6164.6	519.7	52.060
1127		8124.0	20.600	5885.9	587.1	51.734	1177		8507.6	24.36	6170.3	518.5	52.067
1128		8131.6	20.671	5891.6	585.6	51.741	1178		8515.3	24.44	6176.0	517.2	52.074
1129		8139.3	20.741	5897.2	584.1	51.747	1179		8523.0	24.52	6181.7	515.9	52.080
1130	670.33	8146.9	20.812	5902.9	582.7	51.754	1180	720.33	8530.8	24.60	6187.4	514.7	52.087
1131		8154.6	20.883	5908.6	581.2	51.761	1181		8538.5	24.68	6193.2	513.4	52.093
1132		8162.2	20.954	5914.2	579.7	51.768	1182		8546.2	24.77	6198.9	512.2	52.100
1133		8169.9	21.026	5919.9	578.3	51.775	1183		8553.9	24.85	6204.6	510.9	52.106
1134		8177.5	21.097	5925.6	576.8	51.781	1184		8561.6	24.93	6210.3	509.7	52.113
1135	675.33	8185.2	21.169	5931.2	575.4	51.788	1185	725.33	8569.3	25.01	6216.0	508.5	52.119
1136		8192.8	21.241	5936.9	573.9	51.795	1186		8577.0	25.09	6221.8	507.2	52.126
1137		8200.5	21.313	5942.6	572.5	51.802	1187		8584.7	25.18	6227.5	506.0	52.132
1138		8208.2	21.386	5948.2	571.1	51.808	1188		8592.4	25.26	6233.2	504.8	52.139
1139		8215.8	21.458	5953.9	569.6	51.815	1189		8600.1	25.34	6238.9	503.5	52.145
1140	680.33	8223.5	21.531	5959.6	568.2	51.822	1190	730.33	8607.8	25.42	6244.7	502.3	52.152
1141		8231.1	21.604	5965.3	566.8	51.828	1191		8615.6	25.51	6250.4	501.1	52.158
1142		8238.8	21.677	5970.9	565.4	51.835	1192		8623.3	25.59	6256.1	499.9	52.165
1143		8246.5	21.751	5976.6	563.9	51.842	1193		8631.0	25.67	6261.9	498.7	52.171
1144		8254.1	21.824	5982.3	562.5	51.849	1194		8638.7	25.76	6267.6	497.5	52.178
1145	685.33	8261.8	21.898	5988.0	561.1	51.855	1195	735.33	8646.4	25.84	6273.3	496.3	52.184
1146		8269.5	21.972	5993.7	559.7	51.862	1196		8654.1	25.92	6279.1	495.1	52.190
1147		8277.1	22.046	5999.3	558.3	51.869	1197		8661.9	26.01	6284.8	493.9	52.197
1148		8284.8	22.120	6005.0	556.9	51.875	1198		8669.6	26.09	6290.5	492.7	52.203
1149		8292.5	22.195	6010.7	555.6	51.882	1199		8677.3	26.18	6296.3	491.5	52.210

Table 5   Products—200% Theoretical Air (for One Pound-Mole)

T	t	$\bar{h}$	p_r	$\bar{u}$	v_r	$\bar{\phi}$	T	t	$\bar{h}$	p_r	$\bar{u}$	v_r	$\bar{\phi}$
1200	740.33	8685.0	26.26	6302.0	490.3	52.216	1250	790.33	9072.6	30.80	6590.3	435.5	52.533
1201		8692.8	26.35	6307.7	489.1	52.223	1251		9080.4	30.90	6596.1	434.5	52.539
1202		8700.5	26.43	6313.5	488.0	52.229	1252		9088.2	30.99	6601.9	433.5	52.545
1203		8708.2	26.52	6319.2	486.8	52.236	1253		9096.0	31.09	6607.7	432.5	52.551
1204		8715.9	26.61	6325.0	485.6	52.242	1254		9103.7	31.19	6613.5	431.5	52.558
1205	745.33	8723.7	26.69	6330.7	484.5	52.248	1255	795.33	9111.5	31.29	6619.3	430.5	52.564
1206		8731.4	26.78	6336.5	483.3	52.255	1256		9119.3	31.38	6625.1	429.5	52.570
1207		8739.1	26.86	6342.2	482.1	52.261	1257		9127.1	31.48	6630.9	428.5	52.576
1208		8746.9	26.95	6347.9	481.0	52.268	1258		9134.9	31.58	6636.7	427.5	52.582
1209		8754.6	27.04	6353.7	479.8	52.274	1259		9142.7	31.68	6642.5	426.5	52.589
1210	750.33	8762.3	27.13	6359.4	478.7	52.280	1260	800.33	9150.5	31.78	6648.3	425.5	52.595
1211		8770.1	27.21	6365.2	477.6	52.287	1261		9158.3	31.88	6654.1	424.5	52.601
1212		8777.8	27.30	6371.0	476.4	52.293	1262		9166.0	31.98	6659.9	423.5	52.607
1213		8785.5	27.39	6376.7	475.3	52.300	1263		9173.8	32.08	6665.7	422.6	52.613
1214		8793.3	27.48	6382.5	474.1	52.306	1264		9181.6	32.18	6671.5	421.6	52.619
1215	755.33	8801.0	27.57	6388.2	473.0	52.312	1265	805.33	9189.4	32.28	6677.3	420.6	52.626
1216		8808.8	27.65	6394.0	471.9	52.319	1266		9197.2	32.38	6683.1	419.6	52.632
1217		8816.5	27.74	6399.7	470.8	52.325	1267		9205.0	32.48	6688.9	418.7	52.638
1218		8824.3	27.83	6405.5	469.6	52.331	1268		9212.8	32.58	6694.7	417.7	52.644
1219		8832.0	27.92	6411.2	468.5	52.338	1269		9220.6	32.68	6700.6	416.7	52.650
1220	760.33	8839.7	28.01	6417.0	467.4	52.344	1270	810.33	9228.4	32.78	6706.4	415.8	52.656
1221		8847.5	28.10	6422.8	466.3	52.350	1271		9236.2	32.88	6712.2	414.8	52.663
1222		8855.2	28.19	6428.5	465.2	52.357	1272		9244.0	32.98	6718.0	413.9	52.669
1223		8863.0	28.28	6434.3	464.1	52.363	1273		9251.8	33.08	6723.8	412.9	52.675
1224		8870.7	28.37	6440.0	463.0	52.369	1274		9259.6	33.19	6729.6	412.0	52.681
1225	765.33	8878.5	28.46	6445.8	461.9	52.376	1275	815.33	9267.4	33.29	6735.5	411.0	52.687
1226		8886.2	28.55	6451.6	460.8	52.382	1276		9275.2	33.39	6741.3	410.1	52.693
1227		8894.0	28.64	6457.4	459.7	52.388	1277		9283.0	33.49	6747.1	409.1	52.699
1228		8901.8	28.73	6463.1	458.6	52.395	1278		9290.9	33.60	6752.9	408.2	52.705
1229		8909.5	28.83	6468.9	457.5	52.401	1279		9298.7	33.70	6758.7	407.3	52.711
1230	770.33	8917.3	28.92	6474.7	456.5	52.407	1280	820.33	9306.5	33.81	6764.6	406.3	52.718
1231		8925.0	29.01	6480.4	455.4	52.414	1281		9314.3	33.91	6770.4	405.4	52.724
1232		8932.8	29.10	6486.2	454.3	52.420	1282		9322.1	34.01	6776.2	404.5	52.730
1233		8940.5	29.19	6492.0	453.2	52.426	1283		9329.9	34.12	6782.0	403.6	52.736
1234		8948.3	29.29	6497.8	452.2	52.433	1284		9337.7	34.22	6787.9	402.6	52.742
1235	775.33	8956.1	29.38	6503.5	451.1	52.439	1285	825.33	9345.5	34.33	6793.7	401.7	52.748
1236		8963.8	29.47	6509.3	450.1	52.445	1286		9353.4	34.43	6799.5	400.8	52.754
1237		8971.6	29.57	6515.1	449.0	52.451	1287		9361.2	34.54	6805.4	399.9	52.760
1238		8979.4	29.66	6520.9	447.9	52.458	1288		9369.0	34.64	6811.2	399.0	52.766
1239		8987.1	29.75	6526.6	446.9	52.464	1289		9376.8	34.75	6817.0	398.1	52.772
1240	780.33	8994.9	29.85	6532.4	445.8	52.470	1290	830.33	9384.6	34.86	6822.9	397.2	52.778
1241		9002.7	29.94	6538.2	444.8	52.477	1291		9392.5	34.96	6828.7	396.3	52.784
1242		9010.4	30.04	6544.0	443.8	52.483	1292		9400.3	35.07	6834.6	395.4	52.791
1243		9018.2	30.13	6549.8	442.7	52.489	1293		9408.1	35.18	6840.4	394.5	52.797
1244		9026.0	30.23	6555.6	441.7	52.495	1294		9415.9	35.28	6846.2	393.6	52.803
1245	785.33	9033.7	30.32	6561.4	440.6	52.502	1295	835.33	9423.8	35.39	6852.1	392.7	52.809
1246		9041.5	30.42	6567.1	439.6	52.508	1296		9431.6	35.50	6857.9	391.8	52.815
1247		9049.3	30.51	6572.9	438.6	52.514	1297		9439.4	35.61	6863.8	390.9	52.821
1248		9057.1	30.61	6578.7	437.6	52.520	1298		9447.2	35.72	6869.6	390.0	52.827
1249		9064.9	30.70	6584.5	436.5	52.526	1299		9455.1	35.83	6875.4	389.1	52.833

Table 5 Products—200% Theoretical Air (for One Pound-Mole)

T	t	$\bar{h}$	p_r	$\bar{u}$	v_r	$\bar{\phi}$	T	t	$\bar{h}$	p_r	$\bar{u}$	v_r	$\bar{\phi}$
1300	840.33	9462.9	35.93	6881.3	388.2	52.839	1350	890.33	9855.9	41.72	7175.0	347.2	53.135
1301		9470.7	36.04	6887.1	387.4	52.845	1351		9863.8	41.85	7180.9	346.5	53.141
1302		9478.6	36.15	6893.0	386.5	52.851	1352		9871.6	41.97	7186.8	345.7	53.147
1303		9486.4	36.26	6898.8	385.6	52.857	1353		9879.5	42.09	7192.7	345.0	53.153
1304		9494.3	36.37	6904.7	384.7	52.863	1354		9887.4	42.22	7198.6	344.2	53.159
1305	845.33	9502.1	36.48	6910.5	383.9	52.869	1355	895.33	9895.3	42.34	7204.5	343.4	53.165
1306		9509.9	36.59	6916.4	383.0	52.875	1356		9903.2	42.46	7210.4	342.7	53.170
1307		9517.8	36.70	6922.2	382.1	52.881	1357		9911.1	42.59	7216.3	341.9	53.176
1308		9525.6	36.81	6928.1	381.3	52.887	1358		9919.0	42.71	7222.2	341.2	53.182
1309		9533.4	36.93	6934.0	380.4	52.893	1359		9926.9	42.84	7228.1	340.4	53.188
1310	850.33	9541.3	37.04	6939.8	379.6	52.899	1360	900.33	9934.8	42.96	7234.0	339.7	53.194
1311		9549.1	37.15	6945.7	378.7	52.905	1361		9942.7	43.09	7239.9	339.0	53.199
1312		9557.0	37.26	6951.5	377.9	52.911	1362		9950.6	43.22	7245.8	338.2	53.205
1313		9564.8	37.37	6957.4	377.0	52.917	1363		9958.5	43.34	7251.7	337.5	53.211
1314		9572.7	37.49	6963.2	376.2	52.923	1364		9966.4	43.47	7257.7	336.7	53.217
1315	855.33	9580.5	37.60	6969.1	375.3	52.929	1365	905.33	9974.3	43.60	7263.6	336.0	53.223
1316		9588.4	37.71	6975.0	374.5	52.935	1366		9982.2	43.72	7269.5	335.3	53.228
1317		9596.2	37.83	6980.8	373.6	52.941	1367		9990.1	43.85	7275.4	334.5	53.234
1318		9604.1	37.94	6986.7	372.8	52.947	1368		9998.0	43.98	7281.3	333.8	53.240
1319		9611.9	38.05	6992.6	372.0	52.953	1369		10005.9	44.11	7287.2	333.1	53.246
1320	860.33	9619.8	38.17	6998.4	371.1	52.959	1370	910.33	10013.8	44.23	7293.2	332.4	53.252
1321		9627.6	38.28	7004.3	370.3	52.965	1371		10021.7	44.36	7299.1	331.6	53.257
1322		9635.5	38.40	7010.2	369.5	52.970	1372		10029.6	44.49	7305.0	330.9	53.263
1323		9643.3	38.51	7016.0	368.7	52.976	1373		10037.5	44.62	7310.9	330.2	53.269
1324		9651.2	38.63	7021.9	367.8	52.982	1374		10045.4	44.75	7316.8	329.5	53.275
1325	865.33	9659.1	38.74	7027.8	367.0	52.988	1375	915.33	10053.3	44.88	7322.8	328.8	53.280
1326		9666.9	38.86	7033.7	366.2	52.994	1376		10061.3	45.01	7328.7	328.1	53.286
1327		9674.8	38.97	7039.5	365.4	53.000	1377		10069.2	45.14	7334.6	327.3	53.292
1328		9682.6	39.09	7045.4	364.6	53.006	1378		10077.1	45.27	7340.6	326.6	53.298
1329		9690.5	39.21	7051.3	363.8	53.012	1379		10085.0	45.40	7346.5	325.9	53.303
1330	870.33	9698.4	39.32	7057.2	363.0	53.018	1380	920.33	10092.9	45.54	7352.4	325.2	53.309
1331		9706.2	39.44	7063.1	362.1	53.024	1381		10100.8	45.67	7358.4	324.5	53.315
1332		9714.1	39.56	7068.9	361.3	53.030	1382		10108.7	45.80	7364.3	323.8	53.321
1333		9722.0	39.68	7074.8	360.5	53.036	1383		10116.7	45.93	7370.2	323.1	53.326
1334		9729.8	39.80	7080.7	359.7	53.042	1384		10124.6	46.06	7376.2	322.4	53.332
1335	875.33	9737.7	39.91	7086.6	358.9	53.047	1385	925.33	10132.5	46.20	7382.1	321.7	53.338
1336		9745.6	40.03	7092.5	358.1	53.053	1386		10140.4	46.33	7388.0	321.0	53.343
1337		9753.4	40.15	7098.3	357.4	53.059	1387		10148.4	46.46	7394.0	320.3	53.349
1338		9761.3	40.27	7104.2	356.6	53.065	1388		10156.3	46.60	7399.9	319.7	53.355
1339		9769.2	40.39	7110.1	355.8	53.071	1389		10164.2	46.73	7405.9	319.0	53.361
1340	880.33	9777.1	40.51	7116.0	355.0	53.077	1390	930.33	10172.1	46.87	7411.8	318.3	53.366
1341		9784.9	40.63	7121.9	354.2	53.083	1391		10180.1	47.00	7417.7	317.6	53.372
1342		9792.8	40.75	7127.8	353.4	53.089	1392		10188.0	47.14	7423.7	316.9	53.378
1343		9800.7	40.87	7133.7	352.6	53.094	1393		10195.9	47.27	7429.6	316.2	53.383
1344		9808.6	40.99	7139.6	351.9	53.100	1394		10203.9	47.41	7435.6	315.6	53.389
1345	885.33	9816.4	41.11	7145.5	351.1	53.106	1395	935.33	10211.8	47.54	7441.5	314.9	53.395
1346		9824.3	41.23	7151.4	350.3	53.112	1396		10219.7	47.68	7447.5	314.2	53.400
1347		9832.2	41.36	7157.3	349.5	53.118	1397		10227.7	47.82	7453.4	313.5	53.406
1348		9840.1	41.48	7163.2	348.8	53.124	1398		10235.6	47.95	7459.4	312.9	53.412
1349		9848.0	41.60	7169.0	348.0	53.130	1399		10243.5	48.09	7465.3	312.2	53.418

Table 5 Products—200% Theoretical Air (for One Pound-Mole)

T	t	$\bar{h}$	p_r	$\bar{u}$	v_r	$\bar{\phi}$	T	t	$\bar{h}$	p_r	$\bar{u}$	v_r	$\bar{\phi}$
1400	940.33	10251.5	48.23	7471.3	311.5	53.423	1450	990.33	10649.7	55.52	7770.2	280.3	53.703
1401		10259.4	48.37	7477.2	310.9	53.429	1451		10657.7	55.67	7776.2	279.7	53.708
1402		10267.3	48.50	7483.2	310.2	53.435	1452		10665.7	55.82	7782.2	279.1	53.714
1403		10275.3	48.64	7489.1	309.5	53.440	1453		10673.7	55.98	7788.2	278.5	53.719
1404		10283.2	48.78	7495.1	308.9	53.446	1454		10681.6	56.14	7794.2	278.0	53.725
1405	945.33	10291.2	48.92	7501.0	308.2	53.451	1455	995.33	10689.6	56.29	7800.2	277.4	53.730
1406		10299.1	49.06	7507.0	307.6	53.457	1456		10697.6	56.45	7806.2	276.8	53.736
1407		10307.1	49.20	7513.0	306.9	53.463	1457		10705.6	56.60	7812.2	276.2	53.741
1408		10315.0	49.34	7518.9	306.2	53.468	1458		10713.6	56.76	7818.2	275.7	53.747
1409		10323.0	49.48	7524.9	305.6	53.474	1459		10721.6	56.92	7824.3	275.1	53.752
1410	950.33	10330.9	49.62	7530.8	304.9	53.480	1460	1000.33	10729.6	57.07	7830.3	274.5	53.758
1411		10338.8	49.76	7536.8	304.3	53.485	1461		10737.6	57.23	7836.3	274.0	53.763
1412		10346.8	49.90	7542.8	303.6	53.491	1462		10745.6	57.39	7842.3	273.4	53.769
1413		10354.8	50.04	7548.7	303.0	53.497	1463		10753.6	57.55	7848.3	272.8	53.774
1414		10362.7	50.19	7554.7	302.4	53.502	1464		10761.6	57.71	7854.3	272.3	53.780
1415	955.33	10370.7	50.33	7560.7	301.7	53.508	1465	1005.33	10769.6	57.87	7860.4	271.7	53.785
1416		10378.6	50.47	7566.6	301.1	53.513	1466		10777.7	58.02	7866.4	271.1	53.790
1417		10386.6	50.61	7572.6	300.4	53.519	1467		10785.7	58.18	7872.4	270.6	53.796
1418		10394.5	50.76	7578.6	299.8	53.525	1468		10793.7	58.34	7878.4	270.0	53.801
1419		10402.5	50.90	7584.5	299.2	53.530	1469		10801.7	58.50	7884.5	269.5	53.807
1420	960.33	10410.4	51.05	7590.5	298.5	53.536	1470	1010.33	10809.7	58.67	7890.5	268.9	53.812
1421		10418.4	51.19	7596.5	297.9	53.542	1471		10817.7	58.83	7896.5	268.3	53.818
1422		10426.4	51.33	7602.5	297.3	53.547	1472		10825.7	58.99	7902.5	267.8	53.823
1423		10434.3	51.48	7608.4	296.6	53.553	1473		10833.7	59.15	7908.6	267.2	53.829
1424		10442.3	51.62	7614.4	296.0	53.558	1474		10841.7	59.31	7914.6	266.7	53.834
1425	965.33	10450.2	51.77	7620.4	295.4	53.564	1475	1015.33	10849.8	59.48	7920.6	266.1	53.839
1426		10458.2	51.92	7626.4	294.8	53.570	1476		10857.8	59.64	7926.6	265.6	53.845
1427		10466.2	52.06	7632.4	294.1	53.575	1477		10865.8	59.80	7932.7	265.1	53.850
1428		10474.1	52.21	7638.3	293.5	53.581	1478		10873.8	59.96	7938.7	264.5	53.856
1429		10482.1	52.36	7644.3	292.9	53.586	1479		10881.8	60.13	7944.7	264.0	53.861
1430	970.33	10490.1	52.50	7650.3	292.3	53.592	1480	1020.33	10889.8	60.29	7950.8	263.4	53.867
1431		10498.1	52.65	7656.3	291.7	53.597	1481		10897.9	60.46	7956.8	262.9	53.872
1432		10506.0	52.80	7662.3	291.1	53.603	1482		10905.9	60.62	7962.8	262.3	53.877
1433		10514.0	52.95	7668.3	290.5	53.609	1483		10913.9	60.79	7968.9	261.8	53.883
1434		10522.0	53.09	7674.3	289.8	53.614	1484		10921.9	60.95	7974.9	261.3	53.888
1435	975.33	10529.9	53.24	7680.2	289.2	53.620	1485	1025.33	10930.0	61.12	7981.0	260.7	53.894
1436		10537.9	53.39	7686.2	288.6	53.625	1486		10938.0	61.29	7987.0	260.2	53.899
1437		10545.9	53.54	7692.2	288.0	53.631	1487		10946.0	61.45	7993.0	259.7	53.904
1438		10553.9	53.69	7698.2	287.4	53.636	1488		10954.0	61.62	7999.1	259.1	53.910
1439		10561.9	53.84	7704.2	286.8	53.642	1489		10962.1	61.79	8005.1	258.6	53.915
1440	980.33	10569.8	53.99	7710.2	286.2	53.647	1490	1030.33	10970.1	61.96	8011.2	258.1	53.921
1441		10577.8	54.14	7716.2	285.6	53.653	1491		10978.1	62.13	8017.2	257.6	53.926
1442		10585.8	54.29	7722.2	285.0	53.658	1492		10986.2	62.29	8023.3	257.0	53.931
1443		10593.8	54.45	7728.2	284.4	53.664	1493		10994.2	62.46	8029.3	256.5	53.937
1444		10601.8	54.60	7734.2	283.8	53.670	1494		11002.2	62.63	8035.4	256.0	53.942
1445	985.33	10609.7	54.75	7740.2	283.2	53.675	1495	1035.33	11010.3	62.80	8041.4	255.5	53.948
1446		10617.7	54.90	7746.2	282.6	53.681	1496		11018.3	62.97	8047.5	254.9	53.953
1447		10625.7	55.06	7752.2	282.1	53.686	1497		11026.3	63.14	8053.5	254.4	53.958
1448		10633.7	55.21	7758.2	281.5	53.692	1498		11034.4	63.31	8059.6	253.9	53.964
1449		10641.7	55.36	7764.2	280.9	53.697	1499		11042.4	63.49	8065.6	253.4	53.969

Table 5   Products—200% Theoretical Air (for One Pound-Mole)

T	t	$\bar{h}$	p_r	$\bar{u}$	v_r	$\bar{\phi}$	T	t	$\bar{h}$	p_r	$\bar{u}$	v_r	$\bar{\phi}$
1500	1040.33	11050.5	63.66	8071.7	252.9	53.974	**1550**	1090.33	11453.8	72.73	8375.7	228.7	54.239
1501		11058.5	63.83	8077.7	252.4	53.980	**1551**		11461.9	72.92	8381.8	228.3	54.244
1502		11066.6	64.00	8083.8	251.8	53.985	**1552**		11470.0	73.11	8387.9	227.8	54.249
1503		11074.6	64.17	8089.8	251.3	53.990	**1553**		11478.0	73.30	8394.0	227.4	54.255
1504		11082.6	64.35	8095.9	250.8	53.996	**1554**		11486.1	73.49	8400.1	226.9	54.260
1505	1045.33	11090.7	64.52	8102.0	250.3	54.001	**1555**	1095.33	11494.2	73.69	8406.2	226.5	54.265
1506		11098.7	64.69	8108.0	249.8	54.007	**1556**		11502.3	73.88	8412.3	226.0	54.270
1507		11106.8	64.87	8114.1	249.3	54.012	**1557**		11510.4	74.07	8418.5	225.6	54.275
1508		11114.8	65.04	8120.2	248.8	54.017	**1558**		11518.5	74.27	8424.6	225.1	54.281
1509		11122.9	65.22	8126.2	248.3	54.023	**1559**		11526.6	74.46	8430.7	224.7	54.286
1510	1050.33	11130.9	65.39	8132.3	247.8	54.028	**1560**	1100.33	11534.7	74.66	8436.8	224.2	54.291
1511		11139.0	65.57	8138.3	247.3	54.033	**1561**		11542.8	74.85	8442.9	223.8	54.296
1512		11147.0	65.75	8144.4	246.8	54.039	**1562**		11550.9	75.05	8449.0	223.4	54.301
1513		11155.1	65.92	8150.5	246.3	54.044	**1563**		11559.0	75.24	8455.1	222.9	54.307
1514		11163.1	66.10	8156.6	245.8	54.049	**1564**		11567.1	75.44	8461.3	222.5	54.312
1515	1055.33	11171.2	66.28	8162.6	245.3	54.054	**1565**	1105.33	11575.2	75.64	8467.4	222.0	54.317
1516		11179.3	66.45	8168.7	244.8	54.060	**1566**		11583.4	75.84	8473.5	221.6	54.322
1517		11187.3	66.63	8174.8	244.3	54.065	**1567**		11591.5	76.03	8479.6	221.2	54.327
1518		11195.4	66.81	8180.8	243.8	54.070	**1568**		11599.6	76.23	8485.7	220.7	54.332
1519		11203.4	66.99	8186.9	243.3	54.076	**1569**		11607.7	76.43	8491.9	220.3	54.338
1520	1060.33	11211.5	67.17	8193.0	242.8	54.081	**1570**	1110.33	11615.8	76.63	8498.0	219.9	54.343
1521		11219.6	67.35	8199.1	242.4	54.086	**1571**		11623.9	76.83	8504.1	219.4	54.348
1522		11227.6	67.53	8205.1	241.9	54.092	**1572**		11632.0	77.03	8510.2	219.0	54.353
1523		11235.7	67.71	8211.2	241.4	54.097	**1573**		11640.1	77.23	8516.4	218.6	54.358
1524		11243.7	67.89	8217.3	240.9	54.102	**1574**		11648.2	77.43	8522.5	218.1	54.363
1525	1065.33	11251.8	68.07	8223.4	240.4	54.108	**1575**	1115.33	11656.4	77.63	8528.6	217.7	54.369
1526		11259.9	68.25	8229.5	239.9	54.113	**1576**		11664.5	77.83	8534.8	217.3	54.374
1527		11267.9	68.43	8235.5	239.5	54.118	**1577**		11672.6	78.04	8540.9	216.9	54.379
1528		11276.0	68.62	8241.6	239.0	54.123	**1578**		11680.7	78.24	8547.0	216.4	54.384
1529		11284.1	68.80	8247.7	238.5	54.129	**1579**		11688.8	78.44	8553.2	216.0	54.389
1530	1070.33	11292.2	68.98	8253.8	238.0	54.134	**1580**	1120.33	11696.9	78.64	8559.3	215.6	54.394
1531		11300.2	69.17	8259.9	237.5	54.139	**1581**		11705.1	78.85	8565.4	215.2	54.399
1532		11308.3	69.35	8266.0	237.1	54.144	**1582**		11713.2	79.05	8571.6	214.8	54.405
1533		11316.4	69.53	8272.0	236.6	54.150	**1583**		11721.3	79.26	8577.7	214.3	54.410
1534		11324.4	69.72	8278.1	236.1	54.155	**1584**		11729.4	79.46	8583.8	213.9	54.415
1535	1075.33	11332.5	69.90	8284.2	235.7	54.160	**1585**	1125.33	11737.6	79.67	8590.0	213.5	54.420
1536		11340.6	70.09	8290.3	235.2	54.166	**1586**		11745.7	79.87	8596.1	213.1	54.425
1537		11348.7	70.27	8296.4	234.7	54.171	**1587**		11753.8	80.08	8602.3	212.7	54.430
1538		11356.7	70.46	8302.5	234.2	54.176	**1588**		11761.9	80.29	8608.4	212.3	54.435
1539		11364.8	70.65	8308.6	233.8	54.181	**1589**		11770.1	80.49	8614.5	211.8	54.440
1540	1080.33	11372.9	70.83	8314.7	233.3	54.187	**1590**	1130.33	11778.2	80.70	8620.7	211.4	54.446
1541		11381.0	71.02	8320.8	232.9	54.192	**1591**		11786.3	80.91	8626.8	211.0	54.451
1542		11389.1	71.21	8326.9	232.4	54.197	**1592**		11794.5	81.12	8633.0	210.6	54.456
1543		11397.2	71.40	8333.0	231.9	54.202	**1593**		11802.6	81.33	8639.1	210.2	54.461
1544		11405.2	71.59	8339.1	231.5	54.208	**1594**		11810.7	81.54	8645.3	209.8	54.466
1545	1085.33	11413.3	71.77	8345.2	231.0	54.213	**1595**	1135.33	11818.9	81.75	8651.4	209.4	54.471
1546		11421.4	71.96	8351.3	230.5	54.218	**1596**		11827.0	81.96	8657.6	209.0	54.476
1547		11429.5	72.15	8357.4	230.1	54.223	**1597**		11835.1	82.17	8663.7	208.6	54.481
1548		11437.6	72.34	8363.5	229.6	54.228	**1598**		11843.3	82.38	8669.9	208.2	54.486
1549		11445.7	72.53	8369.6	229.2	54.234	**1599**		11851.4	82.59	8676.0	207.8	54.491

Table 5   Products—200% Theoretical Air (for One Pound-Mole)

T	t	$\bar{h}$	p_r	$\bar{u}$	v_r	$\bar{\phi}$	T	t	$\bar{h}$	p_r	$\bar{u}$	v_r	$\bar{\phi}$
1600	1140.33	11859.5	82.80	8682.2	207.37	54.497	1650	1190.33	12267.7	93.97	8991.1	188.44	54.748
1601		11867.7	83.01	8688.3	206.97	54.502	1651		12275.9	94.20	8997.3	188.08	54.753
1602		11875.8	83.23	8694.5	206.57	54.507	1652		12284.1	94.44	9003.5	187.73	54.758
1603		11884.0	83.44	8700.6	206.17	54.512	1653		12292.3	94.67	9009.7	187.37	54.763
1604		11892.1	83.65	8706.8	205.77	54.517	1654		12300.5	94.91	9015.9	187.02	54.768
1605	1145.33	11900.3	83.87	8713.0	205.37	54.522	1655	1195.33	12308.7	95.15	9022.1	186.67	54.773
1606		11908.4	84.08	8719.1	204.98	54.527	1656		12316.9	95.38	9028.3	186.32	54.777
1607		11916.6	84.30	8725.3	204.58	54.532	1657		12325.1	95.62	9034.5	185.96	54.782
1608		11924.7	84.51	8731.4	204.19	54.537	1658		12333.3	95.86	9040.7	185.61	54.787
1609		11932.8	84.73	8737.6	203.80	54.542	1659		12341.4	96.10	9046.9	185.26	54.792
1610	1150.33	11941.0	84.94	8743.8	203.40	54.547	1660	1200.33	12349.6	96.34	9053.1	184.92	54.797
1611		11949.1	85.16	8749.9	203.01	54.552	1661		12357.8	96.58	9059.3	184.57	54.802
1612		11957.3	85.38	8756.1	202.62	54.557	1662		12366.0	96.82	9065.5	184.22	54.807
1613		11965.4	85.59	8762.3	202.23	54.562	1663		12374.2	97.06	9071.8	183.87	54.812
1614		11973.6	85.81	8768.4	201.84	54.568	1664		12382.4	97.30	9078.0	183.53	54.817
1615	1155.33	11981.8	86.03	8774.6	201.45	54.573	1665	1205.33	12390.6	97.54	9084.2	183.18	54.822
1616		11989.9	86.25	8780.8	201.07	54.578	1666		12398.8	97.78	9090.4	182.84	54.827
1617		11998.1	86.47	8786.9	200.68	54.583	1667		12407.0	98.03	9096.6	182.50	54.832
1618		12006.2	86.69	8793.1	200.30	54.588	1668		12415.2	98.27	9102.8	182.15	54.837
1619		12014.4	86.91	8799.3	199.91	54.593	1669		12423.4	98.51	9109.0	181.81	54.842
1620	1160.33	12022.5	87.13	8805.4	199.53	54.598	1670	1210.33	12431.7	98.76	9115.3	181.47	54.846
1621		12030.7	87.35	8811.6	199.15	54.603	1671		12439.9	99.00	9121.5	181.13	54.851
1622		12038.8	87.57	8817.8	198.76	54.608	1672		12448.1	99.25	9127.7	180.79	54.856
1623		12047.0	87.80	8824.0	198.38	54.613	1673		12456.3	99.49	9133.9	180.45	54.861
1624		12055.2	88.02	8830.1	198.00	54.618	1674		12464.5	99.74	9140.2	180.12	54.866
1625	1165.33	12063.3	88.24	8836.3	197.63	54.623	1675	1215.33	12472.7	99.98	9146.4	179.78	54.871
1626		12071.5	88.46	8842.5	197.25	54.628	1676		12480.9	100.23	9152.6	179.44	54.876
1627		12079.7	88.69	8848.7	196.87	54.633	1677		12489.1	100.48	9158.8	179.11	54.881
1628		12087.8	88.91	8854.9	196.49	54.638	1678		12497.3	100.73	9165.1	178.78	54.886
1629		12096.0	89.14	8861.0	196.12	54.643	1679		12505.6	100.98	9171.3	178.44	54.891
1630	1170.33	12104.2	89.36	8867.2	195.75	54.648	1680	1220.33	12513.8	101.22	9177.5	178.11	54.896
1631		12112.3	89.59	8873.4	195.37	54.653	1681		12522.0	101.47	9183.7	177.78	54.900
1632		12120.5	89.82	8879.6	195.00	54.658	1682		12530.2	101.72	9190.0	177.45	54.905
1633		12128.7	90.04	8885.8	194.63	54.663	1683		12538.4	101.97	9196.2	177.11	54.910
1634		12136.8	90.27	8892.0	194.26	54.668	1684		12546.6	102.23	9202.4	176.78	54.915
1635	1175.33	12145.0	90.50	8898.1	193.89	54.673	1685	1225.33	12554.8	102.48	9208.7	176.46	54.920
1636		12153.2	90.72	8904.3	193.52	54.678	1686		12563.1	102.73	9214.9	176.13	54.925
1637		12161.4	90.95	8910.5	193.15	54.683	1687		12571.3	102.98	9221.1	175.80	54.930
1638		12169.5	91.18	8916.7	192.78	54.688	1688		12579.5	103.23	9227.4	175.47	54.935
1639		12177.7	91.41	8922.9	192.42	54.693	1689		12587.7	103.49	9233.6	175.15	54.939
1640	1180.33	12185.9	91.64	8929.1	192.05	54.698	1690	1230.33	12596.0	103.74	9239.9	174.82	54.944
1641		12194.1	91.87	8935.3	191.69	54.703	1691		12604.2	104.00	9246.1	174.50	54.949
1642		12202.3	92.10	8941.5	191.32	54.708	1692		12612.4	104.25	9252.3	174.17	54.954
1643		12210.4	92.33	8947.7	190.96	54.713	1693		12620.6	104.51	9258.6	173.85	54.959
1644		12218.6	92.57	8953.9	190.60	54.718	1694		12628.9	104.76	9264.8	173.53	54.964
1645	1185.33	12226.8	92.80	8960.1	190.23	54.723	1695	1235.33	12637.1	105.02	9271.1	173.21	54.969
1646		12235.0	93.03	8966.3	189.87	54.728	1696		12645.3	105.28	9277.3	172.88	54.973
1647		12243.2	93.26	8972.5	189.51	54.733	1697		12653.6	105.53	9283.6	172.56	54.978
1648		12251.4	93.50	8978.7	189.16	54.738	1698		12661.8	105.79	9289.8	172.25	54.983
1649		12259.5	93.73	8984.8	188.80	54.743	1699		12670.0	106.05	9296.0	171.93	54.988

Table 5   Products—200% Theoretical Air (for One Pound-Mole)

T	t	$\bar{h}$	p_r	$\bar{u}$	v_r	$\bar{\phi}$	T	t	$\bar{h}$	p_r	$\bar{u}$	v_r	$\bar{\phi}$
1700	1240.33	12678.3	106.31	9302.3	171.61	54.993	1750	1290.33	13091.1	119.93	9615.8	156.60	55.232
1701		12686.5	106.57	9308.5	171.29	54.998	1751		13099.3	120.21	9622.1	156.32	55.237
1702		12694.7	106.83	9314.8	170.97	55.003	1752		13107.6	120.50	9628.4	156.03	55.242
1703		12703.0	107.09	9321.0	170.66	55.007	1753		13115.9	120.78	9634.7	155.75	55.246
1704		12711.2	107.35	9327.3	170.34	55.012	1754		13124.2	121.07	9641.0	155.47	55.251
1705	1245.33	12719.4	107.61	9333.5	170.03	55.017	1755	1295.33	13132.5	121.36	9647.3	155.19	55.256
1706		12727.7	107.87	9339.8	169.72	55.022	1756		13140.7	121.65	9653.6	154.91	55.261
1707		12735.9	108.14	9346.0	169.40	55.027	1757		13149.0	121.94	9659.9	154.63	55.265
1708		12744.1	108.40	9352.3	169.09	55.031	1758		13157.3	122.23	9666.2	154.35	55.270
1709		12752.4	108.66	9358.6	168.78	55.036	1759		13165.6	122.52	9672.5	154.07	55.275
1710	1250.33	12760.6	108.93	9364.8	168.47	55.041	1760	1300.33	13173.9	122.81	9678.8	153.80	55.279
1711		12768.9	109.19	9371.1	168.16	55.046	1761		13182.2	123.10	9685.1	153.52	55.284
1712		12777.1	109.46	9377.3	167.85	55.051	1762		13190.5	123.39	9691.4	153.24	55.289
1713		12785.4	109.72	9383.6	167.54	55.056	1763		13198.8	123.69	9697.7	152.97	55.293
1714		12793.6	109.99	9389.8	167.23	55.060	1764		13207.0	123.98	9704.0	152.69	55.298
1715	1255.33	12801.9	110.26	9396.1	166.93	55.065	1765	1305.33	13215.3	124.27	9710.3	152.42	55.303
1716		12810.1	110.52	9402.4	166.62	55.070	1766		13223.6	124.57	9716.6	152.14	55.308
1717		12818.3	110.79	9408.6	166.31	55.075	1767		13231.9	124.86	9722.9	151.87	55.312
1718		12826.6	111.06	9414.9	166.01	55.080	1768		13240.2	125.16	9729.2	151.60	55.317
1719		12834.8	111.33	9421.2	165.70	55.084	1769		13248.5	125.45	9735.5	151.32	55.322
1720	1260.33	12843.1	111.60	9427.4	165.40	55.089	1770	1310.33	13256.8	125.75	9741.8	151.05	55.326
1721		12851.4	111.87	9433.7	165.10	55.094	1771		13265.1	126.05	9748.1	150.78	55.331
1722		12859.6	112.14	9440.0	164.79	55.099	1772		13273.4	126.34	9754.5	150.51	55.336
1723		12867.9	112.41	9446.2	164.49	55.104	1773		13281.7	126.64	9760.8	150.24	55.340
1724		12876.1	112.68	9452.5	164.19	55.108	1774		13290.0	126.94	9767.1	149.97	55.345
1725	1265.33	12884.4	112.95	9458.8	163.89	55.113	1775	1315.33	13298.3	127.24	9773.4	149.70	55.350
1726		12892.6	113.22	9465.0	163.59	55.118	1776		13306.6	127.54	9779.7	149.44	55.354
1727		12900.9	113.50	9471.3	163.29	55.123	1777		13314.9	127.84	9786.0	149.17	55.359
1728		12909.1	113.77	9477.6	162.99	55.128	1778		13323.2	128.14	9792.3	148.90	55.364
1729		12917.4	114.04	9483.9	162.70	55.132	1779		13331.5	128.44	9798.7	148.64	55.368
1730	1270.33	12925.7	114.32	9490.1	162.40	55.137	1780	1320.33	13339.8	128.75	9805.0	148.37	55.373
1731		12933.9	114.59	9496.4	162.10	55.142	1781		13348.1	129.05	9811.3	148.11	55.378
1732		12942.2	114.87	9502.7	161.81	55.147	1782		13356.4	129.35	9817.6	147.84	55.382
1733		12950.5	115.15	9509.0	161.51	55.151	1783		13364.7	129.66	9823.9	147.58	55.387
1734		12958.7	115.42	9515.2	161.22	55.156	1784		13373.0	129.96	9830.3	147.31	55.392
1735	1275.33	12967.0	115.70	9521.5	160.93	55.161	1785	1325.33	13381.4	130.27	9836.6	147.05	55.396
1736		12975.2	115.98	9527.8	160.63	55.166	1786		13389.7	130.57	9842.9	146.79	55.401
1737		12983.5	116.26	9534.1	160.34	55.170	1787		13398.0	130.88	9849.2	146.53	55.406
1738		12991.8	116.54	9540.4	160.05	55.175	1788		13406.3	131.18	9855.6	146.27	55.410
1739		13000.1	116.81	9546.6	159.76	55.180	1789		13414.6	131.49	9861.9	146.01	55.415
1740	1280.33	13008.3	117.10	9552.9	159.47	55.185	1790	1330.33	13422.9	131.80	9868.2	145.75	55.420
1741		13016.6	117.38	9559.2	159.18	55.189	1791		13431.2	132.11	9874.5	145.49	55.424
1742		13024.9	117.66	9565.5	158.89	55.194	1792		13439.5	132.42	9880.9	145.23	55.429
1743		13033.1	117.94	9571.8	158.60	55.199	1793		13447.9	132.73	9887.2	144.97	55.434
1744		13041.4	118.22	9578.1	158.31	55.204	1794		13456.2	133.04	9893.5	144.71	55.438
1745	1285.33	13049.7	118.50	9584.4	158.02	55.208	1795	1335.33	13464.5	133.35	9899.9	144.46	55.443
1746		13058.0	118.79	9590.6	157.74	55.213	1796		13472.8	133.66	9906.2	144.20	55.447
1747		13066.2	119.07	9596.9	157.45	55.218	1797		13481.1	133.97	9912.5	143.95	55.452
1748		13074.5	119.35	9603.2	157.17	55.223	1798		13489.4	134.28	9918.9	143.69	55.457
1749		13082.8	119.64	9609.5	156.88	55.227	1799		13497.8	134.60	9925.2	143.44	55.461

Table 5 Products—200% Theoretical Air (for One Pound-Mole)

T	t	$\bar{h}$	p_r	$\bar{u}$	v_r	$\bar{\phi}$	T	t	$\bar{h}$	p_r	$\bar{u}$	v_r	$\bar{\phi}$
1800	1340.33	13506.1	134.91	9931.5	143.18	55.466	1850	1390.33	13923.2	151.37	10249.4	131.16	55.695
1801		13514.4	135.23	9937.9	142.92	55.471	1851		13931.6	151.71	10255.8	130.93	55.699
1802		13522.7	135.54	9944.2	142.68	55.475	1852		13940.0	152.05	10262.2	130.71	55.704
1803		13531.0	135.86	9950.5	142.42	55.480	1853		13948.3	152.41	10268.5	130.48	55.708
1804		13539.4	136.17	9956.9	142.18	55.484	1854		13956.7	152.75	10274.9	130.26	55.713
1805	1345.33	13547.7	136.49	9963.2	141.92	55.489	1855	1395.33	13965.1	153.10	10281.3	130.03	55.717
1806		13556.0	136.81	9969.6	141.67	55.494	1856		13973.5	153.44	10287.7	129.81	55.722
1807		13564.3	137.12	9975.9	141.42	55.498	1857		13981.8	153.80	10294.1	129.58	55.726
1808		13572.7	137.44	9982.3	141.17	55.503	1858		13990.2	154.14	10300.4	129.36	55.731
1809		13581.0	137.76	9988.6	140.92	55.507	1859		13998.6	154.49	10306.9	129.13	55.735
1810	1350.33	13589.3	138.08	9994.9	140.67	55.512	1860	1400.33	14006.9	154.85	10313.2	128.90	55.740
1811		13597.7	138.40	10001.3	140.42	55.517	1861		14015.3	155.20	10319.6	128.69	55.744
1812		13606.0	138.72	10007.6	140.18	55.521	1862		14023.7	155.55	10326.0	128.46	55.749
1813		13614.3	139.04	10014.0	139.93	55.526	1863		14032.0	155.90	10332.4	128.24	55.753
1814		13622.7	139.37	10020.4	139.68	55.531	1864		14040.5	156.26	10338.8	128.02	55.758
1815	1355.33	13631.0	139.69	10026.7	139.44	55.535	1865	1405.33	14048.8	156.61	10345.2	127.80	55.762
1816		13639.3	140.01	10033.0	139.19	55.540	1866		14057.2	156.96	10351.6	127.58	55.767
1817		13647.7	140.33	10039.3	138.95	55.544	1867		14065.6	157.31	10358.0	127.36	55.771
1818		13656.0	140.66	10045.7	138.70	55.549	1868		14073.9	157.68	10364.3	127.14	55.776
1819		13664.4	140.99	10052.1	138.45	55.554	1869		14082.3	158.03	10370.7	126.92	55.780
1820	1360.33	13672.7	141.31	10058.4	138.22	55.558	1870	1410.33	14090.7	158.39	10377.2	126.70	55.785
1821		13681.1	141.64	10064.8	137.97	55.563	1871		14099.1	158.74	10383.5	126.49	55.789
1822		13689.4	141.96	10071.1	137.73	55.567	1872		14107.5	159.10	10390.0	126.27	55.794
1823		13697.7	142.29	10077.5	137.49	55.572	1873		14115.8	159.46	10396.3	126.05	55.798
1824		13706.1	142.61	10083.9	137.25	55.576	1874		14124.2	159.82	10402.7	125.83	55.803
1825	1365.33	13714.4	142.95	10090.2	137.01	55.581	1875	1415.33	14132.6	160.18	10409.1	125.62	55.807
1826		13722.7	143.28	10096.6	136.76	55.586	1876		14141.0	160.54	10415.5	125.40	55.811
1827		13731.1	143.61	10103.0	136.53	55.590	1877		14149.4	160.90	10422.0	125.19	55.816
1828		13739.4	143.94	10109.3	136.29	55.595	1878		14157.8	161.27	10428.3	124.97	55.820
1829		13747.8	144.27	10115.6	136.05	55.599	1879		14166.1	161.63	10434.7	124.76	55.825
1830	1370.33	13756.2	144.60	10122.0	135.81	55.604	1880	1420.33	14174.6	161.99	10441.1	124.54	55.829
1831		13764.5	144.94	10128.4	135.57	55.608	1881		14182.9	162.35	10447.5	124.33	55.834
1832		13772.8	145.27	10134.7	135.34	55.613	1882		14191.4	162.72	10454.0	124.12	55.838
1833		13781.2	145.60	10141.1	135.10	55.617	1883		14199.7	163.09	10460.3	123.90	55.843
1834		13789.5	145.93	10147.5	134.87	55.622	1884		14208.1	163.46	10466.7	123.69	55.847
1835	1375.33	13797.9	146.27	10153.8	134.63	55.627	1885	1425.33	14216.5	163.83	10473.2	123.48	55.852
1836		13806.3	146.60	10160.2	134.40	55.631	1886		14224.9	164.19	10479.6	123.27	55.856
1837		13814.6	146.94	10166.6	134.16	55.636	1887		14233.3	164.56	10486.0	123.05	55.861
1838		13822.9	147.28	10172.9	133.92	55.640	1888		14241.7	164.93	10492.4	122.85	55.865
1839		13831.3	147.61	10179.3	133.69	55.645	1889		14250.1	165.30	10498.8	122.63	55.869
1840	1380.33	13839.7	147.96	10185.7	133.46	55.649	1890	1430.33	14258.5	165.67	10505.2	122.43	55.874
1841		13848.0	148.29	10192.0	133.23	55.654	1891		14266.9	166.03	10511.6	122.22	55.878
1842		13856.4	148.63	10198.4	132.99	55.658	1892		14275.3	166.41	10518.0	122.01	55.883
1843		13864.7	148.97	10204.8	132.77	55.663	1893		14283.7	166.78	10524.4	121.81	55.887
1844		13873.1	149.31	10211.1	132.53	55.667	1894		14292.1	167.16	10530.9	121.60	55.892
1845	1385.33	13881.5	149.65	10217.6	132.31	55.672	1895	1435.33	14300.5	167.52	10537.3	121.39	55.896
1846		13889.8	149.99	10223.9	132.07	55.676	1896		14308.9	167.90	10543.7	121.18	55.900
1847		13898.1	150.33	10230.3	131.85	55.681	1897		14317.3	168.27	10550.1	120.98	55.905
1848		13906.5	150.68	10236.7	131.62	55.685	1898		14325.7	168.66	10556.5	120.77	55.909
1849		13914.9	151.03	10243.0	131.38	55.690	1899		14334.1	169.03	10563.0	120.57	55.914

Table 5   Products—200% Theoretical Air (for One Pound-Mole)

T	t	$\bar{h}$	p_r	$\bar{u}$	v_r	$\bar{\phi}$	T	t	$\bar{h}$	p_r	$\bar{u}$	v_r	$\bar{\phi}$
1900	1440.33	14342.5	169.41	10569.4	120.36	55.918	1950	1490.33	14763.8	189.15	10891.4	110.64	56.137
1901		14350.9	169.78	10575.8	120.16	55.923	1951		14772.3	189.57	10897.9	110.45	56.141
1902		14359.3	170.17	10582.2	119.95	55.927	1952		14780.7	189.98	10904.3	110.27	56.146
1903		14367.8	170.54	10588.7	119.75	55.931	1953		14789.2	190.40	10910.8	110.08	56.150
1904		14376.1	170.93	10595.1	119.54	55.936	1954		14797.6	190.81	10917.2	109.90	56.154
1905	1445.33	14384.6	171.30	10601.5	119.34	55.940	1955	1495.33	14806.1	191.22	10923.7	109.72	56.159
1906		14393.0	171.69	10607.9	119.14	55.945	1956		14814.5	191.64	10930.1	109.53	56.163
1907		14401.3	172.06	10614.3	118.94	55.949	1957		14823.0	192.05	10936.6	109.35	56.167
1908		14409.8	172.45	10620.8	118.73	55.954	1958		14831.4	192.48	10943.1	109.17	56.172
1909		14418.2	172.83	10627.2	118.53	55.958	1959		14839.9	192.89	10949.6	108.99	56.176
1910	1450.33	14426.6	173.22	10633.6	118.33	55.962	1960	1500.33	14848.3	193.31	10956.0	108.81	56.180
1911		14435.0	173.60	10640.0	118.13	55.967	1961		14856.8	193.74	10962.5	108.62	56.185
1912		14443.5	173.99	10646.5	117.93	55.971	1962		14865.2	194.15	10968.9	108.45	56.189
1913		14451.8	174.37	10652.9	117.73	55.976	1963		14873.7	194.58	10975.4	108.26	56.193
1914		14460.3	174.75	10659.4	117.54	55.980	1964		14882.1	195.00	10981.9	108.09	56.198
1915	1455.33	14468.7	175.15	10665.8	117.33	55.984	1965	1505.33	14890.6	195.42	10988.4	107.91	56.202
1916		14477.1	175.53	10672.2	117.14	55.989	1966		14899.0	195.85	10994.8	107.73	56.206
1917		14485.5	175.93	10678.6	116.94	55.993	1967		14907.5	196.27	11001.3	107.55	56.210
1918		14493.9	176.31	10685.0	116.74	55.997	1968		14915.9	196.69	11007.7	107.38	56.215
1919		14502.4	176.71	10691.5	116.54	56.002	1969		14924.4	197.12	11014.2	107.19	56.219
1920	1460.33	14510.8	177.09	10697.9	116.35	56.006	1970	1510.33	14932.9	197.55	11020.7	107.02	56.223
1921		14519.2	177.49	10704.4	116.15	56.011	1971		14941.3	197.98	11027.2	106.84	56.228
1922		14527.6	177.88	10710.8	115.96	56.015	1972		14949.8	198.40	11033.7	106.66	56.232
1923		14536.1	178.28	10717.3	115.76	56.019	1973		14958.2	198.83	11040.1	106.49	56.236
1924		14544.5	178.66	10723.7	115.57	56.024	1974		14966.7	199.27	11046.6	106.31	56.241
1925	1465.33	14552.9	179.05	10730.1	115.37	56.028	1975	1515.33	14975.2	199.69	11053.1	106.14	56.245
1926		14561.3	179.46	10736.5	115.18	56.033	1976		14983.7	200.13	11059.6	105.96	56.249
1927		14569.8	179.85	10743.0	114.99	56.037	1977		14992.1	200.56	11066.0	105.78	56.253
1928		14578.2	180.25	10749.4	114.79	56.041	1978		15000.6	200.99	11072.6	105.61	56.258
1929		14586.6	180.64	10755.9	114.60	56.046	1979		15009.0	201.43	11079.0	105.43	56.262
1930	1470.33	14595.0	181.05	10762.3	114.40	56.050	1980	1520.33	15017.5	201.86	11085.5	105.26	56.266
1931		14603.5	181.44	10768.8	114.21	56.054	1981		15026.0	202.29	11092.0	105.09	56.270
1932		14611.9	181.84	10775.2	114.02	56.059	1982		15034.4	202.73	11098.5	104.91	56.275
1933		14620.4	182.24	10781.7	113.83	56.063	1983		15042.9	203.17	11105.0	104.74	56.279
1934		14628.8	182.63	10788.1	113.64	56.067	1984		15051.4	203.60	11111.4	104.57	56.283
1935	1475.33	14637.2	183.04	10794.6	113.45	56.072	1985	1525.33	15059.9	204.05	11117.9	104.40	56.288
1936		14645.6	183.44	10801.0	113.26	56.076	1986		15068.3	204.48	11124.4	104.23	56.292
1937		14654.1	183.85	10807.5	113.06	56.081	1987		15076.8	204.93	11130.9	104.05	56.296
1938		14662.5	184.25	10813.9	112.88	56.085	1988		15085.3	205.36	11137.4	103.89	56.300
1939		14671.0	184.66	10820.4	112.68	56.089	1989		15093.8	205.80	11143.9	103.72	56.305
1940	1480.33	14679.4	185.06	10826.8	112.50	56.094	1990	1530.33	15102.3	206.25	11150.4	103.54	56.309
1941		14687.8	185.47	10833.3	112.31	56.098	1991		15110.7	206.69	11156.9	103.37	56.313
1942		14696.3	185.87	10839.7	112.12	56.102	1992		15119.2	207.13	11163.4	103.21	56.317
1943		14704.7	186.28	10846.2	111.94	56.107	1993		15127.7	207.58	11169.8	103.03	56.322
1944		14713.1	186.69	10852.6	111.75	56.111	1994		15136.2	208.02	11176.4	102.87	56.326
1945	1485.33	14721.6	187.09	10859.1	111.56	56.115	1995	1535.33	15144.6	208.46	11182.8	102.70	56.330
1946		14730.0	187.51	10865.5	111.37	56.120	1996		15153.1	208.92	11189.3	102.53	56.334
1947		14738.5	187.92	10872.0	111.19	56.124	1997		15161.6	209.36	11195.9	102.36	56.339
1948		14746.9	188.33	10878.4	111.00	56.128	1998		15170.1	209.80	11202.3	102.20	56.343
1949		14755.4	188.74	10884.9	110.82	56.133	1999		15178.6	210.26	11208.8	102.03	56.347

Table 5   Products—200% Theoretical Air (for One Pound-Mole)

T	t	$\bar{h}$	p_r	$\bar{u}$	v_r	$\bar{\phi}$	T	t	$\bar{h}$	p_r	$\bar{u}$	v_r	$\bar{\phi}$
2000	1540.33	15187.0	210.7	11215.3	101.86	56.351	2050	1590.33	15612.2	234.2	11541.2	93.93	56.561
2001		15195.5	211.2	11221.8	101.70	56.356	2051		15620.7	234.7	11547.7	93.79	56.565
2002		15204.0	211.6	11228.3	101.53	56.360	2052		15629.2	235.2	11554.2	93.64	56.570
2003		15212.5	212.1	11234.8	101.36	56.364	2053		15637.8	235.7	11560.8	93.48	56.574
2004		15221.0	212.5	11241.3	101.20	56.368	2054		15646.2	236.2	11567.3	93.33	56.578
2005	1545.33	15229.5	213.0	11247.8	101.03	56.373	2055	1595.33	15654.8	236.7	11573.8	93.19	56.582
2006		15238.0	213.4	11254.3	100.87	56.377	2056		15663.3	237.1	11580.4	93.04	56.586
2007		15246.4	213.9	11260.8	100.71	56.381	2057		15671.8	237.7	11586.9	92.89	56.590
2008		15254.9	214.3	11267.3	100.54	56.385	2058		15680.4	238.1	11593.5	92.74	56.595
2009		15263.5	214.8	11273.9	100.38	56.390	2059		15688.9	238.6	11600.0	92.59	56.599
2010	1550.33	15271.9	215.2	11280.3	100.21	56.394	2060	1600.33	15697.4	239.1	11606.5	92.45	56.603
2011		15280.4	215.7	11286.9	100.05	56.398	2061		15706.0	239.6	11613.1	92.29	56.607
2012		15288.9	216.2	11293.3	99.89	56.402	2062		15714.5	240.1	11619.7	92.15	56.611
2013		15297.4	216.6	11299.9	99.73	56.406	2063		15723.0	240.6	11626.2	92.00	56.615
2014		15305.9	217.1	11306.3	99.56	56.411	2064		15731.6	241.2	11632.7	91.85	56.619
2015	1555.33	15314.4	217.5	11312.9	99.40	56.415	2065	1605.33	15740.1	241.6	11639.3	91.71	56.623
2016		15322.9	218.0	11319.4	99.24	56.419	2066		15748.6	242.1	11645.8	91.56	56.628
2017		15331.4	218.5	11325.9	99.07	56.423	2067		15757.1	242.6	11652.4	91.42	56.632
2018		15339.9	218.9	11332.4	98.91	56.427	2068		15765.7	243.2	11658.9	91.27	56.636
2019		15348.3	219.4	11338.9	98.76	56.432	2069		15774.2	243.7	11665.4	91.12	56.640
2020	1560.33	15356.9	219.9	11345.4	98.60	56.436	2070	1610.33	15782.8	244.2	11672.0	90.98	56.644
2021		15365.4	220.3	11352.0	98.43	56.440	2071		15791.3	244.7	11678.6	90.84	56.648
2022		15373.8	220.8	11358.4	98.28	56.444	2072		15799.8	245.2	11685.1	90.69	56.652
2023		15382.4	221.3	11365.0	98.12	56.448	2073		15808.4	245.7	11691.7	90.54	56.656
2024		15390.9	221.7	11371.5	97.96	56.453	2074		15816.9	246.2	11698.2	90.40	56.661
2025	1565.33	15399.4	222.2	11378.0	97.80	56.457	2075	1615.33	15825.4	246.7	11704.8	90.26	56.665
2026		15407.9	222.7	11384.5	97.64	56.461	2076		15834.0	247.2	11711.3	90.12	56.669
2027		15416.4	223.1	11391.0	97.49	56.465	2077		15842.5	247.7	11717.9	89.97	56.673
2028		15424.9	223.6	11397.6	97.32	56.470	2078		15851.0	248.3	11724.4	89.83	56.677
2029		15433.4	224.1	11404.0	97.17	56.474	2079		15859.6	248.8	11731.0	89.69	56.681
2030	1570.33	15441.9	224.6	11410.6	97.01	56.478	2080	1620.33	15868.2	249.3	11737.6	89.55	56.685
2031		15450.4	225.0	11417.1	96.85	56.482	2081		15876.7	249.8	11744.2	89.40	56.689
2032		15458.9	225.5	11423.6	96.70	56.486	2082		15885.2	250.3	11750.7	89.26	56.693
2033		15467.4	226.0	11430.2	96.54	56.490	2083		15893.8	250.8	11757.2	89.12	56.698
2034		15475.9	226.5	11436.7	96.39	56.495	2084		15902.4	251.3	11763.8	88.98	56.702
2035	1575.33	15484.4	226.9	11443.2	96.23	56.499	2085	1625.33	15910.9	251.9	11770.3	88.83	56.706
2036		15493.0	227.4	11449.7	96.08	56.503	2086		15919.4	252.4	11776.9	88.69	56.710
2037		15501.4	227.9	11456.2	95.92	56.507	2087		15928.0	252.9	11783.5	88.56	56.714
2038		15510.0	228.4	11462.8	95.76	56.511	2088		15936.5	253.4	11790.0	88.42	56.718
2039		15518.5	228.9	11469.3	95.61	56.516	2089		15945.1	253.9	11796.6	88.28	56.722
2040	1580.33	15527.0	229.3	11475.8	95.46	56.520	2090	1630.33	15953.6	254.5	11803.2	88.13	56.726
2041		15535.5	229.8	11482.4	95.31	56.524	2091		15962.1	255.0	11809.7	88.00	56.730
2042		15544.1	230.3	11488.9	95.15	56.528	2092		15970.7	255.5	11816.3	87.86	56.734
2043		15552.5	230.8	11495.4	95.00	56.532	2093		15979.3	256.1	11822.9	87.72	56.738
2044		15561.1	231.3	11502.0	94.85	56.536	2094		15987.9	256.6	11829.5	87.58	56.743
2045	1585.33	15569.6	231.8	11508.5	94.69	56.541	2095	1635.33	15996.4	257.1	11836.0	87.44	56.747
2046		15578.1	232.2	11515.0	94.54	56.545	2096		16004.9	257.6	11842.6	87.30	56.751
2047		15586.6	232.7	11521.6	94.39	56.549	2097		16013.5	258.2	11849.2	87.17	56.755
2048		15595.1	233.2	11528.1	94.24	56.553	2098		16022.0	258.7	11855.7	87.03	56.759
2049		15603.6	233.7	11534.6	94.08	56.557	2099		16030.6	259.2	11862.3	86.89	56.763

Table 5 Products—200% Theoretical Air (for One Pound-Mole)

T	t	$\bar{h}$	p_r	$\bar{u}$	v_r	$\bar{\phi}$	T	t	$\bar{h}$	p_r	$\bar{u}$	v_r	$\bar{\phi}$
2100	1640.33	16039.2	259.8	11868.9	86.75	56.767	2150	1690.33	16467.9	287.6	12198.3	80.24	56.969
2101		16047.8	260.3	11875.5	86.62	56.771	2151		16476.6	288.1	12205.0	80.11	56.973
2102		16056.3	260.8	11882.0	86.48	56.775	2152		16485.2	288.7	12211.6	79.99	56.977
2103		16064.8	261.4	11888.6	86.35	56.779	2153		16493.8	289.3	12218.2	79.87	56.981
2104		16073.4	261.9	11895.2	86.21	56.783	2154		16502.3	289.9	12224.8	79.75	56.985
2105	1645.33	16081.9	262.4	11901.7	86.07	56.787	2155	1695.33	16510.9	290.4	12231.4	79.62	56.989
2106		16090.5	263.0	11908.3	85.94	56.791	2156		16519.5	291.0	12238.0	79.50	56.993
2107		16099.1	263.5	11914.9	85.81	56.796	2157		16528.1	291.6	12244.6	79.38	56.997
2108		16107.7	264.1	11921.5	85.67	56.800	2158		16536.7	292.2	12251.2	79.25	57.001
2109		16116.2	264.6	11928.0	85.53	56.804	2159		16545.3	292.8	12257.8	79.13	57.005
2110	1650.33	16124.8	265.1	11934.6	85.40	56.808	2160	1700.33	16553.9	293.4	12264.5	79.01	57.009
2111		16133.4	265.7	11941.2	85.27	56.812	2161		16562.5	294.0	12271.1	78.89	57.013
2112		16141.9	266.2	11947.8	85.14	56.816	2162		16571.1	294.6	12277.7	78.77	57.017
2113		16150.5	266.8	11954.3	85.00	56.820	2163		16579.7	295.1	12284.3	78.65	57.021
2114		16159.1	267.3	11960.9	84.87	56.824	2164		16588.3	295.7	12290.9	78.52	57.025
2115	1655.33	16167.6	267.9	11967.5	84.73	56.828	2165	1705.33	16596.9	296.3	12297.5	78.40	57.029
2116		16176.2	268.4	11974.1	84.60	56.832	2166		16605.5	296.9	12304.2	78.28	57.033
2117		16184.7	269.0	11980.7	84.47	56.836	2167		16614.2	297.5	12310.8	78.16	57.037
2118		16193.3	269.5	11987.3	84.34	56.840	2168		16622.7	298.1	12317.4	78.04	57.040
2119		16201.9	270.0	11993.9	84.21	56.844	2169		16631.3	298.7	12324.0	77.92	57.044
2120	1660.33	16210.5	270.6	12000.5	84.08	56.848	2170	1710.33	16640.0	299.3	12330.6	77.81	57.048
2121		16219.0	271.2	12007.0	83.94	56.852	2171		16648.6	299.9	12337.3	77.68	57.052
2122		16227.6	271.7	12013.6	83.81	56.856	2172		16657.1	300.5	12343.8	77.56	57.056
2123		16236.2	272.3	12020.2	83.68	56.860	2173		16665.8	301.1	12350.5	77.44	57.060
2124		16244.8	272.8	12026.8	83.55	56.864	2174		16674.4	301.7	12357.1	77.33	57.064
2125	1665.33	16253.3	273.4	12033.4	83.42	56.868	2175	1715.33	16683.0	302.3	12363.7	77.21	57.068
2126		16261.9	273.9	12040.0	83.29	56.873	2176		16691.6	302.9	12370.4	77.09	57.072
2127		16270.5	274.5	12046.6	83.16	56.877	2177		16700.2	303.5	12377.0	76.97	57.076
2128		16279.1	275.0	12053.2	83.03	56.881	2178		16708.8	304.1	12383.6	76.86	57.080
2129		16287.6	275.6	12059.7	82.90	56.885	2179		16717.4	304.7	12390.2	76.73	57.084
2130	1670.33	16296.2	276.2	12066.4	82.77	56.889	2180	1720.33	16726.0	305.3	12396.9	76.62	57.088
2131		16304.8	276.7	12073.0	82.65	56.893	2181		16734.7	305.9	12403.5	76.50	57.092
2132		16313.4	277.3	12079.6	82.51	56.897	2182		16743.3	306.6	12410.2	76.38	57.096
2133		16322.0	277.8	12086.1	82.39	56.901	2183		16751.9	307.2	12416.7	76.27	57.100
2134		16330.6	278.4	12092.7	82.26	56.905	2184		16760.5	307.8	12423.4	76.15	57.104
2135	1675.33	16339.2	279.0	12099.3	82.13	56.909	2185	1725.33	16769.1	308.4	12430.0	76.04	57.108
2136		16347.8	279.5	12106.0	82.00	56.913	2186		16777.8	309.0	12436.7	75.92	57.112
2137		16356.3	280.1	12112.5	81.88	56.917	2187		16786.4	309.6	12443.3	75.80	57.116
2138		16364.9	280.7	12119.1	81.75	56.921	2188		16794.9	310.2	12449.9	75.69	57.120
2139		16373.5	281.2	12125.7	81.62	56.925	2189		16803.6	310.9	12456.5	75.57	57.124
2140	1680.33	16382.1	281.8	12132.4	81.49	56.929	2190	1730.33	16812.2	311.5	12463.2	75.46	57.128
2141		16390.6	282.4	12138.9	81.37	56.933	2191		16820.8	312.1	12469.8	75.34	57.131
2142		16399.2	282.9	12145.5	81.24	56.937	2192		16829.5	312.7	12476.5	75.23	57.135
2143		16407.8	283.5	12152.1	81.12	56.941	2193		16838.0	313.3	12483.0	75.11	57.139
2144		16416.4	284.1	12158.8	80.99	56.945	2194		16846.7	313.9	12489.7	75.00	57.143
2145	1685.33	16425.0	284.7	12165.3	80.86	56.949	2195	1735.33	16855.3	314.6	12496.3	74.89	57.147
2146		16433.6	285.2	12171.9	80.74	56.953	2196		16863.9	315.2	12503.0	74.77	57.151
2147		16442.2	285.8	12178.5	80.61	56.957	2197		16872.6	315.8	12509.6	74.65	57.155
2148		16450.8	286.4	12185.2	80.49	56.961	2198		16881.2	316.4	12516.3	74.54	57.159
2149		16459.3	287.0	12191.7	80.37	56.965	2199		16889.8	317.1	12522.9	74.43	57.163

Table 5 Products—200% Theoretical Air (for One Pound-Mole)

T	t	$\bar{h}$	p_r	$\bar{u}$	v_r	$\bar{\phi}$	T	t	$\bar{h}$	p_r	$\bar{u}$	v_r	$\bar{\phi}$
2200	1740.33	16898.4	317.7	12529.5	74.32	57.167	2250	1790.33	17330.6	350.4	12862.4	68.92	57.361
2201		16907.1	318.3	12536.2	74.20	57.171	2251		17339.2	351.0	12869.1	68.82	57.365
2202		16915.7	318.9	12542.8	74.09	57.175	2252		17347.9	351.7	12875.8	68.71	57.369
2203		16924.3	319.6	12549.5	73.98	57.179	2253		17356.6	352.4	12882.4	68.61	57.373
2204		16933.0	320.2	12556.1	73.87	57.182	2254		17365.2	353.1	12889.1	68.51	57.377
2205	1745.33	16941.6	320.9	12562.7	73.75	57.186	2255	1795.33	17373.9	353.8	12895.7	68.41	57.380
2206		16950.2	321.5	12569.4	73.64	57.190	2256		17382.5	354.4	12902.4	68.31	57.384
2207		16958.8	322.1	12576.0	73.53	57.194	2257		17391.2	355.1	12909.1	68.20	57.388
2208		16967.5	322.7	12582.7	73.42	57.198	2258		17399.9	355.8	12915.8	68.10	57.392
2209		16976.1	323.4	12589.4	73.31	57.202	2259		17408.5	356.5	12922.5	68.00	57.396
2210	1750.33	16984.7	324.0	12595.9	73.20	57.206	2260	1800.33	17417.2	357.2	12929.2	67.90	57.400
2211		16993.3	324.7	12602.6	73.09	57.210	2261		17425.9	357.9	12935.8	67.80	57.403
2212		17002.0	325.3	12609.3	72.97	57.214	2262		17434.6	358.6	12942.5	67.70	57.407
2213		17010.6	326.0	12615.9	72.86	57.218	2263		17443.2	359.3	12949.2	67.60	57.411
2214		17019.3	326.6	12622.6	72.75	57.222	2264		17451.9	360.0	12955.9	67.50	57.415
2215	1755.33	17027.9	327.2	12629.2	72.64	57.226	2265	1805.33	17460.5	360.6	12962.5	67.40	57.419
2216		17036.6	327.9	12635.9	72.53	57.229	2266		17469.2	361.3	12969.2	67.30	57.422
2217		17045.2	328.5	12642.5	72.42	57.233	2267		17477.9	362.0	12975.9	67.20	57.426
2218		17053.8	329.2	12649.2	72.31	57.237	2268		17486.5	362.7	12982.6	67.10	57.430
2219		17062.5	329.8	12655.8	72.20	57.241	2269		17495.2	363.4	12989.3	67.00	57.434
2220	1760.33	17071.1	330.4	12662.5	72.10	57.245	2270	1810.33	17503.9	364.2	12996.0	66.89	57.438
2221		17079.7	331.1	12669.2	71.99	57.249	2271		17512.6	364.9	13002.7	66.80	57.442
2222		17088.4	331.7	12675.8	71.88	57.253	2272		17521.3	365.6	13009.4	66.70	57.446
2223		17097.0	332.4	12682.4	71.77	57.257	2273		17529.9	366.3	13016.1	66.60	57.449
2224		17105.6	333.1	12689.1	71.66	57.261	2274		17538.6	367.0	13022.8	66.50	57.453
2225	1765.33	17114.3	333.7	12695.7	71.55	57.265	2275	1815.33	17547.3	367.7	13029.5	66.40	57.457
2226		17122.9	334.4	12702.4	71.44	57.268	2276		17555.9	368.4	13036.1	66.30	57.461
2227		17131.6	335.0	12709.1	71.34	57.272	2277		17564.6	369.1	13042.8	66.21	57.465
2228		17140.2	335.7	12715.7	71.23	57.276	2278		17573.3	369.8	13049.5	66.11	57.468
2229		17148.9	336.3	12722.4	71.12	57.280	2279		17582.0	370.5	13056.2	66.01	57.472
2230	1770.33	17157.5	337.0	12729.0	71.02	57.284	2280	1820.33	17590.6	371.2	13062.9	65.91	57.476
2231		17166.1	337.6	12735.7	70.91	57.288	2281		17599.3	371.9	13069.6	65.81	57.480
2232		17174.8	338.3	12742.4	70.80	57.292	2282		17608.0	372.6	13076.3	65.72	57.484
2233		17183.5	339.0	12749.0	70.70	57.295	2283		17616.7	373.4	13083.0	65.62	57.487
2234		17192.1	339.6	12755.7	70.59	57.299	2284		17625.4	374.1	13089.7	65.52	57.491
2235	1775.33	17200.8	340.3	12762.4	70.49	57.303	2285	1825.33	17634.1	374.8	13096.4	65.43	57.495
2236		17209.4	340.9	12769.0	70.38	57.307	2286		17642.8	375.5	13103.1	65.33	57.499
2237		17218.1	341.6	12775.7	70.28	57.311	2287		17651.4	376.2	13109.8	65.23	57.503
2238		17226.7	342.3	12782.3	70.17	57.315	2288		17660.1	376.9	13116.4	65.14	57.506
2239		17235.3	342.9	12789.0	70.07	57.319	2289		17668.7	377.6	13123.1	65.05	57.510
2240	1780.33	17244.0	343.6	12795.7	69.96	57.323	2290	1830.33	17677.4	378.4	13129.8	64.95	57.514
2241		17252.7	344.3	12802.4	69.85	57.327	2291		17686.1	379.1	13136.5	64.86	57.518
2242		17261.3	345.0	12809.0	69.75	57.330	2292		17694.8	379.8	13143.2	64.76	57.521
2243		17270.0	345.6	12815.7	69.64	57.334	2293		17703.5	380.5	13149.9	64.66	57.525
2244		17278.7	346.3	12822.4	69.54	57.338	2294		17712.2	381.3	13156.6	64.57	57.529
2245	1785.33	17287.3	347.0	12829.1	69.43	57.342	2295	1835.33	17720.9	382.0	13163.3	64.47	57.533
2246		17295.9	347.6	12835.7	69.33	57.346	2296		17729.6	382.7	13170.1	64.38	57.537
2247		17304.6	348.3	12842.4	69.23	57.350	2297		17738.3	383.5	13176.8	64.28	57.540
2248		17313.2	349.0	12849.0	69.12	57.353	2298		17747.0	384.2	13183.5	64.19	57.544
2249		17321.9	349.7	12855.7	69.02	57.357	2299		17755.7	384.9	13190.2	64.10	57.548

Table 5　Products—200% Theoretical Air (for One Pound-Mole)

T	t	$\bar{h}$	p_r	$\bar{u}$	v_r	$\bar{\phi}$	T	t	$\bar{h}$	p_r	$\bar{u}$	v_r	$\bar{\phi}$
2300	1840.33	17764.4	385.7	13196.9	64.00	57.552	2350	1890.33	18199.7	423.8	13532.9	59.51	57.739
2301		17773.1	386.4	13203.6	63.91	57.556	2351		18208.4	424.6	13539.6	59.43	57.743
2302		17781.7	387.1	13210.3	63.81	57.559	2352		18217.1	425.4	13546.4	59.34	57.746
2303		17790.4	387.9	13217.0	63.72	57.563	2353		18225.8	426.1	13553.1	59.25	57.750
2304		17799.1	388.6	13223.7	63.63	57.567	2354		18234.5	426.9	13559.8	59.17	57.754
2305	1845.33	17807.8	389.3	13230.4	63.53	57.571	2355	1895.33	18243.3	427.7	13566.6	59.08	57.758
2306		17816.5	390.1	13237.1	63.44	57.574	2356		18252.0	428.5	13573.3	59.00	57.761
2307		17825.2	390.8	13243.8	63.35	57.578	2357		18260.7	429.4	13580.0	58.91	57.765
2308		17833.9	391.6	13250.5	63.25	57.582	2358		18269.4	430.2	13586.8	58.83	57.769
2309		17842.6	392.3	13257.2	63.17	57.586	2359		18278.2	431.0	13593.5	58.74	57.772
2310	1850.33	17851.3	393.0	13263.9	63.07	57.589	2360	1900.33	18286.9	431.8	13600.3	58.66	57.776
2311		17860.0	393.8	13270.6	62.98	57.593	2361		18295.6	432.6	13607.0	58.57	57.780
2312		17868.7	394.5	13277.4	62.89	57.597	2362		18304.4	433.4	13613.8	58.49	57.783
2313		17877.4	395.3	13284.1	62.80	57.601	2363		18313.1	434.2	13620.5	58.41	57.787
2314		17886.1	396.0	13290.8	62.71	57.604	2364		18321.8	435.0	13627.2	58.32	57.791
2315	1855.33	17894.8	396.8	13297.5	62.61	57.608	2365	1905.33	18330.5	435.8	13634.0	58.24	57.794
2316		17903.5	397.5	13304.2	62.52	57.612	2366		18339.3	436.6	13640.7	58.16	57.798
2317		17912.2	398.3	13310.9	62.43	57.616	2367		18348.0	437.4	13647.5	58.07	57.802
2318		17920.9	399.0	13317.7	62.34	57.620	2368		18356.7	438.2	13654.2	57.99	57.806
2319		17929.6	399.8	13324.4	62.25	57.623	2369		18365.5	439.0	13661.0	57.91	57.809
2320	1860.33	17938.3	400.5	13331.1	62.16	57.627	2370	1910.33	18374.2	439.9	13667.7	57.82	57.813
2321		17947.0	401.3	13337.8	62.07	57.631	2371		18382.9	440.7	13674.5	57.74	57.817
2322		17955.7	402.1	13344.5	61.98	57.635	2372		18391.7	441.5	13681.2	57.66	57.820
2323		17964.4	402.8	13351.3	61.89	57.638	2373		18400.4	442.3	13688.0	57.57	57.824
2324		17973.1	403.6	13358.0	61.80	57.642	2374		18409.1	443.1	13694.7	57.49	57.828
2325	1865.33	17981.8	404.3	13364.7	61.71	57.646	2375	1915.33	18417.9	443.9	13701.5	57.41	57.831
2326		17990.5	405.1	13371.4	61.62	57.650	2376		18426.6	444.8	13708.2	57.33	57.835
2327		17999.2	405.9	13378.1	61.53	57.653	2377		18435.3	445.6	13715.0	57.25	57.839
2328		18007.9	406.6	13384.8	61.44	57.657	2378		18444.1	446.4	13721.7	57.16	57.842
2329		18016.6	407.4	13391.5	61.35	57.661	2379		18452.8	447.3	13728.5	57.08	57.846
2330	1870.33	18025.3	408.2	13398.3	61.26	57.665	2380	1920.33	18461.6	448.1	13735.2	57.00	57.850
2331		18034.0	409.0	13405.0	61.17	57.668	2381		18470.3	448.9	13742.0	56.92	57.853
2332		18042.7	409.7	13411.7	61.08	57.672	2382		18479.0	449.8	13748.7	56.84	57.857
2333		18051.5	410.5	13418.4	61.00	57.676	2383		18487.8	450.6	13755.5	56.76	57.861
2334		18060.2	411.2	13425.2	60.91	57.679	2384		18496.5	451.4	13762.2	56.68	57.864
2335	1875.33	18068.9	412.0	13431.9	60.82	57.683	2385	1925.33	18505.3	452.2	13769.0	56.60	57.868
2336		18077.6	412.8	13438.6	60.73	57.687	2386		18513.9	453.1	13775.7	56.51	57.872
2337		18086.3	413.6	13445.4	60.64	57.691	2387		18522.7	453.9	13782.4	56.43	57.875
2338		18095.0	414.4	13452.1	60.55	57.694	2388		18531.4	454.7	13789.2	56.35	57.879
2339		18103.7	415.1	13458.8	60.47	57.698	2389		18540.1	455.6	13795.9	56.27	57.883
2340	1880.33	18112.5	415.9	13465.5	60.38	57.702	2390	1930.33	18549.0	456.4	13802.8	56.19	57.886
2341		18121.2	416.7	13472.3	60.29	57.706	2391		18557.7	457.3	13809.5	56.11	57.890
2342		18129.9	417.5	13479.0	60.20	57.709	2392		18566.4	458.1	13816.3	56.03	57.894
2343		18138.6	418.3	13485.7	60.12	57.713	2393		18575.2	459.0	13823.0	55.95	57.897
2344		18147.3	419.0	13492.5	60.03	57.717	2394		18583.9	459.8	13829.8	55.87	57.901
2345	1885.33	18156.1	419.8	13499.2	59.94	57.720	2395	1935.33	18592.7	460.7	13836.5	55.79	57.905
2346		18164.8	420.6	13505.9	59.85	57.724	2396		18601.4	461.5	13843.3	55.72	57.908
2347		18173.5	421.4	13512.7	59.77	57.728	2397		18610.2	462.3	13850.1	55.64	57.912
2348		18182.2	422.2	13519.4	59.68	57.731	2398		18618.9	463.2	13856.8	55.56	57.916
2349		18190.9	423.0	13526.1	59.60	57.735	2399		18627.7	464.1	13863.6	55.48	57.919

Table 5   Products—200% Theoretical Air (for One Pound-Mole)

T	t	$\bar{h}$	p_r	$\bar{u}$	v_r	$\bar{\phi}$	T	t	$\bar{h}$	p_r	$\bar{u}$	v_r	$\bar{\phi}$
2400	1940.33	18636.4	464.9	13870.4	55.40	57.923	2450	1990.33	19074.6	509.2	14209.3	51.63	58.i04
2401		18645.2	465.8	13877.1	55.32	57.927	2451		19083.4	510.2	14216.0	51.56	58.107
2402		18653.9	466.6	13883.9	55.24	57.930	2452		19092.2	511.0	14222.8	51.49	58.111
2403		18662.7	467.5	13890.6	55.16	57.934	2453		19101.0	512.0	14229.7	51.42	58.114
2404		18671.4	468.4	13897.4	55.08	57.938	2454		19109.8	512.9	14236.5	51.35	58.118
2405	1945.33	18680.2	469.2	13904.2	55.01	57.941	2455	1995.33	19118.6	513.8	14243.3	51.27	58.122
2406		18688.9	470.1	13910.9	54.93	57.945	2456		19127.3	514.8	14250.1	51.20	58.125
2407		18697.7	470.9	13917.7	54.85	57.948	2457		19136.1	515.7	14256.8	51.13	58.129
2408		18706.4	471.8	13924.5	54.77	57.952	2458		19144.9	516.6	14263.6	51.06	58.132
2409		18715.2	472.7	13931.2	54.70	57.956	2459		19153.7	517.5	14270.4	50.99	58.136
2410	1950.33	18723.9	473.5	13938.0	54.62	57.959	2460	2000.33	19162.4	518.5	14277.2	50.92	58.140
2411		18732.7	474.4	13944.8	54.54	57.963	2461		19171.2	519.4	14284.0	50.85	58.143
2412		18741.4	475.3	13951.5	54.46	57.967	2462		19180.0	520.4	14290.8	50.77	58.147
2413		18750.2	476.1	13958.3	54.39	57.970	2463		19188.8	521.3	14297.6	50.71	58.150
2414		18758.9	477.0	13965.1	54.31	57.974	2464		19197.6	522.2	14304.5	50.64	58.154
2415	1955.33	18767.7	477.9	13971.9	54.23	57.978	2465	2005.33	19206.4	523.2	14311.3	50.57	58.157
2416		18776.5	478.7	13978.6	54.16	57.981	2466		19215.2	524.1	14318.1	50.49	58.161
2417		18785.2	479.6	13985.4	54.08	57.985	2467		19224.0	525.0	14324.9	50.42	58.165
2418		18794.0	480.5	13992.2	54.00	57.988	2468		19232.8	526.0	14331.7	50.36	58.168
2419		18802.7	481.4	13998.9	53.93	57.992	2469		19241.5	526.9	14338.4	50.29	58.172
2420	1960.33	18811.5	482.2	14005.7	53.85	57.996	2470	2010.33	19250.3	527.9	14345.2	50.22	58.175
2421		18820.3	483.1	14012.5	53.78	57.999	2471		19259.1	528.8	14352.0	50.15	58.179
2422		18829.1	484.0	14019.3	53.70	58.003	2472		19267.9	529.8	14358.8	50.07	58.182
2423		18837.8	484.9	14026.1	53.63	58.007	2473		19276.7	530.7	14365.6	50.00	58.186
2424		18846.6	485.8	14032.9	53.55	58.010	2474		19285.5	531.6	14372.4	49.94	58.189
2425	1965.33	18855.4	486.7	14039.6	53.47	58.014	2475	2015.33	19294.3	532.6	14379.3	49.87	58.193
2426		18864.1	487.6	14046.4	53.40	58.017	2476		19303.1	533.6	14386.1	49.80	58.196
2427		18872.9	488.4	14053.2	53.33	58.021	2477		19311.9	534.5	14392.9	49.73	58.200
2428		18881.6	489.3	14060.0	53.25	58.025	2478		19320.7	535.5	14399.7	49.66	58.204
2429		18890.4	490.2	14066.7	53.17	58.028	2479		19329.5	536.4	14406.5	49.59	58.207
2430	1970.33	18899.2	491.1	14073.5	53.10	58.032	2480	2020.33	19338.3	537.4	14413.3	49.52	58.211
2431		18907.9	492.0	14080.3	53.02	58.035	2481		19347.0	538.4	14420.1	49.46	58.214
2432		18916.7	492.9	14087.1	52.95	58.039	2482		19355.8	539.3	14426.9	49.39	58.218
2433		18925.5	493.8	14093.9	52.87	58.043	2483		19364.6	540.3	14433.7	49.32	58.221
2434		18934.2	494.7	14100.6	52.80	58.046	2484		19373.5	541.2	14440.6	49.25	58.225
2435	1975.33	18943.0	495.6	14107.4	52.73	58.050	2485	2025.33	19382.3	542.2	14447.4	49.18	58.228
2436		18951.8	496.5	14114.2	52.65	58.053	2486		19391.1	543.2	14454.2	49.12	53.232
2437		18960.5	497.4	14121.0	52.58	58.057	2487		19399.9	544.2	14461.0	49.05	58.236
2438		18969.3	498.3	14127.8	52.51	58.061	2488		19408.6	545.1	14467.8	48.98	58.239
2439		18978.1	499.2	14134.6	52.43	58.064	2489		19417.4	546.1	14474.6	48.91	58.243
2440	1980.33	18986.9	500.1	14141.4	52.36	58.068	2490	2030.33	19426.2	547.1	14481.4	48.85	58.246
2441		18995.7	501.0	14148.2	52.29	58.071	2491		19435.0	548.0	14488.2	48.78	58.250
2442		19004.4	501.9	14155.0	52.21	58.075	2492		19443.8	549.0	14495.1	48.71	58.253
2443		19013.2	502.8	14161.7	52.14	58.079	2493		19452.7	550.0	14501.9	48.65	58.257
2444		19022.0	503.7	14168.5	52.07	58.082	2494		19461.5	551.0	14508.7	48.58	58.260
2445	1985.33	19030.8	504.7	14175.3	51.99	58.086	2495	2035.33	19470.3	551.9	14515.5	48.51	58.264
2446		19039.5	505.5	14182.1	51.92	58.089	2496		19479.1	552.9	14522.4	48.44	58.267
2447		19048.3	506.5	14188.9	51.85	58.093	2497		19487.9	553.9	14529.2	48.38	58.271
2448		19057.1	507.4	14195.7	51.78	58.097	2498		19496.7	554.9	14536.0	48.31	58.274
2449		19065.8	508.3	14202.5	51.70	58.100	2499		19505.5	555.9	14542.8	48.24	58.278

Table 5   Products—200% Theoretical Air (for One Pound-Mole)

T	t	$\bar{h}$	p_r	$\bar{u}$	v_r	$\bar{\phi}$	T	t	$\bar{h}$	p_r	$\bar{u}$	v_r	$\bar{\phi}$
2500	2040.33	19514.3	556.9	14549.6	48.18	58.281	2750	2290.33	21732.0	852.4	16270.8	34.62	59.127
2505		19558.3	561.8	14583.7	47.85	58.299	2755		21776.6	859.4	16305.5	34.40	59.143
2510		19602.4	566.8	14617.9	47.52	58.316	2760		21821.3	866.4	16340.3	34.19	59.159
2515		19646.4	571.8	14652.0	47.20	58.334	2765		21866.0	873.5	16375.1	33.97	59.175
2520		19690.5	576.9	14686.2	46.88	58.352	2770		21910.6	880.6	16409.8	33.76	59.191
2525	2065.33	19734.6	582.0	14720.3	46.56	58.369	2775	2315.33	21955.4	887.8	16444.6	33.54	59.208
2530		19778.7	587.2	14754.5	46.24	58.387	2780		22000.1	895.0	16479.4	33.33	59.224
2535		19822.8	592.3	14788.7	45.93	58.404	2785		22044.8	902.3	16514.2	33.12	59.240
2540		19867.0	597.5	14822.9	45.62	58.421	2790		22089.6	909.6	16549.0	32.92	59.256
2545		19911.1	602.8	14857.1	45.31	58.439	2795		22134.3	916.9	16583.9	32.71	59.272
2550	2090.33	19955.3	608.0	14891.4	45.01	58.456	2800	2340.33	22179.1	924.3	16618.7	32.51	59.288
2555		19999.4	613.4	14925.6	44.70	58.473	2805		22223.8	931.8	16653.5	32.30	59.304
2560		20043.6	618.7	14959.8	44.40	58.491	2810		22268.6	939.3	16688.4	32.10	59.320
2565		20087.8	624.1	14994.1	44.10	58.508	2815		22313.5	946.9	16723.3	31.90	59.336
2570		20132.0	629.5	15028.4	43.81	58.525	2820		22358.3	954.5	16758.2	31.71	59.351
2575	2115.33	20176.3	635.0	15062.7	43.52	58.542	2825	2365.33	22403.1	962.2	16793.1	31.51	59.367
2580		20220.5	640.5	15097.0	43.23	58.559	2830		22447.9	969.9	16827.9	31.31	59.383
2585		20264.8	646.1	15131.3	42.94	58.576	2835		22492.7	977.7	16862.8	31.12	59.399
2590		20309.0	651.7	15165.7	42.65	58.594	2840		22537.6	985.4	16897.8	30.93	59.415
2595		20353.3	657.3	15200.0	42.37	58.611	2845		22582.5	993.3	16932.7	30.74	59.431
2600	2140.33	20397.6	663.0	15234.4	42.09	58.628	2850	2390.33	22627.3	1001.2	16967.6	30.55	59.446
2605		20441.9	668.7	15268.7	41.81	58.645	2855		22672.2	1009.2	17002.6	30.36	59.462
2610		20486.2	674.4	15303.1	41.53	58.662	2860		22717.1	1017.2	17037.6	30.17	59.478
2615		20530.5	680.2	15337.5	41.26	58.679	2865		22762.0	1025.3	17072.6	29.99	59.494
2620		20574.9	686.0	15371.9	40.98	58.696	2870		22807.0	1033.4	17107.5	29.81	59.509
2625	2165.33	20619.2	691.9	15406.3	40.71	58.713	2875	2415.33	22851.9	1041.6	17142.5	29.62	59.525
2630		20663.6	697.8	15440.8	40.45	58.729	2880		22896.8	1049.8	17177.6	29.44	59.540
2635		20708.0	703.8	15475.3	40.18	58.746	2885		22941.8	1058.0	17212.6	29.26	59.556
2640		20752.4	709.8	15509.7	39.92	58.763	2890		22986.7	1066.4	17247.6	29.08	59.572
2645		20796.8	715.8	15544.2	39.66	58.780	2895		23031.7	1074.8	17282.6	28.91	59.587
2650	2190.33	20841.2	721.9	15578.6	39.39	58.797	2900	2440.33	23076.7	1083.2	17317.7	28.73	59.603
2655		20885.6	728.0	15613.1	39.14	58.813	2905		23121.7	1091.7	17352.7	28.56	59.618
2660		20930.0	734.1	15647.6	38.88	58.830	2910		23166.7	1100.2	17387.8	28.38	59.634
2665		20974.5	740.3	15682.2	38.63	58.847	2915		23211.7	1108.9	17422.9	28.21	59.649
2670		21018.9	746.6	15716.7	38.38	58.864	2920		23256.7	1117.5	17458.0	28.04	59.665
2675	2215.33	21063.4	752.9	15751.2	38.13	58.880	2925	2465.33	23301.8	1126.2	17493.1	27.87	59.680
2680		21107.9	759.2	15785.8	37.88	58.897	2930		23346.8	1134.9	17528.2	27.70	59.695
2685		21152.4	765.6	15820.4	37.64	58.913	2935		23391.8	1143.7	17563.3	27.54	59.711
2690		21196.9	772.0	15854.9	37.39	58.930	2940		23437.0	1152.6	17598.5	27.37	59.726
2695		21241.5	778.4	15889.6	37.15	58.947	2945		23482.0	1161.5	17633.7	27.21	59.741
2700	2240.33	21286.0	784.9	15924.1	36.91	58.963	2950	2490.33	23527.1	1170.5	17668.8	27.05	59.757
2705		21330.5	791.5	15958.7	36.68	58.980	2955		23572.2	1179.5	17703.9	26.88	59.772
2710		21375.1	798.1	15993.4	36.44	58.996	2960		23617.3	1188.6	17739.2	26.72	59.787
2715		21419.6	804.7	16028.0	36.21	59.012	2965		23662.4	1197.8	17774.3	26.56	59.802
2720		21464.2	811.4	16062.7	35.98	59.029	2970		23707.5	1207.0	17809.5	26.41	59.818
2725	2265.33	21508.8	818.1	16097.3	35.75	59.045	2975	2515.33	23752.7	1216.3	17844.8	26.25	59.833
2730		21553.4	824.9	16132.0	35.52	59.062	2980		23797.8	1225.6	17880.0	26.09	59.848
2735		21598.0	831.7	16166.7	35.29	59.078	2985		23843.0	1235.0	17915.2	25.94	59.863
2740		21642.7	838.5	16201.4	35.07	59.094	2990		23888.2	1244.4	17950.4	25.78	59.878
2745		21687.3	845.4	16236.1	34.84	59.110	2995		23933.3	1253.9	17985.7	25.63	59.893

Table 6   Products—200% Theoretical Air (for One Pound-Mole)

T	t	Fuel—$(CH_1)_n$			Fuel—$(CH_2)_n$			Fuel—$(CH_3)_n$		
		$\bar{c}_p$	$\bar{c}_v$	$k = \dfrac{\bar{c}_p}{\bar{c}_v}$	$\bar{c}_p$	$\bar{c}_v$	$k = \dfrac{\bar{c}_p}{\bar{c}_v}$	$\bar{c}_p$	$\bar{c}_v$	$k = \dfrac{\bar{c}_p}{\bar{c}_v}$
		Btu	Btu		Btu	Btu		Btu	Btu	
R	F	lb-mole R	lb-mole R		lb-mole R	lb-mole R		lb-mole R	lb-mole R	
300	−159.67	7.009	5.023	1.395	7.030	5.044	1.394	7.044	5.058	1.393
350	−109.67	7.034	5.048	1.393	7.050	5.064	1.392	7.062	5.076	1.391
400	−59.67	7.061	5.076	1.391	7.074	5.088	1.390	7.082	5.096	1.390
450	−9.67	7.091	5.105	1.389	7.098	5.113	1.388	7.104	5.118	1.388
500	40.33	7.121	5.136	1.387	7.125	5.139	1.386	7.127	5.141	1.386
550	90.33	7.153	5.167	1.384	7.152	5.167	1.384	7.152	5.166	1.384
600	140.33	7.186	5.200	1.382	7.182	5.196	1.382	7.179	5.193	1.382
650	190.33	7.221	5.235	1.379	7.214	5.228	1.380	7.209	5.223	1.380
700	240.33	7.258	5.273	1.377	7.249	5.263	1.377	7.242	5.256	1.378
750	290.33	7.298	5.312	1.374	7.286	5.301	1.375	7.278	5.292	1.375
800	340.33	7.341	5.355	1.371	7.327	5.341	1.372	7.317	5.331	1.372
850	390.33	7.386	5.400	1.368	7.370	5.384	1.369	7.359	5.373	1.370
900	440.33	7.433	5.447	1.365	7.416	5.430	1.366	7.404	5.418	1.367
950	490.33	7.482	5.497	1.361	7.464	5.478	1.363	7.451	5.465	1.363
1000	540.33	7.534	5.548	1.358	7.514	5.528	1.359	7.500	5.514	1.360
1100	640.33	7.639	5.654	1.351	7.618	5.632	1.353	7.603	5.617	1.354
1200	740.33	7.748	5.762	1.345	7.725	5.739	1.346	7.709	5.723	1.347
1300	840.33	7.856	5.870	1.338	7.832	5.847	1.340	7.816	5.830	1.341
1400	940.33	7.963	5.977	1.332	7.938	5.952	1.334	7.921	5.936	1.335
1500	1040.33	8.065	6.079	1.327	8.041	6.055	1.328	8.024	6.038	1.329
1600	1140.33	8.163	6.178	1.321	8.140	6.154	1.323	8.123	6.137	1.324
1700	1240.33	8.256	6.271	1.317	8.234	6.248	1.318	8.218	6.232	1.319
1800	1340.33	8.344	6.358	1.312	8.322	6.336	1.313	8.307	6.321	1.314
1900	1440.33	8.426	6.440	1.308	8.406	6.420	1.309	8.392	6.406	1.310
2000	1540.33	8.503	6.517	1.305	8.484	6.498	1.306	8.471	6.485	1.306
2100	1640.33	8.575	6.589	1.301	8.558	6.572	1.302	8.546	6.560	1.303
2200	1740.33	8.641	6.656	1.298	8.626	6.641	1.299	8.616	6.630	1.300
2300	1840.33	8.704	6.718	1.296	8.691	6.705	1.296	8.682	6.696	1.297
2400	1940.33	8.762	6.776	1.293	8.751	6.765	1.294	8.743	6.757	1.294
2500	2040.33	8.816	6.830	1.291	8.807	6.821	1.291	8.800	6.815	1.291
2600	2140.33	8.866	6.880	1.289	8.859	6.873	1.289	8.854	6.868	1.289
2700	2240.33	8.913	6.927	1.287	8.908	6.922	1.287	8.905	6.919	1.287
2800	2340.33	8.957	6.971	1.285	8.954	6.968	1.285	8.952	6.966	1.285
2900	2440.33	8.998	7.012	1.283	8.997	7.011	1.283	8.996	7.011	1.283
3000	2540.33	9.037	7.051	1.282	9.038	7.052	1.282	9.038	7.052	1.282
3200	2740.33	9.107	7.121	1.279	9.112	7.126	1.279	9.115	7.129	1.279
3400	2940.33	9.170	7.184	1.276	9.177	7.192	1.276	9.183	7.197	1.276
3600	3140.33	9.225	7.239	1.274	9.236	7.250	1.274	9.244	7.258	1.274
3800	3340.33	9.275	7.289	1.272	9.289	7.303	1.272	9.299	7.313	1.272
4000	3540.33	9.320	7.334	1.271	9.337	7.351	1.270	9.348	7.362	1.270

Table 7   Products—100% Theoretical Air (for One Pound-Mole)

T	t	$\bar{h}$	p_r	$\bar{u}$	v_r	$\bar{\phi}$	T	t	$\bar{h}$	p_r	$\bar{u}$	v_r	$\bar{\phi}$
300	−159.67	2117.2	.15822	1521.5	20348	42.065	350	−109.67	2474.2	.2754	1779.2	13639	43.165
301		2124.4	.16012	1526.6	20174	42.088	351		2481.4	.2782	1784.3	13538	43.186
302		2131.5	.16203	1531.8	20001	42.112	352		2488.5	.2811	1789.5	13438	43.206
303		2138.6	.16397	1536.9	19831	42.135	353		2495.7	.2840	1794.7	13340	43.226
304		2145.7	.16592	1542.0	19663	42.159	354		2502.9	.2869	1799.9	13241	43.247
305	−154.67	2152.9	.16788	1547.2	19496	42.182	355	−104.67	2510.0	.2898	1805.1	13145	43.267
306		2160.0	.16987	1552.3	19332	42.206	356		2517.2	.2928	1810.2	13048	43.287
307		2167.1	.17187	1557.5	19169	42.229	357		2524.4	.2958	1815.4	12953	43.307
308		2174.2	.17388	1562.6	19009	42.252	358		2531.5	.2988	1820.6	12859	43.327
309		2181.4	.17592	1567.7	18850	42.275	359		2538.7	.3018	1825.8	12766	43.347
310	−149.67	2188.5	.17797	1572.9	18693	42.298	360	−99.67	2545.9	.3048	1830.9	12674	43.367
311		2195.6	.18004	1578.0	18538	42.321	361		2553.0	.3079	1836.1	12582	43.387
312		2202.8	.18212	1583.2	18384	42.344	362		2560.2	.3110	1841.3	12492	43.407
313		2209.9	.18423	1588.3	18232	42.367	363		2567.4	.3141	1846.5	12402	43.426
314		2217.0	.18635	1593.4	18082	42.390	364		2574.5	.3172	1851.7	12313	43.446
315	−144.67	2224.1	.18849	1598.6	17934	42.412	365	−94.67	2581.7	.3204	1856.9	12225	43.466
316		2231.3	.19065	1603.7	17787	42.435	366		2588.9	.3236	1862.1	12138	43.485
317		2238.4	.19283	1608.9	17642	42.457	367		2596.1	.3268	1867.3	12052	43.505
318		2245.5	.19502	1614.0	17499	42.480	368		2603.2	.3300	1872.4	11967	43.525
319		2252.7	.19723	1619.2	17357	42.502	369		2610.4	.3333	1877.6	11882	43.544
320	−139.67	2259.8	.19946	1624.3	17217	42.525	370	−89.67	2617.6	.3365	1882.8	11799	43.563
321		2267.0	.20171	1629.5	17078	42.547	371		2624.8	.3398	1888.0	11716	43.583
322		2274.1	.20398	1634.6	16941	42.569	372		2631.9	.3432	1893.2	11633	43.602
323		2281.2	.20626	1639.8	16805	42.591	373		2639.1	.3465	1898.4	11552	43.621
324		2288.4	.20857	1644.9	16671	42.613	374		2646.3	.3499	1903.6	11471	43.641
325	−134.67	2295.5	.21089	1650.1	16538	42.635	375	−84.67	2653.5	.3533	1908.8	11392	43.660
326		2302.6	.21324	1655.3	16407	42.657	376		2660.7	.3567	1914.0	11312	43.679
327		2309.8	.21560	1660.4	16277	42.679	377		2667.8	.3601	1919.2	11234	43.698
328		2316.9	.21798	1665.6	16148	42.701	378		2675.0	.3636	1924.4	11156	43.717
329		2324.1	.22038	1670.7	16021	42.723	379		2682.2	.3671	1929.6	11080	43.736
330	−129.67	2331.2	.22279	1675.9	15895	42.744	380	−79.67	2689.4	.3706	1934.8	11003	43.755
331		2338.3	.22523	1681.0	15771	42.766	381		2696.6	.3742	1940.0	10928	43.774
332		2345.5	.22769	1686.2	15648	42.788	382		2703.8	.3777	1945.2	10853	43.793
333		2352.6	.23017	1691.3	15526	42.809	383		2710.9	.3813	1950.4	10779	43.811
334		2359.8	.23267	1696.5	15405	42.830	384		2718.1	.3849	1955.6	10706	43.830
335	−124.67	2366.9	.23518	1701.7	15286	42.852	385	−74.67	2725.3	.3886	1960.8	10633	43.849
336		2374.1	.23772	1706.8	15168	42.873	386		2732.5	.3922	1966.0	10561	43.868
337		2381.2	.24027	1712.0	15052	42.894	387		2739.7	.3959	1971.2	10490	43.886
338		2388.4	.24285	1717.2	14936	42.916	388		2746.9	.3996	1976.4	10419	43.905
339		2395.5	.24545	1722.3	14822	42.937	389		2754.1	.4034	1981.6	10349	43.923
340	−119.67	2402.7	.24807	1727.5	14709	42.958	390	−69.67	2761.3	.4072	1986.8	10279	43.942
341		2409.8	.25071	1732.7	14596	42.979	391		2768.5	.4110	1992.0	10211	43.960
342		2417.0	.25336	1737.8	14486	43.000	392		2775.7	.4148	1997.2	10142	43.979
343		2424.1	.25604	1743.0	14376	43.021	393		2782.9	.4186	2002.4	10075	43.997
344		2431.3	.25874	1748.1	14268	43.041	394		2790.1	.4225	2007.6	10008	44.015
345	−114.67	2438.4	.26146	1753.3	14160	43.062	395	−64.67	2797.3	.4264	2012.9	9941	44.033
346		2445.6	.26420	1758.5	14054	43.083	396		2804.5	.4303	2018.1	9876	44.052
347		2452.8	.26697	1763.7	13949	43.104	397		2811.7	.4343	2023.3	9810	44.070
348		2459.9	.26975	1768.8	13845	43.124	398		2818.9	.4383	2028.5	9746	44.088
349		2467.1	.27255	1774.0	13742	43.145	399		2826.1	.4423	2033.7	9682	44.106

Table 7 Products—100% Theoretical Air (for One Pound-Mole)

T	t	$\bar{h}$	p_r	$\bar{u}$	v_r	$\bar{\phi}$	T	t	$\bar{h}$	p_r	$\bar{u}$	v_r	$\bar{\phi}$
400	−59.67	2833.3	.4463	2038.9	9618	44.124	450	−9.67	3194.6	.6851	2300.9	7049	44.975
401		2840.5	.4503	2044.2	9556	44.142	451		3201.8	.6906	2306.2	7008	44.991
402		2847.7	.4544	2049.4	9493	44.160	452		3209.1	.6962	2311.5	6967	45.007
403		2854.9	.4586	2054.6	9431	44.178	453		3216.3	.7019	2316.7	6926	45.023
404		2862.1	.4627	2059.8	9370	44.196	454		3223.6	.7076	2322.0	6886	45.039
405	−54.67	2869.3	.4669	2065.0	9309	44.213	455	−4.67	3230.8	.7133	2327.3	6846	45.055
406		2876.5	.4711	2070.3	9249	44.231	456		3238.1	.7190	2332.5	6806	45.071
407		2883.7	.4753	2075.5	9190	44.249	457		3245.3	.7248	2337.8	6767	45.087
408		2890.9	.4795	2080.7	9130	44.267	458		3252.6	.7306	2343.1	6728	45.103
409		2898.1	.4838	2085.9	9072	44.284	459		3259.8	.7364	2348.3	6689	45.119
410	−49.67	2905.3	.4881	2091.1	9014	44.302	460	.33	3267.1	.7423	2353.6	6650	45.134
411		2912.6	.4925	2096.4	8956	44.319	461		3274.3	.7482	2358.9	6612	45.150
412		2919.8	.4968	2101.6	8899	44.337	462		3281.6	.7542	2364.1	6574	45.166
413		2927.0	.5012	2106.8	8842	44.355	463		3288.9	.7602	2369.4	6536	45.182
414		2934.2	.5057	2112.1	8786	44.372	464		3296.1	.7662	2374.7	6499	45.197
415	−44.67	2941.4	.5101	2117.3	8731	44.389	465	5.33	3303.4	.7722	2380.0	6462	45.213
416		2948.6	.5146	2122.5	8675	44.407	466		3310.6	.7783	2385.2	6425	45.228
417		2955.9	.5191	2127.8	8621	44.424	467		3317.9	.7844	2390.5	6389	45.244
418		2963.1	.5236	2133.0	8567	44.441	468		3325.2	.7906	2395.8	6353	45.260
419		2970.3	.5282	2138.2	8513	44.459	469		3332.4	.7968	2401.1	6317	45.275
420	−39.67	2977.5	.5328	2143.5	8459	44.476	470	10.33	3339.7	.8030	2406.3	6281	45.290
421		2984.7	.5374	2148.7	8407	44.493	471		3347.0	.8093	2411.6	6245	45.306
422		2992.0	.5421	2153.9	8354	44.510	472		3354.2	.8156	2416.9	6210	45.321
423		2999.2	.5468	2159.2	8302	44.527	473		3361.5	.8219	2422.2	6176	45.337
424		3006.4	.5515	2164.4	8251	44.544	474		3368.8	.8283	2427.5	6141	45.352
425	−34.67	3013.6	.5562	2169.6	8200	44.561	475	15.33	3376.0	.8348	2432.8	6107	45.367
426		3020.9	.5610	2174.9	8149	44.578	476		3383.3	.8412	2438.0	6072	45.383
427		3028.1	.5658	2180.1	8099	44.595	477		3390.6	.8477	2443.3	6039	45.398
428		3035.3	.5707	2185.4	8049	44.612	478		3397.9	.8542	2448.6	6005	45.413
429		3042.5	.5755	2190.6	7999	44.629	479		3405.1	.8608	2453.9	5972	45.428
430	−29.67	3049.8	.5804	2195.9	7950	44.646	480	20.33	3412.4	.8674	2459.2	5939	45.444
431		3057.0	.5854	2201.1	7902	44.663	481		3419.7	.8740	2464.5	5906	45.459
432		3064.2	.5903	2206.3	7853	44.679	482		3427.0	.8807	2469.8	5873	45.474
433		3071.5	.5953	2211.6	7805	44.696	483		3434.3	.8874	2475.1	5841	45.489
434		3078.7	.6003	2216.8	7758	44.713	484		3441.5	.8942	2480.4	5809	45.504
435	−24.67	3085.9	.6054	2222.1	7711	44.729	485	25.33	3448.8	.9010	2485.7	5777	45.519
436		3093.2	.6105	2227.3	7664	44.746	486		3456.1	.9078	2491.0	5745	45.534
437		3100.4	.6156	2232.6	7618	44.763	487		3463.4	.9147	2496.3	5714	45.549
438		3107.6	.6208	2237.8	7572	44.779	488		3470.6	.9216	2501.5	5683	45.564
439		3114.9	.6259	2243.1	7527	44.796	489		3477.9	.9285	2506.8	5652	45.579
440	−19.67	3122.1	.6311	2248.3	7481	44.812	490	30.33	3485.2	.9355	2512.2	5621	45.594
441		3129.4	.6364	2253.6	7437	44.829	491		3492.5	.9425	2517.5	5590	45.609
442		3136.6	.6417	2258.9	7392	44.845	492		3499.8	.9496	2522.8	5560	45.623
443		3143.9	.6470	2264.1	7348	44.861	493		3507.1	.9567	2528.1	5530	45.638
444		3151.1	.6523	2269.4	7304	44.878	494		3514.4	.9638	2533.4	5500	45.653
445	−14.67	3158.3	.6577	2274.6	7261	44.894	495	35.33	3521.7	.9710	2538.7	5471	45.668
446		3165.6	.6631	2279.9	7218	44.910	496		3528.9	.9783	2544.0	5441	45.682
447		3172.8	.6685	2285.2	7175	44.926	497		3536.2	.9855	2549.3	5412	45.697
448		3180.1	.6740	2290.4	7133	44.943	498		3543.5	.9928	2554.6	5383	45.712
449		3187.3	.6795	2295.7	7091	44.959	499		3550.8	1.0002	2559.9	5354	45.726

Table 7   Products—100% Theoretical Air (for One Pound-Mole)

T	t	$\bar{h}$	p_r	$\bar{u}$	v_r	$\bar{\phi}$	T	t	$\bar{h}$	p_r	$\bar{u}$	v_r	$\bar{\phi}$
500	40.33	3558.1	1.0075	2565.2	5326	45.741	550	90.33	3924.0	1.4314	2831.7	4124	46.438
501		3565.4	1.0150	2570.5	5297	45.756	551		3931.3	1.4410	2837.1	4103	46.452
502		3572.7	1.0224	2575.8	5269	45.770	552		3938.6	1.4507	2842.5	4083	46.465
503		3580.0	1.0299	2581.1	5241	45.785	553		3946.0	1.4605	2847.8	4063	46.478
504		3587.3	1.0375	2586.4	5213	45.799	554		3953.3	1.4702	2853.2	4044	46.492
505	45.33	3594.6	1.0451	2591.7	5186	45.814	555	95.33	3960.7	1.4801	2858.5	4024	46.505
506		3601.9	1.0527	2597.1	5158	45.828	556		3968.0	1.4900	2863.9	4005	46.518
507		3609.2	1.0603	2602.4	5131	45.842	557		3975.4	1.4999	2869.2	3985	46.531
508		3616.5	1.0680	2607.7	5104	45.857	558		3982.7	1.5099	2874.6	3966	46.544
509		3623.8	1.0758	2613.0	5077	45.871	559		3990.1	1.5199	2880.0	3947	46.558
510	50.33	3631.1	1.0836	2618.3	5051	45.886	560	100.33	3997.4	1.5300	2885.3	3928	46.571
511		3638.4	1.0914	2623.6	5024	45.900	561		4004.8	1.5402	2890.7	3909	46.584
512		3645.7	1.0993	2629.0	4998	45.914	562		4012.1	1.5503	2896.1	3890	46.597
513		3653.0	1.1072	2634.3	4972	45.928	563		4019.5	1.5606	2901.4	3872	46.610
514		3660.3	1.1152	2639.6	4946	45.943	564		4026.8	1.5709	2906.8	3853	46.623
515	55.33	3667.6	1.1232	2644.9	4921	45.957	565	105.33	4034.2	1.5812	2912.2	3835	46.636
516		3674.9	1.1312	2650.2	4895	45.971	566		4041.5	1.5916	2917.5	3816	46.649
517		3682.2	1.1393	2655.6	4870	45.985	567		4048.9	1.6020	2922.9	3798	46.662
518		3689.6	1.1475	2660.9	4845	45.999	568		4056.2	1.6125	2928.3	3780	46.675
519		3696.9	1.1556	2666.2	4820	46.013	569		4063.6	1.6231	2933.6	3762	46.688
520	60.33	3704.2	1.1638	2671.5	4795	46.027	570	110.33	4070.9	1.6337	2939.0	3744	46.701
521		3711.5	1.1721	2676.9	4770	46.041	571		4078.3	1.6443	2944.4	3727	46.714
522		3718.8	1.1804	2682.2	4746	46.056	572		4085.7	1.6550	2949.7	3709	46.727
523		3726.1	1.1888	2687.5	4721	46.070	573		4093.0	1.6658	2955.1	3692	46.739
524		3733.4	1.1972	2692.8	4697	46.083	574		4100.4	1.6766	2960.5	3674	46.752
525	65.33	3740.7	1.2056	2698.2	4673	46.097	575	115.33	4107.7	1.6874	2965.9	3657	46.765
526		3748.1	1.2141	2703.5	4649	46.111	576		4115.1	1.6983	2971.3	3640	46.778
527		3755.4	1.2226	2708.8	4626	46.125	577		4122.5	1.7093	2976.6	3623	46.791
528		3762.7	1.2312	2714.2	4602	46.139	578		4129.8	1.7203	2982.0	3606	46.803
529		3770.0	1.2398	2719.5	4579	46.153	579		4137.2	1.7314	2987.4	3589	46.816
530	70.33	3777.3	1.2484	2724.8	4556	46.167	580	120.33	4144.6	1.7425	2992.8	3572	46.829
531		3784.7	1.2572	2730.2	4533	46.181	581		4151.9	1.7536	2998.2	3555	46.842
532		3792.0	1.2659	2735.5	4510	46.194	582		4159.3	1.7649	3003.5	3539	46.854
533		3799.3	1.2747	2740.9	4487	46.208	583		4166.7	1.7762	3008.9	3522	46.867
534		3806.6	1.2836	2746.2	4465	46.222	584		4174.1	1.7875	3014.3	3506	46.880
535	75.33	3814.0	1.2924	2751.5	4442	46.236	585	125.33	4181.4	1.7989	3019.7	3490	46.892
536		3821.3	1.3014	2756.9	4420	46.249	586		4188.8	1.8103	3025.1	3474	46.905
537		3828.6	1.3104	2762.2	4398	46.263	587		4196.2	1.8218	3030.5	3458	46.917
538		3836.0	1.3194	2767.6	4376	46.277	588		4203.5	1.8334	3035.9	3442	46.930
539		3843.3	1.3285	2772.9	4354	46.290	589		4210.9	1.8450	3041.3	3426	46.942
540	80.33	3850.6	1.3376	2778.2	4332	46.304	590	130.33	4218.3	1.8566	3046.6	3410	46.955
541		3857.9	1.3467	2783.6	4311	46.317	591		4225.7	1.8683	3052.0	3395	46.967
542		3865.3	1.3560	2788.9	4290	46.331	592		4233.1	1.8801	3057.4	3379	46.980
543		3872.6	1.3652	2794.3	4268	46.344	593		4240.4	1.8920	3062.8	3364	46.992
544		3879.9	1.3745	2799.6	4247	46.358	594		4247.8	1.9038	3068.2	3348	47.005
545	85.33	3887.3	1.3839	2805.0	4226	46.371	595	135.33	4255.2	1.9158	3073.6	3333	47.017
546		3894.6	1.3933	2810.3	4205	46.385	596		4262.6	1.9278	3079.0	3318	47.030
547		3901.9	1.4027	2815.7	4185	46.398	597		4270.0	1.9398	3084.4	3303	47.042
548		3909.3	1.4122	2821.0	4164	46.412	598		4277.4	1.9519	3089.8	3288	47.054
549		3916.6	1.4218	2826.4	4144	46.425	599		4284.7	1.9641	3095.2	3273	47.067

Table 7 Products—100% Theoretical Air (for One Pound-Mole)

T	t	$\bar{h}$	p_r	$\bar{u}$	v_r	$\bar{\phi}$	T	t	$\bar{h}$	p_r	$\bar{u}$	v_r	$\bar{\phi}$
600	140.33	4292.1	1.976	3100.6	3258	47.079	650	190.33	4662.6	2.664	3371.8	2618	47.672
601		4299.5	1.989	3106.0	3243	47.091	651		4670.0	2.679	3377.2	2607	47.683
602		4306.9	2.001	3111.4	3229	47.104	652		4677.5	2.695	3382.7	2596	47.695
603		4314.3	2.013	3116.8	3214	47.116	653		4684.9	2.710	3388.1	2585	47.706
604		4321.7	2.026	3122.2	3200	47.128	654		4692.3	2.726	3393.6	2575	47.718
605	145.33	4329.0	2.038	3127.6	3185	47.140	655	195.33	4699.8	2.742	3399.0	2564	47.729
606		4336.4	2.051	3133.0	3171	47.152	656		4707.2	2.757	3404.5	2553	47.740
607		4343.8	2.063	3138.4	3157	47.165	657		4714.6	2.773	3409.9	2542	47.752
608		4351.2	2.076	3143.8	3143	47.177	658		4722.1	2.789	3415.4	2532	47.763
609		4358.6	2.089	3149.2	3129	47.189	659		4729.5	2.805	3420.8	2521	47.774
610	150.33	4366.0	2.102	3154.6	3115	47.201	660	200.33	4737.0	2.821	3426.3	2511	47.786
611		4373.4	2.115	3160.1	3101	47.213	661		4744.4	2.837	3431.7	2500	47.797
612		4380.8	2.127	3165.5	3087	47.225	662		4751.8	2.853	3437.2	2490	47.808
613		4388.2	2.140	3170.9	3073	47.237	663		4759.3	2.869	3442.7	2480	47.819
614		4395.6	2.153	3176.3	3060	47.249	664		4766.7	2.886	3448.1	2469	47.831
615	155.33	4403.0	2.167	3181.7	3046	47.261	665	205.33	4774.2	2.902	3453.6	2459	47.842
616		4410.4	2.180	3187.1	3033	47.274	666		4781.6	2.918	3459.0	2449	47.853
617		4417.8	2.193	3192.5	3019	47.286	667		4789.1	2.935	3464.5	2439	47.864
618		4425.2	2.206	3197.9	3006	47.298	668		4796.5	2.951	3470.0	2429	47.875
619		4432.6	2.220	3203.4	2993	47.309	669		4804.0	2.968	3475.4	2419	47.886
620	160.33	4440.0	2.233	3208.8	2980	47.321	670	210.33	4811.4	2.985	3480.9	2409	47.898
621		4447.4	2.246	3214.2	2967	47.333	671		4818.9	3.001	3486.4	2399	47.909
622		4454.8	2.260	3219.6	2954	47.345	672		4826.3	3.018	3491.8	2389	47.920
623		4462.2	2.274	3225.0	2941	47.357	673		4833.8	3.035	3497.3	2380	47.931
624		4469.6	2.287	3230.5	2928	47.369	674		4841.2	3.052	3502.8	2370	47.942
625	165.33	4477.1	2.301	3235.9	2915	47.381	675	215.33	4848.7	3.069	3508.2	2360	47.953
626		4484.5	2.315	3241.3	2902	47.393	676		4856.2	3.086	3513.7	2351	47.964
627		4491.9	2.328	3246.7	2890	47.405	677		4863.6	3.103	3519.2	2341	47.975
628		4499.3	2.342	3252.2	2877	47.416	678		4871.1	3.121	3524.7	2332	47.986
629		4506.7	2.356	3257.6	2865	47.428	679		4878.5	3.138	3530.1	2322	47.997
630	170.33	4514.1	2.370	3263.0	2852	47.440	680	220.33	4886.0	3.155	3535.6	2313	48.008
631		4521.5	2.384	3268.4	2840	47.452	681		4893.5	3.173	3541.1	2303	48.019
632		4528.9	2.399	3273.9	2828	47.463	682		4900.9	3.190	3546.6	2294	48.030
633		4536.4	2.413	3279.3	2815	47.475	683		4908.4	3.208	3552.1	2285	48.041
634		4543.8	2.427	3284.7	2803	47.487	684		4915.9	3.226	3557.5	2276	48.052
635	175.33	4551.2	2.441	3290.2	2791	47.499	685	225.33	4923.3	3.243	3563.0	2266	48.063
636		4558.6	2.456	3295.6	2779	47.510	686		4930.8	3.261	3568.5	2257	48.074
637		4566.0	2.470	3301.0	2767	47.522	687		4938.3	3.279	3574.0	2248	48.085
638		4573.5	2.485	3306.5	2756	47.534	688		4945.7	3.297	3579.4	2239	48.095
639		4580.9	2.499	3311.9	2744	47.545	689		4953.2	3.315	3584.9	2230	48.106
640	180.33	4588.3	2.514	3317.4	2732	47.557	690	230.33	4960.7	3.333	3590.4	2221	48.117
641		4595.7	2.529	3322.8	2720	47.568	691		4968.1	3.352	3595.9	2213	48.128
642		4603.2	2.543	3328.2	2709	47.580	692		4975.6	3.370	3601.4	2204	48.139
643		4610.6	2.558	3333.7	2697	47.592	693		4983.1	3.388	3606.9	2195	48.149
644		4618.0	2.573	3339.1	2686	47.603	694		4990.6	3.407	3612.4	2186	48.160
645	185.33	4625.4	2.588	3344.5	2674	47.615	695	235.33	4998.0	3.425	3617.9	2178	48.171
646		4632.8	2.603	3350.0	2663	47.626	696		5005.5	3.444	3623.4	2169	48.182
647		4640.3	2.618	3355.4	2652	47.638	697		5013.0	3.462	3628.9	2160	48.192
648		4647.7	2.634	3360.9	2641	47.649	698		5020.5	3.481	3634.4	2152	48.203
649		4655.1	2.649	3366.3	2629	47.661	699		5028.0	3.500	3639.9	2143	48.214

Table 7 Products—100% Theoretical Air (for One Pound-Mole)

T	t	$\bar{h}$	p_r	$\bar{u}$	v_r	$\bar{\phi}$	T	t	$\bar{h}$	p_r	$\bar{u}$	v_r	$\bar{\phi}$
700	240.33	5035.4	3.519	3645.3	2134.7	48.225	750	290.33	5410.8	4.567	3921.4	1762.2	48.742
701		5042.9	3.538	3650.8	2126.3	48.235	751		5418.3	4.590	3926.9	1755.7	48.753
702		5050.4	3.557	3656.3	2118.0	48.246	752		5425.8	4.614	3932.5	1749.2	48.763
703		5057.9	3.576	3661.8	2109.6	48.257	753		5433.4	4.637	3938.0	1742.7	48.773
704		5065.4	3.595	3667.3	2101.4	48.267	754		5440.9	4.660	3943.5	1736.3	48.783
705	245.33	5072.9	3.615	3672.8	2093.1	48.278	755	295.33	5448.4	4.684	3949.1	1729.8	48.793
706		5080.4	3.634	3678.3	2084.9	48.289	756		5456.0	4.707	3954.7	1723.4	48.803
707		5087.8	3.653	3683.8	2076.7	48.299	757		5463.5	4.731	3960.2	1717.1	48.812
708		5095.3	3.673	3689.3	2068.6	48.310	758		5471.0	4.755	3965.8	1710.7	48.822
709		5102.8	3.693	3694.9	2060.6	48.320	759		5478.6	4.779	3971.3	1704.5	48.832
710	250.33	5110.3	3.712	3700.3	2052.5	48.331	760	300.33	5486.1	4.803	3976.9	1698.2	48.842
711		5117.8	3.732	3705.9	2044.5	48.341	761		5493.7	4.827	3982.4	1692.0	48.852
712		5125.3	3.752	3711.4	2036.6	48.352	762		5501.2	4.851	3988.0	1685.7	48.862
713		5132.8	3.772	3716.9	2028.7	48.362	763		5508.8	4.875	3993.5	1679.6	48.872
714		5140.3	3.792	3722.4	2020.8	48.373	764		5516.3	4.899	3999.1	1673.4	48.882
715	255.33	5147.8	3.812	3727.9	2013.0	48.383	765	305.33	5523.9	4.924	4004.7	1667.3	48.892
716		5155.3	3.832	3733.4	2005.2	48.394	766		5531.4	4.948	4010.2	1661.2	48.902
717		5162.8	3.852	3738.9	1997.4	48.404	767		5539.0	4.973	4015.8	1655.1	48.911
718		5170.3	3.873	3744.4	1989.7	48.415	768		5546.5	4.998	4021.4	1649.1	48.921
719		5177.8	3.893	3749.9	1982.0	48.425	769		5554.1	5.023	4026.9	1643.1	48.931
720	260.33	5185.3	3.913	3755.4	1974.4	48.436	770	310.33	5561.6	5.047	4032.5	1637.1	48.941
721		5192.8	3.934	3761.0	1966.8	48.446	771		5569.2	5.072	4038.1	1631.2	48.951
722		5200.3	3.955	3766.5	1959.2	48.456	772		5576.7	5.097	4043.6	1625.3	48.961
723		5207.8	3.975	3772.0	1951.7	48.467	773		5584.3	5.123	4049.2	1619.4	48.970
724		5215.3	3.996	3777.5	1944.2	48.477	774		5591.8	5.148	4054.8	1613.5	48.980
725	265.33	5222.8	4.017	3783.0	1936.8	48.488	775	315.33	5599.4	5.173	4060.4	1607.7	48.990
726		5230.3	4.038	3788.6	1929.4	48.498	776		5606.9	5.199	4065.9	1601.9	49.000
727		5237.8	4.059	3794.1	1922.0	48.508	777		5614.5	5.224	4071.5	1596.1	49.009
728		5245.3	4.080	3799.6	1914.7	48.519	778		5622.1	5.250	4077.1	1590.3	49.019
729		5252.8	4.102	3805.1	1907.4	48.529	779		5629.6	5.276	4082.6	1584.6	49.029
730	270.33	5260.3	4.123	3810.7	1900.1	48.539	780	320.33	5637.2	5.301	4088.2	1578.9	49.038
731		5267.8	4.144	3816.2	1892.9	48.550	781		5644.7	5.327	4093.8	1573.2	49.048
732		5275.4	4.166	3821.7	1885.7	48.560	782		5652.3	5.353	4099.4	1567.6	49.058
733		5282.9	4.187	3827.2	1878.5	48.570	783		5659.9	5.380	4104.9	1562.0	49.068
734		5290.4	4.209	3832.8	1871.4	48.580	784		5667.4	5.406	4110.5	1556.4	49.077
735	275.33	5297.9	4.231	3838.3	1864.3	48.591	785	325.33	5675.0	5.432	4116.1	1550.8	49.087
736		5305.4	4.253	3843.8	1857.3	48.601	786		5682.6	5.458	4121.7	1545.3	49.096
737		5312.9	4.275	3849.4	1850.3	48.611	787		5690.1	5.485	4127.3	1539.8	49.106
738		5320.5	4.297	3854.9	1843.3	48.621	788		5697.7	5.512	4132.8	1534.3	49.116
739		5328.0	4.319	3860.4	1836.4	48.631	789		5705.3	5.538	4138.4	1528.8	49.125
740	280.33	5335.5	4.341	3866.0	1829.5	48.641	790	330.33	5712.9	5.565	4144.0	1523.4	49.135
741		5343.0	4.363	3871.5	1822.6	48.652	791		5720.4	5.592	4149.6	1518.0	49.144
742		5350.5	4.385	3877.0	1815.8	48.662	792		5728.0	5.619	4155.2	1512.6	49.154
743		5358.1	4.408	3882.6	1808.9	48.672	793		5735.6	5.646	4160.8	1507.2	49.164
744		5365.6	4.430	3888.1	1802.2	48.682	794		5743.1	5.673	4166.4	1501.9	49.173
745	285.33	5373.1	4.453	3893.6	1795.4	48.692	795	335.33	5750.7	5.701	4172.0	1496.6	49.183
746		5380.6	4.476	3899.2	1788.7	48.702	796		5758.3	5.728	4177.6	1491.3	49.192
747		5388.2	4.498	3904.7	1782.0	48.712	797		5765.9	5.756	4183.1	1486.0	49.202
748		5395.7	4.521	3910.3	1775.4	48.722	798		5773.5	5.783	4188.7	1480.8	49.211
749		5403.2	4.544	3915.8	1768.8	48.732	799		5781.0	5.811	4194.3	1475.6	49.221

Table 7 Products—100% Theoretical Air (for One Pound-Mole)

T	t	$\bar{h}$	p_r	$\bar{u}$	v_r	$\bar{\phi}$	T	t	$\bar{h}$	p_r	$\bar{u}$	v_r	$\bar{\phi}$
800	340.33	5788.6	5.839	4199.9	1470.4	49.230	850	390.33	6169.1	7.366	4481.1	1238.4	49.692
801		5796.2	5.867	4205.5	1465.2	49.240	851		6176.8	7.399	4486.8	1234.3	49.701
802		5803.8	5.895	4211.1	1460.1	49.249	852		6184.4	7.433	4492.5	1230.2	49.709
803		5811.4	5.923	4216.7	1455.0	49.259	853		6192.0	7.466	4498.1	1226.1	49.718
804		5819.0	5.951	4222.3	1449.9	49.268	854		6199.7	7.500	4503.8	1222.0	49.727
805	345.33	5826.6	5.979	4227.9	1444.8	49.277	855	395.33	6207.3	7.534	4509.4	1217.9	49.736
806		5834.1	6.008	4233.5	1439.7	49.287	856		6215.0	7.568	4515.1	1213.9	49.745
807		5841.7	6.036	4239.1	1434.7	49.296	857		6222.6	7.602	4520.7	1209.8	49.754
808		5849.3	6.065	4244.7	1429.7	49.306	858		6230.3	7.636	4526.4	1205.8	49.763
809		5856.9	6.094	4250.4	1424.7	49.315	859		6237.9	7.670	4532.1	1201.8	49.772
810	350.33	5864.5	6.123	4256.0	1419.7	49.324	860	400.33	6245.6	7.705	4537.7	1197.8	49.781
811		5872.1	6.152	4261.6	1414.8	49.334	861		6253.2	7.739	4543.4	1193.9	49.790
812		5879.7	6.181	4267.2	1409.9	49.343	862		6260.8	7.774	4549.0	1189.9	49.799
813		5887.3	6.210	4272.8	1405.0	49.353	863		6268.5	7.809	4554.7	1186.0	49.808
814		5894.9	6.239	4278.4	1400.1	49.362	864		6276.2	7.844	4560.4	1182.1	49.816
815	355.33	5902.5	6.268	4284.0	1395.3	49.371	865	405.33	6283.8	7.879	4566.0	1178.2	49.825
816		5910.1	6.298	4289.6	1390.4	49.381	866		6291.4	7.914	4571.7	1174.3	49.834
817		5917.7	6.328	4295.2	1385.6	49.390	867		6299.1	7.949	4577.4	1170.4	49.843
818		5925.3	6.357	4300.9	1380.8	49.399	868		6306.8	7.985	4583.0	1166.6	49.852
819		5932.9	6.387	4306.5	1376.1	49.408	869		6314.4	8.020	4588.7	1162.8	49.861
820	360.33	5940.5	6.417	4312.1	1371.3	49.418	870	410.33	6322.1	8.056	4594.4	1159.0	49.869
821		5948.1	6.447	4317.7	1366.6	49.427	871		6329.7	8.092	4600.1	1155.2	49.878
822		5955.7	6.477	4323.3	1361.9	49.436	872		6337.4	8.128	4605.7	1151.4	49.887
823		5963.3	6.507	4329.0	1357.2	49.445	873		6345.1	8.164	4611.4	1147.6	49.896
824		5970.9	6.538	4334.6	1352.6	49.455	874		6352.7	8.200	4617.1	1143.9	49.905
825	365.33	5978.5	6.568	4340.2	1347.9	49.464	875	415.33	6360.4	8.236	4622.8	1140.1	49.913
826		5986.2	6.599	4345.8	1343.3	49.473	876		6368.0	8.272	4628.4	1136.4	49.922
827		5993.8	6.629	4351.5	1338.7	49.482	877		6375.7	8.309	4634.1	1132.7	49.931
828		6001.4	6.660	4357.1	1334.1	49.492	878		6383.4	8.346	4639.8	1129.0	49.940
829		6009.0	6.691	4362.7	1329.6	49.501	879		6391.1	8.382	4645.5	1125.4	49.948
830	370.33	6016.6	6.722	4368.3	1325.1	49.510	880	420.33	6398.7	8.419	4651.2	1121.7	49.957
831		6024.2	6.753	4374.0	1320.5	49.519	881		6406.4	8.456	4656.9	1118.0	49.966
832		6031.8	6.784	4379.6	1316.0	49.528	882		6414.1	8.493	4662.5	1114.4	49.974
833		6039.5	6.816	4385.2	1311.6	49.537	883		6421.7	8.531	4668.2	1110.8	49.983
834		6047.1	6.847	4390.9	1307.1	49.547	884		6429.4	8.568	4673.9	1107.2	49.992
835	375.33	6054.7	6.879	4396.5	1302.7	49.556	885	425.33	6437.1	8.605	4679.6	1103.6	50.000
836		6062.3	6.910	4402.1	1298.3	49.565	886		6444.8	8.643	4685.3	1100.1	50.009
837		6069.9	6.942	4407.8	1293.9	49.574	887		6452.4	8.681	4691.0	1096.5	50.018
838		6077.6	6.974	4413.4	1289.5	49.583	888		6460.1	8.719	4696.7	1093.0	50.026
839		6085.2	7.006	4419.1	1285.1	49.592	889		6467.8	8.757	4702.4	1089.5	50.035
840	380.33	6092.8	7.038	4424.7	1280.8	49.601	890	430.33	6475.5	8.795	4708.1	1086.0	50.044
841		6100.4	7.070	4430.3	1276.5	49.610	891		6483.1	8.833	4713.7	1082.5	50.052
842		6108.1	7.103	4436.0	1272.2	49.619	892		6490.8	8.872	4719.4	1079.0	50.061
843		6115.7	7.135	4441.6	1267.9	49.628	893		6498.5	8.910	4725.1	1075.5	50.070
844		6123.3	7.168	4447.3	1263.6	49.637	894		6506.2	8.949	4730.8	1072.1	50.078
845	385.33	6131.0	7.200	4452.9	1259.4	49.646	895	435.33	6513.9	8.988	4736.6	1068.7	50.087
846		6138.6	7.233	4458.6	1255.1	49.656	896		6521.6	9.027	4742.3	1065.2	50.095
847		6146.2	7.266	4464.2	1250.9	49.665	897		6529.3	9.066	4748.0	1061.8	50.104
848		6153.9	7.299	4469.8	1246.8	49.674	898		6537.0	9.105	4753.7	1058.4	50.113
849		6161.5	7.332	4475.5	1242.6	49.683	899		6544.6	9.144	4759.4	1055.1	50.121

Table 7 Products—100% Theoretical Air (for One Pound-Mole)

T	t	$\bar{h}$	p_r	$\bar{u}$	v_r	$\bar{\phi}$	T	t	$\bar{h}$	p_r	$\bar{u}$	v_r	$\bar{\phi}$
900	440.33	6552.3	9.184	4765.1	1051.7	50.130	950	490.33	6938.4	11.332	5051.8	899.7	50.547
901		6560.0	9.223	4770.8	1048.3	50.138	951		6946.1	11.378	5057.6	896.9	50.555
902		6567.7	9.263	4776.5	1045.0	50.147	952		6953.9	11.425	5063.3	894.2	50.563
903		6575.4	9.303	4782.2	1041.7	50.155	953		6961.6	11.472	5069.1	891.5	50.571
904		6583.1	9.343	4787.9	1038.4	50.164	954		6969.4	11.519	5074.9	888.8	50.580
905	445.33	6590.8	9.383	4793.6	1035.1	50.172	955	495.33	6977.1	11.566	5080.6	886.1	50.588
906		6598.5	9.423	4799.3	1031.8	50.181	956		6984.9	11.614	5086.4	883.4	50.596
907		6606.2	9.464	4805.0	1028.5	50.189	957		6992.6	11.661	5092.2	880.7	50.604
908		6613.9	9.504	4810.8	1025.3	50.198	958		7000.4	11.709	5097.9	878.0	50.612
909		6621.6	9.545	4816.5	1022.0	50.206	959		7008.2	11.757	5103.7	875.4	50.620
910	450.33	6629.3	9.586	4822.2	1018.8	50.215	960	500.33	7015.9	11.805	5109.5	872.7	50.628
911		6637.0	9.626	4827.9	1015.6	50.223	961		7023.7	11.853	5115.3	870.1	50.636
912		6644.7	9.667	4833.6	1012.4	50.232	962		7031.4	11.901	5121.0	867.5	50.644
913		6652.4	9.709	4839.4	1009.2	50.240	963		7039.2	11.950	5126.8	864.8	50.652
914		6660.2	9.750	4845.1	1006.0	50.248	964		7047.0	11.998	5132.6	862.2	50.661
915	455.33	6667.9	9.791	4850.8	1002.8	50.257	965	505.33	7054.7	12.047	5138.4	859.6	50.669
916		6675.6	9.833	4856.5	999.7	50.265	966		7062.5	12.096	5144.2	857.0	50.677
917		6683.3	9.875	4862.2	996.5	50.274	967		7070.3	12.145	5149.9	854.5	50.685
918		6691.0	9.917	4868.0	993.4	50.282	968		7078.0	12.194	5155.7	851.9	50.693
919		6698.7	9.959	4873.7	990.3	50.291	969		7085.8	12.244	5161.5	849.3	50.701
920	460.33	6706.4	10.001	4879.4	987.2	50.299	970	510.33	7093.6	12.293	5167.3	846.8	50.709
921		6714.1	10.043	4885.2	984.1	50.307	971		7101.4	12.343	5173.1	844.3	50.717
922		6721.9	10.086	4890.9	981.0	50.316	972		7109.1	12.393	5178.9	841.7	50.725
923		6729.6	10.128	4896.6	978.0	50.324	973		7116.9	12.443	5184.7	839.2	50.733
924		6737.3	10.171	4902.4	974.9	50.332	974		7124.7	12.493	5190.5	836.7	50.741
925	465.33	6745.0	10.214	4908.1	971.9	50.341	975	515.33	7132.5	12.543	5196.3	834.2	50.749
926		6752.7	10.257	4913.8	968.8	50.349	976		7140.2	12.593	5202.0	831.7	50.757
927		6760.4	10.300	4919.6	965.8	50.357	977		7148.0	12.644	5207.8	829.2	50.765
928		6768.2	10.343	4925.3	962.8	50.366	978		7155.8	12.695	5213.6	826.7	50.773
929		6775.9	10.387	4931.0	959.8	50.374	979		7163.6	12.746	5219.4	824.3	50.781
930	470.33	6783.6	10.430	4936.8	956.9	50.382	980	520.33	7171.4	12.797	5225.2	821.8	50.788
931		6791.4	10.474	4942.5	953.9	50.391	981		7179.1	12.848	5231.0	819.4	50.796
932		6799.1	10.518	4948.3	950.9	50.399	982		7186.9	12.900	5236.8	817.0	50.804
933		6806.8	10.562	4954.0	948.0	50.407	983		7194.7	12.951	5242.6	814.5	50.812
934		6814.5	10.606	4959.7	945.0	50.416	984		7202.5	13.003	5248.4	812.1	50.820
935	475.33	6822.3	10.650	4965.5	942.1	50.424	985	525.33	7210.3	13.055	5254.2	809.7	50.828
936		6830.0	10.695	4971.2	939.2	50.432	986		7218.1	13.107	5260.0	807.3	50.836
937		6837.7	10.739	4977.0	936.3	50.440	987		7225.9	13.159	5265.8	804.9	50.844
938		6845.5	10.784	4982.7	933.4	50.449	988		7233.7	13.212	5271.6	802.5	50.852
939		6853.2	10.829	4988.5	930.6	50.457	989		7241.5	13.264	5277.5	800.2	50.860
940	480.33	6860.9	10.874	4994.2	927.7	50.465	990	530.33	7249.3	13.317	5283.3	797.8	50.868
941		6868.7	10.919	5000.0	924.8	50.473	991		7257.1	13.370	5289.1	795.4	50.875
942		6876.4	10.964	5005.7	922.0	50.482	992		7264.9	13.423	5294.9	793.1	50.883
943		6884.2	11.010	5011.5	919.2	50.490	993		7272.7	13.476	5300.7	790.8	50.891
944		6891.9	11.055	5017.3	916.4	50.498	994		7280.5	13.529	5306.5	788.4	50.899
945	485.33	6899.6	11.101	5023.0	913.5	50.506	995	535.33	7288.3	13.583	5312.3	786.1	50.907
946		6907.4	11.147	5028.8	910.8	50.514	996		7296.1	13.637	5318.1	783.8	50.915
947		6915.1	11.193	5034.5	908.0	50.523	997		7303.9	13.691	5324.0	781.5	50.923
948		6922.9	11.239	5040.3	905.2	50.531	998		7311.7	13.745	5329.8	779.2	50.930
949		6930.6	11.285	5046.0	902.4	50.539	999		7319.5	13.799	5335.6	776.9	50.938

Table 7 Products—100% Theoretical Air (for One Pound-Mole)

T	t	$\bar{h}$	p_r	$\bar{u}$	v_r	$\bar{\phi}$	T	t	$\bar{h}$	p_r	$\bar{u}$	v_r	$\bar{\phi}$
1000	540.33	7327.3	13.853	5341.4	774.7	50.946	1250	790.33	9316.7	33.86	6834.4	396.2	52.720
1005		7366.3	14.128	5370.5	763.4	50.985	1255		9357.3	34.41	6865.1	391.4	52.753
1010		7405.4	14.406	5399.7	752.4	51.024	1260		9397.9	34.98	6895.7	386.6	52.785
1015		7444.5	14.689	5428.9	741.5	51.062	1265		9438.5	35.55	6926.4	381.9	52.817
1020		7483.7	14.977	5458.1	730.9	51.101	1270		9479.2	36.13	6957.2	377.3	52.849
1025	565.33	7522.8	15.268	5487.3	720.4	51.139	1275	815.33	9519.9	36.71	6987.9	372.7	52.881
1030		7562.1	15.564	5516.6	710.2	51.177	1280		9560.6	37.31	7018.7	368.2	52.913
1035		7601.3	15.865	5545.9	700.1	51.215	1285		9601.3	37.91	7049.5	363.8	52.945
1040		7640.5	16.170	5575.2	690.2	51.253	1290		9642.1	38.52	7080.4	359.4	52.977
1045		7679.8	16.480	5604.6	680.5	51.291	1295		9682.9	39.13	7111.2	355.1	53.008
1050	590.33	7719.2	16.795	5634.0	670.9	51.328	1300	840.33	9723.8	39.76	7142.1	350.9	53.040
1055		7758.5	17.114	5663.4	661.5	51.366	1305		9764.6	40.39	7173.1	346.7	53.071
1060		7797.9	17.438	5692.9	652.3	51.403	1310		9805.5	41.03	7204.0	342.6	53.102
1065		7837.3	17.767	5722.3	643.3	51.440	1315		9846.5	41.68	7235.1	338.5	53.134
1070		7876.7	18.100	5751.8	634.4	51.477	1320		9887.4	42.34	7266.1	334.6	53.165
1075	615.33	7916.2	18.439	5781.4	625.7	51.514	1325	865.33	9928.4	43.01	7297.1	330.6	53.196
1080		7955.7	18.782	5811.0	617.1	51.550	1330		9969.4	43.68	7328.2	326.8	53.227
1085		7995.2	19.131	5840.6	608.6	51.587	1335		10010.5	44.36	7359.4	322.9	53.257
1090		8034.8	19.485	5870.2	600.3	51.623	1340		10051.5	45.06	7390.5	319.2	53.288
1095		8074.4	19.843	5899.9	592.2	51.660	1345		10092.6	45.76	7421.7	315.5	53.319
1100	640.33	8114.0	20.207	5929.5	584.2	51.696	1350	890.33	10133.8	46.46	7452.9	311.8	53.349
1105		8153.6	20.577	5959.3	576.3	51.732	1355		10174.9	47.18	7484.1	308.2	53.380
1110		8193.3	20.951	5989.0	568.6	51.768	1360		10216.1	47.91	7515.4	304.6	53.410
1115		8233.0	21.331	6018.8	560.9	51.803	1365		10257.4	48.64	7546.7	301.1	53.440
1120		8272.8	21.717	6048.6	553.5	51.839	1370		10298.6	49.39	7578.0	297.7	53.470
1125	665.33	8312.6	22.108	6078.5	546.1	51.874	1375	915.33	10339.9	50.14	7609.4	294.3	53.500
1130		8352.4	22.504	6108.3	538.9	51.909	1380		10381.2	50.91	7640.7	290.9	53.530
1135		8392.2	22.906	6138.3	531.7	51.945	1385		10422.6	51.68	7672.2	287.6	53.560
1140		8432.1	23.314	6168.2	524.7	51.980	1390		10464.0	52.46	7703.6	284.3	53.590
1145		8472.0	23.728	6198.1	517.9	52.015	1395		10505.4	53.25	7735.1	281.1	53.620
1150	690.33	8511.9	24.147	6228.1	511.1	52.049	1400	940.33	10546.8	54.05	7766.6	278.0	53.650
1155		8551.8	24.573	6258.2	504.4	52.084	1405		10588.3	54.86	7798.1	274.8	53.679
1160		8591.8	25.004	6288.2	497.9	52.119	1410		10629.7	55.68	7829.7	271.7	53.709
1165		8631.8	25.441	6318.3	491.4	52.153	1415		10671.3	56.51	7861.3	268.7	53.738
1170		8671.9	25.884	6348.4	485.1	52.187	1420		10712.8	57.35	7892.9	265.7	53.767
1175	715.33	8711.9	26.333	6378.6	478.8	52.222	1425	965.33	10754.4	58.20	7924.5	262.7	53.797
1180		8752.1	26.789	6408.7	472.7	52.256	1430		10796.0	59.07	7956.2	259.8	53.826
1185		8792.2	27.251	6439.0	466.7	52.290	1435		10837.7	59.94	7988.0	256.9	53.855
1190		8832.4	27.719	6469.2	460.7	52.323	1440		10879.3	60.82	8019.7	254.1	53.884
1195		8872.6	28.193	6499.5	454.9	52.357	1445		10921.0	61.71	8051.5	251.3	53.913
1200	740.33	8912.8	28.674	6529.8	449.1	52.391	1450	990.33	10962.7	62.61	8083.3	248.5	53.942
1205		8953.1	29.162	6560.1	443.4	52.424	1455		11004.5	63.53	8115.1	245.8	53.970
1210		8993.3	29.656	6590.5	437.9	52.458	1460		11046.3	64.45	8146.9	243.1	53.999
1215		9033.7	30.157	6620.8	432.4	52.491	1465		11088.1	65.38	8178.8	240.5	54.028
1220		9074.0	30.664	6651.3	427.0	52.524	1470		11130.0	66.33	8210.7	237.8	54.056
1225	765.33	9114.4	31.178	6681.7	421.6	52.557	1475	1015.33	11171.8	67.29	8242.7	235.3	54.084
1230		9154.8	31.699	6712.2	416.4	52.590	1480		11213.7	68.25	8274.7	232.7	54.113
1235		9195.3	32.228	6742.7	411.2	52.623	1485		11255.7	69.23	8306.7	230.2	54.141
1240		9235.7	32.763	6773.3	406.2	52.655	1490		11297.6	70.22	8338.7	227.7	54.169
1245		9276.2	33.305	6803.8	401.2	52.688	1495		11339.6	71.23	8370.7	225.3	54.197

Table 7   Products—100% Theoretical Air (for One Pound-Mole)

T	t	$\bar{h}$	p_r	$\bar{u}$	v_r	$\bar{\phi}$	T	t	$\bar{h}$	p_r	$\bar{u}$	v_r	$\bar{\phi}$
1500	1040.33	11381.6	72.24	8402.8	222.83	54.226	1750	1290.33	13517.5	140.18	10042.3	133.98	55.542
1505		11423.7	73.26	8434.9	220.45	54.254	1755		13560.9	141.93	10075.7	132.69	55.567
1510		11465.7	74.30	8467.1	218.10	54.281	1760		13604.3	143.71	10109.2	131.43	55.591
1515		11507.8	75.35	8499.3	215.77	54.309	1765		13647.8	145.50	10142.7	130.18	55.616
1520		11550.0	76.41	8531.5	213.48	54.337	1770		13691.2	147.32	10176.3	128.94	55.641
1525	1065.33	11592.1	77.48	8563.7	211.22	54.365	1775	1315.33	13734.7	149.15	10209.8	127.71	55.665
1530		11634.3	78.57	8595.9	208.98	54.392	1780		13778.2	151.00	10243.4	126.50	55.690
1535		11676.5	79.66	8628.2	206.78	54.420	1785		13821.8	152.87	10277.0	125.31	55.714
1540		11718.8	80.77	8660.5	204.60	54.447	1790		13865.4	154.76	10310.7	124.13	55.739
1545		11761.0	81.90	8692.9	202.45	54.475	1795		13909.0	156.66	10344.3	122.96	55.763
1550	1090.33	11803.3	83.03	8725.2	200.33	54.502	1800	1340.33	13952.6	158.59	10378.0	121.80	55.787
1555		11845.6	84.18	8757.6	198.24	54.529	1805		13996.2	160.53	10411.8	120.66	55.811
1560		11888.0	85.34	8790.1	196.17	54.557	1810		14039.8	162.50	10445.4	119.54	55.835
1565		11930.4	86.51	8822.5	194.13	54.584	1815		14083.6	164.48	10479.3	118.42	55.860
1570		11972.8	87.70	8855.0	192.11	54.611	1820		14127.2	166.48	10513.0	117.32	55.884
1575	1115.33	12015.2	88.90	8887.5	190.12	54.638	1825	1365.33	14171.0	168.51	10546.8	116.23	55.908
1580		12057.7	90.11	8920.0	188.16	54.665	1830		14214.8	170.55	10580.7	115.15	55.932
1585		12100.2	91.34	8952.6	186.22	54.691	1835		14258.5	172.62	10614.5	114.08	55.955
1590		12142.7	92.58	8985.2	184.30	54.718	1840		14302.4	174.71	10648.4	113.02	55.979
1595		12185.3	93.84	9017.8	182.41	54.745	1845		14346.3	176.80	10682.4	111.99	56.003
1600	1140.33	12227.8	95.10	9050.5	180.54	54.772	1850	1390.33	14390.0	178.93	10716.2	110.96	56.027
1605		12270.4	96.39	9083.1	178.70	54.798	1855		14433.9	181.08	10750.2	109.93	56.050
1610		12313.1	97.68	9115.8	176.88	54.825	1860		14477.8	183.26	10784.1	108.92	56.074
1615		12355.7	98.99	9148.6	175.08	54.851	1865		14521.8	185.43	10818.2	107.93	56.098
1620		12398.4	100.31	9181.3	173.31	54.878	1870		14565.8	187.65	10852.2	106.94	56.121
1625	1165.33	12441.1	101.65	9214.1	171.55	54.904	1875	1415.33	14609.7	189.87	10886.2	105.98	56.145
1630		12483.9	103.01	9246.9	169.82	54.930	1880		14653.7	192.13	10920.3	105.01	56.168
1635		12526.6	104.37	9279.8	168.11	54.956	1885		14697.8	194.42	10954.4	104.05	56.192
1640		12569.4	105.76	9312.6	166.42	54.982	1890		14741.8	196.71	10988.5	103.11	56.215
1645		12612.2	107.15	9345.5	164.75	55.009	1895		14785.9	199.02	11022.7	102.18	56.238
1650	1190.33	12655.1	108.57	9378.4	163.10	55.035	1900	1440.33	14829.9	201.37	11056.8	101.25	56.261
1655		12698.0	109.99	9411.4	161.47	55.061	1905		14874.1	203.73	11091.1	100.35	56.285
1660		12740.8	111.44	9444.3	159.86	55.086	1910		14918.3	206.14	11125.3	99.43	56.308
1665		12783.8	112.90	9477.3	158.27	55.112	1915		14962.4	208.54	11159.5	98.55	56.331
1670		12826.7	114.37	9510.3	156.70	55.138	1920		15006.5	210.97	11193.7	97.67	56.354
1675	1215.33	12869.7	115.86	9543.4	155.15	55.164	1925	1465.33	15050.8	213.42	11228.1	96.80	56.377
1680		12912.7	117.36	9576.5	153.62	55.189	1930		15095.0	215.92	11262.3	95.92	56.400
1685		12955.7	118.89	9609.6	152.10	55.215	1935		15139.4	218.42	11296.7	95.07	56.423
1690		12998.8	120.42	9642.7	150.60	55.240	1940		15183.6	220.95	11331.0	94.23	56.446
1695		13041.9	121.98	9675.9	149.13	55.266	1945		15227.9	223.50	11365.5	93.39	56.468
1700	1240.33	13085.0	123.55	9709.0	147.67	55.291	1950	1490.33	15272.2	226.07	11399.8	92.57	56.491
1705		13128.1	125.13	9742.2	146.22	55.317	1955		15316.6	228.67	11434.3	91.75	56.514
1710		13171.3	126.73	9775.5	144.80	55.342	1960		15360.9	231.30	11468.6	90.94	56.537
1715		13214.5	128.36	9808.7	143.39	55.367	1965		15405.3	233.95	11503.1	90.14	56.559
1720		13257.7	129.99	9842.0	142.00	55.392	1970		15449.8	236.63	11537.7	89.34	56.582
1725	1265.33	13300.9	131.65	9875.3	140.62	55.417	1975	1515.33	15494.2	239.33	11572.1	88.56	56.604
1730		13344.2	133.32	9908.7	139.26	55.442	1980		15538.6	242.06	11606.6	87.78	56.627
1735		13387.5	135.00	9942.0	137.92	55.467	1985		15583.2	244.82	11641.2	87.01	56.649
1740		13430.8	136.71	9975.4	136.59	55.492	1990		15627.7	247.60	11675.8	86.25	56.672
1745		13474.2	138.43	10008.8	135.27	55.517	1995		15672.1	250.38	11710.3	85.51	56.694

Table 7 Products—100% Theoretical Air (for One Pound-Mole)

T	t	$\bar{h}$	p_r	$\bar{u}$	v_r	$\bar{\phi}$	T	t	$\bar{h}$	p_r	$\bar{u}$	v_r	$\bar{\phi}$
2000	1540.33	15716.7	253.2	11745.0	84.76	56.716	2500	2040.33	20272.0	704.4	15307.4	38.09	58.748
2010		15805.9	258.9	11814.3	83.30	56.761	2510		20365.0	717.6	15380.5	37.54	58.785
2020		15895.1	264.8	11883.7	81.87	56.805	2520		20458.0	731.1	15453.6	36.99	58.822
2030		15984.4	270.7	11953.2	80.47	56.849	2530		20551.0	744.9	15526.7	36.45	58.859
2040		16073.8	276.8	12022.7	79.10	56.893	2540		20644.1	758.7	15600.1	35.93	58.896
2050	1590.33	16163.4	283.0	12092.4	77.75	56.937	2550	2090.33	20737.4	772.8	15673.4	35.41	58.932
2060		16253.0	289.2	12162.1	76.44	56.980	2560		20830.5	787.1	15746.7	34.90	58.969
2070		16342.7	295.6	12232.0	75.14	57.024	2570		20923.8	801.6	15820.2	34.40	59.005
2080		16432.5	302.1	12301.9	73.88	57.067	2580		21017.2	816.4	15893.7	33.91	59.041
2090		16522.4	308.8	12371.9	72.64	57.110	2590		21110.6	831.4	15967.2	33.43	59.077
2100	1640.33	16612.3	315.5	12442.0	71.43	57.153	2600	2140.33	21204.1	846.6	16040.9	32.96	59.113
2110		16702.3	322.4	12512.1	70.24	57.196	2610		21297.6	862.1	16114.5	32.49	59.149
2120		16792.5	329.3	12582.4	69.08	57.238	2620		21391.2	877.8	16188.2	32.03	59.185
2130		16882.6	336.5	12652.7	67.94	57.281	2630		21484.8	893.6	16262.0	31.58	59.221
2140		16972.9	343.7	12723.2	66.82	57.323	2640		21578.6	909.8	16335.9	31.14	59.256
2150	1690.33	17063.2	351.1	12793.6	65.72	57.365	2650	2190.33	21672.3	926.3	16409.8	30.70	59.292
2160		17153.6	358.6	12864.2	64.64	57.407	2660		21766.1	942.9	16483.7	30.28	59.327
2170		17244.2	366.2	12934.8	63.59	57.449	2670		21860.0	959.7	16557.8	29.86	59.362
2180		17334.7	374.0	13005.6	62.55	57.491	2680		21954.0	976.9	16631.9	29.44	59.398
2190		17425.4	381.9	13076.4	61.54	57.532	2690		22047.9	994.3	16706.0	29.03	59.433
2200	1740.33	17516.1	389.9	13147.2	60.55	57.574	2700	2240.33	22142.0	1011.9	16780.2	28.63	59.467
2210		17606.9	398.1	13218.1	59.58	57.615	2710		22236.1	1029.8	16854.5	28.24	59.502
2220		17697.8	406.4	13289.2	58.62	57.656	2720		22330.3	1048.0	16928.8	27.85	59.537
2230		17788.7	414.9	13360.3	57.68	57.697	2730		22424.5	1066.4	17003.1	27.47	59.572
2240		17879.8	423.5	13431.5	56.76	57.738	2740		22518.8	1085.0	17077.6	27.10	59.606
2250	1790.33	17971.0	432.2	13502.8	55.86	57.778	2750	2290.33	22613.1	1103.9	17152.0	26.73	59.640
2260		18062.2	441.1	13574.1	54.98	57.819	2760		22707.5	1123.2	17226.5	26.37	59.675
2270		18153.4	450.2	13645.5	54.11	57.859	2770		22801.9	1142.6	17301.1	26.02	59.709
2280		18244.8	459.4	13717.0	53.26	57.899	2780		22896.5	1162.3	17375.8	25.67	59.743
2290		18336.2	468.7	13788.6	52.43	57.939	2790		22991.0	1182.5	17450.4	25.32	59.777
2300	1840.33	18427.8	478.2	13860.3	51.61	57.979	2800	2340.33	23085.6	1202.6	17525.2	24.99	59.810
2310		18519.3	487.8	13931.9	50.82	58.019	2810		23180.2	1223.2	17599.9	24.65	59.844
2320		18611.0	497.7	14003.8	50.03	58.058	2820		23274.9	1244.1	17674.8	24.33	59.878
2330		18702.6	507.7	14075.5	49.25	58.098	2830		23369.6	1265.3	17749.6	24.00	59.911
2340		18794.4	517.8	14147.5	48.50	58.137	2840		23464.5	1286.8	17824.6	23.69	59.945
2350	1890.33	18886.3	528.1	14219.5	47.75	58.176	2850	2390.33	23559.3	1308.6	17899.6	23.37	59.978
2360		18978.2	538.6	14291.6	47.02	58.215	2860		23654.3	1330.7	17974.7	23.06	60.011
2370		19070.2	549.3	14363.7	46.30	58.254	2870		23749.2	1353.0	18049.8	22.76	60.044
2380		19162.3	560.1	14435.9	45.60	58.293	2880		23844.2	1375.8	18124.9	22.47	60.077
2390		19254.4	571.1	14508.1	44.91	58.332	2890		23939.3	1398.8	18200.1	22.17	60.111
2400	1940.33	19346.5	582.3	14580.4	44.23	58.370	2900	2440.33	24034.4	1422.1	18275.4	21.88	60.143
2410		19438.7	593.7	14652.8	43.56	58.409	2910		24129.5	1445.7	18350.6	21.60	60.176
2420		19531.0	605.2	14725.2	42.91	58.447	2920		24224.8	1469.8	18426.1	21.32	60.209
2430		19623.5	617.0	14797.8	42.27	58.485	2930		24320.0	1494.0	18501.4	21.05	60.241
2440		19716.0	628.8	14870.5	41.64	58.523	2940		24415.4	1518.6	18577.0	20.78	60.274
2450	1990.33	19808.4	641.0	14943.1	41.02	58.561	2950	2490.33	24510.7	1543.6	18652.4	20.51	60.306
2460		19901.0	653.3	15015.8	40.41	58.598	2960		24606.1	1568.8	18728.0	20.25	60.338
2470		19993.7	665.7	15088.6	39.82	58.636	2970		24701.5	1594.5	18803.5	19.99	60.370
2480		20086.4	678.4	15161.5	39.23	58.673	2980		24797.1	1620.5	18879.2	19.73	60.403
2490		20179.2	691.3	15234.4	38.65	58.711	2990		24892.6	1646.9	18954.9	19.48	60.435

Table 7   Products—100% Theoretical Air (for One Pound-Mole)

T	t	$\bar{h}$	p_r	$\bar{u}$	v_r	$\bar{\phi}$	T	t	$\bar{h}$	p_r	$\bar{u}$	v_r	$\bar{\phi}$
3000	2540.33	24988.1	1673.6	19030.5	19.237	60.467	3500	3040.33	29820.7	3543.3	22870.2	10.600	61.956
3010		25083.8	1700.5	19106.4	18.996	60.498	3510		29918.3	3593.5	22948.0	10.482	61.984
3020		25179.5	1727.9	19182.2	18.756	60.530	3520		30015.9	3643.8	23025.7	10.367	62.012
3030		25275.2	1755.8	19258.1	18.519	60.562	3530		30113.7	3695.2	23103.6	10.252	62.040
3040		25371.0	1783.8	19334.0	18.289	60.593	3540		30211.4	3746.8	23181.4	10.139	62.067
3050	2590.33	25466.8	1812.4	19409.9	18.059	60.625	3550	3090.33	30309.1	3799.4	23259.3	10.027	62.095
3060		25562.7	1841.2	19485.9	17.835	60.656	3560		30406.9	3852.2	23337.3	9.917	62.122
3070		25658.5	1870.4	19561.9	17.614	60.687	3570		30504.7	3905.7	23415.1	9.809	62.150
3080		25754.5	1900.2	19638.0	17.394	60.719	3580		30602.6	3960.2	23493.2	9.701	62.177
3090		25850.6	1930.2	19714.2	17.180	60.750	3590		30700.4	4014.9	23571.2	9.596	62.204
3100	2640.33	25946.5	1960.5	19790.4	16.969	60.781	3600	3140.33	30798.3	4070.3	23649.3	9.492	62.232
3110		26042.6	1991.3	19866.6	16.760	60.812	3610		30896.3	4126.3	23727.4	9.389	62.259
3120		26138.8	2022.5	19942.9	16.555	60.843	3620		30994.2	4182.9	23805.4	9.287	62.286
3130		26234.9	2054.1	20019.1	16.352	60.873	3630		31092.3	4240.2	23883.6	9.187	62.313
3140		26331.0	2085.9	20095.4	16.155	60.904	3640		31190.3	4298.2	23961.8	9.088	62.340
3150	2690.33	26427.3	2118.4	20171.9	15.958	60.935	3650	3190.33	31288.4	4356.9	24040.0	8.990	62.367
3160		26523.7	2151.3	20248.4	15.764	60.965	3660		31386.4	4416.2	24118.2	8.894	62.394
3170		26619.9	2184.3	20324.7	15.574	60.996	3670		31484.6	4476.3	24196.5	8.799	62.420
3180		26716.3	2218.1	20401.3	15.385	61.026	3680		31582.8	4537.0	24274.8	8.704	62.447
3190		26812.8	2252.0	20477.9	15.201	61.056	3690		31681.0	4597.9	24353.1	8.613	62.474
3200	2740.33	26909.1	2286.7	20554.4	15.018	61.086	3700	3240.33	31779.2	4660.0	24431.5	8.521	62.500
3210		27005.6	2321.5	20631.0	14.839	61.116	3710		31877.4	4722.9	24509.9	8.430	62.527
3220		27102.2	2356.8	20707.7	14.662	61.146	3720		31975.7	4785.8	24588.3	8.341	62.553
3230		27198.8	2392.8	20784.4	14.486	61.177	3730		32074.0	4850.2	24666.8	8.253	62.580
3240		27295.4	2429.0	20861.2	14.315	61.206	3740		32172.3	4914.6	24745.2	8.167	62.606
3250	2790.33	27392.0	2465.7	20938.0	14.145	61.236	3750	3290.33	32270.7	4980.4	24823.7	8.080	62.632
3260		27488.7	2502.8	21014.8	13.978	61.266	3760		32369.1	5046.3	24902.3	7.996	62.658
3270		27585.4	2540.4	21091.7	13.813	61.295	3770		32467.6	5113.6	24980.9	7.912	62.685
3280		27682.2	2578.5	21168.5	13.651	61.325	3780		32566.0	5181.0	25059.5	7.830	62.711
3290		27779.1	2617.1	21245.6	13.491	61.354	3790		32664.5	5249.1	25138.1	7.748	62.737
3300	2840.33	27875.9	2656.2	21322.5	13.333	61.384	3800	3340.33	32763.1	5318.7	25216.8	7.667	62.763
3310		27972.7	2695.8	21399.5	13.177	61.413	3810		32861.6	5388.5	25295.5	7.588	62.789
3320		28069.7	2735.9	21476.7	13.023	61.443	3820		32960.3	5458.9	25374.3	7.510	62.814
3330		28166.6	2776.1	21553.7	12.873	61.472	3830		33058.8	5530.2	25453.0	7.432	62.840
3340		28263.6	2817.2	21630.8	12.723	61.501	3840		33157.4	5602.3	25531.7	7.356	62.866
3350	2890.33	28360.7	2858.8	21708.1	12.575	61.530	3850	3390.33	33256.1	5675.1	25610.6	7.280	62.892
3360		28457.8	2900.6	21785.3	12.431	61.559	3860		33354.9	5748.8	25689.5	7.206	62.917
3370		28554.9	2943.3	21862.5	12.287	61.588	3870		33453.5	5823.3	25768.2	7.132	62.943
3380		28652.1	2986.1	21939.8	12.147	61.616	3880		33552.3	5898.6	25847.2	7.059	62.968
3390		28749.3	3029.5	22017.2	12.009	61.645	3890		33651.1	5974.0	25926.1	6.988	62.994
3400	2940.33	28846.4	3073.8	22094.5	11.870	61.674	3900	3440.33	33749.9	6051.0	26005.0	6.917	63.019
3410		28943.7	3118.2	22171.9	11.736	61.702	3910		33848.7	6128.8	26084.0	6.846	63.044
3420		29041.0	3163.2	22249.4	11.603	61.731	3920		33947.6	6207.5	26163.0	6.777	63.070
3430		29138.4	3208.8	22326.9	11.471	61.759	3930		34046.5	6286.2	26242.1	6.709	63.095
3440		29235.7	3254.9	22404.4	11.342	61.788	3940		34145.3	6366.6	26321.1	6.641	63.120
3450	2990.33	29333.1	3301.6	22481.9	11.214	61.816	3950	3490.33	34244.3	6447.1	26400.2	6.575	63.145
3460		29430.6	3348.9	22559.5	11.088	61.844	3960		34343.3	6529.3	26479.3	6.509	63.170
3470		29528.0	3396.7	22637.1	10.963	61.872	3970		34442.3	6611.5	26558.4	6.444	63.195
3480		29625.5	3445.1	22714.8	10.840	61.900	3980		34541.3	6695.4	26637.6	6.379	63.220
3490		29723.1	3494.1	22792.4	10.719	61.928	3990		34640.4	6779.4	26716.8	6.316	63.245

Table 8 Products—100% Theoretical Air (for One Pound-Mole)

T	t	$\bar{c}_p$ Btu/lb-mole R	$\bar{c}_v$ Btu/lb-mole R	$k = \dfrac{\bar{c}_p}{\bar{c}_v}$	$\bar{c}_p$ Btu/lb-mole R	$\bar{c}_v$ Btu/lb-mole R	$k = \dfrac{\bar{c}_p}{\bar{c}_v}$	$\bar{c}_p$ Btu/lb-mole R	$\bar{c}_v$ Btu/lb-mole R	$k = \dfrac{\bar{c}_p}{\bar{c}_v}$
R	F	Fuel—$(CH_1)_n$			Fuel—$(CH_2)_n$			Fuel—$(CH_3)_n$		
300	−159.67	7.082	5.096	1.390	7.120	5.135	1.387	7.147	5.161	1.385
350	−109.67	7.130	5.144	1.386	7.160	5.174	1.384	7.180	5.194	1.382
400	−59.67	7.183	5.197	1.382	7.203	5.217	1.381	7.217	5.231	1.380
450	−9.67	7.238	5.252	1.378	7.248	5.262	1.377	7.256	5.270	1.377
500	40.33	7.292	5.306	1.374	7.294	5.308	1.374	7.295	5.309	1.374
550	90.33	7.346	5.360	1.370	7.340	5.354	1.371	7.336	5.350	1.371
600	140.33	7.399	5.413	1.367	7.386	5.400	1.368	7.377	5.391	1.368
650	190.33	7.453	5.467	1.363	7.433	5.447	1.365	7.420	5.434	1.365
700	240.33	7.507	5.521	1.360	7.482	5.496	1.361	7.464	5.479	1.362
750	290.33	7.561	5.575	1.356	7.532	5.546	1.358	7.511	5.525	1.359
800	340.33	7.617	5.631	1.353	7.583	5.597	1.355	7.560	5.574	1.356
850	390.33	7.675	5.689	1.349	7.637	5.651	1.351	7.611	5.625	1.353
900	440.33	7.733	5.747	1.346	7.692	5.706	1.348	7.664	5.678	1.350
950	490.33	7.793	5.807	1.342	7.749	5.763	1.345	7.719	5.733	1.346
1000	540.33	7.854	5.868	1.338	7.808	5.822	1.341	7.776	5.790	1.343
1100	640.33	7.978	5.992	1.331	7.927	5.941	1.334	7.893	5.907	1.336
1200	740.33	8.102	6.117	1.325	8.049	6.063	1.328	8.012	6.026	1.330
1300	840.33	8.226	6.240	1.318	8.170	6.185	1.321	8.132	6.146	1.323
1400	940.33	8.347	6.361	1.312	8.290	6.304	1.315	8.251	6.265	1.317
1500	1040.33	8.463	6.477	1.307	8.406	6.420	1.309	8.367	6.381	1.311
1600	1140.33	8.574	6.588	1.301	8.518	6.532	1.304	8.479	6.493	1.306
1700	1240.33	8.680	6.694	1.297	8.625	6.639	1.299	8.587	6.601	1.301
1800	1340.33	8.779	6.793	1.292	8.726	6.740	1.295	8.690	6.704	1.296
1900	1440.33	8.873	6.887	1.288	8.822	6.836	1.290	8.787	6.801	1.292
2000	1540.33	8.961	6.975	1.285	8.913	6.927	1.287	8.880	6.894	1.288
2100	1640.33	9.043	7.057	1.281	8.998	7.012	1.283	8.967	6.981	1.284
2200	1740.33	9.119	7.133	1.278	9.078	7.092	1.280	9.049	7.064	1.281
2300	1840.33	9.191	7.205	1.276	9.153	7.167	1.277	9.127	7.141	1.278
2400	1940.33	9.257	7.271	1.273	9.223	7.237	1.274	9.200	7.214	1.275
2500	2040.33	9.319	7.333	1.271	9.289	7.303	1.272	9.268	7.282	1.273
2600	2140.33	9.377	7.391	1.269	9.350	7.364	1.270	9.332	7.346	1.270
2700	2240.33	9.431	7.445	1.267	9.408	7.422	1.268	9.392	7.406	1.268
2800	2340.33	9.482	7.496	1.265	9.462	7.476	1.266	9.449	7.463	1.266
2900	2440.33	9.529	7.543	1.263	9.513	7.527	1.264	9.502	7.516	1.264
3000	2540.33	9.573	7.587	1.262	9.561	7.575	1.262	9.552	7.567	1.262
3200	2740.33	9.653	7.667	1.259	9.647	7.661	1.259	9.644	7.658	1.259
3400	2940.33	9.724	7.738	1.257	9.725	7.739	1.257	9.725	7.739	1.257
3600	3140.33	9.786	7.800	1.255	9.793	7.807	1.254	9.797	7.811	1.254
3800	3340.33	9.842	7.856	1.253	9.854	7.868	1.252	9.862	7.876	1.252
4000	3540.33	9.891	7.905	1.251	9.908	7.922	1.251	9.920	7.934	1.250

Table 9 Molecular Weight and Gas Constant for Products of Combustion of Hydrocarbon Fuels of Composition $(CH_x)_n$ with Air

Hydrogen-Carbon Ratio		Percent of Theoretical Fuel										
$\dfrac{\text{lb of H}}{\text{lb of C}}$	$\dfrac{\text{atoms of H}}{\text{atoms of C}}$	0	10	20	30	40	50	60	70	80	90	100
Gas Constant (ft lbf/lbm R)												
0.06	0.715	28.967	29.101	29.234	29.367	29.498	29.629	29.759	29.888	30.016	30.144	30.270
0.08	0.953	28.967	29.069	29.171	29.272	29.372	29.471	29.569	29.667	29.764	29.860	29.955
0.10	1.192	28.967	29.041	29.114	29.186	29.258	29.329	29.399	29.469	29.538	29.606	29.674
0.12	1.430	28.967	29.014	29.062	29.108	29.154	29.200	29.245	29.290	29.334	29.377	29.421
0.14	1.668	28.967	28.991	29.014	29.037	29.060	29.083	29.105	29.127	29.149	29.170	29.192
0.16	1.907	28.967	28.969	28.970	28.972	28.974	28.975	28.977	28.979	28.980	28.982	28.984
0.18	2.145	28.967	28.948	28.930	28.912	28.895	28.877	28.860	28.843	28.826	28.810	28.794
0.20	2.383	28.967	28.930	28.893	28.857	28.822	28.787	28.752	28.718	28.685	28.652	28.620
0.22	2.622	28.967	28.912	28.859	28.806	28.754	28.703	28.653	28.604	28.555	28.507	28.460
0.24	2.860	28.967	28.896	28.827	28.759	28.692	28.626	28.561	28.497	28.435	28.373	28.312
0.26	3.098	28.697	28.881	28.797	28.715	28.634	28.554	28.476	28.399	28.323	28.249	28.175
0.28	3.337	28.967	28.867	28.769	28.674	28.579	28.487	28.396	28.307	28.219	28.133	28.049
0.30	3.575	28.967	28.854	28.744	28.635	28.529	28.424	28.322	28.221	28.123	28.026	27.931
0.32	3.813	28.967	28.842	28.719	28.599	28.481	28.366	28.253	28.141	28.032	27.925	27.821
0.34	4.052	28.967	28.830	28.696	28.565	28.437	28.311	28.187	28.067	27.948	27.832	27.718
Molecular Weight												
0.06	0.715	53.348	53.103	52.861	52.622	52.388	52.156	51.929	51.704	51.483	51.266	51.051
0.08	0.993	53.348	53.160	52.975	52.793	52.613	52.436	52.261	52.089	51.920	51.753	51.588
0.10	1.192	53.348	53.213	53.079	52.948	52.818	52.690	52.564	52.440	52.317	52.197	52.077
0.12	1.430	53.348	53.261	53.175	53.089	53.006	52.923	52.841	52.761	52.681	52.603	52.526
0.14	1.668	53.348	53.305	53.262	53.219	53.177	53.136	53.095	53.055	53.016	52.976	52.938
0.16	1.907	53.348	53.345	53.342	53.339	53.336	53.333	53.330	53.327	53.324	53.321	53.318
0.18	2.145	53.348	53.382	53.416	53.449	53.482	53.514	53.546	53.577	53.608	53.639	53.669
0.20	2.383	53.348	53.417	53.485	53.551	53.617	53.682	53.746	53.810	53.873	53.934	53.996
0.22	2.622	53.348	53.449	53.548	53.646	53.743	53.838	53.933	54.026	54.118	54.209	54.299
0.24	2.860	53.348	53.479	53.607	53.735	53.860	53.984	54.107	54.228	54.347	54.465	54.582
0.26	3.098	53.348	53.507	53.663	53.817	53.969	54.120	54.269	54.416	54.561	54.705	54.847
0.28	3.337	53.348	53.533	53.714	53.894	54.072	54.247	54.421	54.592	54.762	54.929	55.095
0.30	3.575	53.348	53.557	53.763	53.967	54.168	54.367	54.563	54.758	54.950	55.140	55.328
0.32	3.813	53.348	53.580	53.808	54.035	54.258	54.479	54.697	54.913	55.127	55.338	55.547
0.34	4.052	53.348	53.601	53.851	54.099	54.343	54.585	54.824	55.060	55.293	55.524	55.753

Table 10 Enthalpy of Combustion of Hydrocarbons at 25°C and Constant Pressure

Compound	Formula	Enthalpy of Combustion of Liquid $-\Delta H_c^0$, Btu/lb		Enthalpy of Combustion of Gas $-\Delta H_c^0$, Btu/lb	
		Gross[a]	Net[b]	Gross[a]	Net[b]
Paraffins					
Methane	CH_4			23870	21510
Ethane	C_2H_6	22179	20290	22314	20425
Propane	C_3H_8	21474	19758	21636	19920
n-Butane	C_4H_{10}	21123	19495	21285	19657
n-Pentane	C_5H_{12}	20909	19335	21067	19495
n-Hexane	C_6H_{14}	20771	19233	20927	19391
n-Heptane	C_7H_{16}	20668	19157	20825	19314
n-Octane	C_8H_{18}	20587	19098	20744	19253
n-Nonane	C_9H_{20}	20530	19055	20686	19209
n-Decane	$C_{10}H_{22}$	20481	19019	20636	19173
n-Undecane	$C_{11}H_{24}$	20442	18988	20596	19143
n-Dodecane	$C_{12}H_{26}$	20407	18963	20562	19118
n-Tridecane	$C_{13}H_{28}$	20380	18943	20535	19098
n-Tetradecane	$C_{14}H_{30}$	20355	18925	20510	19080
n-Pentadecane	$C_{15}H_{32}$	20334	18909	20488	19064
n-Hexadecane	$C_{16}H_{34}$	20316	18895	20470	19049
Olefins					
Ethene	C_2H_4			21618	20269
Propene	C_3H_6			21031	19682
1-Butene	C_4H_8	20659	19310	20819	19470
1-Pentene	C_5H_{10}	20532	19183	20690	19341
1-Hexene	C_6H_{12}	20452	19103	20609	19260
1-Heptene	C_7H_{14}	20393	19044	20549	19200
1-Octene	C_8H_{16}	20349	19000	20504	19156
1-Nonene	C_9H_{18}	20314	18966	20470	19121
1-Decene	$C_{10}H_{20}$	20287	18938	20442	19093
1-Undecene	$C_{11}H_{22}$	20264	18915	20419	19070
1-Dodecene	$C_{12}H_{24}$	20246	18897	20400	19051
1-Tridecene	$C_{13}H_{26}$	20230	18881	20384	19035
1-Tetradecene	$C_{14}H_{28}$	20216	18867	20370	19021
1-Pentadecene	$C_{15}H_{30}$	20204	18855	20358	19009
1-Hexadecene	$C_{16}H_{32}$	20194	18845	20347	18998
Alkylbenzenes					
Benzene	C_6H_6	17986	17260	18172	17446
Methylbenzene	C_7H_8	18245	17423	18422	17601
Ethylbenzene	C_8H_{10}	18486	17596	18658	17767
n-Propylbenzene	C_9H_{12}	18666	17722	18832	17887
n-Butylbenzene	$C_{10}H_{14}$	18810	17823	18971	17984
n-Pentylbenzene	$C_{11}H_{16}$	18927	17905	19086	18065
n-Hexylbenzene	$C_{12}H_{18}$	19023	17974	19182	18132

[a]Combustion products are H_2O (liq) and CO_2 (g).
[b]Combustion products are H_2O (g) and CO_2 (g).

SYMBOLS AND UNITS USED IN TABLES 11 TO 23

a velocity of sound $= \sqrt{gk\overline{R}T/\overline{m}}$, ft/sec

$\bar{c}_p$ specific heat at constant pressure, Btu/°F lb-mole

$\bar{c}_v$ specific heat at constant volume, Btu/°F lb-mole

g acceleration given to unit mass by unit force $=$ $32.174 \dfrac{\text{ft}}{\text{sec}^2}\dfrac{\text{lbm}}{\text{lbf}}$ [a]

$\bar{h}$ enthalpy per mole, Btu/lb-mole

k $\bar{c}_p/\bar{c}_v$

$\overline{m}$ molecular weight, lb/mole

p_r relative pressure, lbf/in.^{2} [b]

$\overline{R}$ universal gas constant, Btu/R lb-mole

T temperature, R

t temperature, °F

$\bar{u}$ internal energy per mole, Btu/lb-mole

v_r relative volume, ft^3 [b]

$\bar{\phi}$ $\displaystyle\int_{T_0}^{T} \frac{\bar{c}_p}{T}\, dT$, Btu/R lb-mole

[a] lbm and lbf are used for pound mass and pound force, respectively, except that lb is used for either wherever the meaning is obvious.
[b] The ratio of the pressures p_a and p_b corresponding to the temperatures T_a and T_b, respectively, along a given isentropic is equal to the ratio of the relative pressures p_{ra} and p_{rb} as tabulated for T_a and T_b, respectively. Thus

$$\left(\frac{p_a}{p_b}\right)_{s=\text{constant}} = \frac{p_{ra}}{p_{rb}}$$

Similarly

$$\left(\frac{v_a}{v_b}\right)_{s=\text{constant}} = \frac{v_{ra}}{v_{rb}}$$

Table 11 Nitrogen at Low Pressures (for One Pound-Mole)

$\bar{m} = 28.0134$

T	t	$\bar{h}$	p_r	$\bar{u}$	v_r	$\bar{\phi}$	T	t	$\bar{h}$	p_r	$\bar{u}$	v_r	$\bar{\phi}$
300	−159.67	2081.9	.13082	1486.2	24611	41.687	550	90.33	3820.2	1.0923	2728.0	5404	45.901
305	−154.67	2116.7	.13861	1511.0	23614	41.802	555	95.33	3855.0	1.1275	2752.9	5283	45.964
310	−149.67	2151.4	.14673	1535.8	22673	41.915	560	100.33	3889.8	1.1635	2777.7	5165	46.027
315	−144.67	2186.2	.15518	1560.7	21784	42.026	565	105.33	3924.6	1.2003	2802.6	5052	46.089
320	−139.67	2221.0	.16398	1585.5	20943	42.136	570	110.33	3959.4	1.2379	2827.5	4941	46.150
325	−134.67	2255.7	.17312	1610.3	20146	42.243	575	115.33	3994.2	1.2764	2852.3	4834	46.211
330	−129.67	2290.5	.18263	1635.1	19392	42.350	580	120.33	4029.0	1.3157	2877.2	4731	46.271
335	−124.67	2325.2	.19250	1660.0	18676	42.454	585	125.33	4063.8	1.3559	2902.1	4630	46.331
340	−119.67	2360.0	.20274	1684.8	17997	42.557	590	130.33	4098.6	1.3969	2926.9	4532	46.390
345	−114.67	2394.8	.21338	1709.6	17351	42.659	595	135.33	4133.4	1.4389	2951.8	4438	46.449
350	−109.67	2429.5	.22440	1734.5	16738	42.759	600	140.33	4168.2	1.4817	2976.7	4346	46.507
355	−104.67	2464.3	.23582	1759.3	16155	42.857	605	145.33	4203.0	1.5255	3001.6	4256	46.565
360	−99.67	2499.0	.24766	1784.1	15599	42.954	610	150.33	4237.8	1.5701	3026.5	4169	46.622
365	−94.67	2533.8	.25991	1809.0	15071	43.050	615	155.33	4272.6	1.6157	3051.3	4085	46.679
370	−89.67	2568.6	.27259	1833.8	14566	43.145	620	160.33	4307.5	1.6623	3076.2	4003	46.735
375	−84.67	2603.3	.28571	1858.6	14086	43.238	625	165.33	4342.3	1.7097	3101.1	3923	46.791
380	−79.67	2638.1	.29927	1883.5	13627	43.330	630	170.33	4377.1	1.7582	3126.0	3845	46.847
385	−74.67	2672.9	.31328	1908.3	13188	43.421	635	175.33	4412.0	1.8076	3150.9	3770	46.902
390	−69.67	2707.6	.32776	1933.1	12770	43.511	640	180.33	4446.8	1.8581	3175.8	3696	46.956
395	−64.67	2742.4	.34270	1958.0	12369	43.599	645	185.33	4481.6	1.9095	3200.7	3625	47.011
400	−59.67	2777.1	.35813	1982.8	11986	43.687	650	190.33	4516.5	1.9620	3225.7	3555	47.064
405	−54.67	2811.9	.37405	2007.6	11619	43.773	655	195.33	4551.3	2.0155	3250.6	3488	47.118
410	−49.67	2846.7	.39047	2032.5	11268	43.859	660	200.33	4586.2	2.0700	3275.5	3422	47.171
415	−44.67	2881.4	.40740	2057.3	10932	43.943	665	205.33	4621.0	2.1256	3300.4	3357	47.224
420	−39.67	2916.2	.42484	2082.1	10609	44.026	670	210.33	4655.9	2.1822	3325.4	3295	47.276
425	−34.67	2951.0	.44282	2107.0	10300	44.108	675	215.33	4690.8	2.2399	3350.3	3234	47.328
430	−29.67	2985.7	.46132	2131.8	10003	44.190	680	220.33	4725.6	2.2988	3375.3	3175	47.379
435	−24.67	3020.5	.48038	2156.6	9718	44.270	685	225.33	4760.5	2.3587	3400.2	3117	47.430
440	−19.67	3055.2	.49999	2181.5	9444	44.350	690	230.33	4795.4	2.4197	3425.2	3060	47.481
445	−14.67	3090.0	.52017	2206.3	9181	44.428	695	235.33	4830.3	2.4819	3450.1	3005	47.531
450	−9.67	3124.8	.54092	2231.1	8928	44.506	700	240.33	4865.2	2.5452	3475.1	2951	47.581
455	−4.67	3159.5	.56226	2256.0	8684	44.583	705	245.33	4900.1	2.6097	3500.1	2899	47.631
460	.33	3194.3	.58419	2280.8	8450	44.659	710	250.33	4935.0	2.6754	3525.1	2848	47.680
465	5.33	3229.1	.60673	2305.6	8225	44.734	715	255.33	4969.9	2.7422	3550.1	2798	47.729
470	10.33	3263.8	.62988	2330.5	8008	44.808	720	260.33	5004.9	2.8103	3575.1	2749	47.778
475	15.33	3298.6	.65366	2355.3	7798	44.882	725	265.33	5039.8	2.8796	3600.1	2702	47.826
480	20.33	3333.4	.67807	2380.2	7597	44.955	730	270.33	5074.7	2.9501	3625.1	2656	47.874
485	25.33	3368.1	.70313	2405.0	7402	45.027	735	275.33	5109.7	3.0218	3650.1	2610	47.922
490	30.33	3402.9	.72884	2429.8	7215	45.098	740	280.33	5144.7	3.0948	3675.1	2566	47.970
495	35.33	3437.7	.75522	2454.7	7034	45.169	745	285.33	5179.6	3.1690	3700.2	2523	48.017
500	40.33	3472.5	.78227	2479.5	6859	45.238	750	290.33	5214.6	3.2446	3725.2	2481	48.063
505	45.33	3507.2	.81001	2504.4	6691	45.308	755	295.33	5249.6	3.3214	3750.2	2439	48.110
510	50.33	3542.0	.83845	2529.2	6528	45.376	760	300.33	5284.6	3.3996	3775.3	2399	48.156
515	55.33	3576.8	.86759	2554.1	6370	45.444	765	305.33	5319.6	3.4791	3800.4	2360	48.202
520	60.33	3611.6	.89745	2578.9	6218	45.511	770	310.33	5354.6	3.5599	3825.5	2321	48.248
525	65.33	3646.3	.92804	2603.8	6071	45.578	775	315.33	5389.6	3.6421	3850.5	2284	48.293
530	70.33	3681.1	.95937	2628.6	5929	45.644	780	320.33	5424.6	3.7257	3875.6	2247	48.338
535	75.33	3715.9	.99145	2653.5	5791	45.709	785	325.33	5459.7	3.8107	3900.8	2211	48.383
540	80.33	3750.7	1.02429	2678.3	5658	45.774	790	330.33	5494.7	3.8970	3925.9	2175	48.427
545	85.33	3785.5	1.05790	2703.2	5529	45.838	795	335.33	5529.8	3.9848	3951.0	2141	48.472

Table 11 Nitrogen at Low Pressures (for One Pound-Mole)

$\bar{m} = 28.0134$

T	t	$\bar{h}$	p_r	$\bar{u}$	v_r	$\bar{\phi}$	T	t	$\bar{h}$	p_r	$\bar{u}$	v_r	$\bar{\phi}$
800	340.33	5564.8	4.074	3976.1	2107.3	48.516	1300	840.33	9153.9	23.47	6572.2	594.3	51.993
810	350.33	5635.0	4.257	4026.4	2042.0	48.603	1310	850.33	9227.7	24.15	6626.2	582.1	52.050
820	360.33	5705.2	4.446	4076.8	1979.5	48.689	1320	860.33	9301.7	24.85	6680.3	570.1	52.106
830	370.33	5775.5	4.640	4127.2	1919.5	48.774	1330	870.33	9375.7	25.56	6734.5	558.5	52.162
840	380.33	5845.8	4.841	4177.6	1862.0	48.858	1340	880.33	9449.8	26.28	6788.8	547.2	52.218
850	390.33	5916.1	5.049	4228.1	1806.8	48.941	1350	890.33	9524.1	27.02	6843.2	536.2	52.273
860	400.33	5986.5	5.262	4278.7	1753.8	49.024	1360	900.33	9598.4	27.78	6897.6	525.4	52.328
870	410.33	6057.0	5.483	4329.3	1702.9	49.105	1370	910.33	9672.8	28.55	6952.2	515.0	52.382
880	420.33	6127.5	5.710	4379.9	1654.0	49.186	1380	920.33	9747.3	29.34	7006.8	504.8	52.436
890	430.33	6198.1	5.944	4430.6	1606.9	49.266	1390	930.33	9821.9	30.15	7061.5	494.8	52.490
900	440.33	6268.7	6.185	4481.4	1561.7	49.344	1400	940.33	9896.5	30.97	7116.3	485.1	52.544
910	450.33	6339.4	6.433	4532.2	1518.1	49.423	1410	950.33	9971.3	31.81	7171.2	475.7	52.597
920	460.33	6410.1	6.688	4583.1	1476.2	49.500	1420	960.33	10046.1	32.67	7226.2	466.4	52.650
930	470.33	6480.9	6.951	4634.1	1435.8	49.576	1430	970.33	10121.1	33.55	7281.3	457.5	52.702
940	480.33	6551.8	7.221	4685.1	1396.9	49.652	1440	980.33	10196.1	34.44	7336.5	448.7	52.755
950	490.33	6622.7	7.499	4736.1	1359.4	49.727	1450	990.33	10271.2	35.36	7391.7	440.1	52.807
960	500.33	6693.7	7.786	4787.3	1323.2	49.802	1460	1000.33	10346.4	36.29	7447.1	431.8	52.858
970	510.33	6764.8	8.080	4838.5	1288.4	49.875	1470	1010.33	10421.7	37.24	7502.5	423.6	52.910
980	520.33	6835.9	8.382	4889.8	1254.7	49.948	1480	1020.33	10497.1	38.21	7558.1	415.7	52.961
990	530.33	6907.1	8.693	4941.1	1222.2	50.021	1490	1030.33	10572.6	39.20	7613.7	407.9	53.012
1000	540.33	6978.4	9.012	4992.5	1190.8	50.092	1500	1040.33	10648.2	40.21	7669.4	400.3	53.062
1010	550.33	7049.7	9.340	5044.0	1160.5	50.163	1510	1050.33	10723.8	41.24	7725.2	392.9	53.112
1020	560.33	7121.1	9.677	5095.5	1131.2	50.234	1520	1060.33	10799.6	42.29	7781.1	385.7	53.162
1030	570.33	7192.6	10.023	5147.2	1102.8	50.303	1530	1070.33	10875.4	43.37	7837.0	378.6	53.212
1040	580.33	7264.2	10.378	5198.9	1075.4	50.372	1540	1080.33	10951.3	44.46	7893.1	371.7	53.262
1050	590.33	7335.8	10.742	5250.7	1048.9	50.441	1550	1090.33	11027.3	45.57	7949.2	365.0	53.311
1060	600.33	7407.5	11.117	5302.5	1023.3	50.509	1560	1100.33	11103.4	46.71	8005.5	358.4	53.360
1070	610.33	7479.3	11.500	5354.5	998.5	50.576	1570	1110.33	11179.6	47.87	8061.8	352.0	53.408
1080	620.33	7551.2	11.894	5406.5	974.4	50.643	1580	1120.33	11255.9	49.05	8118.2	345.7	53.457
1090	630.33	7623.2	12.298	5458.6	951.1	50.710	1590	1130.33	11332.2	50.26	8174.7	339.5	53.505
1100	640.33	7695.2	12.712	5510.8	928.6	50.775	1600	1140.33	11408.6	51.48	8231.3	333.5	53.553
1110	650.33	7767.3	13.137	5563.0	906.7	50.841	1610	1150.33	11485.1	52.73	8287.9	327.6	53.601
1120	660.33	7839.5	13.573	5615.4	885.6	50.905	1620	1160.33	11561.8	54.01	8344.7	321.9	53.648
1130	670.33	7911.8	14.019	5667.8	865.0	50.970	1630	1170.33	11638.4	55.31	8401.5	316.3	53.695
1140	680.33	7984.2	14.476	5720.3	845.1	51.033	1640	1180.33	11715.2	56.63	8458.4	310.8	53.742
1150	690.33	8056.6	14.945	5772.9	825.8	51.097	1650	1190.33	11792.1	57.98	8515.4	305.4	53.789
1160	700.33	8129.2	15.425	5825.6	807.0	51.159	1660	1200.33	11869.0	59.35	8572.5	300.1	53.835
1170	710.33	8201.8	15.917	5878.3	788.8	51.222	1670	1210.33	11946.0	60.75	8629.6	295.0	53.882
1180	720.33	8274.5	16.421	5931.2	771.2	51.284	1680	1220.33	12023.1	62.18	8686.9	290.0	53.928
1190	730.33	8347.3	16.937	5984.1	754.0	51.345	1690	1230.33	12100.3	63.63	8744.2	285.0	53.973
1200	740.33	8420.2	17.465	6037.1	737.3	51.406	1700	1240.33	12177.6	65.10	8801.6	280.2	54.019
1210	750.33	8493.2	18.006	6090.3	721.2	51.467	1710	1250.33	12254.9	66.61	8859.1	275.5	54.064
1220	760.33	8566.2	18.559	6143.5	705.4	51.527	1720	1260.33	12332.3	68.14	8916.7	270.9	54.110
1230	770.33	8639.4	19.126	6196.7	690.2	51.586	1730	1270.33	12409.8	69.70	8974.3	266.4	54.155
1240	780.33	8712.6	19.706	6250.1	675.3	51.646	1740	1280.33	12487.4	71.29	9032.0	261.9	54.199
1250	790.33	8785.9	20.299	6303.6	660.8	51.705	1750	1290.33	12565.1	72.90	9089.8	257.6	54.244
1260	800.33	8859.3	20.906	6357.1	646.8	51.763	1760	1300.33	12642.8	74.55	9147.7	253.4	54.288
1270	810.33	8932.8	21.526	6410.8	633.1	51.821	1770	1310.33	12720.7	76.22	9205.7	249.2	54.332
1280	820.33	9006.4	22.161	6464.5	619.8	51.879	1780	1320.33	12798.6	77.92	9263.7	245.1	54.376
1290	830.33	9080.1	22.810	6518.3	606.9	51.936	1790	1330.33	12876.5	79.66	9321.9	241.1	54.420

Table 11   Nitrogen at Low Pressures (for One Pound-Mole)

$\bar{m} = 28.0134$

T	t	$\bar{h}$	p_r	$\bar{u}$	v_r	$\bar{\phi}$	T	t	$\bar{h}$	p_r	$\bar{u}$	v_r	$\bar{\phi}$
1800	1340.33	12954.6	81.42	9380.0	237.24	54.463	2800	2340.33	21071.9	493.3	15511.5	60.91	58.041
1820	1360.33	13110.9	85.04	9496.7	229.67	54.550	2820	2360.33	21239.1	508.3	15639.0	59.54	58.100
1840	1380.33	13267.6	88.79	9613.6	222.40	54.635	2840	2380.33	21406.5	523.7	15766.7	58.20	58.159
1860	1400.33	13424.5	92.66	9730.8	215.41	54.720	2860	2400.33	21574.1	539.4	15894.5	56.90	58.218
1880	1420.33	13581.7	96.67	9848.3	208.70	54.804	2880	2420.33	21741.8	555.5	16022.5	55.63	58.277
1900	1440.33	13739.3	100.81	9966.1	202.26	54.887	2900	2440.33	21909.6	572.0	16150.6	54.41	58.335
1920	1460.33	13897.1	105.10	10084.2	196.06	54.970	2920	2460.33	22077.6	588.9	16278.9	53.21	58.392
1940	1480.33	14055.2	109.52	10202.6	190.09	55.052	2940	2480.33	22245.7	606.2	16407.3	52.05	58.450
1960	1500.33	14213.5	114.09	10321.2	184.36	55.133	2960	2500.33	22414.0	623.8	16535.8	50.92	58.507
1980	1520.33	14372.2	118.81	10440.2	178.84	55.214	2980	2520.33	22582.4	641.9	16664.5	49.82	58.564
2000	1540.33	14531.1	123.69	10559.4	173.52	55.294	3000	2540.33	22750.9	660.4	16793.3	48.75	58.620
2020	1560.33	14690.3	128.72	10678.8	168.41	55.373	3020	2560.33	22919.5	679.3	16922.2	47.71	58.676
2040	1580.33	14849.7	133.92	10798.6	163.48	55.451	3040	2580.33	23088.3	698.6	17051.3	46.70	58.732
2060	1600.33	15009.5	139.27	10918.6	158.73	55.529	3060	2600.33	23257.2	718.3	17180.5	45.71	58.787
2080	1620.33	15169.4	144.80	11038.8	154.15	55.606	3080	2620.33	23426.2	738.5	17309.8	44.76	58.842
2100	1640.33	15329.6	150.50	11159.3	149.74	55.683	3100	2640.33	23595.4	759.2	17439.2	43.82	58.897
2120	1660.33	15490.1	156.38	11280.1	145.49	55.759	3120	2660.33	23764.6	780.3	17568.7	42.91	58.951
2140	1680.33	15650.8	162.43	11401.1	141.38	55.835	3140	2680.33	23934.0	801.8	17698.4	42.03	59.005
2160	1700.33	15811.8	168.67	11522.3	137.43	55.910	3160	2700.33	24103.5	823.8	17828.2	41.16	59.059
2180	1720.33	15973.0	175.10	11643.8	133.61	55.984	3180	2720.33	24273.1	846.3	17958.1	40.32	59.113
2200	1740.33	16134.4	181.72	11765.6	129.92	56.058	3200	2740.33	24442.9	869.3	18088.1	39.50	59.166
2220	1760.33	16296.1	188.54	11887.5	126.36	56.131	3220	2760.33	24612.7	892.8	18218.3	38.70	59.219
2240	1780.33	16458.0	195.56	12009.7	122.92	56.203	3240	2780.33	24782.7	916.8	18348.5	37.93	59.271
2260	1800.33	16620.1	202.79	12132.1	119.60	56.275	3260	2800.33	24952.7	941.2	18478.8	37.17	59.324
2280	1820.33	16782.5	210.23	12254.7	116.39	56.347	3280	2820.33	25122.9	966.2	18609.3	36.43	59.376
2300	1840.33	16945.1	217.88	12377.6	113.29	56.418	3300	2840.33	25293.2	991.8	18739.9	35.71	59.428
2320	1860.33	17107.8	225.75	12500.6	110.29	56.488	3320	2860.33	25463.6	1017.8	18870.5	35.01	59.479
2340	1880.33	17270.8	233.84	12623.9	107.39	56.558	3340	2880.33	25634.1	1044.4	19001.3	34.32	59.530
2360	1900.33	17434.0	242.16	12747.4	104.58	56.628	3360	2900.33	25804.7	1071.5	19132.2	33.65	59.581
2380	1920.33	17597.5	250.72	12871.1	101.87	56.697	3380	2920.33	25975.4	1099.2	19263.2	33.00	59.632
2400	1940.33	17761.1	259.51	12995.0	99.25	56.765	3400	2940.33	26146.2	1127.4	19394.2	32.36	59.682
2420	1960.33	17924.9	268.55	13119.1	96.70	56.833	3420	2960.33	26317.1	1156.2	19525.4	31.74	59.732
2440	1980.33	18088.9	277.84	13243.4	94.25	56.901	3440	2980.33	26488.1	1185.6	19656.7	31.14	59.782
2460	2000.33	18253.1	287.37	13367.9	91.86	56.968	3460	3000.33	26659.1	1215.6	19788.1	30.54	59.832
2480	2020.33	18417.5	297.17	13492.6	89.56	57.034	3480	3020.33	26830.3	1246.2	19919.5	29.97	59.881
2500	2040.33	18582.1	307.23	13617.5	87.33	57.100	3500	3040.33	27001.6	1277.4	20051.1	29.40	59.930
2520	2060.33	18746.9	317.55	13742.5	85.16	57.166	3520	3060.33	27173.0	1309.2	20182.7	28.85	59.979
2540	2080.33	18911.9	328.15	13867.8	83.07	57.231	3540	3080.33	27344.4	1341.6	20314.5	28.32	60.027
2560	2100.33	19077.0	339.03	13993.2	81.03	57.296	3560	3100.33	27516.0	1374.6	20446.3	27.79	60.076
2580	2120.33	19242.3	350.19	14118.8	79.06	57.360	3580	3120.33	27687.6	1408.3	20578.2	27.28	60.124
2600	2140.33	19407.8	361.64	14244.6	77.15	57.424	3600	3140.33	27859.3	1442.7	20710.3	26.78	60.172
2620	2160.33	19573.5	373.39	14370.5	75.30	57.488	3620	3160.33	28031.2	1477.6	20842.4	26.29	60.219
2640	2180.33	19739.3	385.43	14496.6	73.50	57.551	3640	3180.33	28203.1	1513.3	20974.5	25.81	60.267
2660	2200.33	19905.3	397.79	14622.9	71.76	57.613	3660	3200.33	28375.1	1549.6	21106.8	25.35	60.314
2680	2220.33	20071.5	410.45	14749.4	70.07	57.676	3680	3220.33	28547.1	1586.7	21239.2	24.89	60.361
2700	2240.33	20237.8	423.43	14876.0	68.43	57.737	3700	3240.33	28719.3	1624.4	21371.6	24.44	60.407
2720	2260.33	20404.3	436.74	15002.8	66.84	57.799	3720	3260.33	28891.5	1662.8	21504.1	24.01	60.454
2740	2280.33	20571.0	450.37	15129.7	65.29	57.860	3740	3280.33	29063.8	1701.9	21636.7	23.58	60.500
2760	2300.33	20737.8	464.34	15256.8	63.79	57.921	3760	3300.33	29236.2	1741.8	21769.4	23.17	60.546
2780	2320.33	20904.7	478.65	15384.1	62.33	57.981	3780	3320.33	29408.7	1782.4	21902.1	22.76	60.592

Table 11   Nitrogen at Low Pressures (for One Pound-Mole)
$\bar{m} = 28.0134$

T	t	$\bar{h}$	p_r	$\bar{u}$	v_r	$\bar{\phi}$	T	t	$\bar{h}$	p_r	$\bar{u}$	v_r	$\bar{\phi}$
3800	3340.33	29581.2	1824	22035.0	22.361	60.637	4600	4140.33	36537.9	4209	27403.0	11.728	62.298
3820	3360.33	29753.9	1866	22167.9	21.971	60.683	4620	4160.33	36713.0	4291	27538.3	11.555	62.336
3840	3380.33	29926.6	1909	22300.9	21.590	60.728	4640	4180.33	36888.2	4373	27673.8	11.386	62.374
3860	3400.33	30099.3	1952	22433.9	21.218	60.772	4660	4200.33	37063.3	4457	27809.2	11.221	62.412
3880	3420.33	30272.2	1997	22567.1	20.854	60.817	4680	4220.33	37238.6	4542	27944.8	11.058	62.449
3900	3440.33	30445.1	2042	22700.3	20.497	60.862	4700	4240.33	37413.9	4628	28080.3	10.898	62.487
3920	3460.33	30618.1	2088	22833.6	20.148	60.906	4720	4260.33	37589.2	4716	28215.9	10.741	62.524
3940	3480.33	30791.2	2135	22966.9	19.807	60.950	4740	4280.33	37764.6	4805	28351.6	10.587	62.561
3960	3500.33	30964.3	2182	23100.3	19.473	60.994	4760	4300.33	37940.0	4895	28487.3	10.436	62.598
3980	3520.33	31137.5	2231	23233.8	19.146	61.037	4780	4320.33	38115.5	4986	28623.1	10.287	62.635
4000	3540.33	31310.8	2280	23367.4	18.826	61.081	4800	4340.33	38291.0	5079	28758.9	10.142	62.671
4020	3560.33	31484.2	2330	23501.0	18.513	61.124	4820	4360.33	38466.5	5173	28894.7	9.998	62.708
4040	3580.33	31657.6	2381	23634.7	18.206	61.167	4840	4380.33	38642.2	5269	29030.6	9.858	62.744
4060	3600.33	31831.1	2433	23768.5	17.906	61.210	4860	4400.33	38817.8	5366	29166.5	9.720	62.780
4080	3620.33	32004.6	2486	23902.3	17.612	61.253	4880	4420.33	38993.5	5464	29302.5	9.584	62.816
4100	3640.33	32178.2	2540	24036.2	17.324	61.295	4900	4440.33	39169.2	5564	29438.5	9.451	62.852
4120	3660.33	32351.9	2594	24170.2	17.042	61.337	4920	4460.33	39345.0	5665	29574.6	9.320	62.888
4140	3680.33	32525.7	2650	24304.2	16.765	61.379	4940	4480.33	39520.8	5768	29710.7	9.191	62.924
4160	3700.33	32699.5	2706	24438.3	16.495	61.421	4960	4500.33	39696.7	5872	29846.8	9.065	62.959
4180	3720.33	32873.3	2764	24572.4	16.230	61.463	4980	4520.33	39872.6	5978	29983.0	8.940	62.995
4200	3740.33	33047.3	2822	24706.7	15.970	61.504	5000	4540.33	40048.6	6085	30119.3	8.818	63.030
4220	3760.33	33221.3	2882	24840.9	15.716	61.546	5020	4560.33	40224.6	6193	30255.5	8.698	63.065
4240	3780.33	33395.3	2942	24975.3	15.466	61.587	5040	4580.33	40400.6	6304	30391.9	8.580	63.100
4260	3800.33	33569.4	3003	25109.7	15.222	61.628	5060	4600.33	40576.7	6415	30528.2	8.465	63.135
4280	3820.33	33743.6	3066	25244.1	14.982	61.669	5080	4620.33	40752.8	6528	30664.6	8.351	63.170
4300	3840.33	33917.8	3129	25378.6	14.748	61.709	5100	4640.33	40928.9	6643	30801.0	8.239	63.204
4320	3860.33	34092.1	3193	25513.2	14.518	61.750	5120	4660.33	41105.1	6759	30937.5	8.129	63.239
4340	3880.33	34266.5	3259	25647.8	14.292	61.790	5140	4680.33	41281.4	6877	31074.0	8.020	63.273
4360	3900.33	34440.9	3325	25782.5	14.071	61.830	5160	4700.33	41457.6	6997	31210.6	7.914	63.307
4380	3920.33	34615.3	3393	25917.3	13.854	61.870	5180	4720.33	41633.9	7118	31347.2	7.809	63.342
4400	3940.33	34789.8	3461	26052.1	13.642	61.910	5200	4740.33	41810.3	7241	31483.8	7.707	63.375
4420	3960.33	34964.4	3531	26186.9	13.433	61.949	5220	4760.33	41986.7	7366	31620.5	7.605	63.409
4440	3980.33	35139.0	3602	26321.8	13.229	61.989	5240	4780.33	42163.1	7492	31757.2	7.506	63.443
4460	4000.33	35313.7	3674	26456.8	13.028	62.028	5260	4800.33	42339.6	7620	31893.9	7.408	63.477
4480	4020.33	35488.4	3747	26591.8	12.832	62.067	5280	4820.33	42516.0	7749	32030.7	7.312	63.510
4500	4040.33	35663.2	3821	26726.9	12.639	62.106	5300	4840.33	42692.6	7881	32167.5	7.217	63.544
4520	4060.33	35838.1	3896	26862.0	12.450	62.145	5320	4860.33	42869.2	8014	32304.4	7.124	63.577
4540	4080.33	36013.0	3973	26997.1	12.264	62.183	5340	4880.33	43045.8	8148	32441.3	7.033	63.610
4560	4100.33	36187.9	4050	27132.4	12.082	62.222	5360	4900.33	43222.4	8285	32578.2	6.943	63.643
4580	4120.33	36362.9	4129	27267.6	11.903	62.260	5380	4920.33	43399.1	8423	32715.1	6.854	63.676

Table 12   Nitrogen

$\bar{m} = 28.0134$

T R	t F	$\bar{c}_p$ Btu lb-mole R	$\bar{c}_v$ Btu lb-mole R	$k = \dfrac{\bar{c}_p}{\bar{c}_v}$	a ft sec
200	−259.67	6.951	4.966	1.400	704.9
250	−209.67	6.952	4.966	1.400	788.1
300	−159.67	6.952	4.966	1.400	863.3
350	−109.67	6.952	4.966	1.400	932.5
400	−59.67	6.952	4.967	1.400	996.9
450	−9.67	6.953	4.967	1.400	1057.3
500	40.33	6.954	4.969	1.400	1114.5
550	90.33	6.957	4.971	1.399	1168.8
600	140.33	6.962	4.976	1.399	1220.6
650	190.33	6.969	4.984	1.398	1270.2
700	240.33	6.981	4.995	1.398	1317.7
750	290.33	6.995	5.010	1.396	1363.4
800	340.33	7.015	5.029	1.395	1407.3
850	390.33	7.038	5.052	1.393	1449.7
900	440.33	7.065	5.079	1.391	1490.6
1000	540.33	7.131	5.145	1.386	1568.4
1100	640.33	7.208	5.222	1.380	1641.6
1200	740.33	7.293	5.307	1.374	1710.8
1300	840.33	7.382	5.396	1.368	1776.6
1400	940.33	7.472	5.486	1.362	1839.6
1500	1040.33	7.561	5.575	1.356	1900.1
1600	1140.33	7.648	5.662	1.351	1958.5
1700	1240.33	7.731	5.745	1.346	2015.0
1800	1340.33	7.809	5.823	1.341	2069.8
1900	1440.33	7.883	5.897	1.337	2123.1
2000	1540.33	7.953	5.967	1.333	2175.1
2100	1640.33	8.017	6.032	1.329	2225.8
2200	1740.33	8.078	6.092	1.326	2275.4
2300	1840.33	8.134	6.148	1.323	2323.9
2400	1940.33	8.186	6.200	1.320	2371.5
2600	2140.33	8.279	6.293	1.316	2463.9
2800	2340.33	8.359	6.374	1.312	2553.0
3000	2540.33	8.429	6.443	1.308	2639.2
3200	2740.33	8.489	6.504	1.305	2722.8
3400	2940.33	8.542	6.556	1.303	2803.9
3600	3140.33	8.589	6.603	1.301	2882.9
3800	3340.33	8.629	6.644	1.299	2959.8
4000	3540.33	8.666	6.680	1.297	3034.8
4200	3740.33	8.698	6.712	1.296	3108.0
4400	3940.33	8.727	6.741	1.295	3179.6
4600	4140.33	8.753	6.767	1.293	3249.6
4800	4340.33	8.777	6.791	1.292	3318.2
5000	4540.33	8.799	6.813	1.291	3385.4
5200	4740.33	8.818	6.832	1.291	3451.3
5400	4940.33	8.836	6.851	1.290	3516.0
5600	5140.33	8.853	6.867	1.289	3579.5
5800	5340.33	8.869	6.883	1.289	3642.0
6000	5540.33	8.883	6.897	1.288	3703.4
6200	5740.33	8.897	6.911	1.287	3763.8
6400	5940.33	8.909	6.923	1.287	3823.2

Table 13 Oxygen at Low Pressures (for One Pound-Mole)

$\bar{m} = 31.9988$

T	t	$\bar{h}$	p_r	$\bar{u}$	v_r	$\bar{\phi}$	T	t	$\bar{h}$	p_r	$\bar{u}$	v_r	$\bar{\phi}$
300	−159.67	2083.0	.6624	1487.3	4860	44.908	550	90.33	3826.7	5.562	2734.5	1061.2	49.134
305	−154.67	2117.8	.7018	1512.1	4664	45.023	555	95.33	3861.8	5.743	2759.7	1037.1	49.197
310	−149.67	2152.5	.7430	1536.9	4478	45.136	560	100.33	3897.0	5.928	2784.9	1013.7	49.260
315	−144.67	2187.3	.7858	1561.8	4302	45.247	565	105.33	3932.2	6.118	2810.2	991.1	49.323
320	−139.67	2222.1	.8303	1586.6	4136	45.357	570	110.33	3967.4	6.312	2835.4	969.1	49.385
325	−134.67	2256.8	.8767	1611.4	3978	45.465	575	115.33	4002.6	6.511	2860.7	947.8	49.447
330	−129.67	2291.6	.9248	1636.3	3829	45.571	580	120.33	4037.8	6.714	2886.0	927.1	49.508
335	−124.67	2326.4	.9748	1661.1	3688	45.675	585	125.33	4073.1	6.921	2911.3	907.0	49.568
340	−119.67	2361.2	1.0267	1686.0	3554	45.778	590	130.33	4108.3	7.134	2936.7	887.5	49.628
345	−114.67	2395.9	1.0806	1710.8	3426	45.880	595	135.33	4143.6	7.351	2962.0	868.6	49.688
350	−109.67	2430.7	1.1364	1735.7	3305	45.980	600	140.33	4178.9	7.573	2987.4	850.2	49.747
355	−104.67	2465.5	1.1943	1760.5	3190	46.079	605	145.33	4214.3	7.800	3012.8	832.4	49.805
360	−99.67	2500.3	1.2543	1785.4	3080	46.176	610	150.33	4249.6	8.032	3038.2	815.0	49.864
365	−94.67	2535.1	1.3164	1810.2	2976	46.272	615	155.33	4285.0	8.269	3063.7	798.1	49.921
370	−89.67	2569.9	1.3807	1835.1	2876	46.367	620	160.33	4320.4	8.511	3089.2	781.7	49.979
375	−84.67	2604.6	1.4471	1859.9	2781	46.460	625	165.33	4355.8	8.759	3114.7	765.8	50.036
380	−79.67	2639.4	1.5159	1884.8	2690	46.552	630	170.33	4391.3	9.012	3140.2	750.2	50.092
385	−74.67	2674.2	1.5869	1909.7	2604	46.643	635	175.33	4426.8	9.270	3165.7	735.1	50.148
390	−69.67	2709.0	1.6604	1934.5	2521	46.733	640	180.33	4462.3	9.533	3191.3	720.4	50.204
395	−64.67	2743.8	1.7362	1959.4	2442	46.822	645	185.33	4497.8	9.803	3216.9	706.1	50.259
400	−59.67	2778.6	1.8144	1984.3	2366	46.909	650	190.33	4533.3	10.077	3242.5	692.2	50.314
405	−54.67	2813.5	1.8952	2009.2	2293	46.996	655	195.33	4568.9	10.358	3268.2	678.6	50.369
410	−49.67	2848.3	1.9785	2034.1	2224	47.081	660	200.33	4604.5	10.644	3293.8	665.4	50.423
415	−44.67	2883.1	2.0645	2059.0	2157	47.166	665	205.33	4640.1	10.936	3319.5	652.5	50.476
420	−39.67	2917.9	2.1530	2083.9	2093	47.249	670	210.33	4675.8	11.235	3345.3	640.0	50.530
425	−34.67	2952.8	2.2443	2108.8	2032	47.331	675	215.33	4711.5	11.539	3371.0	627.8	50.583
430	−29.67	2987.6	2.3383	2133.7	1973	47.413	680	220.33	4747.2	11.849	3396.8	615.9	50.636
435	−24.67	3022.4	2.4351	2158.6	1917	47.494	685	225.33	4782.9	12.166	3422.6	604.2	50.688
440	−19.67	3057.3	2.5348	2183.5	1863	47.573	690	230.33	4818.7	12.489	3448.4	592.9	50.740
445	−14.67	3092.2	2.6374	2208.4	1811	47.652	695	235.33	4854.5	12.818	3474.3	581.9	50.792
450	−9.67	3127.0	2.7429	2233.4	1761	47.730	700	240.33	4890.3	13.154	3500.2	571.1	50.843
455	−4.67	3161.9	2.8515	2258.3	1712	47.807	705	245.33	4926.1	13.496	3526.1	560.6	50.894
460	.33	3196.8	2.9631	2283.3	1666	47.883	710	250.33	4962.0	13.845	3552.1	550.3	50.945
465	5.33	3231.7	3.0779	2308.3	1621	47.959	715	255.33	4997.9	14.201	3578.0	540.3	50.995
470	10.33	3266.6	3.1958	2333.2	1578	48.033	720	260.33	5033.9	14.564	3604.1	530.5	51.045
475	15.33	3301.5	3.3169	2358.2	1537	48.107	725	265.33	5069.8	14.934	3630.1	521.0	51.095
480	20.33	3336.4	3.4414	2383.2	1497	48.180	730	270.33	5105.8	15.310	3656.2	511.7	51.145
485	25.33	3371.4	3.5692	2408.2	1458	48.253	735	275.33	5141.9	15.694	3682.3	502.6	51.194
490	30.33	3406.3	3.7004	2433.2	1421	48.324	740	280.33	5177.9	16.086	3708.4	493.7	51.243
495	35.33	3441.3	3.8350	2458.3	1385	48.395	745	285.33	5214.0	16.484	3734.6	485.0	51.291
500	40.33	3476.2	3.9732	2483.3	1350	48.466	750	290.33	5250.1	16.890	3760.8	476.5	51.340
505	45.33	3511.2	4.1149	2508.4	1317	48.535	755	295.33	5286.3	17.304	3787.0	468.2	51.388
510	50.33	3546.2	4.2603	2533.4	1285	48.604	760	300.33	5322.5	17.725	3813.2	460.1	51.435
515	55.33	3581.2	4.4094	2558.5	1253	48.673	765	305.33	5358.7	18.154	3839.5	452.2	51.483
520	60.33	3616.2	4.5623	2583.6	1223	48.740	770	310.33	5395.0	18.591	3865.8	444.5	51.530
525	65.33	3651.3	4.7190	2608.7	1194	48.807	775	315.33	5431.2	19.036	3892.2	436.9	51.577
530	70.33	3686.3	4.8796	2633.8	1166	48.874	780	320.33	5467.5	19.489	3918.6	429.5	51.624
535	75.33	3721.4	5.0441	2659.0	1138	48.940	785	325.33	5503.9	19.950	3945.0	422.3	51.670
540	80.33	3756.5	5.2127	2684.1	1112	49.005	790	330.33	5540.3	20.420	3971.4	415.2	51.716
545	85.33	3791.6	5.3853	2709.3	1086	49.070	795	335.33	5576.7	20.898	3997.9	408.3	51.762

Table 13   Oxygen at Low Pressures (for One Pound-Mole)

$\bar{m} = 31.9988$

T	t	$\bar{h}$	p_r	$\bar{u}$	v_r	$\bar{\phi}$	T	t	$\bar{h}$	p_r	$\bar{u}$	v_r	$\bar{\phi}$
800	340.33	5613.1	21.38	4024.4	401.5	51.808	1300	840.33	9422.6	136.87	6840.9	101.93	55.495
810	350.33	5686.1	22.38	4077.6	388.4	51.899	1310	850.33	9501.8	141.12	6900.3	99.62	55.555
820	360.33	5759.2	23.42	4130.8	375.8	51.988	1320	860.33	9581.2	145.48	6959.8	97.37	55.616
830	370.33	5832.5	24.49	4184.2	363.7	52.077	1330	870.33	9660.6	149.94	7019.4	95.19	55.676
840	380.33	5905.8	25.60	4237.7	352.2	52.165	1340	880.33	9740.2	154.50	7079.1	93.07	55.735
850	390.33	5979.4	26.74	4291.4	341.1	52.252	1350	890.33	9819.8	159.18	7138.9	91.01	55.795
860	400.33	6053.0	27.93	4345.2	330.5	52.338	1360	900.33	9899.6	163.97	7198.8	89.01	55.853
870	410.33	6126.8	29.15	4399.1	320.2	52.424	1370	910.33	9979.5	168.87	7258.8	87.06	55.912
880	420.33	6200.7	30.42	4453.2	310.4	52.508	1380	920.33	10059.4	173.89	7318.9	85.17	55.970
890	430.33	6274.8	31.73	4507.4	301.0	52.592	1390	930.33	10139.5	179.03	7379.1	83.32	56.028
900	440.33	6349.0	33.08	4561.7	291.9	52.675	1400	940.33	10219.6	184.28	7439.4	81.53	56.085
910	450.33	6423.3	34.48	4616.2	283.2	52.757	1410	950.33	10299.9	189.66	7499.8	79.78	56.142
920	460.33	6497.8	35.92	4670.8	274.8	52.838	1420	960.33	10380.2	195.16	7560.3	78.08	56.199
930	470.33	6572.4	37.41	4725.5	266.8	52.919	1430	970.33	10460.7	200.79	7620.9	76.43	56.256
940	480.33	6647.1	38.95	4780.4	259.0	52.999	1440	980.33	10541.2	206.54	7681.5	74.82	56.312
950	490.33	6722.0	40.53	4835.4	251.5	53.078	1450	990.33	10621.8	212.43	7742.3	73.25	56.368
960	500.33	6797.0	42.17	4890.5	244.3	53.157	1460	1000.33	10702.5	218.44	7803.1	71.73	56.423
970	510.33	6872.1	43.85	4945.8	237.4	53.234	1470	1010.33	10783.3	224.59	7864.1	70.24	56.478
980	520.33	6947.4	45.59	5001.2	230.7	53.312	1480	1020.33	10864.2	230.88	7925.1	68.79	56.533
990	530.33	7022.8	47.38	5056.8	224.2	53.388	1490	1030.33	10945.1	237.31	7986.2	67.38	56.588
1000	540.33	7098.3	49.23	5112.5	218.0	53.464	1500	1040.33	11026.2	243.88	8047.4	66.01	56.642
1010	550.33	7174.0	51.13	5168.3	212.0	53.539	1510	1050.33	11107.3	250.59	8108.7	64.67	56.696
1020	560.33	7249.8	53.09	5224.2	206.2	53.614	1520	1060.33	11188.6	257.45	8170.0	63.36	56.749
1030	570.33	7325.7	55.11	5280.3	200.6	53.688	1530	1070.33	11269.9	264.45	8231.5	62.09	56.803
1040	580.33	7401.8	57.19	5336.5	195.2	53.762	1540	1080.33	11351.3	271.61	8293.0	60.85	56.856
1050	590.33	7478.0	59.33	5392.8	189.9	53.835	1550	1090.33	11432.7	278.92	8354.6	59.64	56.908
1060	600.33	7554.3	61.53	5449.3	184.9	53.907	1560	1100.33	11514.3	286.38	8416.3	58.46	56.961
1070	610.33	7630.8	63.79	5505.9	180.0	53.979	1570	1110.33	11595.9	294.00	8478.1	57.31	57.013
1080	620.33	7707.4	66.12	5562.6	175.3	54.050	1580	1120.33	11677.6	301.78	8540.0	56.19	57.065
1090	630.33	7784.1	68.52	5619.5	170.7	54.121	1590	1130.33	11759.4	309.73	8601.9	55.09	57.116
1100	640.33	7860.9	70.98	5676.4	166.3	54.191	1600	1140.33	11841.3	317.84	8663.9	54.02	57.168
1110	650.33	7937.9	73.52	5733.6	162.0	54.260	1610	1150.33	11923.2	326.12	8726.0	52.98	57.219
1120	660.33	8014.9	76.12	5790.8	157.9	54.330	1620	1160.33	12005.2	334.56	8788.1	51.96	57.270
1130	670.33	8092.1	78.80	5848.1	153.9	54.398	1630	1170.33	12087.3	343.18	8850.4	50.97	57.320
1140	680.33	8169.5	81.55	5905.6	150.0	54.466	1640	1180.33	12169.5	351.98	8912.7	50.00	57.370
1150	690.33	8246.9	84.38	5963.2	146.3	54.534	1650	1190.33	12251.7	360.95	8975.1	49.06	57.420
1160	700.33	8324.5	87.28	6020.9	142.6	54.601	1660	1200.33	12334.0	370.11	9037.5	48.13	57.470
1170	710.33	8402.2	90.26	6078.7	139.1	54.668	1670	1210.33	12416.4	379.44	9100.0	47.23	57.520
1180	720.33	8480.0	93.32	6136.7	135.7	54.734	1680	1220.33	12498.9	388.97	9162.6	46.35	57.569
1190	730.33	8557.9	96.46	6194.7	132.4	54.800	1690	1230.33	12581.4	398.68	9225.3	45.49	57.618
1200	740.33	8636.0	99.69	6252.9	129.2	54.865	1700	1240.33	12664.0	408.58	9288.0	44.65	57.666
1210	750.33	8714.1	103.00	6311.2	126.1	54.930	1710	1250.33	12746.7	418.68	9350.8	43.83	57.715
1220	760.33	8792.4	106.39	6369.6	123.1	54.994	1720	1260.33	12829.4	428.98	9413.7	43.03	57.763
1230	770.33	8870.8	109.88	6428.2	120.1	55.058	1730	1270.33	12912.2	439.47	9476.7	42.25	57.811
1240	780.33	8949.3	113.45	6486.8	117.3	55.122	1740	1280.33	12995.1	450.17	9539.7	41.48	57.859
1250	790.33	9027.9	117.11	6545.6	114.5	55.185	1750	1290.33	13078.0	461.07	9602.7	40.73	57.906
1260	800.33	9106.6	120.87	6604.4	111.9	55.248	1760	1300.33	13161.0	472.18	9665.9	40.00	57.954
1270	810.33	9185.4	124.72	6663.4	109.3	55.310	1770	1310.33	13244.1	483.51	9729.1	39.29	58.001
1280	820.33	9264.4	128.67	6722.5	106.8	55.372	1780	1320.33	13327.2	495.05	9792.3	38.59	58.048
1290	830.33	9343.4	132.72	6781.6	104.3	55.434	1790	1330.33	13410.4	506.80	9855.7	37.90	58.094

Table 13 Oxygen at Low Pressures (for One Pound-Mole)

$\bar{m} = 31.9988$

T	t	$\bar{h}$	p_r	$\bar{u}$	v_r	$\bar{\phi}$	T	t	$\bar{h}$	p_r	$\bar{u}$	v_r	$\bar{\phi}$
1800	1340.33	13493.6	518.8	9919.1	37.235	58.141	2800	2340.33	22060.9	3478	16500.5	8.640	61.919
1820	1360.33	13660.3	543.4	10046.0	35.943	58.233	2820	2360.33	22236.2	3589	16636.1	8.433	61.981
1840	1380.33	13827.2	568.9	10173.3	34.706	58.324	2840	2380.33	22411.6	3702	16771.8	8.232	62.043
1860	1400.33	13994.4	595.4	10300.7	33.523	58.414	2860	2400.33	22587.1	3819	16907.6	8.037	62.105
1880	1420.33	14161.8	622.9	10428.4	32.390	58.504	2880	2420.33	22762.8	3939	17043.5	7.847	62.166
1900	1440.33	14329.5	651.3	10556.3	31.304	58.593	2900	2440.33	22938.6	4061	17179.6	7.663	62.227
1920	1460.33	14497.3	680.8	10684.5	30.264	58.680	2920	2460.33	23114.5	4187	17315.7	7.485	62.287
1940	1480.33	14665.4	711.3	10812.8	29.268	58.768	2940	2480.33	23290.5	4315	17452.1	7.312	62.348
1960	1500.33	14833.7	742.9	10941.4	28.312	58.854	2960	2500.33	23466.7	4447	17588.5	7.143	62.407
1980	1520.33	15002.2	775.6	11070.2	27.395	58.939	2980	2520.33	23643.0	4582	17725.1	6.980	62.467
2000	1540.33	15170.9	809.5	11199.2	26.515	59.024	3000	2540.33	23819.4	4720	17861.8	6.821	62.526
2020	1560.33	15339.9	844.5	11328.4	25.670	59.108	3020	2560.33	23995.9	4862	17998.6	6.666	62.584
2040	1580.33	15509.0	880.6	11457.8	24.859	59.192	3040	2580.33	24172.6	5006	18135.6	6.516	62.643
2060	1600.33	15678.3	918.0	11587.5	24.080	59.274	3060	2600.33	24349.4	5155	18272.6	6.371	62.701
2080	1620.33	15847.9	956.7	11717.3	23.332	59.356	3080	2620.33	24526.3	5306	18409.8	6.229	62.758
2100	1640.33	16017.6	996.6	11847.3	22.612	59.437	3100	2640.33	24703.3	5462	18547.2	6.091	62.815
2120	1660.33	16187.5	1037.9	11977.5	21.921	59.518	3120	2660.33	24880.5	5621	18684.6	5.957	62.872
2140	1680.33	16357.6	1080.5	12107.8	21.255	59.598	3140	2680.33	25057.8	5783	18822.2	5.827	62.929
2160	1700.33	16527.8	1124.4	12238.4	20.615	59.677	3160	2700.33	25235.2	5950	18959.9	5.700	62.985
2180	1720.33	16698.3	1169.8	12369.1	19.999	59.755	3180	2720.33	25412.7	6120	19097.7	5.576	63.041
2200	1740.33	16868.9	1216.6	12500.0	19.406	59.833	3200	2740.33	25590.4	6294	19235.6	5.456	63.097
2220	1760.33	17039.8	1264.9	12631.1	18.835	59.911	3220	2760.33	25768.2	6472	19373.7	5.339	63.152
2240	1780.33	17210.7	1314.7	12762.4	18.285	59.987	3240	2780.33	25946.1	6654	19511.9	5.225	63.208
2260	1800.33	17381.9	1366.0	12893.8	17.755	60.063	3260	2800.33	26124.1	6840	19650.2	5.115	63.262
2280	1820.33	17553.2	1418.9	13025.5	17.244	60.139	3280	2820.33	26302.2	7030	19788.6	5.007	63.317
2300	1840.33	17724.7	1473.5	13157.2	16.751	60.214	3300	2840.33	26480.5	7225	19927.2	4.902	63.371
2320	1860.33	17896.4	1529.6	13289.2	16.277	60.288	3320	2860.33	26658.9	7424	20065.8	4.799	63.425
2340	1880.33	18068.2	1587.5	13421.3	15.818	60.362	3340	2880.33	26837.4	7627	20204.6	4.700	63.479
2360	1900.33	18240.1	1647.1	13553.5	15.376	60.435	3360	2900.33	27016.0	7834	20343.5	4.603	63.532
2380	1920.33	18412.3	1708.4	13685.9	14.950	60.508	3380	2920.33	27194.8	8046	20482.6	4.508	63.585
2400	1940.33	18584.6	1771.6	13818.5	14.538	60.580	3400	2940.33	27373.6	8263	20621.7	4.416	63.638
2420	1960.33	18757.0	1836.6	13951.2	14.140	60.651	3420	2960.33	27552.6	8484	20761.0	4.326	63.690
2440	1980.33	18929.6	1903.5	14084.1	13.756	60.722	3440	2980.33	27731.7	8710	20900.4	4.238	63.742
2460	2000.33	19102.4	1972.3	14217.1	13.385	60.793	3460	3000.33	27911.0	8941	21039.9	4.153	63.794
2480	2020.33	19275.3	2043.0	14350.3	13.027	60.863	3480	3020.33	28090.3	9177	21179.5	4.070	63.846
2500	2040.33	19448.3	2115.8	14483.6	12.680	60.932	3500	3040.33	28269.8	9418	21319.3	3.988	63.897
2520	2060.33	19621.5	2190.6	14617.1	12.345	61.001	3520	3060.33	28449.4	9663	21459.1	3.909	63.949
2540	2080.33	19794.8	2267.5	14750.7	12.021	61.070	3540	3080.33	28629.1	9914	21599.1	3.832	63.999
2560	2100.33	19968.3	2346.5	14884.5	11.708	61.138	3560	3100.33	28808.9	10170	21739.2	3.756	64.050
2580	2120.33	20141.9	2427.7	15018.4	11.405	61.205	3580	3120.33	28988.8	10432	21879.4	3.683	64.101
2600	2140.33	20315.7	2511.2	15152.5	11.111	61.272	3600	3140.33	29168.9	10699	22019.8	3.611	64.151
2620	2160.33	20489.6	2596.8	15286.7	10.827	61.339	3620	3160.33	29349.0	10971	22160.2	3.541	64.201
2640	2180.33	20663.7	2684.8	15421.0	10.552	61.405	3640	3180.33	29529.3	11249	22300.8	3.473	64.250
2660	2200.33	20837.8	2775.2	15555.5	10.286	61.471	3660	3200.33	29709.7	11532	22441.5	3.406	64.300
2680	2220.33	21012.2	2868.0	15690.1	10.028	61.536	3680	3220.33	29890.3	11822	22582.3	3.341	64.349
2700	2240.33	21186.6	2963.2	15824.8	9.778	61.601	3700	3240.33	30070.9	12117	22723.2	3.277	64.398
2720	2260.33	21361.2	3060.9	15959.7	9.536	61.666	3720	3260.33	30251.7	12418	22864.3	3.215	64.447
2740	2280.33	21536.0	3161.1	16094.7	9.302	61.730	3740	3280.33	30432.5	12725	23005.4	3.154	64.495
2760	2300.33	21710.8	3264.0	16229.8	9.074	61.793	3760	3300.33	30613.5	13038	23146.7	3.095	64.543
2780	2320.33	21885.8	3369.5	16365.1	8.854	61.856	3780	3320.33	30794.7	13357	23288.1	3.037	64.591

Table 13 Oxygen at Low Pressures (for One Pound-Mole)

$\bar{m} = 31.9988$

T	t	$\bar{h}$	p_r	$\bar{u}$	v_r	$\bar{\phi}$	T	t	$\bar{h}$	p_r	$\bar{u}$	v_r	$\bar{\phi}$
3800	3340.33	30975.9	13682	23429.6	2.980	64.639	4600	4140.33	38317.9	33073	29182.9	1.4926	66.392
3820	3360.33	31157.2	14014	23571.2	2.925	64.687	4620	4160.33	38503.7	33751	29329.0	1.4690	66.432
3840	3380.33	31338.7	14353	23713.0	2.871	64.734	4640	4180.33	38689.6	34440	29475.2	1.4458	66.472
3860	3400.33	31520.2	14698	23854.8	2.818	64.781	4660	4200.33	38875.6	35141	29621.5	1.4231	66.512
3880	3420.33	31701.9	15049	23996.8	2.767	64.828	4680	4220.33	39061.7	35853	29767.9	1.4008	66.552
3900	3440.33	31883.7	15408	24138.9	2.716	64.875	4700	4240.33	39248.0	36578	29914.4	1.3789	66.592
3920	3460.33	32065.7	15773	24281.1	2.667	64.922	4720	4260.33	39434.3	37314	30061.0	1.3575	66.631
3940	3480.33	32247.7	16145	24423.4	2.619	64.968	4740	4280.33	39620.7	38062	30207.8	1.3364	66.671
3960	3500.33	32429.8	16524	24565.8	2.572	65.014	4760	4300.33	39807.3	38822	30354.6	1.3158	66.710
3980	3520.33	32612.1	16911	24708.4	2.526	65.060	4780	4320.33	39993.9	39594	30501.5	1.2956	66.749
4000	3540.33	32794.5	17304	24851.0	2.481	65.106	4800	4340.33	40180.7	40379	30648.6	1.2757	66.788
4020	3560.33	32977.0	17706	24993.8	2.437	65.151	4820	4360.33	40367.5	41177	30795.7	1.2562	66.827
4040	3580.33	33159.6	18114	25136.7	2.393	65.196	4840	4380.33	40554.5	41988	30942.9	1.2370	66.866
4060	3600.33	33342.3	18530	25279.7	2.351	65.242	4860	4400.33	40741.6	42811	31090.3	1.2183	66.904
4080	3620.33	33525.1	18954	25422.8	2.310	65.286	4880	4420.33	40928.7	43648	31237.7	1.1998	66.943
4100	3640.33	33708.1	19386	25566.0	2.270	65.331	4900	4440.33	41116.0	44497	31385.3	1.1817	66.981
4120	3660.33	33891.1	19826	25709.4	2.230	65.376	4920	4460.33	41303.4	45361	31532.9	1.1640	67.019
4140	3680.33	34074.3	20274	25852.8	2.191	65.420	4940	4480.33	41490.9	46238	31680.7	1.1465	67.057
4160	3700.33	34257.6	20730	25996.4	2.154	65.464	4960	4500.33	41678.4	47129	31828.6	1.1294	67.095
4180	3720.33	34441.0	21194	26140.1	2.117	65.508	4980	4520.33	41866.1	48033	31976.5	1.1126	67.133
4200	3740.33	34624.5	21666	26283.9	2.080	65.552	5000	4540.33	42053.9	48952	32124.6	1.0961	67.171
4220	3760.33	34808.1	22148	26427.8	2.045	65.596	5020	4560.33	42241.7	49885	32272.7	1.0799	67.208
4240	3780.33	34991.8	22637	26571.8	2.010	65.639	5040	4580.33	42429.7	50833	32421.0	1.0640	67.245
4260	3800.33	35175.7	23136	26715.9	1.976	65.682	5060	4600.33	42617.8	51795	32569.3	1.0484	67.283
4280	3820.33	35359.6	23643	26860.2	1.943	65.725	5080	4620.33	42806.0	52772	32717.8	1.0330	67.320
4300	3840.33	35543.7	24160	27004.5	1.910	65.768	5100	4640.33	42994.2	53765	32866.4	1.0180	67.357
4320	3860.33	35727.9	24685	27149.0	1.878	65.811	5120	4660.33	43182.6	54772	33015.0	1.0032	67.394
4340	3880.33	35912.2	25220	27293.5	1.847	65.854	5140	4680.33	43371.1	55795	33163.8	.9886	67.430
4360	3900.33	36096.6	25764	27438.2	1.816	65.896	5160	4700.33	43559.6	56833	33312.6	.9743	67.467
4380	3920.33	36281.1	26318	27583.0	1.786	65.938	5180	4720.33	43748.3	57887	33461.6	.9603	67.504
4400	3940.33	36465.7	26881	27727.9	1.757	65.980	5200	4740.33	43937.1	58957	33610.6	.9465	67.540
4420	3960.33	36650.4	27454	27872.9	1.728	66.022	5220	4760.33	44125.9	60043	33759.7	.9330	67.576
4440	3980.33	36835.3	28037	28018.0	1.699	66.064	5240	4780.33	44314.9	61145	33909.0	.9197	67.612
4460	4000.33	37020.2	28630	28163.3	1.672	66.105	5260	4800.33	44503.9	62264	34058.3	.9066	67.648
4480	4020.33	37205.3	29233	28308.6	1.645	66.147	5280	4820.33	44693.1	63400	34207.7	.8937	67.684
4500	4040.33	37390.4	29846	28454.1	1.618	66.188	5300	4840.33	44882.3	64552	34357.2	.8811	67.720
4520	4060.33	37575.7	30470	28599.6	1.592	66.229	5320	4860.33	45071.6	65722	34506.9	.8687	67.756
4540	4080.33	37761.1	31105	28745.3	1.566	66.270	5340	4880.33	45261.1	66908	34656.6	.8565	67.791
4560	4100.33	37946.6	31750	28891.0	1.541	66.311	5360	4900.33	45450.6	68113	34806.4	.8445	67.827
4580	4120.33	38132.2	32406	29036.9	1.517	66.351	5380	4920.33	45640.2	69334	34956.3	.8327	67.862

Table 14   Oxygen

$\bar{m} = 31.9988$

T R	t F	$\bar{c}_p$ Btu / lb-mole R	$\bar{c}_v$ Btu / lb-mole R	$k = \dfrac{\bar{c}_p}{\bar{c}_v}$	a ft / sec
200	−259.67	6.952	4.966	1.400	659.6
250	−209.67	6.952	4.967	1.400	737.4
300	−159.67	6.953	4.967	1.400	807.8
350	−109.67	6.956	4.970	1.400	872.4
400	−59.67	6.962	4.976	1.399	932.5
450	−9.67	6.975	4.989	1.398	988.7
500	40.33	6.995	5.009	1.396	1041.6
550	90.33	7.025	5.040	1.394	1091.5
600	140.33	7.065	5.079	1.391	1138.8
650	190.33	7.112	5.127	1.387	1183.7
700	240.33	7.167	5.181	1.383	1226.6
750	290.33	7.228	5.242	1.379	1267.6
800	340.33	7.292	5.306	1.374	1307.0
850	390.33	7.358	5.372	1.370	1344.9
900	440.33	7.426	5.440	1.365	1381.6
1000	540.33	7.561	5.575	1.356	1451.6
1100	640.33	7.690	5.704	1.348	1518.0
1200	740.33	7.810	5.824	1.341	1581.2
1300	840.33	7.920	5.934	1.335	1641.9
1400	940.33	8.020	6.034	1.329	1700.3
1500	1040.33	8.110	6.124	1.324	1756.8
1600	1140.33	8.190	6.205	1.320	1811.5
1700	1240.33	8.263	6.277	1.316	1864.7
1800	1340.33	8.328	6.342	1.313	1916.4
1900	1440.33	8.387	6.402	1.310	1966.7
2000	1540.33	8.441	6.455	1.308	2015.8
2100	1640.33	8.491	6.505	1.305	2063.7
2200	1740.33	8.536	6.550	1.303	2110.6
2300	1840.33	8.579	6.593	1.301	2156.4
2400	1940.33	8.618	6.632	1.299	2201.3
2600	2140.33	8.692	6.706	1.296	2288.3
2800	2340.33	8.760	6.774	1.293	2371.9
3000	2540.33	8.824	6.838	1.290	2452.5
3200	2740.33	8.886	6.900	1.288	2530.4
3400	2940.33	8.946	6.961	1.285	2605.8
3600	3140.33	9.006	7.020	1.283	2678.8
3800	3340.33	9.064	7.078	1.281	2749.7
4000	3540.33	9.122	7.136	1.278	2818.6
4200	3740.33	9.178	7.192	1.276	2885.8
4400	3940.33	9.234	7.248	1.274	2951.2
4600	4140.33	9.288	7.302	1.272	3015.2
4800	4340.33	9.340	7.354	1.270	3077.6
5000	4540.33	9.391	7.405	1.268	3138.8
5200	4740.33	9.440	7.455	1.266	3198.7
5400	4940.33	9.488	7.502	1.265	3257.5
5600	5140.33	9.533	7.547	1.263	3315.2
5800	5340.33	9.576	7.590	1.262	3371.9
6000	5540.33	9.617	7.631	1.260	3427.6
6200	5740.33	9.656	7.670	1.259	3482.5
6400	5940.33	9.693	7.707	1.258	3536.4

Table 15 Water Vapor at Low Pressures (for One Pound-Mole)

m̄ = 18.0152

T	t	$\bar{h}$	p_r	$\bar{u}$	v_r	$\bar{\phi}$	T	t	$\bar{h}$	p_r	$\bar{u}$	v_r	$\bar{\phi}$
300	−159.67	2369.1	.06962	1773.3	46245	40.434	550	90.33	4364.9	.7953	3272.7	7422	45.271
305	−154.67	2408.9	.07439	1803.2	44000	40.566	555	95.33	4405.1	.8249	3303.0	7220	45.344
310	−149.67	2448.8	.07940	1833.2	41899	40.695	560	100.33	4445.3	.8554	3333.2	7025	45.416
315	−144.67	2488.6	.08466	1863.1	39931	40.823	565	105.33	4485.5	.8868	3363.5	6837	45.488
320	−139.67	2528.4	.09018	1892.9	38080	40.948	570	110.33	4525.6	.9190	3393.7	6656	45.558
325	−134.67	2568.2	.09597	1922.8	36343	41.072	575	115.33	4565.9	.9521	3424.0	6481	45.629
330	−129.67	2608.0	.10202	1952.7	34712	41.193	580	120.33	4606.2	.9861	3454.4	6312	45.698
335	−124.67	2647.8	.10836	1982.5	33175	41.313	585	125.33	4646.4	1.0211	3484.7	6148	45.768
340	−119.67	2687.6	.11500	2012.4	31729	41.431	590	130.33	4686.7	1.0569	3515.1	5991	45.836
345	−114.67	2727.4	.12192	2042.3	30366	41.547	595	135.33	4727.1	1.0937	3545.5	5838	45.904
350	−109.67	2767.3	.12917	2072.2	29079	41.662	600	140.33	4767.4	1.1315	3575.9	5691	45.971
355	−104.67	2807.1	.13673	2102.1	27863	41.775	605	145.33	4807.7	1.1703	3606.3	5548	46.038
360	−99.67	2846.9	.14462	2132.0	26714	41.886	610	150.33	4848.0	1.2102	3636.7	5409	46.105
365	−94.67	2886.7	.15284	2161.9	25628	41.996	615	155.33	4888.5	1.2510	3667.2	5276	46.171
370	−89.67	2926.5	.16142	2191.8	24598	42.104	620	160.33	4928.8	1.2929	3697.6	5146	46.236
375	−84.67	2966.4	.17036	2221.7	23623	42.211	625	165.33	4969.3	1.3358	3728.1	5021	46.301
380	−79.67	3006.2	.17967	2251.6	22697	42.317	630	170.33	5009.7	1.3800	3758.6	4899	46.366
385	−74.67	3046.1	.18933	2281.6	21822	42.421	635	175.33	5050.2	1.4251	3789.1	4782	46.430
390	−69.67	3086.0	.19940	2311.5	20990	42.524	640	180.33	5090.7	1.4715	3819.7	4668	46.493
395	−64.67	3125.8	.20987	2341.4	20198	42.626	645	185.33	5131.1	1.5190	3850.2	4557	46.556
400	−59.67	3165.7	.22072	2371.3	19448	42.726	650	190.33	5171.6	1.5676	3880.8	4450	46.619
405	−54.67	3205.5	.23202	2401.2	18733	42.825	655	195.33	5212.2	1.6174	3911.5	4346	46.681
410	−49.67	3245.3	.24372	2431.1	18053	42.923	660	200.33	5252.9	1.6685	3942.2	4245	46.743
415	−44.67	3285.2	.25588	2461.0	17405	43.019	665	205.33	5293.5	1.7209	3972.9	4147	46.804
420	−39.67	3325.1	.26849	2491.1	16787	43.115	670	210.33	5334.1	1.7744	4003.6	4052	46.865
425	−34.67	3365.0	.28155	2521.0	16199	43.209	675	215.33	5374.7	1.8291	4034.3	3960	46.925
430	−29.67	3404.9	.29510	2551.0	15637	43.302	680	220.33	5415.4	1.8854	4065.1	3870	46.985
435	−24.67	3444.8	.30914	2581.0	15100	43.395	685	225.33	5456.2	1.9428	4095.8	3784	47.045
440	−19.67	3484.7	.32370	2610.9	14587	43.486	690	230.33	5496.9	2.0016	4126.6	3699	47.104
445	−14.67	3524.6	.33873	2640.9	14098	43.576	695	235.33	5537.7	2.0619	4157.5	3617	47.163
450	−9.67	3564.5	.35429	2670.9	13631	43.665	700	240.33	5578.5	2.1237	4188.4	3537	47.222
455	−4.67	3604.5	.37037	2700.9	13184	43.754	705	245.33	5619.3	2.1866	4219.3	3460	47.280
460	.33	3644.4	.38699	2730.9	12756	43.841	710	250.33	5660.1	2.2512	4250.2	3385	47.338
465	5.33	3684.2	.40416	2760.8	12347	43.927	715	255.33	5700.9	2.3172	4281.1	3311	47.395
470	10.33	3724.2	.42192	2790.8	11955	44.012	720	260.33	5741.9	2.3846	4312.0	3240	47.452
475	15.33	3764.1	.44028	2820.8	11578	44.097	725	265.33	5782.8	2.4536	4343.0	3171	47.509
480	20.33	3804.1	.45925	2850.9	11216	44.181	730	270.33	5823.8	2.5243	4374.1	3103	47.565
485	25.33	3844.1	.47881	2880.9	10870	44.264	735	275.33	5864.8	2.5965	4405.2	3038	47.621
490	30.33	3884.1	.49900	2911.0	10538	44.346	740	280.33	5905.8	2.6702	4436.3	2974	47.677
495	35.33	3924.1	.51988	2941.1	10218	44.427	745	285.33	5946.8	2.7455	4467.3	2912	47.732
500	40.33	3964.2	.54137	2971.3	9911	44.507	750	290.33	5987.9	2.8223	4498.5	2852	47.787
505	45.33	4004.2	.56352	3001.4	9617	44.587	755	295.33	6029.0	2.9010	4529.7	2793	47.841
510	50.33	4044.2	.58634	3031.5	9334	44.666	760	300.33	6070.1	2.9816	4560.9	2735	47.896
515	55.33	4084.3	.60991	3061.6	9062	44.744	765	305.33	6111.3	3.0638	4592.2	2680	47.950
520	60.33	4124.3	.63410	3091.7	8800	44.821	770	310.33	6152.5	3.1479	4623.4	2625	48.003
525	65.33	4164.3	.65905	3121.8	8549	44.898	775	315.33	6193.8	3.2338	4654.8	2572	48.057
530	70.33	4204.5	.68472	3152.0	8307	44.974	780	320.33	6235.1	3.3213	4686.1	2520	48.110
535	75.33	4244.5	.71116	3182.1	8073	45.049	785	325.33	6276.4	3.4105	4717.5	2470	48.162
540	80.33	4284.6	.73841	3212.3	7848	45.124	790	330.33	6317.7	3.5017	4748.8	2421	48.215
545	85.33	4324.8	.76640	3242.5	7631	45.198	795	335.33	6359.0	3.5950	4780.2	2373	48.267

Table 15   Water Vapor at Low Pressures (for One Pound-Mole)

$\bar{m} = 18.0152$

T	t	$\bar{h}$	p_r	$\bar{u}$	v_r	$\bar{\phi}$	T	t	$\bar{h}$	p_r	$\bar{u}$	v_r	$\bar{\phi}$
800	340.33	6400.4	3.690	4811.7	2326.6	48.319	1300	840.33	10718.8	30.26	8137.2	460.98	52.498
810	350.33	6483.3	3.887	4874.7	2236.5	48.422	1310	850.33	10809.1	31.34	8207.6	448.62	52.567
820	360.33	6566.3	4.091	4937.8	2150.9	48.524	1320	860.33	10899.5	32.44	8278.2	436.66	52.636
830	370.33	6649.3	4.304	5001.1	2069.5	48.625	1330	870.33	10990.1	33.58	8348.9	425.08	52.704
840	380.33	6732.6	4.526	5064.5	1991.9	48.724	1340	880.33	11080.9	34.75	8419.9	413.86	52.772
850	390.33	6816.0	4.756	5128.0	1917.9	48.823	1350	890.33	11171.8	35.95	8490.9	402.99	52.840
860	400.33	6899.5	4.996	5191.7	1847.3	48.921	1360	900.33	11262.9	37.19	8562.2	392.47	52.907
870	410.33	6983.2	5.245	5255.5	1779.9	49.017	1370	910.33	11354.1	38.46	8633.5	382.26	52.974
880	420.33	7067.0	5.504	5319.4	1715.7	49.113	1380	920.33	11445.5	39.77	8705.1	372.38	53.040
890	430.33	7150.8	5.774	5383.4	1654.2	49.208	1390	930.33	11537.1	41.12	8776.8	362.79	53.106
900	440.33	7234.9	6.053	5447.6	1595.6	49.302	1400	940.33	11628.9	42.50	8848.7	353.50	53.172
910	450.33	7319.2	6.344	5512.0	1539.4	49.395	1410	950.33	11720.8	43.92	8920.7	344.49	53.238
920	460.33	7403.5	6.645	5576.5	1485.8	49.487	1420	960.33	11812.9	45.39	8992.9	335.75	53.303
930	470.33	7488.0	6.958	5641.0	1434.4	49.578	1430	970.33	11905.1	46.89	9065.4	327.27	53.367
940	480.33	7572.7	7.282	5705.9	1385.2	49.669	1440	980.33	11997.5	48.44	9137.9	319.04	53.432
950	490.33	7657.4	7.619	5770.8	1338.1	49.759	1450	990.33	12090.1	50.02	9210.6	311.06	53.496
960	500.33	7742.3	7.968	5835.8	1293.0	49.848	1460	1000.33	12182.8	51.66	9283.5	303.31	53.560
970	510.33	7827.4	8.329	5901.0	1249.8	49.936	1470	1010.33	12275.7	53.33	9356.5	295.79	53.623
980	520.33	7912.5	8.704	5966.4	1208.3	50.023	1480	1020.33	12368.7	55.05	9429.7	288.49	53.686
990	530.33	7997.9	9.092	6031.9	1168.5	50.110	1490	1030.33	12462.0	56.82	9503.1	281.41	53.749
1000	540.33	8083.4	9.494	6097.6	1130.3	50.196	1500	1040.33	12555.3	58.64	9576.6	274.52	53.811
1010	550.33	8169.1	9.910	6163.3	1093.7	50.281	1510	1050.33	12649.0	60.50	9650.3	267.83	53.873
1020	560.33	8254.9	10.341	6229.3	1058.5	50.365	1520	1060.33	12742.7	62.42	9724.2	261.34	53.935
1030	570.33	8340.8	10.787	6295.4	1024.7	50.449	1530	1070.33	12836.6	64.38	9798.3	255.02	53.997
1040	580.33	8426.8	11.249	6361.5	992.2	50.532	1540	1080.33	12930.7	66.40	9872.5	248.89	54.058
1050	590.33	8513.1	11.726	6428.0	960.9	50.615	1550	1090.33	13024.9	68.47	9946.8	242.93	54.119
1060	600.33	8599.5	12.220	6494.4	930.9	50.697	1560	1100.33	13119.3	70.60	10021.4	237.13	54.180
1070	610.33	8686.0	12.730	6561.1	902.0	50.778	1570	1110.33	13213.9	72.78	10096.1	231.50	54.240
1080	620.33	8772.7	13.257	6627.9	874.2	50.859	1580	1120.33	13308.7	75.02	10171.0	226.02	54.301
1090	630.33	8859.5	13.803	6694.9	847.5	50.939	1590	1130.33	13403.6	77.31	10246.0	220.70	54.360
1100	640.33	8946.5	14.366	6762.0	821.7	51.018	1600	1140.33	13498.7	79.67	10321.3	215.52	54.420
1110	650.33	9033.7	14.948	6829.4	796.9	51.097	1610	1150.33	13593.9	82.09	10396.6	210.48	54.479
1120	660.33	9121.0	15.549	6896.8	773.0	51.175	1620	1160.33	13689.3	84.57	10472.2	205.58	54.538
1130	670.33	9208.4	16.170	6964.3	750.0	51.253	1630	1170.33	13784.9	87.11	10547.9	200.81	54.597
1140	680.33	9296.0	16.811	7032.2	727.7	51.330	1640	1180.33	13880.6	89.72	10623.8	196.17	54.656
1150	690.33	9383.7	17.472	7100.0	706.3	51.407	1650	1190.33	13976.6	92.39	10700.0	191.65	54.714
1160	700.33	9471.7	18.155	7168.1	685.7	51.483	1660	1200.33	14072.7	95.13	10776.2	187.26	54.772
1170	710.33	9559.7	18.859	7236.2	665.8	51.559	1670	1210.33	14169.0	97.94	10852.6	182.98	54.830
1180	720.33	9647.9	19.586	7304.6	646.6	51.634	1680	1220.33	14265.5	100.82	10929.2	178.82	54.888
1190	730.33	9736.3	20.335	7373.1	628.0	51.708	1690	1230.33	14362.0	103.78	11005.9	174.76	54.945
1200	740.33	9824.9	21.108	7441.8	610.1	51.782	1700	1240.33	14458.8	106.80	11082.8	170.81	55.002
1210	750.33	9913.5	21.905	7510.6	592.8	51.856	1710	1250.33	14555.7	109.91	11159.9	166.97	55.059
1220	760.33	10002.4	22.727	7579.6	576.1	51.929	1720	1260.33	14652.9	113.08	11237.2	163.22	55.116
1230	770.33	10091.3	23.573	7648.7	559.9	52.002	1730	1270.33	14750.1	116.34	11314.6	159.58	55.172
1240	780.33	10180.5	24.446	7718.0	544.3	52.074	1740	1280.33	14847.6	119.68	11392.2	156.02	55.228
1250	790.33	10269.8	25.345	7787.5	529.3	52.146	1750	1290.33	14945.2	123.10	11470.0	152.56	55.284
1260	800.33	10359.3	26.272	7857.1	514.7	52.217	1760	1300.33	15043.1	126.60	11548.0	149.18	55.340
1270	810.33	10449.0	27.226	7927.0	500.6	52.288	1770	1310.33	15141.0	130.19	11626.0	145.90	55.395
1280	820.33	10538.7	28.209	7996.8	486.9	52.358	1780	1320.33	15239.1	133.87	11704.3	142.69	55.451
1290	830.33	10628.7	29.221	8066.9	473.8	52.428	1790	1330.33	15337.5	137.63	11782.8	139.57	55.506

Table 15　Water Vapor at Low Pressures (for One Pound-Mole)
$\bar{m} = 18.0152$

T	t	$\bar{h}$	p_r	$\bar{u}$	v_r	$\bar{\phi}$	T	t	$\bar{h}$	p_r	$\bar{u}$	v_r	$\bar{\phi}$
1800	1340.33	15436.0	141.49	11861.4	136.52	55.561	2800	2340.33	26090.1	1495.4	20529.7	20.094	60.243
1820	1360.33	15633.4	149.48	12019.2	130.66	55.670	2820	2360.33	26317.9	1557.7	20717.8	19.428	60.324
1840	1380.33	15831.5	157.86	12177.6	125.09	55.778	2840	2380.33	26546.2	1622.3	20906.4	18.787	60.405
1860	1400.33	16030.3	166.63	12336.6	119.79	55.885	2860	2400.33	26775.1	1689.2	21095.5	18.170	60.485
1880	1420.33	16229.9	175.83	12496.5	114.74	55.992	2880	2420.33	27004.3	1758.5	21285.0	17.575	60.565
1900	1440.33	16430.1	185.46	12656.9	109.94	56.098	2900	2440.33	27234.1	1830.4	21475.1	17.003	60.644
1920	1460.33	16630.9	195.55	12818.1	105.37	56.203	2920	2460.33	27464.4	1904.8	21665.6	16.451	60.724
1940	1480.33	16832.5	206.11	12979.9	101.01	56.308	2940	2480.33	27695.1	1981.8	21856.7	15.920	60.802
1960	1500.33	17034.7	217.15	13142.4	96.86	56.411	2960	2500.33	27926.3	2061.6	22048.2	15.408	60.881
1980	1520.33	17237.5	228.71	13305.5	92.91	56.514	2980	2520.33	28158.0	2144.2	22240.1	14.915	60.959
2000	1540.33	17441.0	240.80	13469.3	89.13	56.616	3000	2540.33	28390.1	2229.7	22432.5	14.439	61.036
2020	1560.33	17645.1	253.44	13633.7	85.54	56.718	3020	2560.33	28622.8	2318.2	22625.5	13.980	61.114
2040	1580.33	17850.0	266.65	13798.8	82.10	56.819	3040	2580.33	28855.9	2409.8	22818.8	13.538	61.191
2060	1600.33	18055.5	280.45	13964.7	78.83	56.919	3060	2600.33	29089.4	2504.5	23012.7	13.112	61.267
2080	1620.33	18261.7	294.88	14131.1	75.70	57.019	3080	2620.33	29323.4	2602.5	23206.9	12.701	61.343
2100	1640.33	18468.5	309.94	14298.2	72.71	57.118	3100	2640.33	29557.8	2703.8	23401.6	12.304	61.419
2120	1660.33	18676.0	325.67	14465.9	69.86	57.216	3120	2660.33	29792.6	2808.6	23596.8	11.921	61.495
2140	1680.33	18884.1	342.10	14634.4	67.13	57.314	3140	2680.33	30028.0	2916.9	23792.4	11.552	61.570
2160	1700.33	19092.9	359.24	14803.4	64.53	57.411	3160	2700.33	30263.8	3029.0	23988.4	11.196	61.645
2180	1720.33	19302.2	377.13	14973.1	62.03	57.507	3180	2720.33	30499.9	3144.8	24184.8	10.852	61.719
2200	1740.33	19512.3	395.79	15143.4	59.65	57.603	3200	2740.33	30736.4	3264.5	24381.7	10.520	61.793
2220	1760.33	19723.0	415.25	15314.4	57.37	57.699	3220	2760.33	30973.5	3388.1	24579.1	10.199	61.867
2240	1780.33	19934.2	435.54	15485.8	55.19	57.793	3240	2780.33	31211.0	3515.9	24776.8	9.889	61.941
2260	1800.33	20146.2	456.70	15658.1	53.11	57.888	3260	2800.33	31448.7	3647.9	24974.8	9.590	62.014
2280	1820.33	20358.7	478.75	15831.0	51.11	57.981	3280	2820.33	31687.1	3784.2	25173.5	9.302	62.087
2300	1840.33	20571.9	501.72	16004.4	49.20	58.074	3300	2840.33	31925.7	3925.0	25372.4	9.023	62.159
2320	1860.33	20785.6	525.65	16178.4	47.36	58.167	3320	2860.33	32164.9	4070.4	25571.8	8.753	62.232
2340	1880.33	21000.0	550.58	16353.1	45.61	58.259	3340	2880.33	32404.3	4220.5	25771.5	8.493	62.303
2360	1900.33	21215.0	576.54	16528.4	43.93	58.350	3360	2900.33	32644.1	4375.4	25971.6	8.241	62.375
2380	1920.33	21430.6	603.57	16704.3	42.32	58.441	3380	2920.33	32884.4	4535.4	26172.2	7.998	62.446
2400	1940.33	21646.8	631.70	16880.8	40.77	58.532	3400	2940.33	33125.0	4700.4	26373.0	7.763	62.517
2420	1960.33	21863.6	660.97	17057.8	39.29	58.622	3420	2960.33	33366.1	4870.7	26574.4	7.535	62.588
2440	1980.33	22081.0	691.43	17235.5	37.87	58.711	3440	2980.33	33607.4	5046.4	26776.1	7.315	62.658
2460	2000.33	22299.0	723.12	17413.8	36.51	58.800	3460	3000.33	33849.3	5227.7	26978.2	7.103	62.729
2480	2020.33	22517.6	756.06	17592.7	35.20	58.889	3480	3020.33	34091.4	5414.7	27180.6	6.897	62.798
2500	2040.33	22736.7	790.32	17772.1	33.95	58.977	3500	3040.33	34333.9	5607.5	27383.4	6.698	62.868
2520	2060.33	22956.4	825.94	17952.0	32.74	59.064	3520	3060.33	34576.8	5806.3	27586.6	6.506	62.937
2540	2080.33	23176.7	862.95	18132.6	31.59	59.151	3540	3080.33	34820.0	6011.3	27790.1	6.320	63.006
2560	2100.33	23397.5	901.42	18313.7	30.48	59.238	3560	3100.33	35063.6	6222.7	27993.9	6.139	63.075
2580	2120.33	23618.9	941.38	18495.4	29.41	59.324	3580	3120.33	35307.6	6440.5	28198.2	5.965	63.143
2600	2140.33	23840.8	982.90	18677.6	28.39	59.410	3600	3140.33	35551.9	6665.1	28402.8	5.796	63.211
2620	2160.33	24063.5	1026.02	18860.5	27.40	59.495	3620	3160.33	35796.5	6896.5	28607.7	5.633	63.279
2640	2180.33	24286.5	1070.79	19043.8	26.46	59.580	3640	3180.33	36041.5	7134.9	28812.9	5.475	63.346
2660	2200.33	24510.1	1117.27	19227.7	25.55	59.664	3660	3200.33	36286.9	7380.5	29018.7	5.322	63.413
2680	2220.33	24734.3	1165.50	19412.2	24.68	59.748	3680	3220.33	36532.6	7633.6	29224.6	5.173	63.480
2700	2240.33	24958.9	1215.57	19597.1	23.84	59.832	3700	3240.33	36778.6	7894.2	29430.9	5.030	63.547
2720	2260.33	25184.1	1267.53	19782.6	23.03	59.915	3720	3260.33	37024.9	8162.6	29637.5	4.891	63.613
2740	2280.33	25409.9	1321.42	19968.6	22.25	59.997	3740	3280.33	37271.6	8439.0	29844.4	4.756	63.680
2760	2300.33	25636.1	1377.31	20155.1	21.50	60.080	3760	3300.33	37518.6	8723.6	30051.8	4.625	63.745
2780	2320.33	25862.9	1435.27	20342.2	20.79	60.162	3780	3320.33	37765.9	9016.7	30259.4	4.499	63.811

Table 15 Water Vapor at Low Pressures (for One Pound-Mole)

$\bar{m} = 18.0152$

T	t	$\bar{h}$	p_r	$\bar{u}$	v_r	$\bar{\phi}$	T	t	$\bar{h}$	p_r	$\bar{u}$	v_r	$\bar{\phi}$
3800	3340.33	38013.5	9318	30467.3	4.3763	63.876	4600	4140.33	48151.7	31512	39016.8	1.5665	66.296
3820	3360.33	38261.5	9629	30675.6	4.2575	63.941	4620	4160.33	48410.3	32415	39235.6	1.5295	66.352
3840	3380.33	38509.9	9948	30884.1	4.1423	64.006	4640	4180.33	48669.2	33340	39454.9	1.4935	66.408
3860	3400.33	38758.4	10277	31093.0	4.0307	64.071	4660	4200.33	48928.3	34289	39674.2	1.4585	66.464
3880	3420.33	39007.3	10615	31302.2	3.9225	64.135	4680	4220.33	49187.5	35261	39893.7	1.4243	66.519
3900	3440.33	39256.5	10963	31511.7	3.8175	64.199	4700	4240.33	49447.1	36257	40113.5	1.3911	66.574
3920	3460.33	39506.0	11321	31721.4	3.7158	64.263	4720	4260.33	49706.8	37278	40333.5	1.3588	66.630
3940	3480.33	39755.9	11690	31931.6	3.6171	64.327	4740	4280.33	49966.8	38324	40553.8	1.3273	66.685
3960	3500.33	40005.9	12068	32141.9	3.5214	64.390	4760	4300.33	50226.9	39396	40774.2	1.2966	66.739
3980	3520.33	40256.3	12458	32352.6	3.4285	64.453	4780	4320.33	50487.3	40493	40994.9	1.2668	66.794
4000	3540.33	40507.0	12858	32563.6	3.3384	64.516	4800	4340.33	50748.0	41618	41215.8	1.2377	66.848
4020	3560.33	40758.0	13270	32774.8	3.2511	64.578	4820	4360.33	51008.7	42770	41436.8	1.2094	66.903
4040	3580.33	41009.2	13693	32986.3	3.1662	64.641	4840	4380.33	51269.7	43950	41658.1	1.1818	66.957
4060	3600.33	41260.7	14128	33198.1	3.0840	64.703	4860	4400.33	51530.8	45158	41879.6	1.1550	67.010
4080	3620.33	41512.5	14575	33410.2	3.0041	64.765	4880	4420.33	51792.3	46395	42101.3	1.1288	67.064
4100	3640.33	41764.7	15035	33622.6	2.9265	64.826	4900	4440.33	52053.9	47662	42323.2	1.1033	67.118
4120	3660.33	42017.1	15507	33835.3	2.8513	64.888	4920	4460.33	52315.7	48959	42545.3	1.0784	67.171
4140	3680.33	42269.8	15992	34048.3	2.7782	64.949	4940	4480.33	52577.7	50287	42767.5	1.0542	67.224
4160	3700.33	42522.7	16490	34261.5	2.7072	65.010	4960	4500.33	52839.8	51647	42990.0	1.0306	67.277
4180	3720.33	42775.8	17002	34475.0	2.6383	65.071	4980	4520.33	53102.2	53038	43212.6	1.0076	67.330
4200	3740.33	43029.3	17528	34688.7	2.5714	65.131	5000	4540.33	53364.9	54463	43435.6	.9852	67.382
4220	3760.33	43283.1	18068	34902.8	2.5064	65.191	5020	4560.33	53627.6	55921	43658.6	.9634	67.435
4240	3780.33	43537.1	18623	35117.0	2.4433	65.251	5040	4580.33	53890.7	57413	43881.9	.9421	67.487
4260	3800.33	43791.3	19193	35331.6	2.3820	65.311	5060	4600.33	54153.8	58940	44105.4	.9213	67.539
4280	3820.33	44045.9	19777	35546.4	2.3224	65.371	5080	4620.33	54417.3	60502	44329.1	.9011	67.591
4300	3840.33	44300.6	20378	35761.4	2.2645	65.430	5100	4640.33	54680.8	62100	44552.9	.8813	67.643
4320	3860.33	44555.7	20994	35976.8	2.2082	65.489	5120	4660.33	54944.5	63736	44776.9	.8621	67.695
4340	3880.33	44811.0	21627	36192.4	2.1535	65.548	5140	4680.33	55208.5	65409	45001.2	.8433	67.746
4360	3900.33	45066.5	22276	36408.2	2.1004	65.607	5160	4700.33	55472.7	67120	45225.6	.8250	67.797
4380	3920.33	45322.4	22943	36624.3	2.0488	65.666	5180	4720.33	55737.1	68871	45450.3	.8071	67.849
4400	3940.33	45578.4	23626	36840.6	1.9986	65.724	5200	4740.33	56001.6	70661	45675.1	.7897	67.900
4420	3960.33	45834.7	24328	37057.2	1.9497	65.782	5220	4760.33	56266.3	72493	45900.1	.7727	67.950
4440	3980.33	46091.2	25047	37274.0	1.9023	65.840	5240	4780.33	56531.2	74366	46125.3	.7562	68.001
4460	4000.33	46347.9	25786	37491.0	1.8561	65.898	5260	4800.33	56796.3	76280	46350.7	.7400	68.051
4480	4020.33	46604.9	26543	37708.2	1.8113	65.955	5280	4820.33	57061.6	78239	46576.3	.7242	68.102
4500	4040.33	46862.1	27320	37925.7	1.7676	66.012	5300	4840.33	57327.1	80241	46802.0	.7088	68.152
4520	4060.33	47119.5	28117	38143.4	1.7251	66.070	5320	4860.33	57592.8	82288	47028.0	.6938	68.202
4540	4080.33	47377.2	28934	38361.4	1.6839	66.126	5340	4880.33	57858.6	84381	47254.1	.6791	68.252
4560	4100.33	47635.1	29772	38579.6	1.6437	66.183	5360	4900.33	58124.6	86520	47480.4	.6648	68.302
4580	4120.33	47893.3	30631	38798.0	1.6046	66.240	5380	4920.33	58390.8	88707	47706.9	.6509	68.351

Table 16   Water Vapor

$\bar{m} = 18.0152$

T R	t F	$\bar{c}_p$ Btu / lb-mole R	$\bar{c}_v$ Btu / lb-mole R	$k = \dfrac{\bar{c}_p}{\bar{c}_v}$	a ft / sec
200	−259.67	7.956	5.970	1.333	857.7
250	−209.67	7.958	5.972	1.333	958.8
300	−159.67	7.961	5.975	1.332	1050.3
350	−109.67	7.964	5.979	1.332	1134.4
400	−59.67	7.971	5.986	1.332	1212.5
450	−9.67	7.984	5.998	1.331	1285.7
500	40.33	8.003	6.017	1.330	1354.7
550	90.33	8.030	6.045	1.329	1420.1
600	140.33	8.066	6.081	1.327	1482.1
650	190.33	8.110	6.125	1.324	1541.3
700	240.33	8.161	6.175	1.322	1597.8
750	290.33	8.218	6.232	1.319	1652.1
800	340.33	8.280	6.294	1.316	1704.2
850	390.33	8.345	6.360	1.312	1754.5
900	440.33	8.414	6.428	1.309	1803.1
1000	540.33	8.557	6.571	1.302	1895.8
1100	640.33	8.706	6.720	1.295	1983.1
1200	740.33	8.861	6.875	1.289	2066.0
1300	840.33	9.020	7.034	1.282	2144.9
1400	940.33	9.182	7.196	1.276	2220.3
1500	1040.33	9.348	7.363	1.270	2292.7
1600	1140.33	9.517	7.531	1.264	2362.2
1700	1240.33	9.687	7.701	1.258	2429.3
1800	1340.33	9.857	7.871	1.252	2494.2
1900	1440.33	10.026	8.040	1.247	2557.1
2000	1540.33	10.193	8.207	1.242	2618.3
2100	1640.33	10.357	8.371	1.237	2677.8
2200	1740.33	10.517	8.532	1.233	2735.8
2300	1840.33	10.674	8.688	1.229	2792.6
2400	1940.33	10.825	8.839	1.225	2848.1
2500	2040.33	10.971	8.985	1.221	2902.5
2600	2140.33	11.112	9.126	1.218	2955.8
2700	2240.33	11.247	9.261	1.214	3008.2
2800	2340.33	11.377	9.391	1.211	3059.7
2900	2440.33	11.501	9.515	1.209	3110.3
3000	2540.33	11.619	9.634	1.206	3160.1
3200	2740.33	11.841	9.855	1.202	3257.5
3400	2940.33	12.042	10.056	1.197	3352.1
3600	3140.33	12.224	10.238	1.194	3444.2
3800	3340.33	12.390	10.404	1.191	3534.0
4000	3540.33	12.542	10.556	1.188	3621.6
4200	3740.33	12.680	10.694	1.186	3707.2
4400	3940.33	12.807	10.822	1.184	3791.0
4600	4140.33	12.925	10.939	1.182	3873.0
4800	4340.33	13.034	11.048	1.180	3953.3
5000	4540.33	13.136	11.150	1.178	4032.0
5200	4740.33	13.232	11.246	1.177	4109.2
5400	4940.33	13.322	11.336	1.175	4184.9
5600	5140.33	13.409	11.423	1.174	4259.3
5800	5340.33	13.491	11.505	1.173	4332.4

Table 17   Carbon Dioxide at Low Pressures (for One Pound-Mole)

$\bar{m} = 44.0098$

T	t	$\bar{h}$	p_r	$\bar{u}$	v_r	$\bar{\phi}$	T	t	$\bar{h}$	p_r	$\bar{u}$	v_r	$\bar{\phi}$
300	−159.67	2106.3	.013766	1510.6	233862	46.361	550	90.33	4145.0	.16183	3052.8	36473	51.255
305	−154.67	2143.2	.014640	1537.6	223578	46.483	555	95.33	4189.8	.16858	3087.7	35329	51.336
310	−149.67	2180.4	.015556	1564.7	213860	46.604	560	100.33	4234.7	.17557	3122.7	34229	51.416
315	−144.67	2217.6	.016516	1592.0	204675	46.723	565	105.33	4279.9	.18281	3157.8	33168	51.497
320	−139.67	2254.9	.017523	1619.4	195972	46.840	570	110.33	4325.2	.19031	3193.2	32143	51.577
325	−134.67	2292.3	.018581	1646.9	187705	46.956	575	115.33	4370.6	.19805	3228.7	31156	51.656
330	−129.67	2329.9	.019687	1674.6	179889	47.071	580	120.33	4416.1	.20609	3264.3	30201	51.735
335	−124.67	2367.8	.020848	1702.6	172445	47.185	585	125.33	4461.9	.21440	3300.1	29282	51.813
340	−119.67	2405.7	.022062	1730.5	165388	47.297	590	130.33	4507.7	.22297	3336.0	28397	51.891
345	−114.67	2443.8	.023332	1758.7	158681	47.409	595	135.33	4553.6	.23186	3372.1	27539	51.969
350	−109.67	2482.1	.024664	1787.0	152289	47.519	600	140.33	4599.8	.24106	3408.3	26711	52.046
355	−104.67	2520.4	.026053	1815.5	146228	47.628	605	145.33	4646.0	.25054	3444.6	25914	52.123
360	−99.67	2559.0	.027507	1844.1	140450	47.736	610	150.33	4692.5	.26038	3481.1	25141	52.199
365	−94.67	2597.7	.029027	1872.8	134942	47.842	615	155.33	4739.0	.27052	3517.7	24397	52.275
370	−89.67	2636.6	.030613	1901.9	129704	47.948	620	160.33	4785.6	.28102	3554.4	23676	52.351
375	−84.67	2675.6	.032273	1930.9	124695	48.053	625	165.33	4832.5	.29188	3591.4	22979	52.426
380	−79.67	2714.8	.034006	1960.2	119919	48.157	630	170.33	4879.5	.30309	3628.4	22306	52.501
385	−74.67	2754.2	.035814	1989.6	115364	48.260	635	175.33	4926.6	.31467	3665.6	21656	52.575
390	−69.67	2793.7	.037703	2019.3	111007	48.362	640	180.33	4973.9	.32663	3703.0	21027	52.649
395	−64.67	2833.4	.039676	2049.0	106840	48.463	645	185.33	5021.3	.33898	3740.4	20420	52.723
400	−59.67	2873.3	.041731	2079.0	102865	48.563	650	190.33	5068.8	.35175	3778.0	19830	52.796
405	−54.67	2913.4	.043870	2109.1	99071	48.663	655	195.33	5116.5	.36494	3815.7	19261	52.870
410	−49.67	2953.5	.046106	2139.3	95431	48.761	660	200.33	5164.1	.37854	3853.4	18711	52.942
415	−44.67	2993.9	.048431	2169.8	91958	48.859	665	205.33	5212.1	.39258	3891.5	18178	53.015
420	−39.67	3034.4	.050858	2200.4	88624	48.956	670	210.33	5260.1	.40709	3929.6	17662	53.087
425	−34.67	3075.1	.053385	2231.2	85433	49.052	675	215.33	5308.2	.42201	3967.8	17165	53.158
430	−29.67	3116.0	.056016	2262.0	82379	49.148	680	220.33	5356.5	.43744	4006.1	16682	53.229
435	−24.67	3157.0	.058758	2293.1	79447	49.243	685	225.33	5404.9	.45334	4044.6	16215	53.300
440	−19.67	3198.2	.061611	2324.4	76640	49.337	690	230.33	5453.4	.46977	4083.2	15763	53.371
445	−14.67	3239.6	.064582	2355.9	73945	49.430	695	235.33	5502.1	.48669	4122.0	15325	53.441
450	−9.67	3281.1	.067669	2387.4	71364	49.523	700	240.33	5551.0	.50418	4160.9	14900	53.511
455	−4.67	3322.8	.070883	2419.2	68886	49.615	705	245.33	5599.9	.52214	4199.9	14490	53.581
460	.33	3364.6	.074220	2451.1	66512	49.707	710	250.33	5648.9	.54068	4238.9	14092	53.650
465	5.33	3406.5	.077691	2483.1	64231	49.797	715	255.33	5698.1	.55977	4278.2	13707	53.719
470	10.33	3448.7	.081299	2515.4	62040	49.888	720	260.33	5747.4	.57948	4317.6	13334	53.788
475	15.33	3491.0	.085058	2547.7	59929	49.977	725	265.33	5796.8	.59976	4357.1	12972	53.856
480	20.33	3533.6	.088955	2580.4	57907	50.066	730	270.33	5846.3	.62069	4396.7	12621	53.924
485	25.33	3576.2	.093003	2613.0	55963	50.155	735	275.33	5895.9	.64229	4436.3	12281	53.992
490	30.33	3619.0	.097216	2645.9	54090	50.243	740	280.33	5945.7	.66450	4476.2	11951	54.060
495	35.33	3662.0	.101579	2679.0	52295	50.330	745	285.33	5995.6	.68734	4516.2	11632	54.127
500	40.33	3705.1	.106116	2712.2	50565	50.417	750	290.33	6045.7	.71090	4556.3	11322	54.194
505	45.33	3748.4	.110823	2745.5	48901	50.503	755	295.33	6095.8	.73512	4596.5	11022	54.260
510	50.33	3791.8	.115704	2779.1	47302	50.588	760	300.33	6146.1	.76009	4636.8	10730	54.327
515	55.33	3835.4	.120764	2812.7	45765	50.673	765	305.33	6196.5	.78583	4677.3	10447	54.393
520	60.33	3879.1	.126007	2846.5	44286	50.758	770	310.33	6247.0	.81227	4717.9	10173	54.458
525	65.33	3923.1	.131464	2880.5	42856	50.842	775	315.33	6297.5	.83952	4758.5	9907	54.524
530	70.33	3967.1	.137103	2914.6	41485	50.925	780	320.33	6348.2	.86752	4799.2	9649	54.589
535	75.33	4011.3	.142956	2948.9	40162	51.008	785	325.33	6399.1	.89635	4840.2	9398	54.654
540	80.33	4055.8	.149028	2983.4	38885	51.091	790	330.33	6450.0	.92606	4881.1	9155	54.719
545	85.33	4100.3	.155312	3018.0	37658	51.173	795	335.33	6501.0	.95655	4922.3	8919	54.783

Table 17 Carbon Dioxide at Low Pressures (for One Pound-Mole)

$\bar{m} = 44.0098$

T	t	$\bar{h}$	p_r	$\bar{u}$	v_r	$\bar{\phi}$	T	t	$\bar{h}$	p_r	$\bar{u}$	v_r	$\bar{\phi}$
800	340.33	6552.1	.9880	4963.4	8689.9	54.847	1300	840.33	12135.6	14.897	9554.0	936.48	60.236
810	350.33	6654.7	1.0535	5046.2	8251.4	54.975	1310	850.33	12255.1	15.602	9653.6	901.07	60.327
820	360.33	6757.7	1.1227	5129.3	7837.8	55.101	1320	860.33	12375.0	16.334	9753.7	867.22	60.418
830	370.33	6861.2	1.1960	5213.0	7447.5	55.227	1330	870.33	12495.1	17.095	9853.9	834.92	60.509
840	380.33	6965.2	1.2733	5297.1	7079.8	55.351	1340	880.33	12615.4	17.889	9954.3	803.86	60.599
850	390.33	7069.6	1.3550	5381.6	6732.1	55.475	1350	890.33	12736.0	18.716	10055.1	774.07	60.689
860	400.33	7174.4	1.4412	5466.5	6403.7	55.597	1360	900.33	12856.8	19.576	10156.0	745.57	60.778
870	410.33	7279.5	1.5323	5551.8	6093.0	55.719	1370	910.33	12977.8	20.468	10257.2	718.29	60.866
880	420.33	7385.1	1.6284	5637.5	5799.5	55.840	1380	920.33	13099.2	21.400	10358.7	692.04	60.955
890	430.33	7491.1	1.7296	5723.7	5522.2	55.959	1390	930.33	13220.8	22.369	10460.4	666.85	61.043
900	440.33	7597.5	1.8362	5810.2	5260.1	56.078	1400	940.33	13342.5	23.373	10562.3	642.80	61.130
910	450.33	7704.3	1.9485	5897.1	5011.8	56.196	1410	950.33	13464.6	24.417	10664.5	619.71	61.217
920	460.33	7811.5	2.0670	5984.5	4776.6	56.313	1420	960.33	13586.8	25.503	10766.9	597.54	61.303
930	470.33	7919.0	2.1917	6072.1	4553.7	56.430	1430	970.33	13709.4	26.631	10869.6	576.25	61.389
940	480.33	8026.9	2.3228	6160.2	4342.8	56.545	1440	980.33	13832.1	27.804	10972.5	555.80	61.475
950	490.33	8135.2	2.4605	6248.6	4143.4	56.659	1450	990.33	13955.0	29.023	11075.5	536.15	61.560
960	500.33	8243.8	2.6056	6337.4	3953.8	56.773	1460	1000.33	14078.3	30.286	11178.9	517.34	61.644
970	510.33	8352.9	2.7585	6426.6	3773.7	56.886	1470	1010.33	14201.7	31.598	11282.5	499.26	61.729
980	520.33	8462.4	2.9188	6516.2	3603.1	56.999	1480	1020.33	14325.3	32.959	11386.2	481.88	61.812
990	530.33	8572.1	3.0872	6606.1	3441.3	57.110	1490	1030.33	14449.1	34.373	11490.2	465.19	61.896
1000	540.33	8682.3	3.2644	6696.4	3287.4	57.221	1500	1040.33	14573.2	35.844	11594.4	449.09	61.979
1010	550.33	8792.8	3.4500	6787.1	3141.7	57.331	1510	1050.33	14697.4	37.363	11698.8	433.71	62.062
1020	560.33	8903.6	3.6451	6878.0	3003.0	57.440	1520	1060.33	14821.9	38.942	11803.4	418.87	62.144
1030	570.33	9014.8	3.8496	6969.4	2871.3	57.548	1530	1070.33	14946.6	40.576	11908.2	404.65	62.225
1040	580.33	9126.3	4.0645	7061.0	2745.9	57.656	1540	1080.33	15071.5	42.274	12013.3	390.94	62.307
1050	590.33	9238.2	4.2895	7153.1	2626.9	57.763	1550	1090.33	15196.6	44.030	12118.5	377.78	62.388
1060	600.33	9350.4	4.5257	7245.4	2513.5	57.870	1560	1100.33	15321.9	45.855	12224.0	365.09	62.468
1070	610.33	9462.9	4.7730	7338.0	2405.8	57.975	1570	1110.33	15447.5	47.745	12329.7	352.88	62.548
1080	620.33	9575.8	5.0318	7431.1	2303.4	58.080	1580	1120.33	15573.2	49.704	12435.5	341.14	62.628
1090	630.33	9689.0	5.3036	7524.4	2205.6	58.185	1590	1130.33	15699.1	51.732	12541.6	329.84	62.708
1100	640.33	9802.5	5.5878	7618.1	2112.6	58.288	1600	1140.33	15825.2	53.838	12647.8	318.93	62.787
1110	650.33	9916.4	5.8854	7712.1	2024.0	58.391	1610	1150.33	15951.5	56.013	12754.3	308.46	62.866
1120	660.33	10030.5	6.1965	7806.3	1939.7	58.494	1620	1160.33	16078.0	58.264	12860.9	298.39	62.944
1130	670.33	10144.9	6.5220	7900.9	1859.3	58.595	1630	1170.33	16204.7	60.593	12967.8	288.69	63.022
1140	680.33	10259.8	6.8625	7995.9	1782.7	58.696	1640	1180.33	16331.6	63.009	13074.8	279.32	63.099
1150	690.33	10374.8	7.2195	8091.1	1709.4	58.797	1650	1190.33	16458.6	65.508	13181.9	270.30	63.177
1160	700.33	10490.1	7.5919	8186.5	1639.7	58.897	1660	1200.33	16585.7	68.100	13289.2	261.59	63.254
1170	710.33	10605.8	7.9811	8282.4	1573.2	58.996	1670	1210.33	16713.1	70.766	13396.8	253.25	63.330
1180	720.33	10721.8	8.3878	8378.5	1509.7	59.095	1680	1220.33	16840.8	73.536	13504.5	245.17	63.406
1190	730.33	10838.0	8.8126	8474.8	1449.1	59.193	1690	1230.33	16968.6	76.400	13612.5	237.39	63.482
1200	740.33	10954.6	9.2561	8571.6	1391.3	59.290	1700	1240.33	17096.6	79.359	13720.6	229.89	63.557
1210	750.33	11071.5	9.7190	8668.6	1336.1	59.387	1710	1250.33	17224.7	82.416	13828.9	222.66	63.633
1220	760.33	11188.5	10.2020	8765.8	1283.3	59.484	1720	1260.33	17352.9	85.582	13937.2	215.68	63.707
1230	770.33	11306.0	10.7057	8863.4	1233.0	59.579	1730	1270.33	17481.4	88.853	14045.9	208.95	63.782
1240	780.33	11423.8	11.2333	8961.3	1184.6	59.675	1740	1280.33	17610.2	92.239	14154.8	202.44	63.856
1250	790.33	11541.6	11.7832	9059.3	1138.4	59.770	1750	1290.33	17739.0	95.735	14263.7	196.17	63.930
1260	800.33	11659.9	12.3552	9157.7	1094.4	59.864	1760	1300.33	17867.9	99.333	14372.8	190.14	64.003
1270	810.33	11778.4	12.9524	9256.4	1052.2	59.958	1770	1310.33	17997.0	103.067	14482.1	184.30	64.077
1280	820.33	11897.2	13.5743	9355.3	1011.9	60.051	1780	1320.33	18126.3	106.920	14591.5	178.66	64.149
1290	830.33	12016.3	14.2218	9454.5	973.4	60.143	1790	1330.33	18255.8	110.894	14701.1	173.22	64.222

Table 17 Carbon Dioxide at Low Pressures (for One Pound-Mole)

$\bar{m} = 44.0098$

T	t	$\bar{h}$	p_r	$\bar{u}$	v_r	$\bar{\phi}$	T	t	$\bar{h}$	p_r	$\bar{u}$	v_r	$\bar{\phi}$
1800	1340.33	18385.1	115.01	14810.5	167.963	64.294	2800	2340.33	31953.5	2333.2	26393.1	12.879	70.272
1820	1360.33	18644.8	123.59	15030.5	158.032	64.437	2820	2360.33	32233.7	2452.8	26633.6	12.338	70.371
1840	1380.33	18905.5	132.82	15251.5	148.669	64.580	2840	2380.33	32513.9	2578.5	26874.1	11.820	70.470
1860	1400.33	19166.2	142.59	15472.5	139.985	64.721	2860	2400.33	32795.1	2710.7	27115.6	11.322	70.570
1880	1420.33	19427.9	152.93	15694.5	131.925	64.860	2880	2420.33	33076.3	2846.9	27357.1	10.856	70.667
1900	1440.33	19689.6	164.02	15916.4	124.314	64.999	2900	2440.33	33357.6	2989.9	27598.6	10.409	70.764
1920	1460.33	19952.2	175.74	16139.4	117.247	65.136	2920	2460.33	33639.7	3140.0	27841.0	9.980	70.861
1940	1480.33	20215.9	188.29	16363.3	110.570	65.273	2940	2480.33	33921.9	3294.4	28083.5	9.577	70.957
1960	1500.33	20479.5	201.54	16587.2	104.366	65.408	2960	2500.33	34204.1	3456.4	28326.0	9.190	71.052
1980	1520.33	20744.1	215.72	16812.1	98.500	65.543	2980	2520.33	34486.3	3626.3	28568.4	8.819	71.147
2000	1540.33	21008.7	230.67	17037.0	93.047	65.676	3000	2540.33	34768.5	3804.6	28810.9	8.462	71.243
2020	1560.33	21274.3	246.41	17262.9	87.975	65.807	3020	2560.33	35051.7	3987.7	29054.4	8.127	71.336
2040	1580.33	21539.9	263.22	17488.7	83.172	65.938	3040	2580.33	35334.8	4179.6	29297.8	7.805	71.429
2060	1600.33	21806.5	280.89	17715.6	78.702	66.068	3060	2600.33	35618.0	4380.8	29541.2	7.496	71.523
2080	1620.33	22074.0	299.76	17943.4	74.465	66.197	3080	2620.33	35901.1	4591.6	29784.7	7.199	71.616
2100	1640.33	22341.5	319.89	18171.2	70.450	66.326	3100	2640.33	36185.3	4807.7	30029.1	6.920	71.707
2120	1660.33	22610.1	340.69	18400.0	66.778	66.451	3120	2660.33	36469.4	5034.0	30273.5	6.651	71.799
2140	1680.33	22878.6	363.21	18628.8	63.229	66.578	3140	2680.33	36752.6	5265.8	30517.0	6.399	71.888
2160	1700.33	23147.1	386.83	18857.6	59.923	66.703	3160	2700.33	37037.7	5513.6	30762.4	6.150	71.979
2180	1720.33	23416.6	411.98	19087.4	56.786	66.828	3180	2720.33	37321.8	5767.4	31006.8	5.917	72.069
2200	1740.33	23686.1	438.33	19317.2	53.862	66.951	3200	2740.33	37606.0	6032.9	31251.2	5.692	72.158
2220	1760.33	23956.6	466.37	19547.9	51.084	67.074	3220	2760.33	37891.1	6304.2	31496.6	5.481	72.246
2240	1780.33	24227.0	496.20	19778.7	48.445	67.198	3240	2780.33	38176.2	6594.4	31742.0	5.273	72.335
2260	1800.33	24498.5	526.88	20010.4	46.031	67.317	3260	2800.33	38461.3	6891.1	31987.4	5.077	72.422
2280	1820.33	24769.9	560.02	20242.2	43.691	67.438	3280	2820.33	38746.4	7201.0	32232.8	4.888	72.510
2300	1840.33	25042.3	594.06	20474.9	41.549	67.555	3300	2840.33	39032.5	7524.9	32479.2	4.706	72.597
2320	1860.33	25314.8	630.16	20707.6	39.509	67.672	3320	2860.33	39318.6	7863.4	32725.6	4.531	72.684
2340	1880.33	25587.2	668.46	20940.3	37.566	67.789	3340	2880.33	39603.7	8208.9	32971.0	4.366	72.770
2360	1900.33	25860.6	709.09	21173.9	35.717	67.906	3360	2900.33	39890.8	8569.6	33218.3	4.208	72.855
2380	1920.33	26134.0	751.43	21407.6	33.990	68.022	3380	2920.33	40176.9	8946.2	33464.7	4.055	72.941
2400	1940.33	26407.4	796.30	21641.3	32.344	68.137	3400	2940.33	40463.0	9339.2	33711.1	3.907	73.026
2420	1960.33	26680.8	843.01	21875.0	30.807	68.250	3420	2960.33	40750.1	9739.8	33958.4	3.768	73.109
2440	1980.33	26956.1	892.46	22110.6	29.340	68.363	3440	2980.33	41037.1	10157.6	34205.8	3.634	73.193
2460	2000.33	27230.5	944.81	22345.3	27.942	68.476	3460	3000.33	41324.2	10593.3	34453.1	3.505	73.276
2480	2020.33	27505.9	999.22	22580.9	26.635	68.588	3480	3020.33	41611.3	11047.7	34700.5	3.380	73.360
2500	2040.33	27781.2	1056.78	22816.6	25.387	68.699	3500	3040.33	41899.3	11510.1	34948.8	3.263	73.441
2520	2060.33	28057.5	1116.53	23053.2	24.221	68.808	3520	3060.33	42186.4	11991.8	35196.2	3.150	73.522
2540	2080.33	28333.9	1179.66	23289.8	23.107	68.917	3540	3080.33	42474.4	12493.7	35444.5	3.041	73.604
2560	2100.33	28610.2	1245.11	23526.4	22.064	69.025	3560	3100.33	42762.5	13016.6	35692.8	2.935	73.685
2580	2120.33	28887.5	1314.19	23764.0	21.068	69.132	3580	3120.33	43050.5	13561.3	35941.1	2.833	73.767
2600	2140.33	29164.8	1387.11	24001.6	20.115	69.239	3600	3140.33	43338.6	14114.8	36189.5	2.737	73.846
2620	2160.33	29442.1	1464.07	24239.1	19.204	69.346	3620	3160.33	43626.6	14690.8	36437.8	2.644	73.926
2640	2180.33	29720.4	1543.76	24477.7	18.352	69.451	3640	3180.33	43915.6	15290.4	36687.1	2.555	74.005
2660	2200.33	29997.7	1627.79	24715.3	17.537	69.557	3660	3200.33	44203.7	15914.4	36935.4	2.468	74.084
2680	2220.33	30276.9	1716.39	24954.8	16.756	69.662	3680	3220.33	44492.7	16563.9	37184.7	2.384	74.164
2700	2240.33	30555.2	1808.00	25193.4	16.026	69.765	3700	3240.33	44781.7	17222.6	37434.0	2.305	74.241
2720	2260.33	30834.5	1904.51	25432.9	15.327	69.869	3720	3260.33	45070.7	17907.6	37683.3	2.229	74.319
2740	2280.33	31113.7	2004.16	25672.5	14.672	69.970	3740	3280.33	45359.8	18619.8	37932.6	2.156	74.396
2760	2300.33	31393.0	2111.13	25912.0	14.030	70.073	3760	3300.33	45648.8	19360.3	38181.9	2.084	74.474
2780	2320.33	31673.2	2219.37	26152.5	13.442	70.172	3780	3320.33	45938.8	20130.2	38432.2	2.015	74.551

Table 17 Carbon Dioxide at Low Pressures (for One Pound-Mole)

$\bar{m} = 44.0098$

T	t	$\bar{h}$	p_r	$\bar{u}$	v_r	$\bar{\phi}$	T	t	$\bar{h}$	p_r	$\bar{u}$	v_r	$\bar{\phi}$
3800	3340.33	46228.8	20931	38682.5	1.9483	74.629	4600	4140.33	57909.7	85219	48774.8	.5793	77.417
3820	3360.33	46518.8	21742	38932.8	1.8855	74.704	4620	4160.33	58204.6	87990	49029.9	.5635	77.480
3840	3380.33	46807.8	22584	39182.1	1.8247	74.780	4640	4180.33	58498.5	90851	49284.1	.5481	77.544
3860	3400.33	47098.8	23458	39433.3	1.7658	74.855	4660	4200.33	58792.4	93806	49538.3	.5331	77.607
3880	3420.33	47388.8	24367	39683.6	1.7088	74.930	4680	4220.33	59087.3	96856	49793.5	.5185	77.671
3900	3440.33	47678.8	25285	39933.9	1.6552	75.004	4700	4240.33	59382.2	100005	50048.6	.5044	77.735
3920	3460.33	47969.7	26265	40185.2	1.6017	75.079	4720	4260.33	59676.1	103154	50302.8	.4910	77.796
3940	3480.33	48259.7	27255	40435.4	1.5514	75.153	4740	4280.33	59971.0	106402	50558.0	.4781	77.858
3960	3500.33	48550.7	28282	40686.7	1.5026	75.226	4760	4300.33	60265.8	109862	50813.1	.4650	77.921
3980	3520.33	48841.7	29348	40938.0	1.4553	75.300	4780	4320.33	60560.7	113321	51068.3	.4527	77.983
4000	3540.33	49132.7	30454	41189.2	1.4095	75.373	4800	4340.33	60855.6	116889	51323.5	.4407	78.044
4020	3560.33	49423.6	31571	41440.5	1.3665	75.445	4820	4360.33	61150.5	120569	51578.6	.4290	78.106
4040	3580.33	49714.6	32761	41691.7	1.3234	75.518	4840	4380.33	61446.3	124241	51834.8	.4181	78.165
4060	3600.33	50005.6	33961	41943.0	1.2829	75.590	4860	4400.33	61741.2	128153	52089.9	.4070	78.227
4080	3620.33	50296.6	35206	42194.2	1.2437	75.661	4880	4420.33	62036.1	132188	52345.1	.3962	78.289
4100	3640.33	50588.5	36497	42446.5	1.2056	75.733	4900	4440.33	62331.9	136214	52601.2	.3860	78.348
4120	3660.33	50879.5	37797	42697.7	1.1698	75.802	4920	4460.33	62626.8	140362	52856.4	.3762	78.408
4140	3680.33	51171.4	39182	42950.0	1.1339	75.874	4940	4480.33	62922.7	144781	53112.5	.3662	78.469
4160	3700.33	51463.4	40578	43202.2	1.1002	75.943	4960	4500.33	63217.6	149190	53367.7	.3568	78.529
4180	3720.33	51755.3	42065	43454.4	1.0664	76.015	4980	4520.33	63513.4	153734	53623.8	.3476	78.588
4200	3740.33	52047.3	43564	43706.7	1.0346	76.084	5000	4540.33	63809.3	158416	53880.0	.3387	78.648
4220	3760.33	52339.2	45115	43958.9	1.0038	76.154	5020	4560.33	64105.1	163077	54136.1	.3303	78.706
4240	3780.33	52631.2	46676	44211.1	.9748	76.221	5040	4580.33	64401.0	168044	54392.3	.3219	78.765
4260	3800.33	52924.1	48338	44464.3	.9458	76.291	5060	4600.33	64696.8	173161	54648.4	.3136	78.825
4280	3820.33	53216.1	50010	44716.6	.9184	76.358	5080	4620.33	64992.7	178257	54904.5	.3058	78.882
4300	3840.33	53509.0	51791	44969.8	.8910	76.428	5100	4640.33	65289.5	183502	55161.6	.2983	78.940
4320	3860.33	53801.9	53583	45223.0	.8652	76.495	5120	4660.33	65585.4	189090	55417.8	.2906	78.999
4340	3880.33	54094.8	55436	45476.2	.8402	76.563	5140	4680.33	65881.2	194654	55673.9	.2834	79.057
4360	3900.33	54387.8	57353	45729.4	.8158	76.630	5160	4700.33	66178.1	200382	55931.0	.2763	79.115
4380	3920.33	54680.7	59336	45982.6	.7922	76.698	5180	4720.33	66474.9	206278	56188.2	.2695	79.172
4400	3940.33	54973.6	61389	46235.8	.7692	76.765	5200	4740.33	66770.8	212135	56444.3	.2631	79.228
4420	3960.33	55267.5	63448	46490.0	.7476	76.831	5220	4760.33	67067.6	218377	56701.4	.2565	79.285
4440	3980.33	55560.5	65643	46743.2	.7259	76.898	5240	4780.33	67364.4	224803	56958.5	.2501	79.343
4460	4000.33	55854.4	67845	46997.4	.7055	76.964	5260	4800.33	67661.3	231186	57215.6	.2442	79.399
4480	4020.33	56147.3	70121	47250.6	.6856	77.030	5280	4820.33	67958.1	237751	57472.8	.2383	79.454
4500	4040.33	56441.2	72474	47504.8	.6663	77.095	5300	4840.33	68254.9	244747	57729.9	.2324	79.512
4520	4060.33	56735.1	74905	47759.0	.6476	77.161	5320	4860.33	68551.8	251696	57987.0	.2268	79.567
4540	4080.33	57029.0	77341	48013.2	.6300	77.224	5340	4880.33	68848.6	258843	58244.1	.2214	79.623
4560	4100.33	57322.9	79936	48267.4	.6122	77.290	5360	4900.33	69145.4	265927	58501.2	.2163	79.677
4580	4120.33	57616.8	82535	48521.6	.5955	77.353	5380	4920.33	69443.2	273479	58759.3	.2111	79.732

Table 18   Carbon Dioxide

$\bar{m} = 44.0098$

T R	t F	$\bar{c}_p$ Btu lb-mole R	$\bar{c}_v$ Btu lb-mole R	$k = \dfrac{\bar{c}_p}{\bar{c}_v}$	a ft sec
200	−259.67	7.003	5.017	1.396	561.6
250	−209.67	7.141	5.155	1.385	625.5
300	−159.67	7.371	5.385	1.369	681.1
350	−109.67	7.665	5.679	1.350	730.5
400	−59.67	7.989	6.003	1.331	775.5
450	−9.67	8.319	6.333	1.314	817.2
500	40.33	8.642	6.656	1.298	856.4
550	90.33	8.950	6.964	1.285	893.6
600	140.33	9.241	7.255	1.274	929.2
650	190.33	9.515	7.529	1.264	963.3
700	240.33	9.772	7.786	1.255	996.2
750	290.33	10.014	8.028	1.247	1028.0
800	340.33	10.241	8.255	1.241	1058.9
850	390.33	10.456	8.470	1.234	1088.8
900	440.33	10.658	8.672	1.229	1117.8
1000	540.33	11.031	9.045	1.220	1173.8
1100	640.33	11.367	9.381	1.212	1227.1
1200	740.33	11.670	9.684	1.205	1278.1
1300	840.33	11.944	9.958	1.199	1327.2
1400	940.33	12.192	10.206	1.195	1374.5
1500	1040.33	12.417	10.431	1.190	1420.3
1600	1140.33	12.620	10.634	1.187	1464.6
1700	1240.33	12.804	10.818	1.184	1507.7
1800	1340.33	12.971	10.985	1.181	1549.5
1900	1440.33	13.123	11.137	1.178	1590.3
2000	1540.33	13.262	11.276	1.176	1630.1
2100	1640.33	13.387	11.401	1.174	1669.0
2200	1740.33	13.502	11.516	1.172	1707.0
2300	1840.33	13.607	11.621	1.171	1744.2
2400	1940.33	13.702	11.717	1.169	1780.7
2500	2040.33	13.790	11.804	1.168	1816.4
2600	2140.33	13.869	11.883	1.167	1851.5
2700	2240.33	13.943	11.957	1.166	1886.0
2800	2340.33	14.012	12.026	1.165	1919.8
2900	2440.33	14.076	12.090	1.164	1953.0
3000	2540.33	14.135	12.149	1.163	1985.7
3200	2740.33	14.239	12.253	1.162	2049.6
3400	2940.33	14.334	12.348	1.161	2111.6
3600	3140.33	14.413	12.428	1.160	2171.8
3800	3340.33	14.485	12.499	1.159	2230.5
4000	3540.33	14.548	12.563	1.158	2287.6
4200	3740.33	14.606	12.620	1.157	2343.4
4400	3940.33	14.658	12.672	1.157	2397.9
4600	4140.33	14.705	12.719	1.156	2451.1
4800	4340.33	14.749	12.763	1.156	2503.3
5000	4540.33	14.789	12.803	1.155	2554.4
5200	4740.33	14.826	12.841	1.155	2604.4
5400	4940.33	14.862	12.876	1.154	2653.5
5600	5140.33	14.896	12.910	1.154	2701.8
5800	5340.33	14.926	12.940	1.153	2749.2

Table 19   Hydrogen at Low Pressures (for One Pound-Mole)

$\bar{m}$ = 2.0158

T	t	$\bar{h}$	p_r	$\bar{u}$	v_r	$\bar{\phi}$	T	t	$\bar{h}$	p_r	$\bar{u}$	v_r	$\bar{\phi}$
300	−159.67	2066.2	.9517	1470.5	3382.9	27.337	550	90.33	3732.2	7.203	2640.0	819.4	31.357
305	−154.67	2097.5	1.0025	1491.8	3265.0	27.441	555	95.33	3766.7	7.430	2664.5	801.6	31.418
310	−149.67	2128.8	1.0549	1513.2	3153.5	27.542	560	100.33	3801.1	7.664	2689.0	784.2	31.480
315	−144.67	2160.4	1.1101	1534.9	3045.0	27.643	565	105.33	3835.7	7.913	2713.7	766.3	31.543
320	−139.67	2192.1	1.1671	1556.6	2942.5	27.742	570	110.33	3870.3	8.154	2738.4	750.2	31.603
325	−134.67	2223.8	1.2269	1578.4	2842.7	27.842	575	115.33	3904.9	8.411	2763.0	733.7	31.665
330	−129.67	2255.7	1.2885	1600.4	2748.4	27.939	580	120.33	3939.5	8.667	2787.7	718.2	31.724
335	−124.67	2287.7	1.3532	1622.4	2656.7	28.036	585	125.33	3974.1	8.931	2812.4	703.0	31.784
340	−119.67	2319.8	1.4198	1644.6	2569.9	28.132	590	130.33	4008.7	9.193	2837.0	688.7	31.841
345	−114.67	2352.0	1.4881	1666.9	2488.0	28.225	595	135.33	4043.4	9.473	2861.8	674.0	31.901
350	−109.67	2384.3	1.5597	1689.3	2408.2	28.318	600	140.33	4078.0	9.752	2886.5	660.3	31.958
355	−104.67	2416.8	1.6331	1711.9	2332.8	28.410	605	145.33	4112.7	10.039	2911.3	646.7	32.016
360	−99.67	2449.4	1.7100	1734.4	2259.3	28.501	610	150.33	4147.4	10.335	2936.0	633.4	32.074
365	−94.67	2482.0	1.7887	1757.1	2189.9	28.590	615	155.33	4182.0	10.628	2960.7	621.0	32.129
370	−89.67	2514.6	1.8710	1779.8	2122.2	28.680	620	160.33	4216.8	10.941	2985.5	608.1	32.187
375	−84.67	2547.4	1.9572	1802.7	2056.2	28.769	625	165.33	4251.4	11.251	3010.3	596.1	32.242
380	−79.67	2580.3	2.0431	1825.7	1995.9	28.855	630	170.33	4286.2	11.571	3035.1	584.3	32.298
385	−74.67	2613.2	2.1350	1848.6	1935.1	28.942	635	175.33	4321.0	11.899	3059.9	572.7	32.354
390	−69.67	2646.3	2.2266	1871.8	1879.6	29.025	640	180.33	4355.7	12.225	3084.8	561.8	32.407
395	−64.67	2679.4	2.3245	1895.0	1823.6	29.111	645	185.33	4390.4	12.560	3109.5	551.1	32.461
400	−59.67	2712.6	2.4242	1918.3	1770.7	29.194	650	190.33	4425.2	12.903	3134.3	540.6	32.514
405	−54.67	2745.9	2.5256	1941.6	1720.9	29.276	655	195.33	4459.9	13.257	3159.2	530.2	32.568
410	−49.67	2779.2	2.6313	1965.0	1672.1	29.357	660	200.33	4494.8	13.606	3184.1	520.6	32.620
415	−44.67	2812.6	2.7415	1988.5	1624.5	29.438	665	205.33	4529.5	13.978	3208.9	510.5	32.673
420	−39.67	2846.1	2.8562	2012.0	1578.1	29.520	670	210.33	4564.3	14.346	3233.8	501.2	32.725
425	−34.67	2879.6	2.9728	2035.6	1534.2	29.599	675	215.33	4599.1	14.724	3258.6	492.0	32.777
430	−29.67	2913.2	3.0910	2059.2	1492.9	29.677	680	220.33	4633.9	15.112	3283.5	482.9	32.828
435	−24.67	2946.8	3.2171	2083.0	1451.0	29.756	685	225.33	4668.7	15.510	3308.4	474.0	32.880
440	−19.67	2980.5	3.3417	2106.8	1413.0	29.832	690	230.33	4703.5	15.919	3333.3	465.2	32.932
445	−14.67	3014.3	3.4746	2130.6	1374.4	29.909	695	235.33	4738.4	16.322	3358.2	457.0	32.981
450	−9.67	3048.1	3.6092	2154.5	1338.0	29.985	700	240.33	4773.1	16.735	3383.0	448.9	33.031
455	−4.67	3082.0	3.7490	2178.4	1302.4	30.060	705	245.33	4808.0	17.158	3408.0	440.9	33.080
460	.33	3115.9	3.8864	2202.4	1270.2	30.131	710	250.33	4842.9	17.593	3432.9	433.1	33.130
465	5.33	3149.7	4.0329	2226.3	1237.4	30.205	715	255.33	4877.7	18.020	3457.8	425.8	33.178
470	10.33	3183.7	4.1849	2250.4	1205.2	30.278	720	260.33	4912.6	18.476	3482.8	418.2	33.227
475	15.33	3217.8	4.3383	2274.5	1175.0	30.350	725	265.33	4947.4	18.925	3507.7	411.1	33.275
480	20.33	3251.9	4.4974	2298.7	1145.4	30.421	730	270.33	4982.3	19.404	3532.6	403.7	33.325
485	25.33	3286.0	4.6622	2322.8	1116.4	30.493	735	275.33	5017.2	19.876	3557.5	396.9	33.372
490	30.33	3320.1	4.8283	2347.1	1089.1	30.562	740	280.33	5052.0	20.338	3582.5	390.5	33.418
495	35.33	3354.2	5.0003	2371.2	1062.4	30.632	745	285.33	5086.9	20.832	3607.4	383.8	33.466
500	40.33	3388.5	5.1784	2395.6	1036.2	30.701	750	290.33	5121.7	21.317	3632.3	377.6	33.511
505	45.33	3422.7	5.3628	2419.8	1010.6	30.771	755	295.33	5156.7	21.813	3657.4	371.4	33.557
510	50.33	3456.9	5.5483	2444.1	986.4	30.838	760	300.33	5191.5	22.342	3682.3	365.0	33.605
515	55.33	3491.2	5.7344	2468.5	963.8	30.904	765	305.33	5226.4	22.862	3707.2	359.1	33.650
520	60.33	3525.6	5.9328	2492.9	940.6	30.971	770	310.33	5261.3	23.394	3732.2	353.2	33.696
525	65.33	3559.9	6.1318	2517.4	918.8	31.037	775	315.33	5296.1	23.939	3757.1	347.4	33.742
530	70.33	3594.4	6.3312	2541.9	898.4	31.101	780	320.33	5331.1	24.471	3782.1	342.1	33.785
535	75.33	3628.8	6.5436	2566.3	877.4	31.166	785	325.33	5365.9	25.015	3807.0	336.8	33.829
540	80.33	3663.2	6.7564	2590.9	857.7	31.230	790	330.33	5400.8	25.572	3832.0	331.5	33.873
545	85.33	3697.7	6.9761	2615.4	838.4	31.293	795	335.33	5435.7	26.141	3856.9	326.4	33.917

Table 19   Hydrogen at Low Pressures (for One Pound-Mole)

$\bar{m} = 2.0158$

T	t	$\bar{h}$	p_r	$\bar{u}$	v_r	$\bar{\phi}$	T	t	$\bar{h}$	p_r	$\bar{u}$	v_r	$\bar{\phi}$
800	340.33	5470.6	26.72	3881.9	321.28	33.960	1300	840.33	8972.7	148.04	6391.1	94.24	37.360
810	350.33	5540.4	27.92	3931.9	311.29	34.048	1310	850.33	9043.1	152.09	6441.7	92.43	37.414
820	360.33	5610.2	29.18	3981.8	301.57	34.135	1320	860.33	9113.6	156.26	6492.3	90.66	37.467
830	370.33	5680.1	30.43	4031.8	292.69	34.218	1330	870.33	9184.1	160.69	6542.9	88.82	37.523
840	380.33	5749.9	31.74	4081.7	284.03	34.302	1340	880.33	9254.6	164.92	6593.6	87.19	37.574
850	390.33	5819.7	33.10	4131.7	275.59	34.385	1350	890.33	9325.1	169.27	6644.2	85.59	37.626
860	400.33	5889.6	34.48	4181.8	267.64	34.467	1360	900.33	9395.6	173.73	6694.9	84.01	37.678
870	410.33	5959.4	35.89	4231.7	260.13	34.546	1370	910.33	9466.2	178.30	6745.6	82.46	37.729
880	420.33	6029.3	37.39	4281.8	252.55	34.627	1380	920.33	9536.7	183.00	6796.2	80.93	37.781
890	430.33	6099.1	38.92	4331.7	245.41	34.707	1390	930.33	9607.3	187.82	6847.0	79.42	37.833
900	440.33	6169.0	40.47	4381.8	238.67	34.784	1400	940.33	9678.0	192.77	6897.8	77.94	37.884
910	450.33	6238.9	42.08	4431.7	232.09	34.862	1410	950.33	9748.6	197.65	6948.5	76.56	37.934
920	460.33	6308.8	43.75	4481.8	225.67	34.939	1420	960.33	9819.9	202.65	7000.0	75.20	37.984
930	470.33	6378.7	45.44	4531.8	219.62	35.015	1430	970.33	9890.2	207.78	7050.4	73.86	38.033
940	480.33	6448.6	47.20	4581.9	213.70	35.090	1440	980.33	9960.5	213.04	7100.9	72.54	38.083
950	490.33	6518.6	48.93	4632.0	208.34	35.162	1450	990.33	10030.8	218.44	7151.3	71.24	38.132
960	500.33	6588.5	50.78	4682.1	202.88	35.235	1460	1000.33	10102.1	223.74	7202.7	70.03	38.180
970	510.33	6658.4	52.69	4732.1	197.55	35.309	1470	1010.33	10172.4	229.41	7253.2	68.77	38.230
980	520.33	6728.3	54.62	4782.2	192.53	35.380	1480	1020.33	10243.7	234.98	7304.6	67.59	38.277
990	530.33	6798.3	56.63	4832.3	187.62	35.452	1490	1030.33	10314.9	240.44	7356.0	66.50	38.323
1000	540.33	6868.4	58.64	4882.5	182.99	35.521	1500	1040.33	10386.2	246.29	7407.4	65.36	38.371
1010	550.33	6938.3	60.73	4932.6	178.47	35.591	1510	1050.33	10456.5	252.27	7457.9	64.24	38.418
1020	560.33	7008.3	62.90	4982.7	174.03	35.660	1520	1060.33	10527.8	258.40	7509.3	63.13	38.466
1030	570.33	7078.3	65.14	5032.9	169.70	35.730	1530	1070.33	10599.1	264.41	7560.7	62.10	38.512
1040	580.33	7148.2	67.32	5082.9	165.78	35.795	1540	1080.33	10670.4	270.56	7612.1	61.08	38.557
1050	590.33	7218.2	69.58	5133.1	161.94	35.861	1550	1090.33	10741.6	276.85	7663.6	60.08	38.603
1060	600.33	7288.3	71.99	5183.3	158.02	35.928	1560	1100.33	10811.9	283.58	7714.0	59.04	38.651
1070	610.33	7358.3	74.40	5233.5	154.33	35.994	1570	1110.33	10883.2	289.89	7765.4	58.12	38.694
1080	620.33	7428.3	76.90	5283.6	150.72	36.059	1580	1120.33	10954.5	296.63	7816.8	57.16	38.740
1090	630.33	7498.3	79.48	5333.8	147.18	36.125	1590	1130.33	11025.8	303.53	7868.3	56.21	38.786
1100	640.33	7568.5	82.06	5384.0	143.85	36.188	1600	1140.33	11097.1	310.28	7919.7	55.34	38.830
1110	650.33	7638.5	84.73	5434.2	140.58	36.252	1610	1150.33	11168.3	317.50	7971.1	54.42	38.875
1120	660.33	7708.6	87.49	5484.4	137.38	36.315	1620	1160.33	11239.6	324.57	8022.5	53.56	38.919
1130	670.33	7778.7	90.24	5534.7	134.38	36.377	1630	1170.33	11310.9	331.79	8073.9	52.72	38.963
1140	680.33	7848.8	93.08	5584.9	131.43	36.439	1640	1180.33	11382.2	339.17	8125.4	51.89	39.006
1150	690.33	7918.8	96.01	5635.1	128.54	36.500	1650	1190.33	11454.4	346.71	8177.8	51.07	39.050
1160	700.33	7989.0	99.04	5685.4	125.70	36.562	1660	1200.33	11525.7	354.42	8229.2	50.26	39.094
1170	710.33	8059.1	102.05	5735.7	123.03	36.621	1670	1210.33	11597.0	362.31	8280.6	49.47	39.137
1180	720.33	8129.3	105.06	5786.0	120.54	36.679	1680	1220.33	11669.2	369.99	8333.0	48.73	39.179
1190	730.33	8199.5	108.25	5836.3	117.97	36.738	1690	1230.33	11740.5	378.23	8384.4	47.95	39.223
1200	740.33	8269.7	111.55	5886.7	115.44	36.798	1700	1240.33	11812.8	386.25	8436.8	47.23	39.264
1210	750.33	8339.9	114.95	5937.0	112.96	36.858	1710	1250.33	11884.1	394.45	8488.2	46.52	39.306
1220	760.33	8410.1	118.21	5987.4	110.75	36.913	1720	1260.33	11956.3	402.82	8540.6	45.82	39.348
1230	770.33	8480.4	121.69	6037.8	108.47	36.971	1730	1270.33	12027.6	411.37	8592.0	45.13	39.390
1240	780.33	8550.7	125.27	6088.3	106.23	37.028	1740	1280.33	12099.8	420.10	8644.4	44.45	39.431
1250	790.33	8621.0	128.96	6138.7	104.02	37.086	1750	1290.33	12171.1	428.59	8695.9	43.82	39.471
1260	800.33	8691.3	132.62	6189.2	101.96	37.142	1760	1300.33	12243.4	437.68	8748.3	43.15	39.513
1270	810.33	8761.7	136.39	6239.7	99.93	37.197	1770	1310.33	12315.6	446.52	8800.7	42.54	39.552
1280	820.33	8832.0	140.12	6290.1	98.03	37.251	1780	1320.33	12386.9	455.54	8852.1	41.93	39.592
1290	830.33	8902.4	144.10	6340.7	96.07	37.306	1790	1330.33	12459.2	465.21	8904.5	41.29	39.634

(111)

Table 19   Hydrogen at Low Pressures (for One Pound-Mole)
$\bar{m} = 2.0158$

T	t	$\bar{h}$	p_r	$\bar{u}$	v_r	$\bar{\phi}$	T	t	$\bar{h}$	p_r	$\bar{u}$	v_r	$\bar{\phi}$
1800	1340.33	12531.4	474.6	8956.9	40.700	39.674	2800	2340.33	20017.7	2501	14457.3	12.014	42.974
1820	1360.33	12675.9	494.0	9061.7	39.539	39.753	2820	2360.33	20172.9	2572	14572.8	11.766	43.030
1840	1380.33	12820.4	514.1	9166.5	38.406	39.832	2840	2380.33	20328.2	2643	14688.3	11.533	43.083
1860	1400.33	12965.0	534.6	9271.3	37.339	39.910	2860	2400.33	20484.4	2718	14804.8	11.294	43.139
1880	1420.33	13110.4	556.4	9377.0	36.260	39.989	2880	2420.33	20640.6	2792	14921.3	11.070	43.192
1900	1440.33	13254.9	578.0	9481.8	35.280	40.065	2900	2440.33	20796.8	2868	15037.9	10.850	43.246
1920	1460.33	13400.4	600.9	9587.6	34.287	40.142	2920	2460.33	20954.0	2947	15155.3	10.634	43.300
1940	1480.33	13545.9	624.2	9693.4	33.353	40.218	2940	2480.33	21111.3	3027	15272.8	10.421	43.353
1960	1500.33	13692.4	648.4	9800.1	32.440	40.293	2960	2500.33	21267.5	3110	15389.3	10.213	43.407
1980	1520.33	13837.9	672.8	9905.9	31.581	40.367	2980	2520.33	21425.7	3192	15507.8	10.018	43.459
2000	1540.33	13984.3	698.2	10012.6	30.741	40.440	3000	2540.33	21582.9	3280	15625.3	9.816	43.512
2020	1560.33	14130.8	724.5	10119.4	29.921	40.514	3020	2560.33	21740.1	3366	15742.8	9.628	43.564
2040	1580.33	14276.3	751.1	10225.1	29.148	40.585	3040	2580.33	21898.3	3458	15861.2	9.434	43.617
2060	1600.33	14423.7	778.6	10332.9	28.393	40.657	3060	2600.33	22056.4	3549	15979.7	9.252	43.669
2080	1620.33	14570.2	807.1	10439.6	27.655	40.728	3080	2620.33	22214.6	3643	16098.2	9.074	43.721
2100	1640.33	14717.6	835.9	10547.3	26.961	40.798	3100	2640.33	22372.8	3739	16216.6	8.898	43.772
2120	1660.33	14865.1	866.5	10655.0	26.255	40.869	3120	2660.33	22531.9	3837	16336.1	8.726	43.824
2140	1680.33	15013.5	896.5	10763.7	25.617	40.937	3140	2680.33	22691.1	3934	16455.5	8.565	43.874
2160	1700.33	15160.9	928.4	10871.5	24.967	41.006	3160	2700.33	22850.3	4038	16574.9	8.398	43.925
2180	1720.33	15309.3	960.5	10980.2	24.356	41.074	3180	2720.33	23009.4	4140	16694.4	8.243	43.975
2200	1740.33	15457.8	993.8	11088.9	23.758	41.141	3200	2740.33	23168.6	4245	16813.8	8.090	44.025
2220	1760.33	15605.2	1028.1	11196.6	23.172	41.209	3220	2760.33	23328.7	4352	16934.2	7.939	44.074
2240	1780.33	15754.6	1062.6	11306.3	22.622	41.274	3240	2780.33	23488.8	4463	17054.7	7.791	44.124
2260	1800.33	15903.0	1099.4	11415.0	22.061	41.342	3260	2800.33	23649.0	4576	17175.1	7.646	44.173
2280	1820.33	16052.4	1136.3	11524.6	21.534	41.407	3280	2820.33	23810.1	4691	17296.5	7.503	44.223
2300	1840.33	16202.8	1174.4	11635.3	21.018	41.473	3300	2840.33	23971.2	4805	17417.9	7.370	44.271
2320	1860.33	16352.2	1212.6	11745.0	20.553	41.536	3320	2860.33	24132.3	4927	17539.3	7.231	44.320
2340	1880.33	16502.5	1253.2	11855.6	20.037	41.602	3340	2880.33	24293.4	5047	17660.6	7.102	44.368
2360	1900.33	16652.9	1294.0	11966.3	19.572	41.665	3360	2900.33	24454.5	5169	17782.0	6.975	44.416
2380	1920.33	16803.3	1337.4	12076.9	19.097	41.731	3380	2920.33	24615.6	5295	17903.4	6.851	44.463
2400	1940.33	16953.6	1380.9	12187.6	18.651	41.794	3400	2940.33	24776.7	5423	18024.8	6.728	44.511
2420	1960.33	17105.0	1424.4	12299.2	18.233	41.856	3420	2960.33	24938.8	5555	18147.2	6.607	44.559
2440	1980.33	17255.4	1470.7	12409.9	17.804	41.920	3440	2980.33	25100.9	5690	18269.6	6.488	44.606
2460	2000.33	17406.7	1517.0	12521.5	17.402	41.981	3460	3000.33	25263.0	5828	18391.9	6.371	44.654
2480	2020.33	17559.0	1564.8	12634.1	17.008	42.043	3480	3020.33	25426.1	5970	18515.3	6.256	44.702
2500	2040.33	17710.4	1614.0	12745.7	16.622	42.104	3500	3040.33	25589.1	6109	18638.6	6.149	44.747
2520	2060.33	17862.7	1663.2	12858.3	16.260	42.164	3520	3060.33	25751.2	6257	18761.0	6.037	44.795
2540	2080.33	18015.0	1713.8	12970.9	15.905	42.223	3540	3080.33	25915.3	6403	18885.3	5.933	44.841
2560	2100.33	18167.3	1766.0	13083.5	15.556	42.283	3560	3100.33	26078.3	6552	19008.7	5.831	44.886
2580	2120.33	18320.6	1818.0	13197.1	15.230	42.341	3580	3120.33	26241.4	6704	19132.0	5.731	44.932
2600	2140.33	18473.9	1873.4	13310.7	14.894	42.400	3600	3140.33	26405.4	6860	19256.3	5.632	44.978
2620	2160.33	18627.2	1930.4	13424.3	14.565	42.460	3620	3160.33	26569.5	7020	19380.6	5.534	45.023
2640	2180.33	18780.5	1987.2	13537.9	14.257	42.517	3640	3180.33	26733.5	7183	19505.0	5.438	45.069
2660	2200.33	18933.8	2047.7	13651.4	13.940	42.577	3660	3200.33	26897.5	7343	19629.3	5.349	45.113
2680	2220.33	19088.1	2108.0	13766.0	13.644	42.634	3680	3220.33	27062.6	7514	19754.6	5.256	45.158
2700	2240.33	19242.4	2170.0	13880.6	13.352	42.692	3700	3240.33	27226.6	7681	19878.9	5.170	45.202
2720	2260.33	19396.7	2233.9	13995.1	13.067	42.750	3720	3260.33	27391.6	7860	20004.2	5.079	45.248
2740	2280.33	19551.9	2297.3	14110.6	12.800	42.805	3740	3280.33	27556.6	8034	20129.5	4.995	45.292
2760	2300.33	19707.2	2364.9	14226.2	12.524	42.863	3760	3300.33	27721.6	8213	20254.8	4.913	45.335
2780	2320.33	19861.4	2432.0	14340.7	12.267	42.918	3780	3320.33	27886.7	8396	20380.1	4.832	45.379

(112)

Table 19 Hydrogen at Low Pressures (for One Pound-Mole)

$\bar{m} = 2.0158$

T	t	$\bar{h}$	p_r	$\bar{u}$	v_r	$\bar{\phi}$	T	t	$\bar{h}$	p_r	$\bar{u}$	v_r	$\bar{\phi}$
3800	3340.33	28052.6	8583	20506.4	4.751	45.423	4600	4140.33	34806.6	19331	25671.6	2.5536	47.035
3820	3360.33	28217.7	8773	20631.7	4.673	45.466	4620	4160.33	34978.4	19702	25803.7	2.5164	47.073
3840	3380.33	28384.6	8960	20758.9	4.599	45.508	4640	4180.33	35151.2	20080	25936.9	2.4798	47.111
3860	3400.33	28550.6	9159	20885.2	4.523	45.552	4660	4200.33	35323.1	20445	26069.0	2.4460	47.146
3880	3420.33	28716.6	9363	21011.5	4.447	45.595	4680	4220.33	35495.9	20837	26202.1	2.4103	47.184
3900	3440.33	28882.6	9571	21137.8	4.373	45.639	4700	4240.33	35668.8	21237	26335.2	2.3750	47.222
3920	3460.33	29048.6	9774	21264.0	4.304	45.681	4720	4260.33	35841.6	21622	26468.3	2.3426	47.258
3940	3480.33	29215.6	9991	21391.3	4.232	45.724	4740	4280.33	36013.4	22037	26600.5	2.3082	47.295
3960	3500.33	29382.5	10204	21518.5	4.165	45.766	4760	4300.33	36186.3	22438	26733.6	2.2766	47.331
3980	3520.33	29549.5	10420	21645.8	4.099	45.808	4780	4320.33	36360.1	22845	26867.6	2.2454	47.367
4000	3540.33	29716.5	10641	21773.0	4.034	45.850	4800	4340.33	36532.9	23260	27000.8	2.2146	47.402
4020	3560.33	29884.4	10867	21901.3	3.970	45.891	4820	4360.33	36706.7	23682	27134.8	2.1841	47.438
4040	3580.33	30052.4	11098	22029.5	3.907	45.933	4840	4380.33	36879.5	24113	27268.0	2.1541	47.474
4060	3600.33	30220.3	11322	22157.7	3.848	45.973	4860	4400.33	37053.3	24551	27402.0	2.1244	47.510
4080	3620.33	30388.3	11562	22285.9	3.787	46.014	4880	4420.33	37227.1	24996	27536.1	2.0951	47.545
4100	3640.33	30556.2	11807	22414.2	3.726	46.056	4900	4440.33	37400.9	25450	27670.2	2.0661	47.581
4120	3660.33	30724.1	12046	22542.4	3.670	46.096	4920	4460.33	37574.7	25913	27804.3	2.0376	47.617
4140	3680.33	30892.1	12302	22670.6	3.612	46.137	4940	4480.33	37749.5	26383	27939.4	2.0094	47.653
4160	3700.33	31061.0	12563	22799.8	3.554	46.179	4960	4500.33	37923.3	26863	28073.4	1.9815	47.688
4180	3720.33	31229.0	12817	22928.1	3.500	46.219	4980	4520.33	38098.1	27350	28208.5	1.9540	47.724
4200	3740.33	31397.9	13089	23057.3	3.444	46.261	5000	4540.33	38271.9	27819	28342.6	1.9288	47.758
4220	3760.33	31566.8	13353	23186.5	3.392	46.300	5020	4560.33	38446.7	28325	28477.7	1.9019	47.794
4240	3780.33	31736.7	13623	23316.6	3.340	46.340	5040	4580.33	38621.5	28810	28612.7	1.8773	47.827
4260	3800.33	31905.6	13898	23445.9	3.289	46.380	5060	4600.33	38796.2	29304	28747.8	1.8530	47.861
4280	3820.33	32074.5	14179	23575.1	3.239	46.419	5080	4620.33	38971.0	29837	28882.8	1.8272	47.897
4300	3840.33	32244.4	14465	23705.2	3.190	46.459	5100	4640.33	39146.8	30348	29018.9	1.8034	47.931
4320	3860.33	32414.3	14757	23835.4	3.142	46.499	5120	4660.33	39321.6	30868	29153.9	1.7800	47.964
4340	3880.33	32584.2	15055	23965.6	3.094	46.539	5140	4680.33	39497.3	31429	29290.0	1.7551	48.000
4360	3900.33	32754.1	15344	24095.8	3.049	46.576	5160	4700.33	39673.1	31968	29426.0	1.7322	48.034
4380	3920.33	32925.0	15654	24226.9	3.003	46.616	5180	4720.33	39848.8	32516	29562.1	1.7096	48.068
4400	3940.33	33095.9	15954	24358.1	2.960	46.654	5200	4740.33	40024.6	33074	29698.1	1.6873	48.102
4420	3960.33	33265.8	16277	24488.3	2.914	46.694	5220	4760.33	40200.3	33641	29834.1	1.6652	48.135
4440	3980.33	33436.6	16589	24619.4	2.872	46.731	5240	4780.33	40376.1	34217	29970.2	1.6434	48.169
4460	4000.33	33607.5	16907	24750.6	2.831	46.769	5260	4800.33	40552.8	34804	30107.2	1.6219	48.203
4480	4020.33	33778.4	17249	24881.7	2.787	46.809	5280	4820.33	40728.6	35401	30243.2	1.6006	48.237
4500	4040.33	33949.3	17580	25012.9	2.747	46.846	5300	4840.33	40905.3	36008	30380.3	1.5796	48.270
4520	4060.33	34120.1	17917	25144.1	2.707	46.884	5320	4860.33	41081.1	36589	30516.3	1.5604	48.302
4540	4080.33	34292.0	18260	25276.2	2.668	46.922	5340	4880.33	41257.8	37216	30653.3	1.5398	48.336
4560	4100.33	34462.9	18629	25407.3	2.627	46.962	5360	4900.33	41435.5	37816	30791.3	1.5211	48.368
4580	4120.33	34634.7	18987	25539.5	2.589	46.999	5380	4920.33	41612.2	38464	30928.3	1.5010	48.401

Table 20   Hydrogen

$\bar{m} = 2.0158$

T R	t F	$\bar{c}_p$ Btu lb-mole R	$\bar{c}_v$ Btu lb-mole R	$k = \dfrac{\bar{c}_p}{\bar{c}_v}$	a ft sec
200	−259.67	5.547	3.561	1.558	2772.0
250	−209.67	5.930	3.944	1.504	3044.8
300	−159.67	6.240	4.254	1.467	3294.5
350	−109.67	6.474	4.488	1.442	3528.8
400	−59.67	6.645	4.659	1.426	3751.2
450	−9.67	6.766	4.780	1.415	3963.6
500	40.33	6.845	4.859	1.409	4168.0
550	90.33	6.899	4.913	1.404	4364.5
600	140.33	6.933	4.947	1.401	4554.1
650	190.33	6.952	4.967	1.400	4737.3
700	240.33	6.968	4.983	1.399	4913.9
750	290.33	6.976	4.990	1.398	5085.2
800	340.33	6.980	4.994	1.398	5251.4
850	390.33	6.984	4.998	1.397	5412.4
900	440.33	6.988	5.002	1.397	5568.7
1000	540.33	6.998	5.012	1.396	5868.3
1100	640.33	7.006	5.020	1.396	6153.3
1200	740.33	7.022	5.036	1.394	6424.0
1300	840.33	7.040	5.054	1.393	6683.0
1400	940.33	7.064	5.078	1.391	6930.7
1500	1040.33	7.095	5.110	1.389	7167.7
1600	1140.33	7.129	5.143	1.386	7395.9
1700	1240.33	7.171	5.185	1.383	7615.0
1800	1340.33	7.215	5.229	1.380	7826.7
1900	1440.33	7.262	5.276	1.376	8031.2
2000	1540.33	7.314	5.328	1.373	8228.9
2100	1640.33	7.368	5.382	1.369	8420.7
2200	1740.33	7.425	5.439	1.365	8606.5
2300	1840.33	7.483	5.497	1.361	8787.6
2400	1940.33	7.540	5.554	1.358	8964.3
2500	2040.33	7.600	5.614	1.354	9136.3
2600	2140.33	7.657	5.672	1.350	9304.9
2700	2240.33	7.715	5.729	1.347	9469.8
2800	2340.33	7.773	5.787	1.343	9631.2
2900	2440.33	7.826	5.840	1.340	9790.2
3000	2540.33	7.880	5.894	1.337	9946.0
3200	2740.33	7.987	6.001	1.331	10249.1
3400	2940.33	8.092	6.107	1.325	10541.8
3600	3140.33	8.190	6.204	1.320	10826.6
3800	3340.33	8.279	6.293	1.316	11104.1
4000	3540.33	8.364	6.379	1.311	11374.2
4200	3740.33	8.444	6.458	1.308	11638.1
4400	3940.33	8.521	6.535	1.304	11895.4
4600	4140.33	8.595	6.609	1.300	12146.9
4800	4340.33	8.664	6.678	1.297	12393.3
5000	4540.33	8.730	6.744	1.294	12634.7
5200	4740.33	8.791	6.806	1.292	12871.7
5400	4940.33	8.853	6.867	1.289	13103.6
5600	5140.33	8.915	6.929	1.287	13330.7
5800	5340.33	8.974	6.988	1.284	13553.8

Table 21   Carbon Monoxide at Low Pressures (for One Pound-Mole)

$\bar{m} = 28.0104$

T	t	$\bar{h}$	p_r	$\bar{u}$	v_r	$\bar{\phi}$	T	t	$\bar{h}$	p_r	$\bar{u}$	v_r	$\bar{\phi}$
300	−159.67	2082.1	.2708	1486.3	11890	43.132	550	90.33	3820.7	2.262	2728.5	2609.6	47.347
305	−154.67	2116.8	.2869	1511.1	11408	43.247	555	95.33	3855.5	2.335	2753.4	2551.1	47.410
310	−149.67	2151.6	.3037	1536.0	10953	43.360	560	100.33	3890.3	2.409	2778.2	2494.4	47.472
315	−144.67	2186.4	.3212	1560.8	10524	43.471	565	105.33	3925.1	2.486	2803.1	2439.5	47.534
320	−139.67	2221.1	.3394	1585.6	10117	43.580	570	110.33	3960.0	2.564	2828.0	2386.2	47.596
325	−134.67	2255.9	.3584	1610.5	9733	43.688	575	115.33	3994.8	2.643	2852.9	2334.5	47.656
330	−129.67	2290.6	.3780	1635.3	9368	43.794	580	120.33	4029.6	2.725	2877.8	2284.3	47.717
335	−124.67	2325.4	.3985	1660.1	9022	43.899	585	125.33	4064.5	2.808	2902.7	2235.7	47.776
340	−119.67	2360.2	.4197	1685.0	8694	44.002	590	130.33	4099.3	2.893	2927.6	2188.5	47.836
345	−114.67	2394.9	.4417	1709.8	8382	44.103	595	135.33	4134.1	2.980	2952.5	2142.6	47.895
350	−109.67	2429.7	.4645	1734.6	8086	44.203	600	140.33	4169.0	3.069	2977.5	2098.1	47.953
355	−104.67	2464.4	.4882	1759.5	7804	44.302	605	145.33	4203.8	3.160	3002.4	2054.8	48.011
360	−99.67	2499.2	.5126	1784.3	7536	44.399	610	150.33	4238.7	3.252	3027.3	2012.8	48.068
365	−94.67	2534.0	.5380	1809.1	7281	44.495	615	155.33	4273.6	3.347	3052.3	1972.0	48.125
370	−89.67	2568.7	.5643	1834.0	7037	44.590	620	160.33	4308.4	3.443	3077.2	1932.3	48.182
375	−84.67	2603.5	.5914	1858.8	6805	44.683	625	165.33	4343.3	3.542	3102.1	1893.7	48.238
380	−79.67	2638.3	.6195	1883.6	6583	44.775	630	170.33	4378.2	3.642	3127.1	1856.1	48.293
385	−74.67	2673.0	.6485	1908.5	6371	44.866	635	175.33	4413.1	3.745	3152.0	1819.6	48.348
390	−69.67	2707.8	.6785	1933.3	6169	44.956	640	180.33	4448.0	3.850	3177.0	1784.1	48.403
395	−64.67	2742.5	.7094	1958.1	5975	45.044	645	185.33	4482.9	3.956	3202.0	1749.5	48.457
400	−59.67	2777.3	.7413	1983.0	5790	45.132	650	190.33	4517.8	4.065	3227.0	1715.8	48.511
405	−54.67	2812.1	.7743	2007.8	5613	45.218	655	195.33	4552.7	4.176	3252.0	1683.1	48.565
410	−49.67	2846.8	.8083	2032.6	5443	45.303	660	200.33	4587.6	4.290	3276.9	1651.2	48.618
415	−44.67	2881.6	.8433	2057.5	5281	45.388	665	205.33	4622.5	4.405	3301.9	1620.1	48.671
420	−39.67	2916.4	.8794	2082.3	5125	45.471	670	210.33	4657.5	4.523	3327.0	1589.8	48.723
425	−34.67	2951.1	.9167	2107.1	4976	45.553	675	215.33	4692.4	4.643	3352.0	1560.3	48.775
430	−29.67	2985.9	.9550	2132.0	4832	45.635	680	220.33	4727.4	4.765	3377.0	1531.5	48.827
435	−24.67	3020.7	.9944	2156.8	4694	45.715	685	225.33	4762.4	4.889	3402.0	1503.5	48.878
440	−19.67	3055.4	1.0350	2181.7	4562	45.794	690	230.33	4797.3	5.016	3427.1	1476.2	48.929
445	−14.67	3090.2	1.0768	2206.5	4435	45.873	695	235.33	4832.3	5.145	3452.1	1449.5	48.979
450	−9.67	3125.0	1.1198	2231.3	4313	45.951	700	240.33	4867.3	5.277	3477.2	1423.5	49.029
455	−4.67	3159.8	1.1640	2256.2	4195	46.028	705	245.33	4902.3	5.411	3502.3	1398.2	49.079
460	.33	3194.5	1.2094	2281.0	4082	46.104	710	250.33	4937.3	5.548	3527.4	1373.4	49.129
465	5.33	3229.3	1.2560	2305.9	3973	46.179	715	255.33	4972.3	5.687	3552.4	1349.3	49.178
470	10.33	3264.1	1.3040	2330.7	3868	46.253	720	260.33	5007.4	5.828	3577.6	1325.7	49.227
475	15.33	3298.8	1.3532	2355.6	3767	46.327	725	265.33	5042.4	5.972	3602.7	1302.7	49.275
480	20.33	3333.6	1.4038	2380.4	3670	46.400	730	270.33	5077.5	6.119	3627.8	1280.3	49.323
485	25.33	3368.4	1.4556	2405.3	3576	46.472	735	275.33	5112.5	6.268	3652.9	1258.3	49.371
490	30.33	3403.2	1.5089	2430.1	3485	46.543	740	280.33	5147.6	6.420	3678.1	1236.9	49.419
495	35.33	3438.0	1.5635	2455.0	3398	46.614	745	285.33	5182.7	6.575	3703.2	1216.0	49.466
500	40.33	3472.7	1.6195	2479.8	3313	46.684	750	290.33	5217.8	6.732	3728.4	1195.5	49.513
505	45.33	3507.5	1.6770	2504.7	3232	46.753	755	295.33	5252.9	6.892	3753.6	1175.6	49.560
510	50.33	3542.3	1.7359	2529.5	3153	46.821	760	300.33	5288.0	7.055	3778.8	1156.0	49.606
515	55.33	3577.1	1.7962	2554.4	3077	46.889	765	305.33	5323.2	7.221	3804.0	1136.9	49.652
520	60.33	3611.9	1.8581	2579.3	3003	46.956	770	310.33	5358.3	7.389	3829.2	1118.3	49.698
525	65.33	3646.7	1.9215	2604.1	2932	47.023	775	315.33	5393.5	7.561	3854.5	1100.0	49.743
530	70.33	3681.5	1.9864	2629.0	2863	47.089	780	320.33	5428.7	7.735	3879.7	1082.2	49.789
535	75.33	3716.3	2.0528	2653.9	2797	47.154	785	325.33	5463.9	7.912	3905.0	1064.7	49.834
540	80.33	3751.1	2.1208	2678.7	2732	47.219	790	330.33	5499.1	8.092	3930.2	1047.7	49.878
545	85.33	3785.9	2.1905	2703.6	2670	47.283	795	335.33	5534.3	8.275	3955.5	1031.0	49.923

Table 21 Carbon Monoxide at Low Pressures (for One Pound-Mole)
$\bar{m} = 28.0104$

T	t	$\bar{h}$	p_r	$\bar{u}$	v_r	$\bar{\phi}$	T	t	$\bar{h}$	p_r	$\bar{u}$	v_r	$\bar{\phi}$
800	340.33	5569.5	8.461	3980.8	1014.6	49.967	1300	840.33	9194.7	49.59	6613.0	281.34	53.478
810	350.33	5640.0	8.843	4031.5	983.0	50.055	1310	850.33	9269.5	51.04	6668.1	275.43	53.536
820	360.33	5710.6	9.237	4082.2	952.7	50.141	1320	860.33	9344.5	52.53	6723.2	269.68	53.593
830	370.33	5781.3	9.644	4133.0	923.6	50.227	1330	870.33	9419.6	54.05	6778.4	264.08	53.649
840	380.33	5852.0	10.064	4183.8	895.7	50.311	1340	880.33	9494.8	55.60	6833.7	258.62	53.706
850	390.33	5922.7	10.498	4234.7	868.9	50.395	1350	890.33	9570.1	57.19	6889.2	253.31	53.762
860	400.33	5993.5	10.945	4285.7	843.2	50.478	1360	900.33	9645.4	58.82	6944.7	248.14	53.817
870	410.33	6064.5	11.406	4336.8	818.5	50.560	1370	910.33	9720.9	60.48	7000.3	243.10	53.873
880	420.33	6135.4	11.882	4387.9	794.8	50.641	1380	920.33	9796.5	62.17	7056.0	238.19	53.928
890	430.33	6206.5	12.372	4439.1	772.0	50.721	1390	930.33	9872.2	63.91	7111.8	233.41	53.982
900	440.33	6277.6	12.877	4490.3	750.1	50.801	1400	940.33	9947.9	65.68	7167.7	228.75	54.037
910	450.33	6348.8	13.397	4541.7	728.9	50.880	1410	950.33	10023.8	67.49	7223.7	224.20	54.091
920	460.33	6420.1	13.933	4593.1	708.6	50.957	1420	960.33	10099.7	69.34	7279.8	219.77	54.144
930	470.33	6491.4	14.485	4644.6	689.0	51.035	1430	970.33	10175.8	71.23	7336.0	215.45	54.198
940	480.33	6562.8	15.053	4696.1	670.1	51.111	1440	980.33	10251.9	73.16	7392.3	211.24	54.251
950	490.33	6634.3	15.638	4747.8	651.9	51.187	1450	990.33	10328.1	75.13	7448.6	207.13	54.303
960	500.33	6705.9	16.239	4799.5	634.4	51.262	1460	1000.33	10404.4	77.14	7505.1	203.12	54.356
970	510.33	6777.6	16.858	4851.3	617.5	51.336	1470	1010.33	10480.9	79.19	7561.7	199.21	54.408
980	520.33	6849.4	17.495	4903.2	601.1	51.409	1480	1020.33	10557.4	81.29	7618.3	195.39	54.460
990	530.33	6921.2	18.149	4955.2	585.4	51.482	1490	1030.33	10634.0	83.42	7675.0	191.67	54.511
1000	540.33	6993.2	18.822	5007.3	570.1	51.555	1500	1040.33	10710.7	85.61	7731.9	188.03	54.563
1010	550.33	7065.2	19.514	5059.5	555.4	51.626	1510	1050.33	10787.5	87.84	7788.8	184.49	54.614
1020	560.33	7137.3	20.225	5111.7	541.2	51.697	1520	1060.33	10864.3	90.11	7845.8	181.02	54.665
1030	570.33	7209.5	20.955	5164.0	527.5	51.768	1530	1070.33	10941.3	92.43	7902.9	177.64	54.715
1040	580.33	7281.8	21.705	5216.5	514.2	51.838	1540	1080.33	11018.3	94.79	7960.1	174.34	54.765
1050	590.33	7354.1	22.475	5269.0	501.4	51.907	1550	1090.33	11095.5	97.21	8017.4	171.11	54.815
1060	600.33	7426.6	23.266	5321.6	488.9	51.976	1560	1100.33	11172.7	99.67	8074.8	167.96	54.865
1070	610.33	7499.2	24.078	5374.3	476.9	52.044	1570	1110.33	11250.0	102.18	8132.2	164.89	54.914
1080	620.33	7571.8	24.912	5427.1	465.2	52.111	1580	1120.33	11327.5	104.74	8189.8	161.88	54.963
1090	630.33	7644.6	25.768	5480.0	454.0	52.178	1590	1130.33	11405.0	107.35	8247.4	158.94	55.012
1100	640.33	7717.4	26.645	5533.0	443.0	52.245	1600	1140.33	11482.5	110.01	8305.2	156.07	55.061
1110	650.33	7790.4	27.546	5586.1	432.4	52.311	1610	1150.33	11560.2	112.73	8363.0	153.27	55.109
1120	660.33	7863.4	28.470	5639.2	422.2	52.376	1620	1160.33	11638.0	115.49	8420.9	150.53	55.157
1130	670.33	7936.5	29.417	5692.5	412.2	52.441	1630	1170.33	11715.8	118.31	8478.8	147.85	55.205
1140	680.33	8009.8	30.388	5745.9	402.6	52.506	1640	1180.33	11793.7	121.19	8536.9	145.23	55.253
1150	690.33	8083.1	31.384	5799.3	393.2	52.570	1650	1190.33	11871.7	124.12	8595.1	142.66	55.300
1160	700.33	8156.5	32.405	5852.9	384.2	52.634	1660	1200.33	11949.8	127.10	8653.3	140.16	55.348
1170	710.33	8230.0	33.452	5906.6	375.3	52.697	1670	1210.33	12028.0	130.14	8711.6	137.71	55.395
1180	720.33	8303.6	34.524	5960.3	366.8	52.759	1680	1220.33	12106.3	133.24	8770.0	135.31	55.441
1190	730.33	8377.3	35.622	6014.2	358.5	52.822	1690	1230.33	12184.6	136.40	8828.5	132.97	55.488
1200	740.33	8451.2	36.748	6068.1	350.4	52.883	1700	1240.33	12263.0	139.61	8887.0	130.67	55.534
1210	750.33	8525.1	37.900	6122.2	342.6	52.945	1710	1250.33	12341.5	142.89	8945.7	128.43	55.580
1220	760.33	8599.1	39.081	6176.3	335.0	53.006	1720	1260.33	12420.1	146.22	9004.4	126.23	55.626
1230	770.33	8673.2	40.290	6230.6	327.6	53.066	1730	1270.33	12498.8	149.62	9063.2	124.09	55.671
1240	780.33	8747.4	41.527	6284.9	320.4	53.126	1740	1280.33	12577.5	153.08	9122.1	121.98	55.717
1250	790.33	8821.7	42.794	6339.4	313.5	53.186	1750	1290.33	12656.3	156.60	9181.1	119.93	55.762
1260	800.33	8896.1	44.091	6393.9	306.7	53.245	1760	1300.33	12735.2	160.18	9240.1	117.91	55.807
1270	810.33	8970.6	45.418	6448.5	300.1	53.304	1770	1310.33	12814.2	163.83	9299.2	115.94	55.852
1280	820.33	9045.2	46.776	6503.3	293.7	53.362	1780	1320.33	12893.2	167.55	9358.4	114.01	55.896
1290	830.33	9119.9	48.165	6558.1	287.4	53.421	1790	1330.33	12972.4	171.33	9417.7	112.12	55.941

(116)

Table 21 Carbon Monoxide at Low Pressures (for One Pound-Mole)

$\bar{m} = 28.0104$

T	t	$\bar{h}$	p_r	$\bar{u}$	v_r	$\bar{\phi}$	T	t	$\bar{h}$	p_r	$\bar{u}$	v_r	$\bar{\phi}$
1800	1340.33	13051.6	175.2	9477.0	110.27	55.985	2800	2340.33	21269.9	1085.8	15709.5	27.674	59.607
1820	1360.33	13210.2	183.1	9595.9	106.68	56.072	2820	2360.33	21438.8	1119.2	15838.7	27.041	59.668
1840	1380.33	13369.1	191.3	9715.1	103.24	56.159	2840	2380.33	21607.9	1153.3	15968.1	26.425	59.727
1860	1400.33	13528.3	199.7	9834.6	99.93	56.245	2860	2400.33	21777.2	1188.3	16097.6	25.828	59.787
1880	1420.33	13687.8	208.5	9954.4	96.76	56.331	2880	2420.33	21946.5	1224.2	16227.2	25.247	59.846
1900	1440.33	13847.6	217.6	10074.5	93.71	56.415	2900	2440.33	22116.0	1260.9	16357.0	24.682	59.904
1920	1460.33	14007.7	227.0	10194.8	90.79	56.499	2920	2460.33	22285.6	1298.4	16486.9	24.134	59.963
1940	1480.33	14168.0	236.6	10315.5	87.97	56.582	2940	2480.33	22455.3	1336.9	16616.9	23.600	60.021
1960	1500.33	14328.6	246.7	10436.3	85.27	56.664	2960	2500.33	22625.2	1376.2	16747.0	23.082	60.078
1980	1520.33	14489.5	257.0	10557.5	82.67	56.746	2980	2520.33	22795.2	1416.4	16877.3	22.578	60.135
2000	1540.33	14650.7	267.7	10678.9	80.17	56.827	3000	2540.33	22965.3	1457.6	17007.7	22.087	60.192
2020	1560.33	14812.0	278.8	10800.6	77.76	56.907	3020	2560.33	23135.5	1499.7	17138.2	21.610	60.249
2040	1580.33	14973.7	290.2	10922.5	75.44	56.987	3040	2580.33	23305.8	1542.8	17268.8	21.146	60.305
2060	1600.33	15135.6	301.9	11044.7	73.21	57.066	3060	2600.33	23476.3	1586.8	17399.5	20.695	60.361
2080	1620.33	15297.7	314.1	11167.2	71.07	57.144	3080	2620.33	23646.8	1631.8	17530.4	20.255	60.416
2100	1640.33	15460.1	326.6	11289.8	69.00	57.222	3100	2640.33	23817.5	1677.8	17661.3	19.828	60.472
2120	1660.33	15622.8	339.6	11412.7	67.00	57.299	3120	2660.33	23988.3	1724.9	17792.4	19.411	60.527
2140	1680.33	15785.6	352.9	11535.9	65.08	57.375	3140	2680.33	24159.2	1773.0	17923.6	19.006	60.581
2160	1700.33	15948.7	366.6	11659.3	63.23	57.451	3160	2700.33	24330.2	1822.1	18054.8	18.611	60.635
2180	1720.33	16112.0	380.8	11782.9	61.44	57.527	3180	2720.33	24501.3	1872.3	18186.2	18.227	60.689
2200	1740.33	16275.6	395.4	11906.7	59.71	57.601	3200	2740.33	24672.5	1923.6	18317.7	17.852	60.743
2220	1760.33	16439.3	410.4	12030.7	58.05	57.675	3220	2760.33	24843.8	1976.0	18449.3	17.488	60.796
2240	1780.33	16603.3	425.9	12155.0	56.44	57.749	3240	2780.33	25015.2	2029.5	18581.0	17.132	60.850
2260	1800.33	16767.5	441.8	12279.5	54.89	57.822	3260	2800.33	25186.7	2084.2	18712.8	16.786	60.902
2280	1820.33	16931.9	458.2	12404.1	53.39	57.894	3280	2820.33	25358.3	2140.0	18844.7	16.449	60.955
2300	1840.33	17096.5	475.1	12529.0	51.95	57.966	3300	2840.33	25530.0	2197.0	18976.7	16.120	61.007
2320	1860.33	17261.3	492.5	12654.1	50.55	58.038	3320	2860.33	25701.8	2255.1	19108.8	15.799	61.059
2340	1880.33	17426.3	510.4	12779.4	49.20	58.108	3340	2880.33	25873.7	2314.5	19240.9	15.486	61.110
2360	1900.33	17591.5	528.8	12904.9	47.89	58.179	3360	2900.33	26045.7	2375.1	19373.2	15.181	61.162
2380	1920.33	17756.9	547.7	13030.6	46.63	58.248	3380	2920.33	26217.8	2437.0	19505.6	14.884	61.213
2400	1940.33	17922.5	567.1	13156.5	45.41	58.318	3400	2940.33	26389.9	2500.1	19638.0	14.594	61.264
2420	1960.33	18088.3	587.1	13282.5	44.23	58.386	3420	2960.33	26562.2	2564.5	19770.5	14.311	61.314
2440	1980.33	18254.3	607.7	13408.8	43.09	58.455	3440	2980.33	26734.5	2630.3	19903.2	14.035	61.364
2460	2000.33	18420.4	628.8	13535.2	41.98	58.523	3460	3000.33	26907.0	2697.3	20035.9	13.766	61.414
2480	2020.33	18586.7	650.5	13661.8	40.91	58.590	3480	3020.33	27079.5	2765.7	20168.7	13.503	61.464
2500	2040.33	18753.2	672.8	13788.6	39.88	58.657	3500	3040.33	27252.1	2835.4	20301.6	13.247	61.514
2520	2060.33	18919.9	695.6	13915.5	38.88	58.723	3520	3060.33	27424.8	2906.5	20434.5	12.997	61.563
2540	2080.33	19086.7	719.1	14042.6	37.90	58.789	3540	3080.33	27597.5	2979.1	20567.6	12.752	61.612
2560	2100.33	19253.7	743.2	14169.9	36.96	58.855	3560	3100.33	27770.4	3053.0	20700.7	12.514	61.660
2580	2120.33	19420.9	768.0	14297.4	36.05	58.920	3580	3120.33	27943.3	3128.4	20833.9	12.281	61.709
2600	2140.33	19588.2	793.4	14425.0	35.17	58.984	3600	3140.33	28116.3	3205.2	20967.2	12.053	61.757
2620	2160.33	19755.7	819.4	14552.7	34.31	59.048	3620	3160.33	28289.4	3283.6	21100.6	11.831	61.805
2640	2180.33	19923.3	846.2	14680.7	33.48	59.112	3640	3180.33	28462.6	3363.4	21234.0	11.614	61.853
2660	2200.33	20091.1	873.6	14808.8	32.68	59.176	3660	3200.33	28635.8	3444.8	21367.6	11.402	61.900
2680	2220.33	20259.1	901.7	14937.0	31.90	59.238	3680	3220.33	28809.1	3527.7	21501.2	11.195	61.947
2700	2240.33	20427.2	930.5	15065.4	31.14	59.301	3700	3240.33	28982.5	3612.1	21634.9	10.993	61.994
2720	2260.33	20595.4	960.1	15193.9	30.40	59.363	3720	3260.33	29156.0	3698.2	21768.6	10.795	62.041
2740	2280.33	20763.8	990.3	15322.6	29.69	59.425	3740	3280.33	29329.5	3785.9	21902.4	10.602	62.088
2760	2300.33	20932.4	1021.4	15451.4	29.00	59.486	3760	3300.33	29503.1	3875.1	22036.3	10.413	62.134
2780	2320.33	21101.1	1053.2	15580.4	28.33	59.547	3780	3320.33	29676.8	3966.1	22170.3	10.228	62.180

Table 21 Carbon Monoxide at Low Pressures (for One Pound-Mole)

$\bar{m} = 28.0104$

T	t	$\bar{h}$	p_r	$\bar{u}$	v_r	$\bar{\phi}$	T	t	$\bar{h}$	p_r	$\bar{u}$	v_r	$\bar{\phi}$
3800	3340.33	29850.6	4059	22304.3	10.047	62.226	4600	4140.33	36850.0	9416	27715.0	5.242	63.897
3820	3360.33	30024.4	4153	22438.4	9.871	62.272	4620	4160.33	37026.0	9599	27851.4	5.165	63.935
3840	3380.33	30198.3	4249	22572.6	9.698	62.317	4640	4180.33	37202.1	9785	27987.7	5.089	63.973
3860	3400.33	30372.3	4347	22706.8	9.530	62.362	4660	4200.33	37378.3	9973	28124.2	5.014	64.011
3880	3420.33	30546.3	4446	22841.1	9.364	62.407	4680	4220.33	37554.4	10164	28260.6	4.941	64.049
3900	3440.33	30720.4	4548	22975.5	9.203	62.452	4700	4240.33	37730.7	10359	28397.1	4.869	64.087
3920	3460.33	30894.5	4651	23110.0	9.045	62.496	4720	4260.33	37906.9	10556	28533.7	4.799	64.124
3940	3480.33	31068.7	4756	23244.5	8.890	62.541	4740	4280.33	38083.2	10756	28670.3	4.729	64.161
3960	3500.33	31243.0	4863	23379.0	8.739	62.585	4760	4300.33	38259.6	10959	28806.9	4.661	64.198
3980	3520.33	31417.4	4972	23513.7	8.591	62.629	4780	4320.33	38436.0	11165	28943.6	4.595	64.235
4000	3540.33	31591.8	5082	23648.3	8.446	62.672	4800	4340.33	38612.4	11374	29080.3	4.529	64.272
4020	3560.33	31766.3	5195	23783.1	8.305	62.716	4820	4360.33	38788.9	11586	29217.0	4.465	64.309
4040	3580.33	31940.8	5309	23917.9	8.166	62.759	4840	4380.33	38965.4	11801	29353.8	4.401	64.345
4060	3600.33	32115.4	5426	24052.8	8.030	62.802	4860	4400.33	39141.9	12019	29490.7	4.339	64.382
4080	3620.33	32290.0	5544	24187.7	7.897	62.845	4880	4420.33	39318.5	12241	29627.5	4.278	64.418
4100	3640.33	32464.7	5665	24322.7	7.767	62.888	4900	4440.33	39495.2	12465	29764.5	4.218	64.454
4120	3660.33	32639.5	5787	24457.8	7.640	62.931	4920	4460.33	39671.9	12693	29901.4	4.160	64.490
4140	3680.33	32814.3	5912	24592.9	7.515	62.973	4940	4480.33	39848.6	12924	30038.4	4.102	64.526
4160	3700.33	32989.2	6039	24728.1	7.393	63.015	4960	4500.33	40025.3	13159	30175.4	4.045	64.562
4180	3720.33	33164.2	6168	24863.3	7.273	63.057	4980	4520.33	40202.1	13397	30312.5	3.989	64.597
4200	3740.33	33339.2	6299	24998.6	7.155	63.099	5000	4540.33	40378.9	13638	30449.6	3.934	64.633
4220	3760.33	33514.2	6432	25133.9	7.041	63.140	5020	4560.33	40555.8	13883	30586.8	3.881	64.668
4240	3780.33	33689.3	6568	25269.3	6.928	63.182	5040	4580.33	40732.7	14131	30724.0	3.828	64.703
4260	3800.33	33864.5	6705	25404.7	6.818	63.223	5060	4600.33	40909.6	14382	30861.2	3.776	64.738
4280	3820.33	34039.7	6845	25540.2	6.710	63.264	5080	4620.33	41086.6	14637	30998.4	3.724	64.773
4300	3840.33	34215.0	6988	25675.8	6.604	63.305	5100	4640.33	41263.6	14896	31135.7	3.674	64.808
4320	3860.33	34390.3	7132	25811.4	6.500	63.345	5120	4660.33	41440.7	15158	31273.1	3.625	64.843
4340	3880.33	34565.7	7279	25947.0	6.398	63.386	5140	4680.33	41617.7	15424	31410.4	3.576	64.877
4360	3900.33	34741.1	7429	26082.7	6.298	63.426	5160	4700.33	41794.9	15693	31547.8	3.529	64.911
4380	3920.33	34916.6	7580	26218.5	6.201	63.466	5180	4720.33	41972.0	15966	31685.3	3.482	64.946
4400	3940.33	35092.1	7735	26354.3	6.105	63.506	5200	4740.33	42149.2	16243	31822.7	3.436	64.980
4420	3960.33	35267.7	7891	26490.2	6.011	63.546	5220	4760.33	42326.4	16524	31960.2	3.390	65.014
4440	3980.33	35443.3	8050	26626.1	5.919	63.586	5240	4780.33	42503.7	16808	32097.8	3.346	65.048
4460	4000.33	35619.0	8212	26762.0	5.828	63.625	5260	4800.33	42681.0	17097	32235.4	3.302	65.082
4480	4020.33	35794.7	8376	26898.0	5.740	63.665	5280	4820.33	42858.3	17389	32373.0	3.259	65.115
4500	4040.33	35970.5	8543	27034.1	5.653	63.704	5300	4840.33	43035.7	17685	32510.6	3.216	65.149
4520	4060.33	36146.3	8712	27170.2	5.568	63.743	5320	4860.33	43213.0	17985	32648.3	3.174	65.182
4540	4080.33	36322.1	8884	27306.3	5.484	63.782	5340	4880.33	43390.5	18289	32786.0	3.133	65.215
4560	4100.33	36498.0	9059	27442.5	5.402	63.820	5360	4900.33	43567.9	18597	32923.7	3.093	65.249
4580	4120.33	36674.0	9236	27578.8	5.321	63.859	5380	4920.33	43745.4	18909	33061.5	3.053	65.282

Table 22   Carbon Monoxide

$\bar{m}$ = 28.0104

T R	t F	$\bar{c}_p$ Btu lb-mole R	$\bar{c}_v$ Btu lb-mole R	$k = \dfrac{\bar{c}_p}{\bar{c}_v}$	a ft sec
200	−259.67	6.952	4.966	1.400	705.0
250	−209.67	6.952	4.966	1.400	788.2
300	−159.67	6.952	4.966	1.400	863.4
350	−109.67	6.952	4.967	1.400	932.6
400	−59.67	6.953	4.967	1.400	996.9
450	−9.67	6.954	4.968	1.400	1057.4
500	40.33	6.957	4.971	1.399	1114.5
550	90.33	6.962	4.976	1.399	1168.7
600	140.33	6.970	4.984	1.398	1220.4
650	190.33	6.982	4.997	1.397	1269.8
700	240.33	6.999	5.014	1.396	1317.1
750	290.33	7.021	5.036	1.394	1362.4
800	340.33	7.048	5.063	1.392	1406.1
850	390.33	7.080	5.094	1.390	1448.1
900	440.33	7.116	5.130	1.387	1488.6
1000	540.33	7.198	5.212	1.381	1565.7
1100	640.33	7.289	5.303	1.374	1638.2
1200	740.33	7.386	5.400	1.368	1706.8
1300	840.33	7.484	5.498	1.361	1772.3
1400	940.33	7.581	5.595	1.355	1835.0
1500	1040.33	7.674	5.688	1.349	1895.3
1600	1140.33	7.763	5.777	1.344	1953.5
1700	1240.33	7.846	5.860	1.339	2010.0
1800	1340.33	7.924	5.938	1.334	2064.8
1900	1440.33	7.996	6.011	1.330	2118.2
2000	1540.33	8.063	6.078	1.327	2170.2
2100	1640.33	8.125	6.140	1.323	2221.1
2200	1740.33	8.183	6.197	1.320	2270.8
2300	1840.33	8.235	6.250	1.318	2319.4
2400	1940.33	8.284	6.298	1.315	2367.1
2500	2040.33	8.329	6.343	1.313	2413.9
2600	2140.33	8.370	6.385	1.311	2459.8
2700	2240.33	8.409	6.423	1.309	2504.8
2800	2340.33	8.444	6.459	1.307	2549.2
2900	2440.33	8.477	6.491	1.306	2592.7
3000	2540.33	8.508	6.522	1.304	2635.6
3200	2740.33	8.563	6.577	1.302	2719.4
3400	2940.33	8.611	6.625	1.300	2800.7
3600	3140.33	8.652	6.667	1.298	2879.8
3800	3340.33	8.689	6.703	1.296	2956.9
4000	3540.33	8.722	6.736	1.295	3032.0
4200	3740.33	8.751	6.765	1.294	3105.4
4400	3940.33	8.777	6.792	1.292	3177.1
4600	4140.33	8.801	6.815	1.291	3247.2
4800	4340.33	8.823	6.837	1.290	3315.8
5000	4540.33	8.842	6.856	1.290	3383.1
5200	4740.33	8.860	6.874	1.289	3449.1
5400	4940.33	8.877	6.891	1.288	3513.9
5600	5140.33	8.892	6.906	1.288	3577.5
5800	5340.33	8.907	6.921	1.287	3639.9

(119)

Table 23 Monatomic Gases, He, Ar, Hg, etc., at Low Pressures (for One Pound-Mole)

Molecular weight, $\bar{m}$: He 4.0026, Ar 39.948, Hg 200.59

T	t	$\bar{h}$	p_r	$\bar{u}$	v_r	$\bar{\phi}$
R	F	Btu lb-mole		Btu lb-mole		Btu lb-mole R
100	−359.67	496.5	.18120	297.9	5922.4	28.616
200	−259.67	992.9	1.0250	595.8	2093.9	32.057
300	−159.67	1489.4	2.8247	893.6	1139.8	34.070
400	−59.67	1985.9	5.7985	1191.5	740.3	35.499
500	40.33	2482.3	10.1295	1489.4	529.7	36.606
600	140.33	2978.8	15.979	1787.3	402.97	37.512
700	240.33	3475.3	23.491	2085.2	319.78	38.277
800	340.33	3971.7	32.801	2383.0	261.73	38.940
900	440.33	4468.2	44.032	2680.9	219.35	39.525
1000	540.33	4964.6	57.301	2978.8	187.28	40.048
1100	640.33	5461.1	72.72	3276.7	162.33	40.521
1200	740.33	5957.6	90.39	3574.5	142.47	40.953
1300	840.33	6454.0	110.41	3872.4	126.35	41.350
1400	940.33	6950.5	132.89	4170.3	113.06	41.718
1500	1040.33	7447.0	157.90	4468.2	101.94	42.061
1600	1140.33	7943.4	185.55	4766.1	92.54	42.381
1700	1240.33	8439.9	215.92	5063.9	84.49	42.682
1800	1340.33	8936.4	249.08	5361.8	77.55	42.966
1900	1440.33	9432.8	285.13	5659.7	71.51	43.234
2000	1540.33	9929.3	324.14	5957.6	66.21	43.489
2100	1640.33	10425.8	366.19	6255.5	61.54	43.731
2200	1740.33	10922.2	411.36	6553.3	57.39	43.962
2300	1840.33	11418.7	459.71	6851.2	53.69	44.183
2400	1940.33	11915.2	511.32	7149.1	50.37	44.394
2500	2040.33	12411.6	566.26	7447.0	47.38	44.597
2600	2140.33	12908.1	624.59	7744.9	44.67	44.791
2700	2240.33	13404.6	686.39	8042.7	42.21	44.979
2800	2340.33	13901.0	751.73	8340.6	39.97	45.159
2900	2440.33	14397.5	820.65	8638.5	37.92	45.334
3000	2540.33	14893.9	893.24	8936.4	36.04	45.502
3200	2740.33	15886.9	1049.6	9532.1	32.72	45.822
3400	2940.33	16879.8	1221.4	10127.9	29.87	46.123
3600	3140.33	17872.7	1409.0	10723.6	27.42	46.407
3800	3340.33	18865.7	1613.0	11319.4	25.28	46.675
4000	3540.33	19858.6	1833.6	11915.2	23.41	46.930
4200	3740.33	20851.5	2071.5	12510.9	21.758	47.172
4400	3940.33	21844.5	2327.0	13106.7	20.292	47.403
4600	4140.33	22837.4	2600.5	13702.4	18.983	47.624
4800	4340.33	23830.3	2892.5	14298.2	17.809	47.835
5000	4540.33	24823.2	3203.2	14893.9	16.751	48.038
5200	4740.33	25816.2	3533.2	15489.7	15.794	48.233
5400	4940.33	26809.1	3882.8	16085.5	14.925	48.420
5600	5140.33	27802.0	4252.4	16681.2	14.132	48.601
5800	5340.33	28795.0	4642.3	17277.0	13.408	48.775
6000	5540.33	29787.9	5052.9	17872.7	12.743	48.943

SYMBOLS FOR POLYTROPIC FUNCTIONS AND COMPRESSIBLE-FLOW FUNCTIONS IN TABLES 24 TO 59

A cross-sectional area

a velocity of sound

c_p specific heat at constant pressure

c_v specific heat at constant volume

D diameter of duct

F wall-force function

$$= pA + \frac{\rho A V^2}{g} = pA(1 + kM^2)$$

f coefficient of friction $= \dfrac{2g\tau_w}{V^2}$

G flow per unit area or mass velocity $= w/A$

g acceleration given to unit mass by unit force $= 32.174\,\dfrac{\text{ft}}{\text{sec}^2}\,\dfrac{\text{lbm}}{\text{lbf}}$

h enthalpy per unit mass

k ratio of specific heats $= c_p/c_v$

L length of duct

L_{max} maximum length of duct for steady adiabatic constant-area flow with friction

M Mach number $= V/a$

$\bar{m}$ molecular weight

n exponent for polytropic process

p static pressure

$\bar{R}$ universal gas constant $= 1545.34 \dfrac{\text{ft lb}}{\text{R lb-mole}}$

R gas constant $= \bar{R}/\bar{m}$

r pressure ratio

s entropy per unit mass

T absolute thermodynamic temperature

u internal energy per unit mass

V velocity

v specific volume

w mass rate of flow

α Mach angle $=$ arc sin $(1/M)$

α' angle between two-dimensional shock and incident flow, see sketch on p. 171

ρ density

θ $= \alpha' - \omega'$

τ_w shear stress at wall

ω vector angle of characteristic curve in hodograph plane, see sketch p. 166

ω' wedge angle for two-dimensional shock, see sketch on p. 171

ω'_m maximum wedge angle

ω'_a wedge angle for sonic flow downstream of two-dimensional shock

Superscript * refers to conditions in which $M = 1$; in particular, M^* denotes the ratio of the velocity divided by the velocity of sound at the position at which $M = 1$.

Subscripts

0 refers to isentropic stagnation conditions

x and y refer to conditions upstream and downstream of a shock, respectively

[a]lbm and lbf are used for pound mass and pound force, respectively, except that lb is used for either wherever the meaning is obvious.

Formulas in Which the Quantities of Tables 25 to 29 Appear

General:

$$pv = \frac{\bar{R}T}{\bar{m}}. \quad r = \frac{p}{p_0}.$$

The Polytropic Process:

$$\frac{v_0}{v} = r^{\frac{1}{n}}. \qquad \frac{T}{T_0} = r^{\frac{n-1}{n}}. \qquad \int_{p_0}^{p} p\,dv = \frac{T_0}{\bar{m}}\frac{\bar{R}}{n-1}\left(1 - r^{\frac{n-1}{n}}\right) = \frac{T}{\bar{m}}\frac{\bar{R}}{n-1}\left(r^{-\left(\frac{n-1}{n}\right)} - 1\right).$$

$$\int_{p_0}^{p} v\,dp = -\frac{T_0}{\bar{m}}\frac{n\bar{R}}{n-1}\left(1 - r^{\frac{n-1}{n}}\right) = -\frac{T_n}{\bar{m}}\frac{\bar{R}}{n-1}\left(r^{-\left(\frac{n-1}{n}\right)} - 1\right).$$

Steady Flow through a Nozzle or a Diffuser:

$$\frac{V^2 - V_0^2}{2g} = \frac{T_0}{\bar{m}}\frac{n\bar{R}}{n-1}\left(1 - r^{\frac{n-1}{n}}\right) = \frac{T}{\bar{m}}\frac{n\bar{R}}{n-1}\left(r^{-\left(\frac{n-1}{n}\right)} - 1\right).$$

$$V = \sqrt{\frac{T_0}{\bar{m}}}\sqrt{\frac{2gn\bar{R}}{n-1}}\sqrt{\left(1 - r^{\frac{n-1}{n}}\right)} = \sqrt{\frac{T}{\bar{m}}}\sqrt{\frac{2gn\bar{R}}{n-1}}\sqrt{r^{-\left(\frac{n-1}{n}\right)} - 1} \quad \text{for } V_0 = 0.$$

$$G = \frac{w}{A} = p_0\sqrt{\frac{\bar{m}}{T_0}}\sqrt{\frac{2g}{\bar{R}}\frac{n}{n-1}}\, r^{\frac{1}{n}}\sqrt{1 - r^{\frac{n-1}{n}}} = p_0\sqrt{\frac{\bar{m}}{T}}\sqrt{\frac{2g}{\bar{R}}\frac{n}{n-1}}\, r^{\frac{1}{n}}\sqrt{r^{\frac{n-1}{n}} - r^{2\left(\frac{n-1}{n}\right)}} \quad \text{for } V_0 = 0.$$

Reversible Adiabatic Process:

$$n = k = c_p/c_v. \qquad u_0 - u = \int_{p_0}^{p} p\,dv = \frac{T_0}{\bar{m}}\frac{\bar{R}}{k-1}\left(1 - r^{\frac{k-1}{k}}\right) = \frac{T}{\bar{m}}\frac{\bar{R}}{k-1}\left(r^{-\left(\frac{k-1}{k}\right)} - 1\right).$$

$$h_0 - h = -\int_{p_0}^{p} v\,dp = \frac{T_0}{\bar{m}}\frac{k\bar{R}}{k-1}\left(1 - r^{\frac{k-1}{k}}\right) = \frac{T}{\bar{m}}\frac{k\bar{R}}{k-1}\left(r^{-\left(\frac{k-1}{k}\right)} - 1\right). \qquad M = \sqrt{\frac{2}{k-1}\left(r^{-\left(\frac{k-1}{k}\right)} - 1\right)} \quad \text{for } M_0 = 0.$$

Table 24   Functions of n in the Formulas Above

n	$\frac{1}{n}$	$\frac{1}{n-1}$	$\frac{n}{n-1}$	$\frac{\bar{R}}{n-1}$ $\frac{\text{ft lb}}{\text{lb-mole °F}}$	$\frac{n\bar{R}}{n-1}$ $\frac{\text{ft lb}}{\text{lb-mole °F}}$	$\sqrt{\frac{2gn\bar{R}}{n-1}}$ $\frac{\text{ft}}{\text{sec}}\left(\frac{\text{lbf}}{\text{lb-mole °F}}\right)^{1/2}$	$\sqrt{\frac{2g}{\bar{R}}\frac{n}{n-1}}$ $\left(\frac{\text{lb-mole R}}{\text{lbf sec}^2}\right)^{1/2}$	$\frac{p^*}{p_0} =$ $\left(\frac{2}{n+1}\right)^{\frac{n}{n-1}}$	n
1.05	0.95238	20.000	21.000	30906.	32452.	1445.1	0.93512	0.5954	1.05
1.10	.90909	10.000	11.000	15453.	16999.	1045.9	.67679	.5847	1.10
1.15	.86957	6.6667	7.6667	10302.	11847.	873.13	.56502	.5744	1.15
1.20	.83333	5.0000	6.0000	7726.6	9271.9	772.42	.49984	.5645	1.20
1.25	.80000	4.0000	5.0000	6181.3	7726.6	705.12	.45629	.5549	1.25
1.30	.76923	3.3333	4.3333	5151.1	6696.4	656.43	.42479	.5457	1.30
1.35	.74074	2.8571	3.8571	4415.2	5960.5	619.31	.40077	.5369	1.35
1.40	.71429	2.5000	3.5000	3863.3	5408.6	589.94	.38176	.5283	1.40
1.45	.68966	2.2222	3.2222	3434.0	4979.4	566.05	.36630	.5200	1.45
1.50	.66667	2.0000	3.0000	3090.6	4636.0	546.18	.35344	.5120	1.50
1.55	.64516	1.8182	2.8182	2809.7	4355.0	529.37	.34256	.5043	1.55
1.60	.62500	1.6667	2.6667	2575.5	4120.9	514.95	.33323	.4968	1.60
1.65	.60606	1.5385	2.5385	2377.4	3922.7	502.41	.32512	.4895	1.65
1.70	.58824	1.4286	2.4286	2207.6	3752.9	491.42	.31800	.4825	1.70
1.75	.57143	1.3333	2.3333	2060.4	3605.7	481.69	.31171	.4757	1.75
1.80	.55556	1.2500	2.2500	1931.7	3477.0	473.01	.30609	.4690	1.80

Table 25 Functions of Pressure Ratio

n = 1.4

r	$\dfrac{1}{r}$	$\dfrac{1}{r^n}$	$r^{\frac{n-1}{n}}$	$\sqrt{1-r^{\frac{n-1}{n}}}$	$\dfrac{1}{r^n}\sqrt{1-r^{\frac{n-1}{n}}}$	r	$\dfrac{1}{r}$	$\dfrac{1}{r^n}$	$r^{\frac{n-1}{n}}$	$\sqrt{1-r^{\frac{n-1}{n}}}$	$\dfrac{1}{r^n}\sqrt{1-r^{\frac{n-1}{n}}}$
1	1	1	1	0	0	.930	1.075	.94948	.97948	.14325	.13601
0.999	1.001	.99929	.99971	.01690	.01689	.928	1.078	.94803	.97888	.14534	.13778
.998	1.002	.99857	.99943	.02391	.02388	.926	1.080	.94657	.97827	.14740	.13952
.997	1.003	.99786	.99914	.02931	.02924	.924	1.082	.94510	.97767	.14943	.14123
.996	1.004	.99714	.99886	.03383	.03373	.922	1.085	.94364	.97706	.15144	.14291
.995	1.005	.99643	.99857	.03783	.03770	.920	1.087	.94218	.97646	.15343	.14456
.994	1.006	.99571	.99828	.04145	.04127	.918	1.089	.94072	.97585	.15540	.14618
.993	1.007	.99500	.99800	.04478	.04455	.916	1.092	.93925	.97524	.15734	.14778
.992	1.008	.99428	.99771	.04788	.04760	.914	1.094	.93779	.97463	.15927	.14936
.991	1.009	.99356	.99742	.05079	.05046	.912	1.096	.93632	.97402	.16117	.15091
.990	1.010	.99285	.99713	.05355	.05316	.910	1.099	.93485	.97341	.16305	.15243
.989	1.011	.99213	.99684	.05617	.05573	.908	1.101	.93339	.97280	.16492	.15393
.988	1.012	.99141	.99656	.05868	.05818	.906	1.104	.93192	.97219	.16676	.15541
.987	1.013	.99070	.99627	.06109	.06052	.904	1.106	.93045	.97158	.16859	.15687
.986	1.014	.98998	.99598	.06341	.06277	.902	1.109	.92898	.97096	.17041	.15830
.985	1.015	.98926	.99569	.06564	.06494	.900	1.111	.92750	.97035	.17220	.15972
.984	1.016	.98855	.99540	.06780	.06703	.895	1.117	.92382	.96880	.17663	.16317
.983	1.017	.98783	.99511	.06990	.06906	.890	1.124	.92013	.96725	.18096	.16651
.982	1.018	.98711	.99482	.07195	.07102	.885	1.130	.91644	.96570	.18521	.16973
.981	1.019	.98639	.99453	.07393	.07293	.880	1.136	.91274	.96414	.18938	.17285
.980	1.020	.98567	.99424	.07586	.07478	.875	1.143	.90903	.96257	.19347	.17588
.979	1.021	.98495	.99395	.07775	.07658	.870	1.149	.90531	.96099	.19750	.17881
.978	1.022	.98424	.99366	.07960	.07834	.865	1.156	.90159	.95941	.20146	.18165
.977	1.024	.98352	.99337	.08140	.08006	.860	1.163	.89787	.95782	.20537	.18440
.976	1.025	.98280	.99308	.08317	.08174	.855	1.170	.89414	.95623	.20922	.18707
.975	1.026	.98208	.99279	.08490	.08338	.850	1.176	.89040	.95463	.21301	.18966
.974	1.027	.98136	.99250	.08659	.08498	.845	1.183	.88665	.95302	.21674	.19218
.973	1.028	.98064	.99221	.08826	.08655	.840	1.190	.88290	.95141	.22043	.19463
.972	1.029	.97992	.99192	.08990	.08809	.835	1.198	.87914	.94979	.22408	.19701
.971	1.030	.97920	.99163	.09150	.08960	.830	1.205	.87538	.94816	.22768	.19932
.970	1.031	.97848	.99134	.09308	.09108	.825	1.212	.87161	.94652	.23125	.20157
.969	1.032	.97776	.99104	.09464	.09253	.820	1.220	.86784	.94488	.23478	.20375
.968	1.033	.97704	.99075	.09617	.09396	.815	1.227	.86406	.94323	.23826	.20588
.967	1.034	.97632	.99046	.09768	.09537	.810	1.235	.86027	.94157	.24171	.20795
.966	1.035	.97559	.99017	.09917	.09675	.805	1.242	.85647	.93990	.24513	.20996
.965	1.036	.97487	.98987	.10064	.09811	.800	1.250	.85267	.93823	.24852	.21191
.964	1.037	.97415	.98958	.10208	.09944	.795	1.258	.84886	.93656	.25188	.21381
.963	1.038	.97343	.98929	.10351	.10076	.790	1.266	.84504	.93487	.25520	.21566
.962	1.040	.97271	.98899	.10492	.10205	.785	1.274	.84121	.93317	.25850	.21746
.961	1.041	.97199	.98870	.10631	.10333	.780	1.282	.83738	.93147	.26178	.21921
.960	1.042	.97126	.98840	.10768	.10459	.775	1.290	.83354	.92976	.26502	.22091
.958	1.044	.96982	.98782	.11038	.10705	.770	1.299	.82970	.92804	.26824	.22256
.956	1.046	.96837	.98723	.11302	.10945	.765	1.307	.82585	.92631	.27144	.22417
.954	1.048	.96692	.98664	.11561	.11178	.760	1.316	.82199	.92458	.27462	.22573
.952	1.050	.96547	.98604	.11814	.11406	.755	1.324	.81812	.92284	.27777	.22725
.950	1.053	.96402	.98545	.12062	.11628	.750	1.333	.81425	.92109	.28090	.22873
.948	1.055	.96257	.98486	.12305	.11844	.745	1.342	.81037	.91933	.28401	.23016
.946	1.057	.96112	.98426	.12544	.12056	.740	1.351	.80648	.91757	.28711	.23155
.944	1.059	.95967	.98367	.12779	.12264	.735	1.361	.80258	.91580	.29019	.23290
.942	1.062	.95822	.98307	.13010	.12467	.730	1.370	.79868	.91401	.29325	.23421
.940	1.064	.95677	.98248	.13238	.12665	.725	1.379	.79477	.91221	.29629	.23548
.938	1.066	.95531	.98188	.13462	.12860	.720	1.389	.79085	.91041	.29931	.23671
.936	1.068	.95386	.98128	.13683	.13051	.715	1.399	.78692	.90860	.30233	.23791
.934	1.071	.95240	.98068	.13900	.13238	.710	1.408	.78299	.90678	.30532	.23906
.932	1.073	.95094	.98008	.14114	.13421	.705	1.418	.77905	.90495	.30830	.24018

Table 25   n = 1.4

r	$\frac{1}{r}$	$\frac{1}{r^n}$	$r^{\frac{n-1}{n}}$	$\sqrt{1-r^{\frac{n-1}{n}}}$	$\frac{1}{r^n}\sqrt{1-r^{\frac{n-1}{n}}}$	r	$\frac{1}{r}$	$\frac{1}{r^n}$	$r^{\frac{n-1}{n}}$	$\sqrt{1-r^{\frac{n-1}{n}}}$	$\frac{1}{r^n}\sqrt{1-r^{\frac{n-1}{n}}}$
.700	1.429	.77510	.90311	.31127	.24126	.425	2.353	.54271	.78311	.46571	.25274
.695	1.439	.77114	.90126	.31422	.24231	.420	2.381	.53814	.78047	.46854	.25214
.690	1.449	.76717	.89940	.31716	.24332	.415	2.410	.53355	.77781	.47138	.25150
.685	1.460	.76319	.89753	.32009	.24429	.410	2.439	.52895	.77512	.47422	.25084
.680	1.471	.75921	.89566	.32301	.24523	.405	2.469	.52434	.77240	.47707	.25015
.675	1.481	.75522	.89377	.32592	.24614	.400	2.500	.51971	.76967	.47993	.24942
.670	1.493	.75122	.89188	.32882	.24701	.395	2.532	.51506	.76691	.48280	.24867
.665	1.504	.74721	.88998	.33170	.24785	.390	2.564	.51039	.76412	.48567	.24789
.660	1.515	.74320	.88806	.33458	.24866	.385	2.597	.50571	.76131	.48856	.24707
.655	1.527	.73917	.88613	.33745	.24943	.380	2.632	.50101	.75847	.49146	.24622
.650	1.538	.73514	.88419	.34031	.25017	.375	2.667	49629	.75560	.49436	.24535
.645	1.550	.73109	.88224	.34316	.25088	.370	2.703	.49155	.75271	.49728	.24444
.640	1.562	.72704	.88028	.34600	.25155	.365	2.740	.48680	.74979	.50021	.24350
.635	1.575	.72298	.87831	.34883	.25220	.360	2.778	.48203	.74684	.50315	.24253
.630	1.587	.71891	.87633	.35166	.25281	.355	2.817	.47724	.74386	.50610	.24153
.625	1.600	.71483	.87434	.35448	.25340	.350	2.857	.47243	.74086	.50906	.24049
.620	1.613	.71074	.87233	.35730	.25395	.345	2.896	.46760	.73782	.51204	.23943
.615	1.626	.70664	.87032	.36011	.25447	.340	2.941	.46275	.73475	.51503	.23833
.610	1.639	.70253	.86829	.36291	.25496	.335	2.985	.45788	.73164	.51803	.23719
.605	1.653	.69841	.86625	.36571	.25542	.330	3.030	.45299	.72850	.52105	.23603
.600	1.667	.69428	.86420	.36851	.25585	.325	3.077	.44807	.72533	.52409	.23483
.595	1.681	.69015	.86214	.37130	.25625	.320	3.125	.44313	.72213	.52714	.23359
.590	1.695	.68600	.86006	.37409	.25662	.315	3.175	.43818	.71889	.53020	.23232
.585	1.709	.68184	.85797	.37687	.25696	.310	3.226	.43320	.71561	.53328	.23102
.580	1.724	.67767	.85587	.37965	.25727	.305	3.279	.42820	.71229	.53638	.22968
.575	1.739	.67350	.85376	.38242	.25756	.300	3.333	.42317	.70893	.53950	.22830
.570	1.754	.66931	.85163	.38519	.25781	.295	3.390	.41812	.70554	.54264	.22689
.565	1.770	.66511	.84949	.38796	.25803	.290	3.448	.41305	.70210	.54580	.22544
.560	1.786	.66090	.84733	.39073	.25823	.285	3.509	.40795	.69862	.54898	.22396
.555	1.802	.65668	.84516	.39350	.25840	.280	3.571	.40282	.69510	.55218	.22243
.550	1.818	.65245	.84298	.39626	.25854	.275	3.636	.39767	.69153	.55540	.22087
.545	1.835	.64820	.84078	.39902	.25865	.270	3.704	.39249	.68791	.55865	.21927
.540	1.852	.64395	.83857	.40178	.25873	.265	3.774	.38728	.68425	.56192	.21762
.535	1.869	.63969	.83635	.40454	.25878	.260	3.846	.38205	.68053	.56521	.21594
.530	1.887	.63541	.83411	.40730	.25880	.255	3.922	.37679	.67677	.56853	.21422
.525	1.905	.63112	.83185	.41006	.25880	.250	4.000	.37150	.67295	.57188	.21245
.520	1.923	.62682	.82958	.41282	.25877	.245	4.082	.36618	.66908	.57526	.21065
.515	1.942	.62251	.82729	.41558	.25870	.240	4.167	.36082	.66515	.57867	.20880
.510	1.961	.61819	.82499	.41834	.25861	.235	4.255	.35544	.66116	.58211	.20690
.505	1.980	.61386	.82267	.42110	.25850	.230	4.348	.35002	.65711	.58558	.20496
.500	2.000	.60951	.82034	.42387	.25835	.225	4.444	.34457	.65299	.58908	.20298
.495	2.020	.60515	.81798	.42663	.25818	.220	4.545	.33908	.64881	.59261	.20094
.490	2.041	.60078	.81561	.42940	.25797	.215	4.651	.33356	.64457	.59619	.19886
.485	2.062	.59639	.81323	.43217	.25774	.210	4.762	.32800	.64025	.59980	.19673
.480	2.083	.59199	.81082	.43494	.25748	.205	4.878	.32240	.63585	.60345	.19455
.475	2.105	.58758	.80840	.43772	.25720	.200	5.000	.31676	.63139	.60714	.19232
.470	2.128	.58316	.80596	.44050	.25688	.198	5.051	.31450	.62957	.60863	.19141
.465	2.151	.57872	.80350	.44328	.25654	.196	5.102	.31223	.62775	.61012	.19050
.460	2.174	.57427	.80102	.44607	.25616	.194	5.155	.30995	.62591	.61162	.18957
.455	2.198	.56980	.79853	.44886	.25576	.192	5.208	.30766	.62406	.61313	.18864
.450	2.222	.56532	.79601	.45165	.25533	.190	5.263	.30537	.62220	.61465	.18770
.445	2.247	.56083	.79347	.45445	.25487	.188	5.319	.30307	.62032	.61618	.18675
.440	2.273	.55632	.79091	.45726	.25438	.186	5.376	.30076	.61843	.61771	.18579
.435	2.299	.55180	.78834	.46007	.25386	.184	5.435	.29845	.61652	.61926	.18482
.430	2.326	.54726	.78574	.46289	.25332	.182	5.495	.29613	.61460	.62081	.18384

Table 25 n = 1.4

r	$\dfrac{1}{r}$	$r^{\frac{1}{n}}$	$r^{\frac{n-1}{n}}$	$\sqrt{1-r^{\frac{n-1}{n}}}$	$r^{\frac{1}{n}}\sqrt{1-r^{\frac{n-1}{n}}}$	r	$\dfrac{1}{r}$	$r^{\frac{1}{n}}$	$r^{\frac{n-1}{n}}$	$\sqrt{1-r^{\frac{n-1}{n}}}$	$r^{\frac{1}{n}}\sqrt{1-r^{\frac{n-1}{n}}}$
.180	5.556	.29380	.61266	.62237	.18285	.070	14.3	.14965	.46777	.72954	.10918
.178	5.618	.29146	.61071	.62394	.18185	.069	14.5	.14812	.46585	.73086	.10825
.176	5.682	.28912	.60874	.62551	.18085	.068	14.7	.14658	.46391	.73218	.10732
.174	5.747	.28677	.60676	.62709	.17983	.067	14.9	.14504	.46195	.73352	.10639
.172	5.814	.28441	.60476	.62869	.17881	.066	15.2	.14349	.45997	.73487	.10545
.170	5.882	.28205	.60274	.63029	.17777	.065	15.4	.14193	.45797	.73623	.10450
.168	5.952	.27967	.60070	.63190	.17672	.064	15.6	.14037	.45594	.73760	.10354
.166	6.024	.27729	.59865	.63352	.17567	.063	15.9	.13880	.45389	.73899	.10257
.164	6.098	.27490	.59658	.63516	.17460	.062	16.1	.13722	.45182	.74039	.10160
.162	6.173	.27250	.59449	.63680	.17353	.061	16.4	.13564	.44973	.74180	.10062
.160	6.250	.27009	.59239	.63845	.17244	.060	16.7	.13405	.44761	.74323	.09963
.158	6.329	.26768	.59027	.64010	.17134	.059	16.9	.13245	.44547	.74467	.09863
.156	6.410	.26525	.58812	.64177	.17023	.058	17.2	.13084	.44330	.74613	.09762
.154	6.494	.26282	.58595	.64346	.16911	.057	17.5	.12922	.44110	.74760	.09660
.152	6.579	.26038	.58376	.64516	.16798	.056	17.9	.12760	.43887	.74908	.09558
.150	6.667	.25793	.58156	.64687	.16684	.055	18.2	.12597	.43662	.75059	.09455
.148	6.757	.25546	.57934	.64858	.16569	.054	18.5	.12433	.43434	.75211	.09351
.146	6.849	.25299	.57709	.65031	.16453	.053	18.9	.12268	.43202	.75364	.09246
.144	6.944	.25051	.57482	.65205	.16335	.052	19.2	.12102	.42968	.75520	.09139
.142	7.042	.24802	.57253	.65381	.16216	.051	19.6	.11935	.42730	.75677	.09032
.140	7.143	.24552	.57021	.65558	.16096	.050	20.0	.11768	.42489	.75836	.08924
.138	7.246	.24301	.56787	.65736	.15975	.049	20.4	.11599	.42245	.75997	.08815
.136	7.353	.24049	.56551	.65916	.15852	.048	20.8	.11430	.41997	.76160	.08705
.134	7.463	.23796	.56312	.66097	.15728	.047	21.3	.11259	.41745	.76325	.08593
.132	7.576	.23542	.56071	.66279	.15603	.046	21.7	.11087	.41489	.76493	.08481
.130	7.692	.23286	.55827	.66463	.15477	.045	22.2	.10915	.41229	.76662	.08367
.128	7.812	.23030	.55580	.66648	.15349	.044	22.7	.10741	.40965	.76834	.08253
.126	7.937	.22773	.55330	.66835	.15220	.043	23.3	.10566	.40697	.77008	.08137
.124	8.065	.22514	.55078	.67024	.15090	.042	23.8	.10390	.40424	.77185	.08019
.122	8.197	.22254	.54823	.67214	.14958	.041	24.4	.10212	.40147	.77364	.07901
.120	8.333	.21992	.54564	.67406	.14824	.0400	25.00	.10034	.39865	.77547	.07781
.118	8.475	.21730	.54303	.67600	.14689	.0395	25.32	.09944	.39722	.77639	.07721
.116	8.621	.21466	.54038	.67795	.14553	.0390	25.64	.09854	.39577	.77732	.07660
.114	8.772	.21201	.53770	.67992	.14415	.0385	25.97	.09764	.39432	.77826	.07599
.112	8.929	.20935	.53499	.68191	.14276	.0380	26.32	.09673	.39285	.77920	.07537
.110	9.091	.20667	.53224	.68393	.14135	.0375	26.67	.09582	.39136	.78015	.07475
.108	9.259	.20398	.52946	.68596	.13992	.0370	27.03	.09490	.38986	.78111	.07413
.106	9.434	.20128	.52664	.68801	.13848	.0365	27.40	.09399	.38835	.78208	.07350
.104	9.615	.19856	.52379	.69008	.13702	.0360	27.78	.09307	.38682	.78305	.07288
.102	9.804	.19582	.52089	.69218	.13554	.0355	28.17	.09214	.38528	.78404	.07224
.100	10.0	.19307	.51795	.69430	.13405	.0350	28.57	.09121	.38372	.78503	.07160
.098	10.2	.19030	.51497	.69644	.13254	.0345	28.99	.09028	.38215	.78603	.07096
.096	10.4	.18752	.51194	.69861	.13100	.0340	29.41	.08934	.38056	.78704	.07032
.094	10.6	.18472	.50887	.70080	.12945	.0335	29.85	.08840	.37895	.78807	.06967
.092	10.9	.18191	.50576	.70302	.12788	.0330	30.30	.08746	.37733	.78910	.06901
.090	11.1	.17907	.50259	.70527	.12630	.0325	30.77	.08651	.37568	.79014	.06835
.088	11.4	.17622	.49938	.70755	.12469	.0320	31.25	.08556	.37402	.79119	.06769
.086	11.6	.17335	.49611	.70986	.12305	.0315	31.75	.08460	.37235	.79225	.06702
.084	11.9	.17046	.49278	.71219	.12140	.0310	32.26	.08364	.37065	.79332	.06635
.082	12.2	.16755	.48940	.71456	.11973	.0305	32.79	.08267	.36893	.79440	.06567
.080	12.5	.16462	.48596	.71697	.11803	.0300	33.33	.08170	.36719	.79549	.06499
.078	12.8	.16167	.48246	.71941	.11631	.0295	33.89	.08073	.36543	.79660	.06431
.076	13.2	.15870	.47889	.72188	.11456	.0290	34.48	.07975	.36365	.79772	.06362
.074	13.5	.15571	.47525	.72439	.11279	.0285	35.09	.07876	.36185	.79884	.06292
.072	13.9	.15269	.47155	.72695	.11100	.0280	35.71	.07777	.36002	.79998	.06222

Table 25   n = 1.4

r	$\frac{1}{r}$	$\frac{1}{r^n}$	$r^{\frac{n-1}{n}}$	$\sqrt{1-r^{\frac{n-1}{n}}}$	$\frac{1}{r^n}\sqrt{1-r^{\frac{n-1}{n}}}$	r	$\frac{1}{r}$	$\frac{1}{r^n}$	$r^{\frac{n-1}{n}}$	$\sqrt{1-r^{\frac{n-1}{n}}}$	$\frac{1}{r^n}\sqrt{1-r^{\frac{n-1}{n}}}$
.0275	36.36	.07678	.35817	.80114	.06151	.0060	167	.02588	.23184	.87645	.02268
.0270	37.04	.07578	.35630	.80231	.06080	.0058	172	.02526	.22960	.87772	.02217
.0265	37.74	.07477	.35440	.80349	.06008	.0056	179	.02464	.22731	.87902	.02166
.0260	38.46	.07376	.35248	.80469	.05936	.0054	185	.02400	.22496	.88036	.02113
.0255	39.22	.07275	.35053	.80590	.05863	.0052	192	.02337	.22255	.88173	.02060
.0250	40.00	.07173	.34855	.80712	.05789	.0050	200	.02272	.22007	.88314	.02006
.0245	40.82	.07070	.34655	.80836	.05715	.0048	208	.02207	.21752	.88458	.01952
.0240	41.67	.06966	.34451	.80962	.05640	.0046	217	.02141	.21489	.88606	.01897
.0235	42.55	.06862	.34244	.81090	.05565	.0044	227	.02074	.21218	.88759	.01841
.0230	43.48	.06758	.34035	.81219	.05489	.0042	238	.02006	.20938	.88917	.01784
.0225	44.44	.06653	.33822	.81350	.05412	.0040	250	.01937	.20648	.89080	.01726
.0220	45.45	.06547	.33605	.81483	.05334	.0039	256	.01903	.20499	.89163	.01696
.0215	46.51	.06440	.33385	.81618	.05256	.0038	263	.01868	.20347	.89248	.01667
.0210	47.62	.06333	.33162	.81755	.05177	.0037	270	.01832	.20193	.89335	.01637
.0205	48.78	.06225	.32934	.81894	.05098	.0036	278	.01797	.20036	.89423	.01607
.0200	50.00	.06116	.32702	.82035	.05017	.0035	286	.01761	.19875	.89513	.01576
.0195	51.28	.06006	.32467	.82179	.04936	.0034	294	.01725	.19711	.89604	.01546
.0190	52.63	.05896	.32227	.82325	.04854	.0033	303	.01688	.19544	.89698	.01515
.0185	54.05	.05784	.31982	.82473	.04771	.0032	312	.01652	.19373	.89793	.01483
.0180	55.56	.05672	.31733	.82624	.04687	.0031	323	.01615	.19198	.89890	.01452
.0175	57.14	.05559	.31479	.82778	.04602	.0030	333	.01577	.19019	.89989	.01420
.0170	58.82	.05445	.31219	.82934	.04516	.0029	345	.01540	.18835	.90091	.01387
.0165	60.61	.05331	.30953	.83094	.04429	.0028	357	.01502	.18647	.90196	.01354
.0160	62.50	.05215	.30682	.83257	.04342	.0027	370	.01463	.18455	.90303	.01321
.0155	64.52	.05098	.30405	.83423	.04253	.0026	385	.01424	.18257	.90412	.01288
.0150	66.67	.04980	.30122	.83593	.04163	.0025	400	.01385	.18053	.90524	.01254
.0145	68.97	.04861	.29832	.83767	.04072	.0024	417	.01345	.17844	.90640	.01219
.0140	71.43	.04740	.29534	.83944	.03979	.0023	435	.01305	.17628	.90759	.01184
.0135	74.07	.04619	.29229	.84126	.03886	.0022	455	.01264	.17406	.90881	.01149
.0130	76.92	.04496	.28915	.84312	.03791	.0021	476	.01223	.17176	.91008	.01113
.0125	80.00	.04372	.28593	.84503	.03694	.0020	500	.01181	.16938	.91138	.01076
.0120	83.33	.04246	.28261	.84699	.03596	.0019	526	.01138	.16692	.91273	.01039
.0115	86.96	.04119	.27920	.84900	.03497	.0018	556	.01095	.16436	.91413	.01001
.0110	90.91	.03990	.27568	.85107	.03396	.0017	588	.01051	.16170	.91559	.00963
.0105	95.24	.03860	.27204	.85321	.03293	.0016	625	.01007	.15892	.91710	.00923
.0100	100	.03728	.26827	.85541	.03189	.0015	667	.00961	.15602	.91868	.00883
.0098	102	.03674	.26673	.85631	.03146	.0014	714	.00915	.15298	.92034	.00842
.0096	104	.03620	.26516	.85723	.03104	.0013	769	.00868	.14977	.92208	.00800
.0094	106	.03566	.26357	.85816	.03061	.0012	833	.00820	.14638	.92392	.00757
.0092	109	.03512	.26195	.85910	.03017	.0011	909	.00770	.14279	.92586	.00713
.0090	111	.03457	.26031	.86005	.02974	.0010	1000	.00720	.13895	.92793	.00668
.0088	114	.03402	.25865	.86102	.02929	.0009	1111	.00668	.13483	.93015	.00621
.0086	116	.03347	.25696	.86200	.02885	.0008	1250	.00614	.13037	.93254	.00572
.0084	119	.03291	.25523	.86300	.02840	.0007	1429	.00558	.12549	.93515	.00522
.0082	122	.03235	.25348	.86401	.02795	.0006	1667	.00500	.12008	.93804	.00469
.0080	125	.03178	.25170	.86504	.02749	.0005	2000	.00439	.11399	.94128	.00413
.0078	128	.03121	.24989	.86609	.02703	.0004	2500	.00374	.10694	.94502	.00354
.0076	132	.03064	.24804	.86716	.02657	.0003	3333	.00304	.09851	.94947	.00289
.0074	135	.03006	.24616	.86824	.02610	.0002	5000	.00228	.08773	.95513	.00218
.0072	139	.02948	.24424	.86935	.02563	.0001	10,000	.00139	.07197	.96334	.00134
.0070	143	.02889	.24228	.87047	.02515	0	∞	0	0	1	0
.0068	147	.02830	.24028	.87162	.02467						
.0066	152	.02770	.23824	.87279	.02418						
.0064	156	.02710	.23616	.87398	.02369						
.0062	161	.02649	.23402	.87520	.02319						

Table 26 $r^{\frac{1}{n}}$

Values of n

r	$\frac{1}{r}$	1.05	1.10	1.15	1.20	1.25	1.30	1.35	1.40	1.45	1.50	1.55	1.60	1.65	1.70	1.75	1.80	r
1	1	1	1	1	1	1	1	1	1	1	1	1	1	1	1	1	1	1
0.98	1.02	0.9809	.9818	.9826	.9833	.9840	.9846	.9851	.9857	.9862	.9866	.9871	.9875	.9878	.9882	.9885	.9888	0.98
.96	1.04	9619	9636	9651	9666	9679	9691	9702	9713	9722	9732	9740	9748	9756	9763	9769	9776	.96
.94	1.06	9428	9453	9476	9497	9517	9535	9552	9568	9582	9596	9609	9621	9632	9643	9653	9662	.94
.92	1.09	9237	9270	9301	9329	9355	9379	9401	9422	9441	9459	9476	9492	9507	9521	9535	9547	.92
.90	1.11	9045	9087	9125	9159	9192	9222	9249	9275	9299	9322	9343	9363	9381	9399	9416	9431	.90
.88	1.14	8854	8903	8948	8990	9028	9063	9097	9127	9156	9183	9208	9232	9255	9276	9296	9314	.88
.86	1.16	8662	8719	8771	8819	8863	8905	8943	8979	9012	9043	9073	9100	9126	9151	9174	9196	.86
.84	1.19	8470	8534	8593	8648	8698	8745	8788	8829	8867	8903	8936	8968	8997	9025	9052	9077	.84
.82	1.22	8278	8349	8415	8476	8532	8584	8633	8678	8721	8761	8798	8834	8867	8898	8928	8956	.82
.80	1.25	8085	8164	8236	8303	8365	8423	8476	8527	8574	8618	8659	8698	8735	8770	8803	8834	.80
.78	1.28	7893	7978	8057	8130	8197	8260	8319	8374	8425	8474	8519	8562	8602	8640	8676	8711	.78
.76	1.32	7700	7792	7877	7956	8029	8097	8160	8220	8276	8328	8377	8424	8468	8509	8549	8586	.76
.74	1.35	7507	7605	7696	7781	7859	7932	8001	8065	8125	8181	8234	8285	8332	8377	8419	8460	.74
.72	1.39	7314	7418	7515	7605	7689	7767	7840	7909	7973	8033	8090	8144	8195	8243	8288	8332	.72
.70	1.43	7120	7231	7333	7429	7518	7601	7678	7751	7819	7884	7944	8002	8056	8107	8156	8202	.70
.68	1.47	6926	7043	7151	7251	7345	7433	7515	7592	7665	7733	7797	7858	7916	7970	8022	8071	.68
.66	1.52	6732	6854	6968	7073	7172	7264	7351	7432	7508	7580	7649	7713	7774	7832	7886	7939	.66
.64	1.56	6537	6665	6784	6894	6998	7094	7185	7270	7351	7427	7498	7566	7630	7691	7749	7804	.64
.62	1.61	6343	6475	6599	6714	6822	6923	7018	7107	7192	7271	7346	7417	7485	7549	7610	7668	.62
.60	1.67	6148	6285	6413	6533	6645	6751	6850	6943	7031	7114	7192	7267	7337	7405	7468	7529	.60
.58	1.72	5952	6094	6227	6351	6468	6577	6680	6777	6868	6955	7037	7114	7188	7258	7325	7389	.58
.56	1.79	5757	5903	6040	6168	6289	6402	6508	6609	6704	6794	6879	6960	7037	7110	7180	7246	.56
.54	1.85	5561	5711	5852	5984	6108	6225	6335	6440	6538	6631	6720	6804	6884	6960	7032	7101	.54
.52	1.92	5364	5519	5663	5799	5927	6047	6161	6268	6370	6466	6558	6645	6728	6807	6882	6954	.52
.50	2.00	5168	5325	5473	5612	5743	5867	5984	6095	6200	6300	6394	6484	6570	6652	6730	6804	.50
.48	2.08	4971	5131	5282	5425	5559	5686	5806	5920	6028	6130	6228	6321	6409	6494	6574	6651	.48
.46	2.17	4773	4936	5090	5236	5373	5503	5626	5743	5854	5959	6059	6155	6246	6333	6416	6496	.46
.44	2.27	4575	4741	4897	5045	5185	5318	5444	5563	5677	5785	5888	5986	6080	6170	6255	6338	.44
.42	2.38	4377	4545	4703	4853	4996	5131	5259	5381	5498	5608	5714	5815	5911	6003	6091	6176	.42
.40	2.50	4178	4347	4508	4660	4804	4942	5073	5197	5316	5429	5537	5640	5739	5833	5924	6011	.40
.38	2.63	3979	4149	4311	4465	4611	4751	4883	5010	5131	5246	5357	5462	5563	5660	5753	5842	.38
.36	2.78	3779	3950	4113	4268	4416	4557	4692	4820	4943	5061	5173	5281	5384	5483	5578	5669	.36
.34	2.94	3579	3750	3914	4070	4219	4361	4497	4627	4752	4871	4986	5095	5201	5302	5399	5492	.34
.32	3.12	3378	3549	3713	3869	4019	4162	4300	4431	4557	4678	4794	4906	5013	5116	5215	5310	.32
.30	3.33	3177	3347	3510	3667	3817	3961	4099	4232	4359	4481	4599	4712	4821	4925	5026	5123	.30
.28	3.57	2975	3144	3306	3462	3612	3756	3895	4028	4157	4280	4399	4513	4623	4729	4832	4930	.28
.26	3.85	2772	2939	3099	3254	3404	3548	3687	3821	3949	4074	4193	4309	4420	4528	4631	4731	.26
.24	4.17	2569	2732	2891	3044	3193	3336	3475	3608	3737	3862	3982	4099	4211	4319	4424	4526	.24
.22	4.55	2364	2525	2680	2832	2978	3120	3258	3391	3520	3644	3765	3882	3995	4104	4210	4312	.22
.20	5.00	2159	2315	2467	2615	2759	2900	3036	3168	3296	3420	3540	3657	3770	3880	3986	4090	.20
.18	5.56	1953	2104	2251	2395	2536	2674	2808	2938	3065	3188	3308	3424	3537	3647	3754	3857	.18
.16	6.25	1746	1890	2032	2172	2308	2442	2573	2701	2826	2947	3066	3181	3293	3403	3509	3613	.16
.14	7.14	1537	1674	1809	1943	2074	2204	2331	2455	2577	2696	2813	2926	3037	3146	3251	3354	.14
.12	8.33	1327	1455	1582	1709	1834	1957	2079	2199	2317	2433	2546	2658	2766	2873	2977	3079	.12
.10	10.0	1116	1233	1350	1468	1585	1701	1817	1931	2043	2154	2264	2371	2477	2581	2683	2783	.10
.08	12.5	0902	1006	1112	1219	1326	1433	1540	1646	1752	1857	1960	2063	2164	2263	2362	2458	.08
.06	16.7	0686	0775	0866	0959	1053	1148	1244	1340	1437	1533	1628	1723	1818	1911	2004	2095	.06
.04	25.0	0466	0536	0609	0684	0761	0841	0921	1003	1086	1170	1253	1337	1422	1506	1590	1673	.04
.02	50.0	0241	0285	0333	0384	0437	0493	0551	0612	0673	0737	0801	0867	0934	1001	1069	1138	.02
0	∞	0	0	0	0	0	0	0	0	0	0	0	0	0	0	0	0	0

Table 27 $r^{\frac{n-1}{n}}$

Values of n

r	$\frac{1}{r}$	1.05	1.10	1.15	1.20	1.25	1.30	1.35	1.40	1.45	1.50	1.55	1.60	1.65	1.70	1.75	1.80	r
1	1	1	1	1	1	1	1	1	1	1	1	1	1	1	1	1	1	1
0.98	1.02	0.9990	.9982	.9974	.9966	.9960	.9953	.9948	.9942	.9938	.9933	.9929	.9925	.9921	.9917	.9914	.9911	0.98
.96	1.04	9981	9963	9947	9932	9919	9906	9895	9884	9874	9865	9856	9848	9840	9833	9827	9820	.96
.94	1.06	9971	9944	9920	9897	9877	9858	9841	9825	9810	9796	9783	9771	9759	9748	9738	9729	.94
.92	1.09	9960	9924	9892	9862	9835	9809	9786	9765	9745	9726	9708	9692	9677	9662	9649	9636	.92
.90	1.11	9950	9905	9864	9826	9791	9760	9731	9703	9678	9655	9633	9613	9593	9575	9558	9543	.90
.88	1.14	9939	9884	9835	9789	9748	9709	9674	9641	9611	9583	9557	9532	9509	9487	9467	9448	.88
.86	1.16	9928	9864	9805	9752	9703	9658	9617	9578	9543	9510	9479	9450	9423	9398	9374	9352	.86
.84	1.19	9917	9843	9775	9714	9657	9606	9558	9514	9473	9435	9400	9367	9336	9307	9280	9254	.84
.82	1.22	9906	9821	9744	9675	9611	9552	9499	9449	9403	9360	9320	9283	9248	9215	9185	9156	.82
.80	1.25	9894	9799	9713	9635	9564	9498	9438	9382	9331	9283	9239	9197	9158	9122	9088	9056	.80
.78	1.28	9882	9777	9681	9594	9515	9443	9376	9315	9258	9205	9156	9110	9068	9028	8990	8955	.78
.76	1.32	9870	9754	9648	9553	9466	9386	9313	9246	9184	9126	9072	9022	8975	8931	8890	8852	.76
.74	1.35	9858	9730	9615	9511	9416	9329	9249	9176	9108	9045	8987	8932	8881	8834	8789	8747	.74
.72	1.39	9845	9706	9581	9467	9364	9270	9184	9104	9031	8963	8900	8841	8786	8735	8687	8642	.72
.70	1.43	9832	9681	9545	9423	9312	9210	9117	9031	8952	8879	8811	8748	8689	8634	8582	8534	.70
.68	1.47	9818	9655	9509	9377	9258	9148	9048	8957	8872	8794	8721	8653	8591	8532	8477	8425	.68
.66	1.52	9804	9629	9472	9331	9203	9086	8979	8881	8790	8707	8629	8557	8490	8427	8369	8314	.66
.64	1.56	9790	9602	9435	9283	9146	9021	8907	8803	8707	8618	8535	8459	8388	8321	8259	8201	.64
.62	1.61	9775	9575	9396	9234	9088	8956	8834	8723	8621	8527	8440	8359	8284	8213	8148	8086	.62
.60	1.67	9760	9546	9355	9184	9029	8888	8760	8642	8534	8434	8342	8257	8177	8103	8034	7969	.60
.58	1.72	9744	9517	9314	9132	8968	8819	8683	8559	8445	8340	8242	8152	8069	7991	7918	7850	.58
.56	1.79	9728	9487	9272	9079	8905	8748	8604	8473	8353	8243	8140	8046	7958	7876	7800	7728	.56
.54	1.85	9711	9455	9228	9024	8841	8675	8524	8386	8259	8143	8036	7937	7845	7759	7679	7604	.54
.52	1.92	9693	9423	9182	8967	8774	8599	8441	8296	8163	8041	7929	7825	7729	7639	7556	7478	.52
.50	2.00	9675	9389	9136	8909	8706	8522	8355	8203	8064	7937	7820	7711	7610	7517	7430	7349	.50
.48	2.08	9657	9355	9087	8849	8635	8442	8267	8108	7963	7830	7707	7594	7489	7392	7301	7217	.48
.46	2.17	9637	9318	9037	8786	8562	8359	8176	8010	7858	7719	7592	7474	7365	7263	7169	7081	.46
.44	2.27	9617	9281	8984	8721	8486	8274	8083	7909	7751	7606	7473	7350	7237	7132	7034	6943	.44
.42	2.38	9595	9242	8930	8654	8407	8186	7986	7805	7640	7489	7350	7223	7105	6996	6895	6801	.42
.40	2.50	9573	9201	8873	8584	8326	8094	7886	7697	7525	7368	7224	7092	6970	6857	6752	6655	.40
.38	2.63	9550	9158	8814	8511	8241	7999	7781	7585	7406	7243	7094	6957	6831	6714	6606	6505	.38
.36	2.78	9525	9113	8752	8434	8152	7900	7673	7468	7283	7114	6959	6817	6687	6566	6454	6350	.36
.34	2.94	9499	9066	8687	8354	8059	7796	7560	7347	7155	6980	6819	6673	6538	6413	6298	6191	.34
.32	3.12	9472	9016	8619	8270	7962	7688	7442	7221	7021	6840	6674	6523	6384	6255	6137	6027	.32
.30	3.33	9443	8963	8547	8182	7860	7574	7319	7089	6882	6694	6523	6367	6223	6091	5969	5856	.30
.28	3.57	9412	8907	8470	8088	7752	7455	7189	6951	6736	6542	6366	6204	6056	5921	5795	5679	.28
.26	3.85	9379	8847	8389	7989	7638	7328	7052	6805	6583	6383	6200	6034	5882	5743	5614	5494	.26
.24	4.17	9343	8783	8302	7883	7518	7194	6907	6651	6422	6214	6027	5856	5700	5556	5425	5303	.24
.22	4.55	9304	8714	8208	7770	7387	7051	6753	6488	6251	6037	5843	5668	5507	5361	5226	5102	.22
.20	5.00	9262	8639	8106	7647	7248	6898	6589	6314	6068	5848	5649	5469	5305	5155	5017	4890	.20
.18	5.56	9216	8557	7996	7514	7097	6732	6411	6127	5873	5646	5442	5257	5089	4936	4795	4667	.18
.16	6.25	9164	8465	7874	7368	6931	6551	6218	5924	5662	5429	5219	5030	4858	4702	4559	4429	.16
.14	7.14	9106	8363	7738	7206	6749	6353	6007	5702	5433	5192	4978	4784	4609	4450	4306	4174	.14
.12	8.33	9040	8247	7584	7023	6544	6131	5771	5456	5179	4932	4713	4515	4338	4177	4031	3897	.12
.10	10.0	8962	8111	7406	6813	6310	5878	5505	5179	4894	4642	4417	4217	4037	3875	3728	3594	.10
.08	12.5	8867	7948	7193	6564	6034	5583	5195	4860	4566	4309	4081	3878	3697	3535	3388	3255	.08
.06	16.7	8746	7743	6928	6257	5697	5224	4822	4476	4176	3915	3685	3482	3301	3140	2995	2864	.06
.04	25.0	8579	7463	6571	5848	5253	4758	4341	3986	3683	3420	3192	2991	2814	2657	2517	2392	.04
.02	50.0	8300	7007	6003	5210	4573	4054	3627	3270	2970	2714	2495	2306	2141	1997	1870	1758	.02
0	∞	0	0	0	0	0	0	0	0	0	0	0	0	0	0	0	0	0

Table 28 $\sqrt{1-r^{\frac{n-1}{n}}}$

Values of n

r	$\frac{1}{r}$	1.05	1.10	1.15	1.20	1.25	1.30	1.35	1.40	1.45	1.50	1.55	1.60	1.65	1.70	1.75	1.80	r
1	1	0	0	0	0	0	0	0	0	0	0	0	0	0	0	0	0	1
0.98	1.02	0.0310	.0428	.0513	.0580	.0635	.0682	.0723	.0759	.0791	.0819	.0845	.0869	.0890	.0910	.0928	.0945	.98
.96	1.04	0441	0609	0749	0823	0902	0968	1026	1077	1122	1163	1199	1233	1263	1291	1317	1341	.96
.94	1.06	0542	0749	0897	1013	1109	1191	1261	1324	1379	1429	1474	1514	1552	1586	1618	1647	.94
.92	1.09	0630	0869	1040	1175	1286	1381	1462	1534	1598	1656	1707	1755	1798	1837	1874	1907	.92
.90	1.11	0707	0976	1168	1319	1444	1550	1642	1722	1794	1858	1916	1968	2016	2060	2101	2139	.90
.88	1.14	0779	1075	1286	1452	1589	1705	1806	1894	1972	2042	2106	2163	2216	2264	2309	2350	.88
.86	1.16	0846	1167	1396	1576	1724	1849	1958	2054	2138	2214	2283	2345	2402	2454	2502	2546	.86
.84	1.19	0909	1254	1500	1692	1851	1986	2102	2204	2295	2376	2449	2516	2576	2632	2683	2731	.84
.82	1.22	0970	1337	1599	1804	1973	2116	2239	2348	2444	2530	2608	2678	2742	2801	2855	2906	.82
.80	1.25	1028	1417	1694	1911	2089	2240	2371	2485	2587	2677	2759	2833	2901	2963	3020	3073	.80
.78	1.28	1085	1494	1786	2014	2202	2361	2498	2618	2724	2819	2905	2983	3054	3118	3178	3233	.78
.76	1.32	1139	1570	1875	2114	2311	2477	2621	2746	2857	2957	3046	3127	3201	3269	3331	3389	.76
.74	1.35	1193	1643	1962	2212	2418	2591	2740	2871	2987	3090	3183	3268	3344	3415	3479	3539	.74
.72	1.39	1246	1715	2048	2308	2522	2702	2857	2993	3113	3221	3317	3404	3484	3557	3624	3686	.72
.70	1.43	1298	1786	2132	2402	2624	2811	2972	3113	3237	3348	3448	3538	3621	3696	3765	3829	.70
.68	1.47	1349	1856	2215	2495	2725	2918	3085	3230	3359	3473	3576	3669	3754	3832	3903	3969	.68
.66	1.52	1400	1925	2297	2587	2824	3024	3196	3346	3478	3596	3703	3798	3886	3966	4039	4106	.66
.64	1.56	1450	1994	2378	2677	2922	3128	3305	3460	3596	3718	3827	3926	4015	4097	4172	4242	.64
.62	1.61	1500	2062	2459	2767	3020	3232	3414	3573	3713	3838	3950	4051	4143	4227	4304	4375	.62
.60	1.67	1550	2130	2539	2857	3116	3335	3522	3685	3829	3957	4072	4175	4269	4355	4434	4507	.60
.58	1.72	1600	2198	2619	2946	3213	3437	3629	3796	3944	4075	4192	4298	4395	4482	4563	4637	.58
.56	1.79	1650	2266	2699	3035	3309	3539	3736	3907	4058	4192	4312	4421	4519	4609	4691	4766	.56
.54	1.85	1700	2334	2779	3124	3405	3641	3842	4018	4172	4309	4432	4542	4642	4734	4818	4895	.54
.52	1.92	1751	2402	2859	3213	3501	3743	3949	4128	4286	4426	4551	4663	4766	4859	4944	5022	.52
.50	2.00	1802	2471	2940	3303	3598	3845	4056	4239	4399	4542	4670	4784	4888	4983	5070	5149	.50
.48	2.08	1853	2541	3022	3393	3695	3947	4163	4349	4513	4659	4788	4905	5011	5107	5195	5276	.48
.46	2.17	1905	2611	3104	3484	3793	4050	4270	4461	4628	4776	4908	5026	5134	5231	5321	5402	.46
.44	2.27	1958	2682	3187	3576	3891	4154	4379	4573	4743	4893	5027	5148	5257	5356	5446	5529	.44
.42	2.38	2012	2754	3271	3669	3991	4259	4488	4685	4858	5011	5147	5270	5380	5481	5572	5656	.42
.40	2.50	2066	2827	3356	3763	4092	4366	4598	4799	4975	5130	5268	5393	5505	5606	5699	5784	.40
.38	2.63	2122	2902	3443	3859	4195	4473	4710	4915	5093	5251	5391	5516	5630	5733	5826	5912	.38
.36	2.78	2179	2978	3532	3957	4299	4583	4824	5031	5213	5372	5514	5642	5756	5860	5955	6041	.36
.34	2.94	2238	3056	3623	4057	4405	4695	4939	5150	5334	5496	5640	5768	5884	5989	6084	6172	.34
.32	3.12	2298	3137	3716	4159	4514	4809	5057	5271	5458	5621	5767	5897	6014	6120	6216	6304	.32
.30	3.33	2361	3220	3812	4264	4626	4925	5178	5395	5584	5749	5896	6028	6146	6252	6349	6437	.30
.28	3.57	2425	3306	3911	4372	4741	5045	5302	5522	5713	5880	6029	6161	6280	6387	6484	6573	.28
.26	3.85	2493	3395	4014	4484	4860	5169	5429	5652	5845	6015	6164	6298	6417	6525	6623	6712	.26
.24	4.17	2563	3488	4121	4601	4982	5297	5561	5787	5982	6153	6303	6438	6558	6666	6764	6853	.24
.22	4.55	2637	3586	4233	4723	5111	5430	5698	5926	6123	6295	6447	6582	6703	6811	6909	6999	.22
.20	5.00	2716	3689	4352	4851	5246	5570	5841	6071	6270	6444	6596	6731	6852	6961	7059	7148	.20
.18	5.56	2800	3799	4477	4986	5388	5717	5991	6224	6424	6598	6751	6887	7008	7116	7214	7303	.18
.16	6.25	2891	3917	4611	5130	5539	5872	6150	6384	6586	6761	6914	7050	7171	7279	7376	7464	.16
.14	7.14	2990	4046	4756	5286	5702	6039	6319	6556	6758	6934	7087	7222	7342	7450	7546	7633	.14
.12	8.33	3099	4187	4915	5456	5879	6220	6503	6741	6943	7119	7271	7406	7525	7631	7726	7812	.12
.10	10.0	3223	4346	5093	5645	6075	6420	6705	6943	7146	7320	7472	7605	7722	7826	7920	8004	.10
.08	12.5	3366	4529	5298	5862	6297	6646	6932	7170	7371	7544	7693	7824	7939	8041	8132	8213	.08
.06	16.7	3541	4751	5542	6118	6560	6911	7196	7432	7631	7801	7947	8074	8185	8283	8370	8448	.06
.04	25.0	3770	5037	5855	6444	6890	7240	7523	7755	7948	8112	8252	8372	8477	8569	8650	8723	.04
.02	50.0	4123	5471	6322	6921	7367	7711	7983	8204	8385	8536	8663	8771	8865	8946	9017	9079	.02
0	∞	1	1	1	1	1	1	1	1	1	1	1	1	1	1	1	1	0

Table 29 $\quad r^{\frac{1}{n}}\sqrt{1-r^{\frac{n-1}{n}}}$

Values of n

r	$\frac{1}{r}$	1.05	1.10	1.15	1.20	1.25	1.30	1.35	1.40	1.45	1.50	1.55	1.60	1.65	1.70	1.75	1.80	r
1	1	0	0	0	0	0	0	0	0	0	0	0	0	0	0	0	0	1
0.98	1.02	0.0304	.0421	.0504	.0570	.0625	.0672	.0713	.0749	.0780	.0808	.0834	.0858	.0880	.0900	.0918	.0935	0.98
.96	1.04	0424	0586	0703	0796	0873	0938	0995	1046	1091	1131	1168	1201	1232	1260	1287	1311	.96
.94	1.06	0511	0708	0850	0962	1055	1135	1205	1267	1322	1371	1416	1457	1495	1529	1561	1591	.94
.92	1.09	0581	0806	0967	1096	1203	1295	1375	1446	1509	1566	1618	1665	1709	1749	1786	1821	.92
.90	1.11	0640	0887	1066	1208	1327	1429	1518	1597	1668	1732	1790	1843	1892	1937	1978	2017	.90
.88	1.14	0690	0957	1151	1305	1434	1545	1642	1729	1806	1876	1939	1997	2051	2100	2146	2189	.88
.86	1.16	0733	1017	1224	1389	1528	1647	1751	1844	1927	2002	2071	2134	2192	2246	2295	2342	.86
.84	1.19	0770	1070	1289	1463	1610	1737	1848	1946	2035	2116	2189	2256	2318	2375	2429	2479	.84
.82	1.22	0803	1116	1345	1529	1683	1816	1934	2038	2131	2216	2294	2366	2432	2493	2549	2602	.82
.80	1.25	0831	1157	1395	1586	1748	1887	2010	2119	2218	2307	2389	2464	2534	2598	2658	2714	.80
.78	1.28	0856	1192	1439	1637	1805	1950	2078	2192	2295	2389	2475	2554	2627	2694	2758	2817	.78
.76	1.32	0877	1223	1477	1682	1855	2006	2139	2257	2365	2463	2552	2634	2711	2782	2848	2909	.76
.74	1.35	0896	1250	1510	1721	1900	2055	2192	2316	2427	2528	2621	2707	2787	2860	2929	2994	.74
.72	1.39	0911	1272	1539	1755	1939	2099	2240	2367	2482	2587	2684	2773	2855	2932	3004	3071	.72
.70	1.43	0924	1292	1563	1785	1973	2136	2282	2413	2532	2640	2739	2831	2917	2997	3071	3141	.70
.68	1.47	0934	1307	1584	1809	2001	2169	2318	2452	2574	2686	2789	2884	2972	3054	3131	3203	.68
.66	1.52	0942	1320	1600	1830	2025	2197	2349	2487	2612	2726	2832	2930	3021	3106	3185	3260	.66
.64	1.56	0948	1329	1613	1846	2045	2219	2375	2516	2644	2761	2870	2971	3064	3151	3233	3310	.64
.62	1.61	0952	1335	1622	1858	2060	2237	2396	2539	2670	2791	2902	3005	3101	3191	3275	3355	.62
.60	1.67	0953	1339	1628	1866	2071	2251	2412	2558	2692	2815	2928	3034	3133	3225	3312	3393	.60
.58	1.72	0952	1340	1631	1871	2078	2260	2424	2573	2709	2834	2950	3058	3159	3253	3342	3426	.58
.56	1.79	0950	1338	1630	1872	2081	2266	2431	2582	2721	2848	2967	3077	3180	3277	3368	3454	.56
.54	1.85	0946	1333	1626	1869	2080	2266	2434	2587	2728	2857	2978	3090	3196	3295	3388	3476	.54
.52	1.92	0939	1326	1619	1863	2075	2263	2433	2588	2730	2862	2985	3099	3206	3307	3402	3492	.52
.50	2.00	0931	1316	1609	1854	2066	2256	2427	2584	2728	2861	2986	3103	3212	3315	3412	3503	.50
.48	2.08	0921	1304	1596	1841	2054	2244	2417	2575	2721	2856	2982	3101	3212	3316	3415	3509	.48
.46	2.17	0909	1289	1580	1824	2038	2229	2402	2562	2709	2846	2974	3094	3207	3313	3414	3509	.46
.44	2.27	0896	1271	1561	1804	2018	2209	2384	2544	2692	2831	2960	3082	3196	3304	3407	3504	.44
.42	2.38	0881	1252	1538	1781	1994	2185	2360	2521	2671	2810	2941	3064	3180	3290	3394	3493	.42
.40	2.50	0863	1229	1513	1754	1966	2157	2333	2494	2645	2785	2917	3041	3159	3270	3376	3476	.40
.38	2.63	0844	1204	1484	1723	1934	2125	2300	2462	2613	2755	2888	3013	3132	3245	3352	3454	.38
.36	2.78	0824	1176	1453	1689	1898	2089	2263	2425	2577	2719	2853	2979	3099	3213	3321	3425	.36
.34	2.94	0801	1146	1418	1651	1859	2047	2221	2383	2535	2677	2812	2939	3060	3175	3285	3389	.34
.32	3.12	0776	1113	1380	1609	1814	2002	2175	2336	2487	2630	2765	2893	3015	3131	3241	3347	.32
.30	3.33	0750	1078	1338	1563	1766	1951	2123	2283	2434	2577	2712	2840	2963	3079	3191	3298	.30
.28	3.57	0721	1039	1293	1514	1712	1895	2065	2224	2375	2517	2652	2781	2904	3021	3133	3241	.28
.26	3.85	0691	0998	1244	1459	1654	1834	2002	2159	2308	2450	2585	2713	2836	2954	3067	3176	.26
.24	4.17	0658	0953	1191	1401	1591	1767	1932	2088	2236	2376	2510	2638	2761	2879	2993	3102	.24
.22	4.55	0624	0905	1135	1337	1522	1694	1856	2009	2155	2294	2427	2555	2677	2795	2909	3018	.22
.20	5.00	0587	0854	1074	1269	1448	1615	1773	1923	2066	2204	2335	2462	2584	2701	2814	2923	.20
.18	5.56	0547	0799	1008	1194	1367	1529	1682	1829	1969	2104	2233	2358	2479	2595	2708	2817	.18
.16	6.25	0505	0740	0937	1114	1279	1434	1582	1724	1861	1993	2120	2243	2362	2477	2588	2697	.16
.14	7.14	0460	0677	0861	1027	1183	1331	1473	1610	1742	1869	1993	2113	2230	2343	2454	2561	.14
.12	8.33	0411	0609	0778	0932	1078	1218	1352	1482	1609	1732	1852	1968	2082	2192	2300	2405	.12
.10	10.0	0360	0536	0688	0829	0963	1092	1218	1340	1460	1577	1691	1803	1913	2020	2125	2227	.10
.08	12.5	0304	0456	0589	0714	0835	0952	1067	1180	1291	1401	1508	1614	1718	1820	1920	2019	.08
.06	16.7	0243	0368	0480	0587	0691	0794	0895	0996	1096	1196	1294	1391	1488	1583	1677	1770	.06
.04	25.0	0176	0270	.0356	0441	0525	0609	0693	0778	0863	0949	1034	1120	1205	1290	1375	1459	.04
.02	50.0	0099	0156	0211	0266	0322	0380	0440	0502	0565	0629	0694	0761	0828	0896	0964	1033	.02
0	∞	0	0	0	0	0	0	0	0	0	0	0	0	0	0	0	0	0

Table 30 One-Dimensional Isentropic Compressible-Flow Functions

For a perfect gas with constant specific heat and molecular weight
k = 1.4

M	M*	$\frac{A}{A^*}$	$\frac{p}{p_0}$	$\frac{\rho}{\rho_0}$	$\frac{T}{T_0}$	$\frac{F}{F^*}$	$\left(\frac{A}{A^*}\right)\left(\frac{p}{p_0}\right)$	M	M*	$\frac{A}{A^*}$	$\frac{p}{p_0}$	$\frac{\rho}{\rho_0}$	$\frac{T}{T_0}$	$\frac{F}{F^*}$	$\left(\frac{A}{A^*}\right)\left(\frac{p}{p_0}\right)$
0	0	∞	1.00000	1.00000	1.00000	∞	∞	0.40	.43133	1.5901	.89562	.92428	.96899	1.3749	1.4241
0.01	0.01096	57.874	.99993	.99995	.99998	45.650	57.870	.41	.44177	1.5587	.89071	.92066	.96747	1.3527	1.3883
.02	.02191	28.942	.99972	.99980	.99992	22.834	28.934	.42	.45218	1.5289	.88572	.91697	.96592	1.3318	1.3542
.03	.03286	19.300	.99937	.99955	.99982	15.232	19.288	.43	.46256	1.5007	.88065	.91322	.96434	1.3122	1.3216
.04	.04381	14.482	.99888	.99920	.99968	11.435	14.465	.44	.47292	1.4740	.87550	.90940	.96272	1.2937	1.2905
.05	.05476	11.592	.99825	.99875	.99950	9.1584	11.571	.45	.48326	1.4487	.87027	.90552	.96108	1.2763	1.2607
.06	.06570	9.6659	.99748	.99820	.99928	7.6428	9.6415	.46	.49357	1.4246	.86496	.90157	.95940	1.2598	1.2322
.07	.07664	8.2915	.99658	.99755	.99902	6.5620	8.2631	.47	.50385	1.4018	.85958	.89756	.95769	1.2443	1.2050
.08	.08758	7.2616	.99553	.99680	.99872	5.7529	7.2291	.48	.51410	1.3801	.85413	.89349	.95595	1.2296	1.1788
.09	.09851	6.4613	.99435	.99596	.99838	5.1249	6.4248	.49	.52432	1.3594	.84861	.88936	.95418	1.2158	1.1537
.10	.10943	5.8218	.99303	.99502	.99800	4.6236	5.7812	.50	.53452	1.3398	.84302	.88517	.95238	1.2027	1.12951
.11	.12035	5.2992	.99157	.99398	.99758	4.2146	5.2546	.51	.54469	1.3212	.83737	.88092	.95055	1.1903	1.10631
.12	.13126	4.8643	.98998	.99284	.99714	3.8747	4.8157	.52	.55482	1.3034	.83166	.87662	.94869	1.1786	1.08397
.13	.14216	4.4968	.98826	.99160	.99664	3.5880	4.4440	.53	.56493	1.2864	.82589	.87227	.94681	1.1675	1.06245
.14	.15306	4.1824	.98640	.99027	.99610	3.3432	4.1255	.54	.57501	1.2703	.82005	.86788	.94489	1.1571	1.04173
.15	.16395	3.9103	.98441	.98884	.99552	3.1317	3.8493	.55	.58506	1.2550	.81416	.86342	.94295	1.1472	1.02174
.16	.17483	3.6727	.98228	.98731	.99490	2.9474	3.6076	.56	.59508	1.2403	.80822	.85892	.94098	1.1378	1.00244
.17	.18569	3.4635	.98003	.98569	.99425	2.7855	3.3943	.57	.60506	1.2263	.80224	.85437	.93898	1.1289	.98381
.18	.19654	3.2779	.97765	.98398	.99356	2.6422	3.2046	.58	.61500	1.2130	.79621	.84977	.93696	1.1205	.96581
.19	.20738	3.1122	.97514	.98217	.99283	2.5146	3.0348	.59	.62491	1.2003	.79012	.84513	.93491	1.1126	.94839
.20	.21822	2.9635	.97250	.98027	.99206	2.4004	2.8820	.60	.63480	1.1882	.78400	.84045	.93284	1.10504	.93155
.21	.22904	2.8293	.96973	.97828	.99125	2.2976	2.7437	.61	.64466	1.1766	.77784	.83573	.93074	1.09793	.91525
.22	.23984	2.7076	.96685	.97621	.99041	2.2046	2.6178	.62	.65448	1.1656	.77164	.83096	.92861	1.09120	.89946
.23	.25063	2.5968	.96383	.97403	.98953	2.1203	2.5029	.63	.66427	1.1551	.76540	.82616	.92646	1.08485	.88416
.24	.26141	2.4956	.96070	.97177	.98861	2.0434	2.3975	.64	.67402	1.1451	.75913	.82132	.92428	1.07883	.86932
.25	.27216	2.4027	.95745	.96942	.98765	1.9732	2.3005	.65	.68374	1.1356	.75283	.81644	.92208	1.07314	.85493
.26	.28291	2.3173	.95408	.96699	.98666	1.9088	2.2109	.66	.69342	1.1265	.74650	.81153	.91986	1.06777	.84096
.27	.29364	2.2385	.95060	.96446	.98563	1.8496	2.1279	.67	.70307	1.1178	.74014	.80659	.91762	1.06271	.82740
.28	.30435	2.1656	.94700	.96185	.98456	1.7950	2.0508	.68	.71268	1.1096	.73376	.80162	.91535	1.05792	.81421
.29	.31504	2.0979	.94329	.95916	.98346	1.7446	1.9789	.69	.72225	1.1018	.72735	.79662	.91306	1.05340	.80141
.30	.32572	2.0351	.93947	.95638	.98232	1.6979	1.9119	.70	.73179	1.09437	.72092	.79158	.91075	1.04915	.78896
.31	.33638	1.9765	.93554	.95352	.98114	1.6546	1.8491	.71	.74129	1.08729	.71448	.78652	.90842	1.04514	.77685
.32	.34701	1.9218	.93150	.95058	.97993	1.6144	1.7902	.72	.75076	1.08057	.70802	.78143	.90606	1.04137	.76507
.33	.35762	1.8707	.92736	.94756	.97868	1.5769	1.7348	.73	.76019	1.07419	.70155	.77632	.90368	1.03783	.75360
.34	.36821	1.8229	.92312	.94446	.97740	1.5420	1.6828	.74	.76958	1.06814	.69507	.77119	.90129	1.03450	.74243
.35	.37879	1.7780	.91877	.94128	.97608	1.5094	1.6336	.75	.77893	1.06242	.68857	.76603	.89888	1.03137	.73155
.36	.38935	1.7358	.91433	.93803	.97473	1.4789	1.5871	.76	.78825	1.05700	.68207	.76086	.89644	1.02844	.72095
.37	.39988	1.6961	.90979	.93470	.97335	1.4503	1.5431	.77	.79753	1.05188	.67556	.75567	.89399	1.02570	.71062
.38	.41039	1.6587	.90516	.93129	.97193	1.4236	1.5014	.78	.80677	1.04705	.66905	.75046	.89152	1.02314	.70054
.39	.42087	1.6234	.90044	.92782	.97048	1.3985	1.4618	.79	.81597	1.04250	.66254	.74524	.88903	1.02075	.69070

See page 121 for definitions of symbols.

Table 30 One-Dimensional Isentropic Compressible-Flow Functions

For a perfect gas with constant specific heat and molecular weight
k = 1.4

M	M*	$\frac{A}{A^*}$	$\frac{p}{p_0}$	$\frac{\rho}{\rho_0}$	$\frac{T}{T_0}$	$\frac{F}{F^*}$	$\left(\frac{A}{A^*}\right)\left(\frac{p}{p_0}\right)$	M	M*	$\frac{A}{A^*}$	$\frac{p}{p_0}$	$\frac{\rho}{\rho_0}$	$\frac{T}{T_0}$	$\frac{F}{F^*}$	$\left(\frac{A}{A^*}\right)\left(\frac{p}{p_0}\right)$
0.80	.82514	1.03823	.65602	.74000	.88652	1.01853	.68110	1.20	1.1583	1.03044	.41238	.53114	.77640	1.01082	.42493
.81	.83426	1.03422	.64951	.73474	.88400	1.01646	.67173	1.21	1.1658	1.03344	.40702	.52620	.77350	1.01178	.42063
.82	.84334	1.03046	.64300	.72947	.88146	1.01455	.66259	1.22	1.1732	1.03657	.40171	.52129	.77061	1.01278	.41640
.83	.85239	1.02696	.63650	.72419	.87890	1.01278	.65366	1.23	1.1806	1.03983	.39645	.51640	.76771	1.01381	.41224
.84	.86140	1.02370	.63000	.71890	.87633	1.01115	.64493	1.24	1.1879	1.04323	.39123	.51154	.76481	1.01486	.40814
.85	.87037	1.02067	.62351	.71361	.87374	1.00966	.63640	1.25	1.1952	1.04676	.38606	.50670	.76190	1.01594	.40411
.86	.87929	1.01787	.61703	.70831	.87114	1.00829	.62806	1.26	1.2025	1.05041	.38094	.50189	.75900	1.01705	.40014
.87	.88817	1.01530	.61057	.70300	.86852	1.00704	.61991	1.27	1.2097	1.05419	.37586	.49710	.75610	1.01818	.39622
.88	.89702	1.01294	.60412	.69769	.86589	1.00591	.61193	1.28	1.2169	1.05810	.37083	.49234	.75319	1.01933	.39237
.89	.90583	1.01080	.59768	.69237	.86324	1.00490	.60413	1.29	1.2240	1.06214	.36585	.48761	.75029	1.02050	.38858
.90	.91460	1.00886	.59126	.68704	.86058	1.00399	.59650	1.30	1.2311	1.06631	.36092	.48291	.74738	1.02170	.38484
.91	.92333	1.00713	.58486	.68171	.85791	1.00318	.58903	1.31	1.2382	1.07060	.35603	.47823	.74448	1.02292	.38116
.92	.93201	1.00560	.57848	.67639	.85523	1.00248	.58171	1.32	1.2452	1.07502	.35119	.47358	.74158	1.02415	.37754
.93	.94065	1.00426	.57212	.67107	.85253	1.00188	.57455	1.33	1.2522	1.07957	.34640	.46895	.73867	1.02540	.37397
.94	.94925	1.00311	.56578	.66575	.84982	1.00136	.56754	1.34	1.2591	1.08424	.34166	.46436	.73577	1.02666	.37044
.95	.95781	1.00214	.55946	.66044	.84710	1.00093	.56066	1.35	1.2660	1.08904	.33697	.45980	.73287	1.02794	.36697
.96	.96633	1.00136	.55317	.65513	.84437	1.00059	.55392	1.36	1.2729	1.09397	.33233	.45527	.72997	1.02924	.36355
.97	.97481	1.00076	.54691	.64982	.84162	1.00033	.54732	1.37	1.2797	1.09902	.32774	.45076	.72707	1.03056	.36018
.98	.98325	1.00033	.54067	.64452	.83887	1.00014	.54085	1.38	1.2865	1.10420	.32319	.44628	.72418	1.03189	.35686
.99	.99165	1.00008	.53446	.63923	.83611	1.00003	.53450	1.39	1.2932	1.10950	.31869	.44183	.72128	1.03323	.35359
1.00	1.00000	1.00000	.52828	.63394	.83333	1.00000	.52828	1.40	1.2999	1.1149	.31424	.43742	.71839	1.03458	.35036
1.01	1.00831	1.00008	.52213	.62866	.83055	1.00003	.52218	1.41	1.3065	1.1205	.30984	.43304	.71550	1.03595	.34717
1.02	1.01658	1.00033	.51602	.62339	.82776	1.00013	.51619	1.42	1.3131	1.1262	.30549	.42869	.71261	1.03733	.34403
1.03	1.02481	1.00074	.50994	.61813	.82496	1.00030	.51031	1.43	1.3197	1.1320	.30119	.42436	.70973	1.03872	.34093
1.04	1.03300	1.00130	.50389	.61288	.82215	1.00053	.50454	1.44	1.3262	1.1379	.29693	.42007	.70685	1.04012	.33787
1.05	1.04114	1.00202	.49787	.60765	.81933	1.00082	.49888	1.45	1.3327	1.1440	.29272	.41581	.70397	1.04153	.33486
1.06	1.04924	1.00290	.49189	.60243	.81651	1.00116	.49332	1.46	1.3392	1.1502	.28856	.41158	.70110	1.04295	.33189
1.07	1.05730	1.00394	.48595	.59722	.81368	1.00155	.48787	1.47	1.3456	1.1565	.28445	.40738	.69823	1.04438	.32896
1.08	1.06532	1.00512	.48005	.59203	.81084	1.00200	.48251	1.48	1.3520	1.1629	.28039	.40322	.69537	1.04581	.32607
1.09	1.07330	1.00645	.47418	.58685	.80800	1.00250	.47724	1.49	1.3583	1.1695	.27637	.39909	.69251	1.04725	.32321
1.10	1.08124	1.00793	.46835	.58169	.80515	1.00305	.47206	1.50	1.3646	1.1762	.27240	.39498	.68965	1.04870	.32039
1.11	1.08914	1.00955	.46256	.57655	.80230	1.00365	.46698	1.51	1.3708	1.1830	.26848	.39091	.68680	1.05016	.31761
1.12	1.09699	1.01131	.45682	.57143	.79944	1.00429	.46199	1.52	1.3770	1.1899	.26461	.38687	.68396	1.05162	.31487
1.13	1.10480	1.01322	.45112	.56632	.79657	1.00497	.45708	1.53	1.3832	1.1970	.26078	.38287	.68112	1.05309	.31216
1.14	1.11256	1.01527	.44545	.56123	.79370	1.00569	.45225	1.54	1.3894	1.2042	.25700	.37890	.67828	1.05456	.30948
1.15	1.1203	1.01746	.43983	.55616	.79083	1.00646	.44751	1.55	1.3955	1.2115	.25326	.37496	.67545	1.05604	.30685
1.16	1.1280	1.01978	.43425	.55112	.78795	1.00726	.44284	1.56	1.4016	1.2190	.24957	.37105	.67262	1.05752	.30424
1.17	1.1356	1.02224	.42872	.54609	.78507	1.00810	.43825	1.57	1.4076	1.2266	.24593	.36717	.66980	1.05900	.30167
1.18	1.1432	1.02484	.42323	.54108	.78218	1.00897	.43374	1.58	1.4135	1.2343	.24233	.36332	.66699	1.06049	.29913
1.19	1.1508	1.02757	.41778	.53610	.77929	1.00988	.42930	1.59	1.4195	1.2422	.23878	.35951	.66418	1.06198	.29662

Table 30 One-Dimensional Isentropic Compressible-Flow Functions

For a perfect gas with constant specific heat and molecular weight
k = 1.4

M	M*	$\frac{A}{A^*}$	$\frac{p}{p_0}$	$\frac{\rho}{\rho_0}$	$\frac{T}{T_0}$	$\frac{F}{F^*}$	$\left(\frac{A}{A^*}\right)\left(\frac{p}{p_0}\right)$	M	M*	$\frac{A}{A^*}$	$\frac{p}{p_0}$	$\frac{\rho}{\rho_0}$	$\frac{T}{T_0}$	$\frac{F}{F^*}$	$\left(\frac{A}{A^*}\right)\left(\frac{p}{p_0}\right)$
1.60	1.4254	1.2502	.23527	.35573	.66138	1.06348	.29414	2.00	1.6330	1.6875	.12780	.23005	.55556	1.1227	.21567
1.61	1.4313	1.2583	.23181	.35198	.65858	1.06498	.29169	2.01	1.6375	1.7017	.12583	.22751	.55310	1.1241	.21412
1.62	1.4371	1.2666	.22839	.34826	.65579	1.06648	.28928	2.02	1.6420	1.7160	.12389	.22499	.55064	1.1255	.21259
1.63	1.4429	1.2750	.22501	.34458	.65301	1.06798	.28690	2.03	1.6465	1.7305	.12198	.22250	.54819	1.1269	.21107
1.64	1.4487	1.2835	.22168	.34093	.65023	1.06948	.28454	2.04	1.6509	1.7452	.12009	.22004	.54576	1.1283	.20957
1.65	1.4544	1.2922	.21839	.33731	.64746	1.07098	.28221	2.05	1.6553	1.7600	.11823	.21760	.54333	1.1297	.20808
1.66	1.4601	1.3010	.21515	.33372	.64470	1.07249	.27991	2.06	1.6597	1.7750	.11640	.21519	.54091	1.1311	.20661
1.67	1.4657	1.3099	.21195	.33016	.64194	1.07399	.27764	2.07	1.6640	1.7902	.11460	.21281	.53850	1.1325	.20515
1.68	1.4713	1.3190	.20879	.32664	.63919	1.07550	.27540	2.08	1.6683	1.8056	.11282	.21045	.53611	1.1339	.20371
1.69	1.4769	1.3282	.20567	.32315	.63645	1.07701	.27318	2.09	1.6726	1.8212	.11107	.20811	.53373	1.1352	.20228
1.70	1.4825	1.3376	.20259	.31969	.63372	1.07851	.27099	2.10	1.6769	1.8369	.10935	.20580	.53135	1.1366	.20087
1.71	1.4880	1.3471	.19955	.31626	.63099	1.08002	.26882	2.11	1.6811	1.8529	.10766	.20352	.52898	1.1380	.19947
1.72	1.4935	1.3567	.19656	.31286	.62827	1.08152	.26668	2.12	1.6853	1.8690	.10599	.20126	.52663	1.1393	.19809
1.73	1.4989	1.3665	.19361	.30950	.62556	1.08302	.26457	2.13	1.6895	1.8853	.10434	.19902	.52428	1.1407	.19672
1.74	1.5043	1.3764	.19070	.30617	.62286	1.08453	.26248	2.14	1.6936	1.9018	.10272	.19681	.52194	1.1420	.19537
1.75	1.5097	1.3865	.18782	.30287	.62016	1.08603	.26042	2.15	1.6977	1.9185	.10113	.19463	.51962	1.1434	.19403
1.76	1.5150	1.3967	.18499	.29959	.61747	1.08753	.25838	2.16	1.7018	1.9354	.09956	.19247	.51730	1.1447	.19270
1.77	1.5203	1.4071	.18220	.29635	.61479	1.08903	.25636	2.17	1.7059	1.9525	.09802	.19033	.51499	1.1460	.19138
1.78	1.5256	1.4176	.17944	.29314	.61211	1.09053	.25436	2.18	1.7099	1.9698	.09650	.18821	.51269	1.1474	.19008
1.79	1.5308	1.4282	.17672	.28997	.60945	1.09202	.25239	2.19	1.7139	1.9873	.09500	.18612	.51041	1.1487	.18879
1.80	1.5360	1.4390	.17404	.28682	.60680	1.09352	.25044	2.20	1.7179	2.0050	.09352	.18405	.50813	1.1500	.18751
1.81	1.5412	1.4499	.17140	.28370	.60415	1.09500	.24851	2.21	1.7219	2.0229	.09207	.18200	.50586	1.1513	.18624
1.82	1.5463	1.4610	.16879	.28061	.60151	1.09649	.24660	2.22	1.7258	2.0409	.09064	.17998	.50361	1.1526	.18499
1.83	1.5514	1.4723	.16622	.27756	.59888	1.09798	.24472	2.23	1.7297	2.0592	.08923	.17798	.50136	1.1539	.18375
1.84	1.5564	1.4837	.16369	.27453	.59626	1.09946	.24286	2.24	1.7336	2.0777	.08784	.17600	.49912	1.1552	.18252
1.85	1.5614	1.4952	.16120	.27153	.59365	1.1009	.24102	2.25	1.7374	2.0964	.08648	.17404	.49689	1.1565	.18130
1.86	1.5664	1.5069	.15874	.26857	.59105	1.1024	.23919	2.26	1.7412	2.1154	.08514	.17211	.49468	1.1578	.18009
1.87	1.5714	1.5188	.15631	.26563	.58845	1.1039	.23739	2.27	1.7450	2.1345	.08382	.17020	.49247	1.1590	.17890
1.88	1.5763	1.5308	.15392	.26272	.58586	1.1054	.23561	2.28	1.7488	2.1538	.08252	.16830	.49027	1.1603	.17772
1.89	1.5812	1.5429	.15156	.25984	.58329	1.1068	.23385	2.29	1.7526	2.1734	.08123	.16643	.48809	1.1616	.17655
1.90	1.5861	1.5552	.14924	.25699	.58072	1.1083	.23211	2.30	1.7563	2.1931	.07997	.16458	.48591	1.1629	.17539
1.91	1.5909	1.5677	.14695	.25417	.57816	1.1097	.23039	2.31	1.7600	2.2131	.07873	.16275	.48374	1.1641	.17424
1.92	1.5957	1.5804	.14469	.25138	.57561	1.1112	.22868	2.32	1.7637	2.2333	.07751	.16095	.48158	1.1653	.17310
1.93	1.6005	1.5932	.14247	.24862	.57307	1.1126	.22699	2.33	1.7673	2.2537	.07631	.15916	.47944	1.1666	.17197
1.94	1.6052	1.6062	.14028	.24588	.57054	1.1141	.22532	2.34	1.7709	2.2744	.07513	.15739	.47730	1.1678	.17085
1.95	1.6099	1.6193	.13813	.24317	.56802	1.1155	.22367	2.35	1.7745	2.2953	.07396	.15564	.47517	1.1690	.16975
1.96	1.6146	1.6326	.13600	.24049	.56551	1.1170	.22204	2.36	1.7781	2.3164	.07281	.15391	.47305	1.1703	.16866
1.97	1.6193	1.6461	.13390	.23784	.56301	1.1184	.22042	2.37	1.7817	2.3377	.07168	.15220	.47095	1.1715	.16757
1.98	1.6239	1.6597	.13184	.23522	.56051	1.1198	.21882	2.38	1.7852	2.3593	.07057	.15052	.46885	1.1727	.16649
1.99	1.6285	1.6735	.12981	.23262	.55803	1.1213	.21724	2.39	1.7887	2.3811	.06948	.14885	.46676	1.1739	.16543

Table 30   One-Dimensional Isentropic Compressible-Flow Functions

For a perfect gas with constant specific heat and molecular weight
k = 1.4

M	M*	$\dfrac{A}{A^*}$	$\dfrac{p}{p_0}$	$\dfrac{\rho}{\rho_0}$	$\dfrac{T}{T_0}$	$\dfrac{F}{F^*}$	$\left(\dfrac{A}{A^*}\right)\left(\dfrac{p}{p_0}\right)$
2.40	1.7922	2.4031	.06840	.14720	.46468	1.1751	.16437
2.41	1.7957	2.4254	.06734	.14557	.46262	1.1763	.16332
2.42	1.7991	2.4479	.06630	.14395	.46056	1.1775	.16229
2.43	1.8025	2.4706	.06527	.14235	.45851	1.1786	.16126
2.44	1.8059	2.4936	.06426	.14078	.45647	1.1798	.16024
2.45	1.8093	2.5168	.06327	.13922	.45444	1.1810	.15923
2.46	1.8126	2.5403	.06229	.13768	.45242	1.1821	.15823
2.47	1.8159	2.5640	.06133	.13616	.45041	1.1833	.15724
2.48	1.8192	2.5880	.06038	.13465	.44841	1.1844	.15626
2.49	1.8225	2.6122	.05945	.13316	.44642	1.1856	.15528
2.50	1.8258	2.6367	.05853	.13169	.44444	1.1867	.15432
2.51	1.8290	2.6615	.05763	.13023	.44247	1.1879	.15337
2.52	1.8322	2.6865	.05674	.12879	.44051	1.1890	.15242
2.53	1.8354	2.7117	.05586	.12737	.43856	1.1901	.15148
2.54	1.8386	2.7372	.05500	.12597	.43662	1.1912	.15055
2.55	1.8417	2.7630	.05415	.12458	.43469	1.1923	.14963
2.56	1.8448	2.7891	.05332	.12321	.43277	1.1934	.14871
2.57	1.8479	2.8154	.05250	.12185	.43085	1.1945	.14780
2.58	1.8510	2.8420	.05169	.12051	.42894	1.1956	.14691
2.59	1.8541	2.8689	.05090	.11918	.42705	1.1967	.14601
2.60	1.8572	2.8960	.05012	.11787	.42517	1.1978	.14513
2.61	1.8602	2.9234	.04935	.11658	.42330	1.1989	.14426
2.62	1.8632	2.9511	.04859	.11530	.42143	1.2000	.14339
2.63	1.8662	2.9791	.04784	.11403	.41957	1.2011	.14253
2.64	1.8692	3.0074	.04711	.11278	.41772	1.2021	.14168
2.65	1.8721	3.0359	.04639	.11154	.41589	1.2031	.14083
2.66	1.8750	3.0647	.04568	.11032	.41406	1.2042	.13999
2.67	1.8779	3.0938	.04498	.10911	.41224	1.2052	.13916
2.68	1.8808	3.1233	.04429	.10792	.41043	1.2062	.13834
2.69	1.8837	3.1530	.04361	.10674	.40863	1.2073	.13752
2.70	1.8865	3.1830	.04295	.10557	.40684	1.2083	.13671
2.71	1.8894	3.2133	.04230	.10442	.40505	1.2093	.13591
2.72	1.8922	3.2440	.04166	.10328	.40327	1.2103	.13511
2.73	1.8950	3.2749	.04102	.10215	.40151	1.2113	.13432
2.74	1.8978	3.3061	.04039	.10104	.39976	1.2123	.13354
2.75	1.9005	3.3376	.03977	.09994	.39801	1.2133	.13276
2.76	1.9032	3.3695	.03917	.09885	.39627	1.2143	.13199
2.77	1.9060	3.4017	.03858	.09777	.39454	1.2153	.13123
2.78	1.9087	3.4342	.03800	.09671	.39282	1.2163	.13047
2.79	1.9114	3.4670	.03742	.09566	.39111	1.2173	.12972
2.80	1.9140	3.5001	.03685	.09462	.38941	1.2182	.12897
2.81	1.9167	3.5336	.03629	.09360	.38771	1.2192	.12823
2.82	1.9193	3.5674	.03574	.09259	.38603	1.2202	.12750
2.83	1.9220	3.6015	.03520	.09158	.38435	1.2211	.12678
2.84	1.9246	3.6359	.03467	.09059	.38268	1.2221	.12605
2.85	1.9271	3.6707	.03415	.08962	.38102	1.2230	.12534
2.86	1.9297	3.7058	.03363	.08865	.37937	1.2240	.12463
2.87	1.9322	3.7413	.03312	.08769	.37773	1.2249	.12393
2.88	1.9348	3.7771	.03262	.08674	.37610	1.2258	.12323
2.89	1.9373	3.8133	.03213	.08581	.37448	1.2268	.12254
2.90	1.9398	3.8498	.03165	.08489	.37286	1.2277	.12185
2.91	1.9423	3.8866	.03118	.08398	.37125	1.2286	.12117
2.92	1.9448	3.9238	.03071	.08308	.36965	1.2295	.12049
2.93	1.9472	3.9614	.03025	.08218	.36806	1.2304	.11982
2.94	1.9497	3.9993	.02980	.08130	.36648	1.2313	.11916
2.95	1.9521	4.0376	.02935	.08043	.36490	1.2322	.11850
2.96	1.9545	4.0763	.02891	.07957	.36333	1.2331	.11785
2.97	1.9569	4.1153	.02848	.07872	.36177	1.2340	.11720
2.98	1.9593	4.1547	.02805	.07788	.36022	1.2348	.11656
2.99	1.9616	4.1944	.02764	.07705	.35868	1.2357	.11591
3.00	1.9640	4.2346	.02722	.07623	.35714	1.2366	.11528
3.10	1.9866	4.6573	.02345	.06852	.34223	1.2450	.10921
3.20	2.0079	5.1210	.02023	.06165	.32808	1.2530	.10359
3.30	2.0279	5.6287	.01748	.05554	.31466	1.2605	.09837
3.40	2.0466	6.1837	.01512	.05009	.30193	1.2676	.09353
3.50	2.0642	6.7896	.01311	.04523	.28986	1.2743	.08902
3.60	2.0808	7.4501	.01138	.04089	.27840	1.2807	.08482
3.70	2.0964	8.1691	.00990	.03702	.26752	1.2867	.08090
3.80	2.1111	8.9506	.00863	.03355	.25720	1.2924	.07723
3.90	2.1250	9.7990	.00753	.03044	.24740	1.2978	.07380
4.00	2.1381	10.719	.00658	.02766	.23810	1.3029	.07059
4.10	2.1505	11.715	.00577	.02516	.22925	1.3077	.06758
4.20	2.1622	12.792	.00506	.02292	.22085	1.3123	.06475
4.30	2.1732	13.955	.00445	.02090	.21286	1.3167	.06209
4.40	2.1837	15.210	.00392	.01909	.20525	1.3208	.05959
4.50	2.1936	16.562	.00346	.01745	.19802	1.3247	.05723
4.60	2.2030	18.018	.00305	.01597	.19113	1.3284	.05500
4.70	2.2119	19.583	.00270	.01463	.18457	1.3320	.05289
4.80	2.2204	21.264	.00240	.01343	.17832	1.3354	.05091
4.90	2.2284	23.067	.00213	.01233	.17235	1.3386	.04904
5.00	2.2361	25.000	$189(10)^{-5}$	.01134	.16667	1.3416	.04725
6.00	2.2953	53.180	$633(10)^{-6}$	.00519	.12195	1.3655	.03368
7.00	2.3333	104.143	$242(10)^{-6}$	.00261	.09259	1.3810	.02516
8.00	2.3591	190.109	$102(10)^{-6}$	.00141	.07246	1.3915	.01947
9.00	2.3772	327.189	$474(10)^{-7}$	.000815	.05814	1.3989	.01550
10.00	2.3904	535.938	$236(10)^{-7}$	.000495	.04762	1.4044	.01263
∞	2.4495	∞	0	0	0	1.4289	0

Table 31   One-Dimensional Isentropic Compressible-Flow Functions

For a perfect gas with constant specific heat and molecular weight
k = 1.0

M	M*	$\dfrac{A}{A^*}$	$\dfrac{p}{p_0}=\dfrac{\rho}{\rho_0}$	$\dfrac{T}{T_0}$	$\dfrac{F}{F^*}$	$\left(\dfrac{A}{A^*}\right)\left(\dfrac{p}{p_0}\right)$	M	M*	$\dfrac{A}{A^*}$	$\dfrac{p}{p_0}=\dfrac{\rho}{\rho_0}$	$\dfrac{T}{T_0}$	$\dfrac{F}{F^*}$	$\left(\dfrac{A}{A^*}\right)\left(\dfrac{p}{p_0}\right)$
0	0	∞	1.0000	1.000	∞	∞	1.75	1.75	1.603	.2163	1.000	1.161	.3466
0.05	.05	12.146	.9989		10.025	12.136	1.80	1.80	1.703	.1979		1.178	.3370
.10	.10	6.096	.9951		5.050	6.065	1.85	1.85	1.815	.1806		1.195	.3279
.15	.15	4.089	.9888		3.408	4.044	1.90	1.90	1.941	.1645		1.213	.3192
.20	.20	3.094	.9802		2.600	3.033	1.95	1.95	2.082	.1494		1.231	.3110
.25	.25	2.503	.9693		2.125	2.426	2.00	2.00	2.241	.1353		1.250	.3033
.30	.30	2.115	.9561		1.817	2.022	2.05	2.05	2.419	.1223		1.269	.2959
.35	.35	1.842	.9406		1.604	1.733	2.10	2.10	2.620	.1102		1.288	.2888
.40	.40	1.643	.9231		1.450	1.516	2.15	2.15	2.846	.09914		1.307	.2821
.45	.45	1.491	.9037		1.336	1.348	2.20	2.20	3.100	.08892		1.327	.2757
.50	.50	1.375	.8825		1.250	1.2131	2.25	2.25	3.388	.07956		1.347	.2696
.55	.55	1.283	.8597		1.184	1.1028	2.30	2.30	3.714	.07100		1.367	.2637
.60	.60	1.210	.8353		1.133	1.0109	2.35	2.35	4.083	.06321		1.387	.2581
.65	.65	1.153	.8096		1.0942	.9331	2.40	2.40	4.502	.05614		1.408	.2527
.70	.70	1.107	.7827		1.0643	.8665	2.45	2.45	4.979	.04973		1.429	.2475
.75	.75	1.0714	.7549		1.0417	.8087	2.50	2.50	5.522	.04394		1.450	.2426
.80	.80	1.0441	.7262		1.0250	.7582	2.55	2.55	6.142	.03873		1.471	.2379
.85	.85	1.0240	.6968		1.0132	.7136	2.60	2.60	6.852	.03405		1.492	.2333
.90	.90	1.0104	.6670		1.0056	.6739	2.65	2.65	7.665	.02986		1.513	.2289
.95	.95	1.0025	.6369		1.0013	.6385	2.70	2.70	8.600	.02612		1.535	.2247
1.00	1.00	1.0000	.6065		1.0000	.6065	2.75	2.75	9.676	.02279		1.557	.2206
1.05	1.05	1.0025	.5762		1.0012	.5777	2.80	2.80	10.92	.01984		1.579	.2166
1.10	1.10	1.0097	.5461		1.0045	.5514	2.85	2.85	12.35	.01723		1.600	.2128
1.15	1.15	1.0217	.5163		1.0098	.5274	2.90	2.90	14.02	.01492		1.622	.2092
1.20	1.20	1.0384	.4868		1.0167	.5054	2.95	2.95	15.95	.01289		1.644	.2056
1.25	1.25	1.0598	.4578		1.0250	.4852	3.00	3.00	18.20	$1111(10)^{-5}$		1.667	.2022
1.30	1.30	1.0861	.4295		1.0346	.4666	3.50	3.50	79.22	$219(10)^{-5}$		1.893	.1733
1.35	1.35	1.117	.4020		1.0453	.4493	4.00	4.00	452.0	$335(10)^{-6}$		2.125	.1516
1.40	1.40	1.154	.3753		1.0571	.4332	4.50	4.50	3364	$401(10)^{-7}$		2.361	.1348
1.45	1.45	1.197	.3495		1.0698	.4183	5.00	5.00	32550	$373(10)^{-8}$		2.600	.1213
1.50	1.50	1.245	.3247		1.0833	.4044	6.00	6.00	$664(10)^{4}$	$152(10)^{-10}$		3.083	.1010
1.55	1.55	1.300	.3008		1.0976	.3913	7.00	7.00	$378(10)^{7}$	$229(10)^{-13}$		3.571	.08665
1.60	1.60	1.363	.2780		1.113	.3791	8.00	8.00	$599(10)^{10}$	$127(10)^{-16}$		4.062	.07578
1.65	1.65	1.434	.2563		1.128	.3676	9.00	9.00	$262(10)^{14}$	$258(10)^{-20}$		4.556	.06741
1.70	1.70	1.514	.2357		1.144	.3568	10.00	10.00	$314(10)^{18}$	$193(10)^{-24}$		5.050	.06065
							∞	∞	∞	0	1.000	∞	0

Table 32 One-Dimensional Isentropic Compressible-Flow Functions

For a perfect gas with constant specific heat and molecular weight
$k = 1.1$

M	M*	$\frac{A}{A^*}$	$\frac{p}{p_0}$	$\frac{\rho}{\rho_0}$	$\frac{T}{T_0}$	$\frac{F}{F^*}$	$\left(\frac{A}{A^*}\right)\left(\frac{p}{p_0}\right)$
0	0	∞	1.0000	1.0000	1.0000	∞	∞
0.05	0.05123	11.999	.9986	.9988	.9999	9.785	11.982
.10	.1024	6.023	.9945	.9950	.9995	4.931	5.990
.15	.1536	4.042	.9877	.9888	.9989	3.332	3.992
.20	.2047	3.059	.9783	.9802	.9980	2.545	2.993
.25	.2558	2.476	.9663	.9693	.9969	2.083	2.393
.30	.3067	2.094	.9518	.9561	.9955	1.784	1.993
.35	.3575	1.825	.9350	.9408	.9939	1.577	1.707
.40	.4082	1.628	.9161	.9234	.9921	1.429	1.492
.45	.4588	1.480	.8951	.9042	.9900	1.319	1.325
.50	.5092	1.365	.8723	.8832	.9877	1.237	1.1908
.55	.5594	1.275	.8478	.8606	.9851	1.174	1.0812
.60	.6094	1.204	.8218	.8366	.9823	1.125	.9897
.65	.6592	1.148	.7945	.8113	.9793	1.0882	.9121
.70	.7087	1.104	.7662	.7850	.9761	1.0599	.8456
.75	.7579	1.0689	.7370	.7578	.9727	1.0387	.7878
.80	.8069	1.0425	.7071	.7298	.9690	1.0231	.7372
.85	.8557	1.0231	.6768	.7013	.9651	1.0122	.6924
.90	.9041	1.0100	.6462	.6724	.9610	1.0051	.6526
.95	.9522	1.0024	.6154	.6431	.9568	1.0012	.6169
1.00	1.0000	1.0000	.5847	.6139	.9524	1.0000	.5847
1.05	1.0474	1.0023	.5542	.5847	.9478	1.0011	.5555
1.10	1.0945	1.0092	.5240	.5557	.9430	1.0041	.5289
1.15	1.141	1.0204	.4944	.5271	.9380	1.0087	.5046
1.20	1.188	1.0360	.4654	.4989	.9328	1.0148	.4822
1.25	1.234	1.0559	.4371	.4713	.9275	1.0221	.4616
1.30	1.279	1.0801	.4097	.4443	.9221	1.0305	.4425
1.35	1.324	1.1088	.3832	.4180	.9165	1.0397	.4249
1.40	1.369	1.1421	.3576	.3926	.9107	1.0497	.4084
1.45	1.413	1.1802	.3330	.3680	.9048	1.0604	.3930
1.50	1.457	1.223	.3095	.3443	.8989	1.0717	.3787
1.55	1.501	1.272	.2871	.3216	.8928	1.0836	.3652
1.60	1.544	1.326	.2658	.2999	.8865	1.0958	.3526
1.65	1.586	1.387	.2456	.2791	.8801	1.108	.3407
1.70	1.628	1.454	.2266	.2593	.8737	1.121	.3294

M	M*	$\frac{A}{A^*}$	$\frac{p}{p_0}$	$\frac{\rho}{\rho_0}$	$\frac{T}{T_0}$	$\frac{F}{F^*}$	$\left(\frac{A}{A^*}\right)\left(\frac{p}{p_0}\right)$
1.75	1.670	1.528	.2086	.2406	.8672	1.134	.3188
1.80	1.711	1.610	.1917	.2228	.8606	1.147	.3088
1.85	1.752	1.701	.1759	.2060	.8539	1.161	.2993
1.90	1.792	1.801	.1612	.1902	.8471	1.175	.2902
1.95	1.832	1.911	.1474	.1754	.8402	1.189	.2816
2.00	1.871	2.032	.1346	.1615	.8333	1.203	.2735
2.05	1.910	2.165	.1227	.1485	.8264	1.217	.2657
2.10	1.948	2.312	.1117	.1363	.8194	1.231	.2582
2.15	1.986	2.473	.1015	.1250	.8123	1.245	.2511
2.20	2.023	2.651	.09218	.1145	.8052	1.259	.2444
2.25	2.060	2.846	.08357	.10473	.7980	1.273	.2379
2.30	2.096	3.061	.07566	.09568	.7908	1.286	.2317
2.35	2.132	3.299	.06842	.08731	.7836	1.300	.2257
2.40	2.167	3.560	.06179	.07959	.7764	1.314	.2200
2.45	2.202	3.848	.05574	.07247	.7692	1.328	.2145
2.50	2.236	4.165	.05022	.06592	.7619	1.342	.2092
2.55	2.270	4.515	.04520	.05990	.7546	1.356	.2041
2.60	2.303	4.902	.04064	.05438	.7473	1.369	.1992
2.65	2.336	5.328	.03650	.04932	.7401	1.382	.1945
2.70	2.368	5.799	.03276	.04470	.7329	1.395	.1900
2.75	2.400	6.320	.02936	.04047	.7256	1.408	.1856
2.80	2.432	6.895	.02630	.03661	.7184	1.422	.1814
2.85	2.463	7.532	.02354	.03310	.7112	1.435	.1773
2.90	2.493	8.237	.02104	.02989	.7040	1.448	.1733
2.95	2.523	9.016	.01880	.02698	.6968	1.460	.1695
3.0	2.553	9.880	$1679(10)^{-5}$	.02434	.6897	1.472	.1658
3.5	2.824	25.83	$522(10)^{-5}$	.008414	.6202	1.589	.1348
4.0	3.055	71.75	$156(10)^{-5}$	.002801	.5556	1.691	.1116
4.5	3.250	205.8	$456(10)^{-6}$	.000918	.4969	1.779	.09385
5.0	3.416	597.7	$134(10)^{-6}$	.000301	.4444	1.854	.07988
6.0	3.674	4949	$121(10)^{-7}$	$338(10)^{-7}$	.3571	1.973	.05967
7.0	3.862	37976	$121(10)^{-8}$	$419(10)^{-8}$	.2899	2.060	.04608
8.0	4.000	$262(10)^3$	$139(10)^{-9}$	$585(10)^{-9}$	.2381	2.125	.03654
9.0	4.104	$161(10)^4$	$184(10)^{-10}$	$927(10)^{-10}$	.1980	2.174	.02962
10.0	4.183	$887(10)^4$	$275(10)^{-11}$	$165(10)^{-10}$	.1667	2.211	.02446
∞	4.583	∞	0	0	0	2.400	0

Table 33 One-Dimensional Isentropic Compressible-Flow Functions

For a perfect gas with constant specific heat and molecular weight
$k = 1.2$

M	M*	$\frac{A}{A^*}$	$\frac{p}{p_0}$	$\frac{\rho}{\rho_0}$	$\frac{T}{T_0}$	$\frac{F}{F^*}$	$\left(\frac{A}{A^*}\right)\left(\frac{p}{p_0}\right)$	M	M*	$\frac{A}{A^*}$	$\frac{p}{p_0}$	$\frac{\rho}{\rho_0}$	$\frac{T}{T_0}$	$\frac{F}{F^*}$	$\left(\frac{A}{A^*}\right)\left(\frac{p}{p_0}\right)$
0	0	∞	1.0000	1.0000	1.0000	∞	∞	1.75	1.606	1.470	.2013	.2629	.7656	1.114	.2960
0.05	0.05243	11.857	.9985	.9988	.9998	9.562	11.839	1.80	1.641	1.539	.1856	.2458	.7553	1.125	.2858
.10	.1048	5.953	.9940	.9950	.9991	4.822	5.917	1.85	1.675	1.615	.1710	.2295	.7450	1.136	.2762
.15	.1571	3.996	.9866	.9888	.9978	3.260	3.942	1.90	1.708	1.698	.1573	.2141	.7347	1.147	.2671
.20	.2093	3.026	.9763	.9802	.9960	2.493	2.954	1.95	1.741	1.787	.1446	.1996	.7245	1.158	.2584
.25	.2614	2.451	.9633	.9693	.9938	2.044	2.361	2.00	1.773	1.884	.1328	.1859	.7143	1.168	.2502
.30	.3133	2.073	.9477	.9562	.9911	1.753	1.965	2.05	1.804	1.989	.1218	.1730	.7041	1.179	.2423
.35	.3649	1.809	.9296	.9409	.9879	1.553	1.681	2.10	1.835	2.103	.1117	.1609	.6940	1.190	.2348
.40	.4162	1.615	.9092	.9236	.9843	1.409	1.468	2.15	1.865	2.226	.1023	.1496	.6839	1.201	.2277
.45	.4672	1.469	.8867	.9046	.9802	1.304	1.302	2.20	1.894	2.359	.09362	.1390	.6739	1.212	.2209
.50	.5179	1.356	.8623	.8839	.9756	1.224	1.170	2.25	1.923	2.504	.08563	.1290	.6639	1.222	.2144
.55	.5683	1.268	.8363	.8616	.9706	1.164	1.0606	2.30	1.951	2.660	.07826	.1197	.6540	1.232	.2082
.60	.6183	1.199	.8088	.8379	.9653	1.118	.9694	2.35	1.978	2.829	.07148	.1110	.6442	1.242	.2022
.65	.6678	1.144	.7801	.8131	.9595	1.0826	.8922	2.40	2.005	3.011	.06526	.1029	.6345	1.252	.1965
.70	.7168	1.100	.7505	.7873	.9533	1.0559	.8258	2.45	2.031	3.208	.05955	.09529	.6249	1.262	.1910
75	.7654	1.0666	.7201	.7606	.9467	1.0360	.7681	2.50	2.057	3.421	.05431	.08825	.6154	1.272	.1858
.80	.8134	1.0410	.6892	.7333	.9398	1.0214	.7174	2.55	2.082	3.650	.04951	.08170	.6060	1.282	.1808
.85	.8609	1.0222	.6580	.7055	.9326	1.0112	.6726	2.60	2.106	3.898	.04512	.07562	.5967	1.291	.1759
.90	.9078	1.0096	.6267	.6774	.9251	1.0047	.6327	2.65	2.130	4.166	.04110	.06997	.5875	1.300	.1712
.95	.9542	1.0023	.5954	.6492	.9172	1.0011	.5968	2.70	2.154	4.455	.03743	.06472	.5784	1.309	.1667
1.00	1.0000	1.0000	.5644	.6209	.9091	1.0000	.5644	2.75	2.177	4.767	.03408	.05985	.5694	1.318	.1624
1.05	1.0451	1.0022	.5339	.5928	.9007	1.0010	.5351	2.80	2.199	5.103	.03102	.05534	.5605	1.327	.1583
1.10	1.0896	1.0087	.5039	.5649	.8921	1.0037	.5083	2.85	2.220	5.466	.02823	.05116	.5518	1.335	.1543
1.15	1.134	1.0194	.4746	.5374	.8832	1.0079	.4838	2.90	2.241	5.858	.02569	.04729	.5432	1.343	.1505
1.20	1.177	1.0340	.4461	.5104	.8741	1.0133	.4613	2.95	2.262	6.280	.02337	.04370	.5347	1.352	.1467
1.25	1.219	1.0525	.4185	.4839	.8648	1.0197	.4405	3.00	2.283	6.735	.02126	.04039	.5263	1.360	.1432
1.30	1.261	1.0749	.3918	.4581	.8554	1.0270	.4212	3.5	2.461	13.76	.008242	.01834	.4494	1.434	.1134
1.35	1.302	1.101	.3662	.4330	.8458	1.0350	.4033	4.0	2.602	28.35	.003237	.008417	.3846	1.493	.09179
1.40	1.342	1.131	.3417	.4087	.8361	1.0437	.3867	4.5	2.714	57.96	.001305	.003948	.3306	1.541	.07564
1.45	1.382	1.166	.3182	.3852	.8263	1.0529	.3712	5.0	2.803	116.3	.000544	.001904	.2857	1.580	.06329
1.50	1.421	1.205	.2959	.3625	.8163	1.0625	.3566	6.0	2.934	435.9	$106(10)^{-6}$	$486(10)^{-6}$	.2174	1.637	.04601
1.55	1.459	1.248	.2747	.3407	.8063	1.0724	.3430	7.0	3.023	1469	$237(10)^{-7}$	$140(10)^{-6}$	.1695	1.677	.03482
1.60	1.497	1.296	.2547	.3199	.7962	1.0826	.3302	8.0	3.084	4467	$609(10)^{-8}$	$451(10)^{-7}$	.1351	1.704	.02720
1.65	1.534	1.349	.2358	.3000	.7860	1.0930	.3181	9.0	3.129	12383	$176(10)^{-8}$	$160(10)^{-7}$	.1099	1.724	.02181
1.70	1.570	1.407	.2180	.2810	.7758	1.1036	.3067	10.0	3.162	31623	$564(10)^{-9}$	$621(10)^{-8}$	.09091	1.739	.01785
								∞	3.317	∞	0	0	0	1.809	0

Table 34   One-Dimensional Isentropic Compressible-Flow Functions

For a perfect gas with constant specific heat and molecular weight
k = 1.3

M	M*	$\frac{A}{A^*}$	$\frac{p}{p_0}$	$\frac{\rho}{\rho_0}$	$\frac{T}{T_0}$	$\frac{F}{F^*}$	$\left(\frac{A}{A^*}\right)\left(\frac{p}{p_0}\right)$
0	0	∞	1.0000	1.0000	1.0000	∞	∞
0.05	0.0536	11.721	.9984	.9988	.9996	9.354	11.702
.10	.1072	5.885	.9936	.9951	.9985	4.720	5.848
.15	.1606	3.952	.9855	.9889	.9966	3.194	3.895
.20	.2138	2.994	.9744	.9803	.9940	2.445	2.917
.25	.2668	2.426	.9603	.9694	.9907	2.007	2.330
.30	.3195	2.054	.9435	.9563	.9867	1.724	1.938
.35	.3719	1.793	.9241	.9411	.9820	1.530	1.657
.40	.4239	1.602	.9023	.9240	.9766	1.391	1.446
.45	.4754	1.459	.8784	.9051	.9705	1.289	1.281
.50	.5264	1.348	.8526	.8845	.9638	1.213	1.1491
.55	.5769	1.261	.8251	.8625	.9566	1.155	1.0407
.60	.6267	1.193	.7962	.8392	.9488	1.111	.9501
.65	.6759	1.139	.7662	.8148	.9404	1.0777	.8731
.70	.7245	1.0972	.7354	.7895	.9315	1.0524	.8069
.75	.7724	1.0644	.7040	.7634	.9222	1.0336	.7493
.80	.8195	1.0395	.6723	.7367	.9124	1.0199	.6988
.85	.8658	1.0214	.6403	.7096	.9022	1.0104	.6540
.90	.9113	1.0092	.6084	.6823	.8917	1.0043	.6140
.95	.9561	1.0022	.5768	.6549	.8808	1.0010	.5781
1.00	1.0000	1.0000	.5457	.6276	.8696	1.0000	.5457
1.05	1.0430	1.0021	.5152	.6004	.8581	1.0009	.5163
1.10	1.0852	1.0083	.4854	.5735	.8464	1.0034	.4895
1.15	1.127	1.0183	.4565	.5470	.8345	1.0072	.4649
1.20	1.167	1.0321	.4285	.5210	.8224	1.0120	.4423
1.25	1.206	1.0495	.4015	.4956	.8102	1.0177	.4214
1.30	1.245	1.0704	.3756	.4709	.7978	1.0241	.4021
1.35	1.283	1.0948	.3509	.4468	.7853	1.0312	.3842
1.40	1.320	1.123	.3273	.4235	.7728	1.0388	.3675
1.45	1.356	1.154	.3049	.4010	.7603	1.0467	.3519
1.50	1.391	1.189	.2836	.3793	.7477	1.0549	.3374
1.55	1.425	1.228	.2635	.3585	.7351	1.0634	.3237
1.60	1.458	1.271	.2446	.3385	.7225	1.0720	.3109
1.65	1.491	1.318	.2268	.3194	.7100	1.0808	.2989
1.70	1.523	1.369	.2101	.3011	.6976	1.0897	.2875
1.75	1.554	1.424	.1944	.2836	.6852	1.0986	.2768
1.80	1.584	1.484	.1797	.2670	.6729	1.108	.2667
1.85	1.613	1.549	.1660	.2513	.6607	1.116	.2571
1.90	1.641	1.618	.1533	.2364	.6487	1.125	.2481
1.95	1.669	1.693	.1415	.2222	.6368	1.134	.2395
2.00	1.696	1.773	.1305	.2087	.6250	1.143	.2313
2.05	1.722	1.859	.1203	.1960	.6134	1.152	.2236
2.10	1.747	1.951	.1108	.1841	.6019	1.160	.2162
2.15	1.772	2.050	.1020	.1728	.5905	1.168	.2092
2.20	1.796	2.156	.0939	.1621	.5793	1.176	.2025
2.25	1.819	2.268	.08645	.1521	.5684	1.184	.1961
2.30	1.842	2.388	.07955	.1427	.5576	1.192	.1900
2.35	1.864	2.517	.07318	.1338	.5470	1.200	.1842
2.40	1.885	2.654	.06731	.1254	.5365	1.208	.1786
2.45	1.906	2.799	.06190	.1176	.5262	1.216	.1733
2.50	1.926	2.954	.05692	.1103	.5161	1.223	.1682
2.55	1.946	3.119	.05234	.1034	.5062	1.230	.1633
2.60	1.965	3.295	.04813	.09693	.4965	1.237	.1586
2.65	1.983	3.482	.04426	.09087	.4870	1.244	.1541
2.70	2.001	3.681	.04070	.08520	.4777	1.250	.1498
2.75	2.019	3.892	.03743	.07988	.4686	1.257	.1457
2.80	2.036	4.116	.03442	.07490	.4596	1.264	.1417
2.85	2.052	4.354	.03166	.07024	.4508	1.270	.1379
2.90	2.068	4.607	.02913	.06587	.4422	1.276	.1342
2.95	2.084	4.875	.02680	.06178	.4338	1.282	.1307
3.0	2.099	5.160	.02466	.05796	.4255	1.288	.12725
3.5	2.228	9.110	.01090	.03092	.3524	1.338	.09926
4.0	2.326	15.94	.00498	.01692	.2941	1.378	.07935
4.5	2.402	27.39	.00236	.00954	.2477	1.409	.06472
5.0	2.460	45.96	.00117	.00555	.2105	1.433	.05370
6.0	2.543	120.1	$321(10)^{-6}$	.00206	.15625	1.468	.03856
7.0	2.598	285.3	$101(10)^{-6}$	$847(10)^{-6}$	.11976	1.491	.02893
8.0	2.635	623.1	$361(10)^{-7}$	$382(10)^{-6}$	.09434	1.507	.02247
9.0	2.662	1266	$142(10)^{-7}$	$186(10)^{-6}$	.07605	1.519	.01793
10.0	2.681	2416	$606(10)^{-8}$	$969(10)^{-7}$	.06250	1.527	.01463
∞	2.769	∞	0	0	0	1.565	0

Table 35 One-Dimensional Isentropic Compressible-Flow Functions

For a perfect gas with constant specific heat and molecular weight
k = 1.67

M	M*	$\frac{A}{A^*}$	$\frac{p}{p_0}$	$\frac{\rho}{\rho_0}$	$\frac{T}{T_0}$	$\frac{F}{F^*}$	$\left(\frac{A}{A^*}\right)\left(\frac{p}{p_0}\right)$
0	0	∞	1.0000	1.0000	1.0000	0	∞
0.05	0.05775	11.265	.9979	.9988	.9992	8.687	11.242
.10	.1154	5.661	.9917	.9950	.9967	4.392	5.614
.15	.1727	3.805	.9815	.9888	.9925	2.982	3.735
.20	.2296	2.887	.9674	.9803	.9868	2.293	2.793
.25	.2859	2.344	.9497	.9695	.9795	1.892	2.226
.30	.3415	1.989	.9286	.9566	.9708	1.635	1.847
.35	.3963	1.741	.9046	.9417	.9606	1.460	1.575
.40	.4502	1.560	.8780	.9250	.9491	1.336	1.370
.45	.5031	1.424	.8491	.9067	.9364	1.245	1.209
.50	.5549	1.320	.8184	.8869	.9227	1.178	1.0803
.55	.6055	1.239	.7862	.8658	.9080	1.128	.9742
.60	.6548	1.176	.7529	.8437	.8924	1.0909	.8853
.65	.7029	1.126	.7190	.8207	.8760	1.0628	.8097
.70	.7496	1.0874	.6847	.7970	.8590	1.0422	.7445
.75	.7949	1.0576	.6503	.7728	.8414	1.0265	.6877
.80	.8388	1.0351	.6162	.7483	.8234	1.0155	.6378
.85	.8812	1.0189	.5826	.7236	.8051	1.0080	.5936
.90	.9222	1.0080	.5497	.6988	.7866	1.0033	.5541
.95	.9618	1.0019	.5177	.6742	.7679	1.0008	.5187
1.00	1.0000	1.0000	.4867	.6497	.7491	1.0000	.4867
1.05	1.0368	1.0018	.4568	.6255	.7303	1.0007	.4576
1.10	1.0721	1.0071	.4282	.6017	.7116	1.0024	.4312
1.15	1.106	1.0154	.4009	.5784	.6930	1.0051	.4070
1.20	1.139	1.0266	.3749	.5557	.6746	1.0085	.3849
1.25	1.170	1.0406	.3502	.5335	.6564	1.0124	.3646
1.30	1.200	1.0573	.3269	.5119	.6385	1.0167	.3457
1.35	1.229	1.0765	.3049	.4910	.6209	1.0213	.3282
1.40	1.257	1.0981	.2842	.4707	.6036	1.0262	.3121
1.45	1.283	1.122	.2647	.4511	.5867	1.0313	.2972
1.50	1.309	1.148	.2465	.4323	.5702	1.0364	.2830
1.55	1.333	1.176	.2295	.4142	.5541	1.0416	.2700
1.60	1.356	1.207	.2136	.3968	.5383	1.0468	.2579
1.65	1.379	1.240	.1988	.3801	.5230	1.0520	.2465
1.70	1.400	1.275	.1850	.3640	.5081	1.0572	.2358
1.75	1.420	1.312	.1721	.3486	.4936	1.0623	.2257
1.80	1.440	1.351	.1601	.3339	.4795	1.0673	.2163
1.85	1.459	1.392	.1490	.3198	.4658	1.0722	.2075
1.90	1.477	1.436	.1386	.3063	.4526	1.0770	.1991
1.95	1.494	1.482	.1290	.2934	.4398	1.0817	.1912
2.00	1.511	1.530	.1201	.2811	.4274	1.0863	.1838
2.05	1.527	1.580	.1119	.2694	.4153	1.0908	.1768
2.10	1.542	1.632	.1042	.2582	.4036	1.0952	.1701
2.15	1.556	1.687	.09712	.2475	.3923	1.0994	.1638
2.20	1.570	1.744	.09053	.2373	.3814	1.1035	.1579
2.25	1.583	1.803	.08442	.2276	.3709	1.107	.1522
2.30	1.596	1.865	.07875	.2183	.3607	1.111	.1468
2.35	1.608	1.929	.07349	.2094	.3508	1.115	.1417
2.40	1.620	1.995	.06862	.2010	.3413	1.119	.1369
2.45	1.631	2.064	.06410	.1930	.3321	1.123	.1323
2.50	1.642	2.135	.05990	.1853	.3232	1.126	.1279
2.55	1.653	2.209	.05601	.1780	.3146	1.129	.1237
2.60	1.663	2.285	.05239	.1710	.3063	1.132	.1197
2.65	1.673	2.364	.04903	.1644	.2983	1.135	.1159
2.70	1.682	2.445	.04591	.1581	.2905	1.138	.1123
2.75	1.691	2.529	.04301	.1520	.2830	1.141	.1088
2.80	1.699	2.616	.04032	.1462	.2757	1.144	.1055
2.85	1.707	2.705	.03781	.1407	.2687	1.146	.1023
2.90	1.715	2.797	.03547	.1354	.2620	1.149	.09924
2.95	1.723	2.892	.03330	.1304	.2554	1.152	.09633
3.0	1.730	2.990	.03128	.12560	.2491	1.154	.09354
3.5	1.790	4.134	.01720	.08779	.1959	1.174	.07111
4.0	1.833	5.608	.009939	.06321	.1572	1.189	.05574
4.5	1.864	7.456	.006007	.04676	.1285	1.200	.04479
5.0	1.887	9.721	.003779	.03542	.1067	1.208	.03673
6.0	1.918	15.68	$165(10)^{-5}$	.02160	.07657	1.220	.02593
7.0	1.938	23.85	$807(10)^{-6}$	.01406	.05742	1.227	.01925
8.0	1.951	34.58	$429(10)^{-6}$	.00963	.04456	1.232	.01484
9.0	1.960	48.24	$244(10)^{-6}$	.00687	.03554	1.235	.01178
10.0	1.967	65.18	$147(10)^{-6}$	.00507	.02898	1.238	.00958
∞	1.996	∞	0	0	0	1.249	0

Table 36   Rayleigh Line—One-Dimensional Compressible-Flow Functions for Stagnation-Temperature Change in the Absence of Friction and Area Change

For a perfect gas with constant specific heat and molecular weight
k = 1.4

M	$\dfrac{T_0}{T_0*}$	$\dfrac{T}{T*}$	$\dfrac{p}{p*}$	$\dfrac{p_0}{p_0*}$	$\dfrac{V}{V*}$	M	$\dfrac{T_0}{T_0*}$	$\dfrac{T}{T*}$	$\dfrac{p}{p*}$	$\dfrac{p_0}{p_0*}$	$\dfrac{V}{V*}$
0	0	0	2.4000	1.2679	0	0.40	.52903	.61515	1.9608	1.1566	.31372
0.01	.000480	.000576	2.3997	1.2678	.000240	.41	.54651	.63448	1.9428	1.1523	.32658
.02	.00192	.00230	2.3987	1.2675	.000959	.42	.56376	.65345	1.9247	1.1480	.33951
.03	.00431	.00516	2.3970	1.2671	.00216	.43	.58075	.67205	1.9065	1.1437	.35251
.04	.00765	.00917	2.3946	1.2665	.00383	.44	.59748	.69025	1.8882	1.1394	.36556
.05	.01192	.01430	2.3916	1.2657	.00598	.45	.61393	.70803	1.8699	1.1351	.37865
.06	.01712	.02053	2.3880	1.2647	.00860	.46	.63007	.72538	1.8515	1.1308	.39178
.07	.02322	.02784	2.3837	1.2636	.01168	.47	.64589	.74228	1.8331	1.1266	.40493
.08	.03021	.03621	2.3787	1.2623	.01522	.48	.66139	.75871	1.8147	1.1224	.41810
.09	.03807	.04562	2.3731	1.2608	.01922	.49	.67655	.77466	1.7962	1.1182	.43127
.10	.04678	.05602	2.3669	1.2591	.02367	.50	.69136	.79012	1.7778	1.1140	.44445
.11	.05630	.06739	2.3600	1.2573	.02856	.51	.70581	.80509	1.7594	1.1099	.45761
.12	.06661	.07970	2.3526	1.2554	.03388	.52	.71990	.81955	1.7410	1.1059	.47075
.13	.07768	.09290	2.3445	1.2533	.03962	.53	.73361	.83351	1.7226	1.1019	.48387
.14	.08947	.10695	2.3359	1.2510	.04578	.54	.74695	.84695	1.7043	1.0979	.49696
.15	.10196	.12181	2.3267	1.2486	.05235	.55	.75991	.85987	1.6860	1.09397	.51001
.16	.11511	.13743	2.3170	1.2461	.05931	.56	.77248	.87227	1.6678	1.09010	.52302
.17	.12888	.15377	2.3067	1.2434	.06666	.57	.78467	.88415	1.6496	1.08630	.53597
.18	.14324	.17078	2.2959	1.2406	.07438	.58	.79647	.89552	1.6316	1.08255	.54887
.19	.15814	.18841	2.2845	1.2377	.08247	.59	.80789	.90637	1.6136	1.07887	.56170
.20	.17355	.20661	2.2727	1.2346	.09091	.60	.81892	.91670	1.5957	1.07525	.57447
.21	.18943	.22533	2.2604	1.2314	.09969	.61	.82956	.92653	1.5780	1.07170	.58716
.22	.20574	.24452	2.2477	1.2281	.10879	.62	.83982	.93585	1.5603	1.06821	.59978
.23	.22244	.26413	2.2345	1.2248	.11820	.63	.84970	.94466	1.5427	1.06480	.61232
.24	.23948	.28411	2.2209	1.2213	.12792	.64	.85920	.95298	1.5253	1.06146	.62477
.25	.25684	.30440	2.2069	1.2177	.13793	.65	.86833	.96081	1.5080	1.05820	.63713
.26	.27446	.32496	2.1925	1.2140	.14821	.66	.87709	.96816	1.4908	1.05502	.64941
.27	.29231	.34573	2.1777	1.2102	.15876	.67	.88548	.97503	1.4738	1.05192	.66159
.28	.31035	.36667	2.1626	1.2064	.16955	.68	.89350	.98144	1.4569	1.04890	.67367
.29	.32855	.38773	2.1472	1.2025	.18058	.69	.90117	.98739	1.4401	1.04596	.68564
.30	.34686	.40887	2.1314	1.1985	.19183	.70	.90850	.99289	1.4235	1.04310	.69751
.31	.36525	.43004	2.1154	1.1945	.20329	.71	.91548	.99796	1.4070	1.04033	.70927
.32	.38369	.45119	2.0991	1.1904	.21494	.72	.92212	1.00260	1.3907	1.03764	.72093
.33	.40214	.47228	2.0825	1.1863	.22678	.73	.92843	1.00682	1.3745	1.03504	.73248
.34	.42057	.49327	2.0657	1.1821	.23879	.74	.93442	1.01062	1.3585	1.03253	.74392
.35	.43894	.51413	2.0487	1.1779	.25096	.75	.94009	1.01403	1.3427	1.03010	.75525
.36	.45723	.53482	2.0314	1.1737	.26327	.76	.94546	1.01706	1.3270	1.02776	.76646
.37	.47541	.55530	2.0140	1.1695	.27572	.77	.95052	1.01971	1.3115	1.02552	.77755
.38	.49346	.57553	1.9964	1.1652	.28828	.78	.95528	1.02198	1.2961	1.02337	.78852
.39	.51134	.59549	1.9787	1.1609	.30095	.79	.95975	1.02390	1.2809	1.02131	.79938

See page 121 for definitions of symbols.

Table 36   Rayleigh Line—One-Dimensional Compressible-Flow Functions for Stagnation-Temperature Change in the Absence of Friction and Area Change

For a perfect gas with constant specific heat and molecular weight
k = 1.4

M	$\dfrac{T_0}{T_0*}$	$\dfrac{T}{T*}$	$\dfrac{p}{p*}$	$\dfrac{p_0}{p_0*}$	$\dfrac{V}{V*}$	M	$\dfrac{T_0}{T_0*}$	$\dfrac{T}{T*}$	$\dfrac{p}{p*}$	$\dfrac{p_0}{p_0*}$	$\dfrac{V}{V*}$
0.80	.96394	1.02548	1.2658	1.01934	.81012	1.20	.97872	.91185	.79576	1.01941	1.1459
.81	.96786	1.02672	1.2509	1.01746	.82075	1.21	.97685	.90671	.78695	1.02140	1.1522
.82	.97152	1.02763	1.2362	1.01569	.83126	1.22	.97492	.90153	.77827	1.02348	1.1584
.83	.97492	1.02823	1.2217	1.01399	.84164	1.23	.97294	.89632	.76971	1.02566	1.1645
.84	.97807	1.02853	1.2073	1.01240	.85190	1.24	.97092	.89108	.76127	1.02794	1.1705
.85	.98097	1.02854	1.1931	1.01091	.86204	1.25	.96886	.88581	.75294	1.03032	1.1764
.86	.98363	1.02826	1.1791	1.00951	.87206	1.26	.96675	.88052	.74473	1.03280	1.1823
.87	.98607	1.02771	1.1652	1.00819	.88196	1.27	.96461	.87521	.73663	1.03536	1.1881
.88	.98828	1.02690	1.1515	1.00698	.89175	1.28	.96243	.86988	.72865	1.03803	1.1938
.89	.99028	1.02583	1.1380	1.00587	.90142	1.29	.96022	.86453	.72078	1.04080	1.1994
.90	.99207	1.02451	1.1246	1.04485	.91097	1.30	.95798	.85917	.71301	1.04365	1.2050
.91	.99366	1.02297	1.1114	1.00393	.92039	1.31	.95571	.85380	.70535	1.04661	1.2105
.92	.99506	1.02120	1.09842	1.00310	.92970	1.32	.95341	.84843	.69780	1.04967	1.2159
.93	.99627	1.01921	1.08555	1.00237	.93889	1.33	.95108	.84305	.69035	1.05283	1.2212
.94	.99729	1.01702	1.07285	1.00174	.94796	1.34	.94873	.83766	.68301	1.05608	1.2264
.95	.99814	1.01463	1.06030	1.00121	.95692	1.35	.94636	.83227	.67577	1.05943	1.2316
.96	.99883	1.01205	1.04792	1.00077	.96576	1.36	.94397	.82698	.66863	1.06288	1.2367
.97	.99935	1.00929	1.03570	1.00043	.97449	1.37	.94157	.82151	.66159	1.06642	1.2417
.98	.99972	1.00636	1.02364	1.00019	.98311	1.38	.93915	.81613	.65464	1.07006	1.2467
.99	.99993	1.00326	1.01174	1.00004	.99161	1.39	.93671	.81076	.64778	1.07380	1.2516
1.00	1.00000	1.00000	1.00000	1.00000	1.00000	1.40	.93425	.80540	.64102	1.07765	1.2564
1.01	.99993	.99659	.98841	1.00004	1.00828	1.41	.93178	.80004	.63436	1.08159	1.2612
1.02	.99973	.99304	.97697	1.00019	1.01644	1.42	.92931	.79469	.62779	1.08563	1.2659
1.03	.99940	.98936	.96569	1.00043	1.02450	1.43	.92683	.78936	.62131	1.08977	1.2705
1.04	.99895	.98553	.95456	1.00077	1.03246	1.44	.92434	.78405	.61491	1.09400	1.2751
1.05	.99838	.98161	.94358	1.00121	1.04030	1.45	.92184	.77875	.60860	1.0983	1.2796
1.06	.99769	.97755	.93275	1.00175	1.04804	1.46	.91933	.77346	.60237	1.1028	1.2840
1.07	.99690	.97339	.92206	1.00238	1.05567	1.47	.91682	.76819	.59623	1.1073	1.2884
1.08	.99600	.96913	.91152	1.00311	1.06320	1.48	.91431	.76294	.59018	1.1120	1.2927
1.09	.99501	.96477	.90112	1.00394	1.07062	1.49	.91179	.75771	.58421	1.1167	1.2970
1.10	.99392	.96031	.89086	1.00486	1.07795	1.50	.90928	.75250	.57831	1.1215	1.3012
1.11	.99274	.95577	.88075	1.00588	1.08518	1.51	.90676	.74731	.57250	1.1264	1.3054
1.12	.99148	.95115	.87078	1.00699	1.09230	1.52	.90424	.74215	.56677	1.1315	1.3095
1.13	.99013	.94646	.86094	1.00820	1.09933	1.53	.90172	.73701	.56111	1.1367	1.3135
1.14	.98871	.94169	.85123	1.00951	1.10626	1.54	.89920	.73189	.55553	1.1420	1.3175
1.15	.98721	.93685	.84166	1.01092	1.1131	1.55	.89669	.72680	.55002	1.1473	1.3214
1.16	.98564	.93195	.83222	1.01243	1.1198	1.56	.89418	.72173	.54458	1.1527	1.3253
1.17	.98400	.92700	.82292	1.01403	1.1264	1.57	.89167	.71669	.53922	1.1582	1.3291
1.18	.98230	.92200	.81374	1.01572	1.1330	1.58	.88917	.71168	.53393	1.1639	1.3329
1.19	.98054	.91695	.80468	1.01752	1.1395	1.59	.88668	.70669	.52871	1.1697	1.3366

Table 36 Rayleigh Line—One-Dimensional Compressible-Flow Functions for Stagnation-Temperature Change in the Absence of Friction and Area Change

For a perfect gas with constant specific heat and molecular weight
k = 1.4

M	$\dfrac{T_0}{T_0*}$	$\dfrac{T}{T*}$	$\dfrac{p}{p*}$	$\dfrac{p_0}{p_0*}$	$\dfrac{V}{V*}$	M	$\dfrac{T_0}{T_0*}$	$\dfrac{T}{T*}$	$\dfrac{p}{p*}$	$\dfrac{p_0}{p_0*}$	$\dfrac{V}{V*}$
1.60	.88419	.70173	.52356	1.1756	1.3403	2.00	.79339	.52893	.36364	1.5031	1.4545
1.61	.88170	.69680	.51848	1.1816	1.3439	2.01	.79139	.52526	.36057	1.5138	1.4567
1.62	.87922	.69190	.51346	1.1877	1.3475	2.02	.78941	.52161	.35754	1.5246	1.4589
1.63	.87675	.68703	.50851	1.1939	1.3511	2.03	.78744	.51800	.35454	1.5356	1.4610
1.64	.87429	.68219	.50363	1.2002	1.3546	2.04	.78549	.51442	.35158	1.5467	1.4631
1.65	.87184	.67738	.49881	1.2066	1.3580	2.05	.78355	.51087	.34866	1.5579	1.4652
1.66	.86940	.67259	.49405	1.2131	1.3614	2.06	.78162	.50735	.34577	1.5693	1.4673
1.67	.86696	.66784	.48935	1.2197	1.3648	2.07	.77971	.50386	.34291	1.5808	1.4694
1.68	.86453	.66312	.48471	1.2264	1.3681	2.08	.77781	.50040	.34009	1.5924	1.4714
1.69	.86211	.65843	.48014	1.2332	1.3713	2.09	.77593	.49697	.33730	1.6042	1.4734
1.70	.85970	.65377	.47563	1.2402	1.3745	2.10	.77406	.49356	.33454	1.6161	1.4753
1.71	.85731	.64914	.47117	1.2473	1.3777	2.11	.77221	.49018	.33181	1.6282	1.4773
1.72	.85493	.64455	.46677	1.2545	1.3809	2.12	.77037	.48683	.32912	1.6404	1.4792
1.73	.85256	.63999	.46242	1.2618	1.3840	2.13	.76854	.48351	.32646	1.6528	1.4811
1.74	.85020	.63546	.45813	1.2692	1.3871	2.14	.76673	.48022	.32383	1.6653	1.4830
1.75	.84785	.63096	.45390	1.2767	1.3901	2.15	.76493	.47696	.32122	1.6780	1.4849
1.76	.84551	.62649	.44972	1.2843	1.3931	2.16	.76314	.47373	.31864	1.6908	1.4867
1.77	.84318	.62205	.44559	1.2920	1.3960	2.17	.76137	.47052	.31610	1.7037	1.4885
1.78	.84087	.61765	.44152	1.2998	1.3989	2.18	.75961	.46734	.31359	1.7168	1.4903
1.79	.83857	.61328	.43750	1.3078	1.4018	2.19	.75787	.46419	.31110	1.7300	1.4921
1.80	.83628	.60894	.43353	1.3159	1.4046	2.20	.75614	.46106	.30864	1.7434	1.4939
1.81	.83400	.60463	.42960	1.3241	1.4074	2.21	.75442	.45796	.30621	1.7570	1.4956
1.82	.83174	.60036	.42573	1.3324	1.4102	2.22	.75271	.45489	.30381	1.7707	1.4973
1.83	.82949	.59612	.42191	1.3408	1.4129	2.23	.75102	.45184	.30143	1.7846	1.4990
1.84	.82726	.59191	.41813	1.3494	1.4156	2.24	.74934	.44882	.29908	1.7986	1.5007
1.85	.82504	.58773	.41440	1.3581	1.4183	2.25	.74767	.44582	.29675	1.8128	1.5024
1.86	.82283	.58359	.41072	1.3669	1.4209	2.26	.74602	.44285	.29445	1.8271	1.5040
1.87	.82064	.57948	.40708	1.3758	1.4235	2.27	.74438	.43990	.29218	1.8416	1.5056
1.88	.81846	.57540	.40349	1.3848	1.4261	2.28	.74275	.43698	.28993	1.8562	1.5072
1.89	.81629	.57135	.39994	1.3940	1.4286	2.29	.74114	.43409	.28771	1.8710	1.5088
1.90	.81414	.56734	.39643	1.4033	1.4311	2.30	.73954	.43122	.28551	1.8860	1.5104
1.91	.81200	.56336	.39297	1.4127	1.4336	2.31	.73795	.42837	.28333	1.9012	1.5119
1.92	.80987	.55941	.38955	1.4222	1.4360	2.32	.73638	.42555	.28118	1.9165	1.5134
1.93	.80776	.55549	.38617	1.4319	1.4384	2.33	.73482	.42276	.27905	1.9320	1.5150
1.94	.80567	.55160	.38283	1.4417	1.4408	2.34	.73327	.41999	.27695	1.9476	1.5165
1.95	.80359	.54774	.37954	1.4516	1.4432	2.35	.73173	.41724	.27487	1.9634	1.5180
1.96	.80152	.54391	.37628	1.4616	1.4455	2.36	.73020	.41451	.27281	1.9794	1.5195
1.97	.79946	.54012	.37306	1.4718	1.4478	2.37	.72868	.41181	.27077	1.9955	1.5209
1.98	.79742	.53636	.36988	1.4821	1.4501	2.38	.72718	.40913	.26875	2.0118	1.5223
1.99	.79540	.53263	.36674	1.4925	1.4523	2.39	.72569	.40647	.26675	2.0283	1.5237

Table 36 Rayleigh Line—One-Dimensional Compressible-Flow Functions for Stagnation-Temperature Change in the Absence of Friction and Area Change

For a perfect gas with constant specific heat and molecular weight
k = 1.4

M	$\dfrac{T_0}{T_0{}^*}$	$\dfrac{T}{T^*}$	$\dfrac{p}{p^*}$	$\dfrac{p_0}{p_0{}^*}$	$\dfrac{V}{V^*}$	M	$\dfrac{T_0}{T_0{}^*}$	$\dfrac{T}{T^*}$	$\dfrac{p}{p^*}$	$\dfrac{p_0}{p_0{}^*}$	$\dfrac{V}{V^*}$
2.40	.72421	.40383	.26478	2.0450	1.5252	2.80	.67380	.31486	.20040	2.8731	1.5711
2.41	.72274	.40122	.26283	2.0619	1.5266	2.81	.67273	.31299	.19909	2.8982	1.5721
2.42	.72129	.39863	.26090	2.0789	1.5279	2.82	.67167	.31114	.19780	2.9236	1.5730
2.43	.71985	.39606	.25899	2.0961	1.5293	2.83	.67062	.30931	.19652	2.9493	1.5739
2.44	.71842	.39352	.25710	2.1135	1.5306	2.84	.66958	.30749	.19525	2.9752	1.5748
2.45	.71700	.39100	.25523	2.1311	1.5320	2.85	.66855	.30568	.19399	3.0013	1.5757
2.46	.71559	.38850	.25337	2.1489	1.5333	2.86	.66752	.30389	.19274	3.0277	1.5766
2.47	.71419	.38602	.25153	2.1669	1.5346	2.87	.66650	.30211	.19151	3.0544	1.5775
2.48	.71280	.38356	.24972	2.1850	1.5359	2.88	.66549	.30035	.19029	3.0813	1.5784
2.49	.71142	.38112	.24793	2.2033	1.5372	2.89	.66449	.29860	.18908	3.1084	1.5792
2.50	.71005	.37870	.24616	2.2218	1.5385	2.90	.66350	.29687	.18788	3.1358	1.5801
2.51	.70870	.37630	.24440	2.2405	1.5398	2.91	.66252	.29515	.18669	3.1635	1.5809
2.52	.70736	.37392	.24266	2.2594	1.5410	2.92	.66154	.29344	.18551	3.1914	1.5818
2.53	.70603	.37157	.24094	2.2785	1.5422	2.93	.66057	.29175	.18435	3.2196	1.5826
2.54	.70471	.36923	.23923	2.2978	1.5434	2.94	.65961	.29007	.18320	3.2481	1.5834
2.55	.70340	.36691	.23754	2.3173	1.5446	2.95	.65865	.28841	.18205	3.2768	1.5843
2.56	.70210	.36461	.23587	2.3370	1.5458	2.96	.65770	.28676	.18091	3.3058	1.5851
2.57	.70081	.36233	.23422	2.3569	1.5470	2.97	.65676	.28512	.17978	3.3351	1.5859
2.58	.69953	.36007	.23258	2.3770	1.5482	2.98	.65583	.28349	.17867	3.3646	1.5867
2.59	.69825	.35783	.23096	2.3972	1.5494	2.99	.65490	.28188	.17757	3.3944	1.5875
2.60	.69699	.35561	.22936	2.4177	1.5505	3.00	.65398	.28028	.17647	3.4244	1.5882
2.61	.69574	.35341	.22777	2.4384	1.5516	3.50	.61580	.21419	.13223	5.3280	1.6198
2.62	.69450	.35123	.22620	2.4593	1.5527	4.00	.58909	.16831	.10256	8.2268	1.6410
2.63	.69327	.34906	.22464	2.4804	1.5538	4.50	.56983	.13540	.08177	12.502	1.6559
2.64	.69205	.34691	.22310	2.5017	1.5549	5.00	.55555	.11111	.06667	18.634	1.6667
2.65	.69084	.34478	.22158	2.5233	1.5560	6.00	.53633	.07849	.04669	38.946	1.6809
2.66	.68964	.34267	.22007	2.5451	1.5571	7.00	.52437	.05826	.03448	75.414	1.6897
2.67	.68845	.34057	.21857	2.5671	1.5582	8.00	.51646	.04491	.02649	136.62	1.6954
2.68	.68727	.33849	.21709	2.5892	1.5593	9.00	.51098	.03565	.02098	233.88	1.6993
2.69	.68610	.33643	.21562	2.6116	1.5603	10.00	.50702	.02897	.01702	381.61	1.7021
2.70	.68494	.33439	.21417	2.6342	1.5613	∞	.48980	0	0	∞	1.7143
2.71	.68378	.33236	.21273	2.6571	1.5623						
2.72	.68263	.33035	.21131	2.6802	1.5633						
2.73	.68150	.32836	.20990	2.7035	1.5644						
2.74	.68038	.32638	.20850	2.7270	1.5654						
2.75	.67926	.32442	.20712	2.7508	1.5663						
2.76	.67815	.32248	.20575	2.7748	1.5673						
2.77	.67704	.32055	.20439	2.7990	1.5683						
2.78	.67595	.31864	.20305	2.8235	1.5692						
2.79	.67487	.31674	.20172	2.8482	1.5702						

Table 37 Rayleigh Line—One-Dimensional Compressible-Flow Functions for Stagnation-Temperature Change in the Absence of Friction and Area Change

For a perfect gas with constant specific heat and molecular weight
k = 1.0

M	$\frac{T_0}{T_0*}=\frac{T}{T*}$	$\frac{p}{p*}$	$\frac{p_0}{p_0*}$	$\frac{V}{V*}$
0	0	2.000	1.213	0
0.05	.00995	1.995	1.212	.00499
.10	.03921	1.980	1.207	.01980
.15	.08608	1.956	1.200	.04401
.20	.14793	1.923	1.190	.07692
.25	.2215	1.882	1.178	.1176
.30	.3030	1.835	1.164	.1651
.35	.3889	1.782	1.149	.2183
.40	.4756	1.724	1.133	.2758
.45	.5602	1.663	1.116	.3368
.50	.6400	1.600	1.0997	.4000
.55	.7132	1.536	1.0834	.4645
.60	.7785	1.471	1.0679	.5294
.65	.8352	1.406	1.0534	.5940
.70	.8828	1.342	1.0402	.6577
.75	.9216	1.280	1.0285	.7200
.80	.9518	1.220	1.0186	.7805
.85	.9740	1.161	1.0107	.8389
.90	.9890	1.105	1.0048	.8950
.95	.9974	1.0512	1.0012	.9488
1.00	1.0000	1.0000	1.0000	1.0000
1.05	.9976	.9512	1.0013	1.0488
1.10	.9910	.9049	1.0052	1.0951
1.15	.9807	.8611	1.0118	1.1389
1.20	.9675	.8197	1.0214	1.1802
1.25	.9518	.7805	1.0340	1.220
1.30	.9342	.7435	1.0498	1.257
1.35	.9151	.7086	1.0690	1.291
1.40	.8948	.6757	1.0919	1.324
1.45	.8737	.6447	1.1187	1.355
1.50	.8521	.6154	1.150	1.384
1.55	.8301	.5878	1.186	1.412
1.60	.8080	.5618	1.226	1.438
1.65	.7859	.5373	1.271	1.463
1.70	.7639	.5141	1.323	1.486

M	$\frac{T_0}{T_0*}=\frac{T}{T*}$	$\frac{p}{p*}$	$\frac{p_0}{p_0*}$	$\frac{V}{V*}$
1.75	.7422	.4923	1.381	1.508
1.80	.7209	.4717	1.446	1.528
1.85	.7000	.4522	1.519	1.547
1.90	.6795	.4338	1.600	1.566
1.95	.6595	.4164	1.691	1.584
2.00	.6400	.4000	1.793	1.601
2.05	.6211	.3844	1.907	1.616
2.10	.6027	.3697	2.034	1.630
2.15	.5849	.3557	2.176	1.644
2.20	.5677	.3425	2.336	1.657
2.25	.5510	.3299	2.515	1.670
2.30	.5348	.3179	2.716	1.682
2.35	.5192	.3066	2.942	1.693
2.40	.5042	.2959	3.197	1.704
2.45	.4897	.2857	3.484	1.714
2.50	.4757	.2759	3.808	1.724
2.55	.4621	.2666	4.175	1.733
2.60	.4490	.2577	4.591	1.742
2.65	.4364	.2493	5.064	1.751
2.70	.4243	.2413	5.602	1.759
2.75	.4126	.2336	6.215	1.766
2.80	.4013	.2262	6.916	1.774
2.85	.3904	.2192	7.719	1.781
2.90	.3799	.2125	8.640	1.787
2.95	.3698	.2061	9.699	1.794
3.00	.3600	.2000	10.92	1.800
3.50	.2791	.1509	41.85	1.849
4.00	.2215	.1176	212.71	1.882
4.50	.1794	.09412	1425	1.906
5.00	.1479	.07692	12519	1.923
6.00	.10519	.05405	$215(10)^4$	1.946
7.00	.07840	.04000	$106(10)^7$	1.960
8.00	.06059	.03077	$147(10)^{10}$	1.969
9.00	.04818	.02439	$574(10)^{13}$	1.976
10.00	.03921	.01980	$623(10)^{17}$	1.980
∞	0	0	∞	2.000

Table 38 Rayleigh Line—One-Dimensional Compressible-Flow Functions for Stagnation-Temperature Change in the Absence of Friction and Area Change

For a perfect gas with constant specific heat and molecular weight
k = 1.1

M	$\dfrac{T_0}{T_0^*}$	$\dfrac{T}{T^*}$	$\dfrac{p}{p^*}$	$\dfrac{p_0}{p_0^*}$	$\dfrac{V}{V^*}$	M	$\dfrac{T_0}{T_0^*}$	$\dfrac{T}{T^*}$	$\dfrac{p}{p^*}$	$\dfrac{p_0}{p_0^*}$	$\dfrac{V}{V^*}$
0	0	0	2.100	1.228	0	1.75	.7771	.7076	.4807	1.347	1.472
0.05	.01044	.01097	2.094	1.226	.00524	1.80	.7591	.6859	.4601	1.403	1.491
.10	.04111	.04315	2.077	1.221	.02077	1.85	.7415	.6648	.4407	1.465	1.508
.15	.09009	.09449	2.049	1.213	.04611	1.90	.7243	.6443	.4224	1.532	1.525
.20	.15444	.16184	2.011	1.203	.08046	1.95	.7076	.6243	.4052	1.607	1.541
.25	.2305	.2413	1.965	1.190	.1228	2.00	.6914	.6049	.3889	1.689	1.556
.30	.3144	.3286	1.911	1.174	.1720	2.05	.6756	.5862	.3735	1.780	1.570
.35	.4020	.4195	1.851	1.157	.2267	2.10	.6603	.5681	.3589	1.879	1.583
.40	.4898	.5102	1.786	1.140	.2857	2.15	.6456	.5506	.3451	1.987	1.595
.45	.5746	.5973	1.717	1.122	.3478	2.20	.6313	.5337	.3321	2.106	1.607
.50	.6540	.6782	1.647	1.1040	.4118	2.25	.6175	.5174	.3197	2.237	1.618
.55	.7261	.7510	1.576	1.0867	.4766	2.30	.6042	.5017	.3079	2.380	1.629
.60	.7898	.8147	1.504	1.0702	.5416	2.35	.5914	.4866	.2968	2.537	1.639
.65	.8446	.8684	1.434	1.0550	.6057	2.40	.5790	.4720	.2863	2.709	1.649
.70	.8902	.9123	1.365	1.0412	.6686	2.45	.5671	.4580	.2763	2.897	1.658
.75	.9270	.9467	1.297	1.0291	.7297	2.50	.5556	.4444	.2667	3.104	1.667
.80	.9554	.9720	1.232	1.0189	.7887	2.55	.5445	.4314	.2576	3.332	1.675
.85	.9761	.9892	1.170	1.0109	.8453	2.60	.5338	.4189	.2489	3.581	1.683
.90	.9899	.9989	1.111	1.0050	.8995	2.65	.5235	.4068	.2406	3.855	1.690
.95	.9976	1.0023	1.0538	1.0013	.9511	2.70	.5136	.3952	.2328	4.156	1.697
1.00	1.0000	1.0000	1.0000	1.0000	1.0000	2.75	.5041	.3840	.2253	4.487	1.704
1.05	.9979	.9930	.9490	1.0013	1.0463	2.80	.4949	.3733	.2182	4.851	1.711
1.10	.9919	.9821	.9009	1.0051	1.0901	2.85	.4860	.3629	.2114	5.251	1.717
1.15	.9827	.9679	.8555	1.0116	1.1314	2.90	.4775	.3529	.2049	5.692	1.723
1.20	.9710	.9511	.8127	1.0209	1.1703	2.95	.4693	.3433	.1986	6.176	1.729
1.25	.9572	.9322	.7724	1.0331	1.207	3.00	.4613	.3341	.1927	6.710	1.734
1.30	.9418	.9118	.7345	1.0483	1.241	3.50	.3960	.2578	.1451	16.26	1.777
1.35	.9251	.8902	.6989	1.0665	1.273	4.00	.3496	.2040	.1129	42.42	1.806
1.40	.9074	.8678	.6654	1.0880	1.304	4.50	.3160	.1648	.0902	115.70	1.827
1.45	.8892	.8449	.6339	1.1130	1.333	5.00	.2909	.1357	.0737	322.33	1.842
1.50	.8706	.8217	.6043	1.141	1.360	6.00	.2568	.09631	.05172	2508	1.862
1.55	.8518	.7984	.5765	1.173	1.385	7.00	.2356	.07169	.03825	18430	1.874
1.60	.8329	.7753	.5503	1.210	1.409	8.00	.2215	.05536	.02941	$123(10)^3$	1.882
1.65	.8141	.7524	.5257	1.251	1.431	9.00	.2116	.04400	.02331	$743(10)^3$	1.888
1.70	.7955	.7298	.5025	1.297	1.452	10.00	.2045	.03579	.01892	$401(10)^4$	1.892
						∞	.1736	0	0	∞	1.909

Table 39 Rayleigh Line—One-Dimensional Compressible-Flow Functions for Stagnation-Temperature Change in the Absence of Friction and Area Change

For a perfect gas with constant specific heat and molecular weight
k = 1.2

M	$\dfrac{T_0}{T_0{}^*}$	$\dfrac{T}{T^*}$	$\dfrac{p}{p^*}$	$\dfrac{p_0}{p_0{}^*}$	$\dfrac{V}{V^*}$	M	$\dfrac{T_0}{T_0{}^*}$	$\dfrac{T}{T^*}$	$\dfrac{p}{p^*}$	$\dfrac{p_0}{p_0{}^*}$	$\dfrac{V}{V^*}$
0	0	0	2.200	1.242	0	1.75	.8054	.6782	.4706	1.320	1.441
0.05	.01094	.01203	2.193	1.239	.00548	1.80	.7900	.6563	.4501	1.369	1.458
.10	.04301	.04726	2.173	1.234	.02174	1.85	.7750	.6351	.4308	1.422	1.474
.15	.09408	.10325	2.141	1.226	.04820	1.90	.7604	.6146	.4126	1.480	1.490
.20	.16089	.17627	2.099	1.214	.08397	1.95	.7462	.5947	.3955	1.543	1.504
.25	.2395	.2618	2.047	1.199	.1279	2.00	.7325	.5755	.3793	1.612	1.517
.30	.3255	.3548	1.986	1.183	.1787	2.05	.7192	.5570	.3641	1.687	1.530
.35	.4147	.4507	1.918	1.165	.2350	2.10	.7063	.5391	.3497	1.767	1.542
.40	.5034	.5450	1.846	1.146	.2953	2.15	.6939	.5219	.3360	1.854	1.553
.45	.5884	.6343	1.770	1.127	.3584	2.20	.6819	.5054	.3231	1.948	1.564
.50	.6672	.7160	1.692	1.1078	.4231	2.25	.6703	.4895	.3109	2.050	1.574
.55	.7381	.7881	1.614	1.0895	.4884	2.30	.6591	.4742	.2994	2.159	1.584
.60	.8003	.8497	1.536	1.0722	.5531	2.35	.6484	.4595	.2884	2.277	1.593
.65	.8531	.9004	1.460	1.0563	.6168	2.40	.6381	.4453	.2780	2.405	1.602
.70	.8969	.9405	1.385	1.0420	.6788	2.45	.6281	.4317	.2682	2.542	1.610
.75	.9318	.9704	1.313	1.0296	.7388	2.50	.6185	.4187	.2588	2.690	1.618
.80	.9585	.9910	1.244	1.0191	.7964	2.55	.6093	.4062	.2499	2.849	1.625
.85	.9779	1.0032	1.178	1.0109	.8514	2.60	.6004	.3941	.2414	3.021	1.632
.90	.9907	1.0081	1.115	1.0049	.9037	2.65	.5918	.3825	.2334	3.205	1.639
.95	.9978	1.0067	1.0562	1.0012	.9532	2.70	.5836	.3713	.2257	3.403	1.645
1.00	1.0000	1.0000	1.0000	1.0000	1.0000	2.75	.5757	.3606	.2184	3.617	1.651
1.05	.9981	.9888	.9471	1.0013	1.0441	2.80	.5681	.3503	.2114	3.847	1.657
1.10	.9927	.9741	.8972	1.0050	1.0856	2.85	.5608	.3404	.2047	4.094	1.663
1.15	.9845	.9564	.8504	1.0114	1.1247	2.90	.5537	.3309	.1983	4.359	1.668
1.20	.9740	.9365	.8065	1.0204	1.1613	2.95	.5469	.3217	.1923	4.644	1.673
1.25	.9617	.9149	.7653	1.0322	1.196	3.00	.5404	.3128	.1864	4.951	1.678
1.30	.9481	.8921	.7266	1.0467	1.228	3.50	.4865	.2405	.1401	9.597	1.717
1.35	.9334	.8685	.6903	1.0640	1.258	4.00	.4486	.1898	.1089	18.99	1.743
1.40	.9180	.8443	.6563	1.0843	1.286	4.50	.4211	.1531	.08696	37.61	1.761
1.45	.9021	.8199	.6245	1.1077	1.313	5.00	.4006	.1259	.07097	73.64	1.774
1.50	.8859	.7955	.5946	1.134	1.338	6.00	.3730	.08919	.04977	266.2	1.792
1.55	.8695	.7712	.5666	1.164	1.361	7.00	.3557	.06632	.03679	875.9	1.803
1.60	.8532	.7473	.5403	1.197	1.383	8.00	.3443	.05118	.02828	2621	1.810
1.65	.8370	.7237	.5156	1.234	1.404	9.00	.3363	.04065	.02240	7181	1.815
1.70	.8211	.7007	.4924	1.275	1.423	10.00	.3306	.03306	.01818	18182	1.818
						∞	.3056	0	0	∞	1.833

Table 40 Rayleigh Line—One-Dimensional Compressible-Flow Functions for Stagnation-Temperature Change in the Absence of Friction and Area Change

For a perfect gas with constant specific heat and molecular weight
k = 1.3

M	$\dfrac{T_0}{T_0*}$	$\dfrac{T}{T*}$	$\dfrac{p}{p*}$	$\dfrac{p_0}{p_0*}$	$\dfrac{V}{V*}$	M	$\dfrac{T_0}{T_0*}$	$\dfrac{T}{T*}$	$\dfrac{p}{p*}$	$\dfrac{p_0}{p_0*}$	$\dfrac{V}{V*}$
0	0	0	2.300	1.255	0	1.75	.8285	.6529	.4617	1.296	1.414
0.05	.01143	.01314	2.293	1.253	.00573	1.80	.8153	.6309	.4413	1.340	1.430
.10	.04489	.05155	2.270	1.247	.02270	1.85	.8024	.6097	.4221	1.387	1.445
.15	.09803	.11236	2.234	1.237	.05028	1.90	.7898	.5892	.4040	1.438	1.459
.20	.16726	.19120	2.186	1.224	.08745	1.95	.7776	.5695	.3870	1.493	1.472
.25	.2482	.2828	2.127	1.209	.1329	2.00	.7659	.5505	.3710	1.552	1.484
.30	.3363	.3816	2.059	1.191	.1853	2.05	.7545	.5322	.3559	1.615	1.495
.35	.4270	.4822	1.934	1.172	.2430	2.10	.7435	.5146	.3416	1.683	1.506
.40	.5165	.5800	1.904	1.152	.3046	2.15	.7329	.4977	.3281	1.755	1.517
.45	.6015	.6713	1.821	1.131	.3687	2.20	.7227	.4815	.3154	1.832	1.527
.50	.6796	.7533	1.736	1.1112	.4340	2.25	.7129	.4659	.3034	1.915	1.536
.55	.7494	.8244	1.651	1.0919	.4994	2.30	.7034	.4510	.2920	2.003	1.545
.60	.8099	.8837	1.567	1.0739	.5640	2.35	.6943	.4367	.2812	2.097	1.553
.65	.8611	.9312	1.485	1.0574	.6272	2.40	.6855	.4229	.2710	2.197	1.561
.70	.9029	.9673	1.405	1.0426	.6885	2.45	.6771	.4097	.2613	2.303	1.568
.75	.9361	.9928	1.328	1.0299	.7473	2.50	.6690	.3971	.2521	2.416	1.575
.80	.9614	1.0088	1.255	1.0193	.8035	2.55	.6612	.3850	.2433	2.536	1.582
.85	.9795	1.0163	1.186	1.0109	.8569	2.60	.6537	.3733	.2350	2.664	1.588
.90	.9914	1.0166	1.120	1.0049	.9075	2.65	.6465	.3621	.2271	2.800	1.594
.95	.9980	1.0108	1.0583	1.0012	.9552	2.70	.6396	.3513	.2195	2.944	1.600
1.00	1.0000	1.0000	1.0000	1.0000	1.0000	2.75	.6329	.3410	.2123	3.096	1.606
1.05	.9982	.9851	.9452	1.0012	1.0421	2.80	.6265	.3311	.2055	3.258	1.611
1.10	.9933	.9669	.8939	1.0049	1.0816	2.85	.6203	.3216	.1990	3.429	1.616
1.15	.9859	.9461	.8458	1.0111	1.1186	2.90	.6144	.3124	.1928	3.611	1.621
1.20	.9765	.9235	.8008	1.0199	1.1532	2.95	.6087	.3036	.1868	3.804	1.626
1.25	.9656	.8996	.7588	1.0312	1.186	3.00	.6032	.2952	.1811	4.007	1.630
1.30	.9534	.8747	.7194	1.0451	1.216	3.50	.5582	.2262	.1359	6.806	1.665
1.35	.9404	.8493	.6826	1.0617	1.244	4.00	.5265	.1781	.1055	11.57	1.688
1.40	.9268	.8237	.6483	1.0809	1.270	4.50	.5037	.1435	.08417	19.44	1.704
1.45	.9128	.7980	.6161	1.1028	1.295	5.00	.4867	.1178	.06866	32.06	1.716
1.50	.8986	.7726	.5860	1.128	1.318	6.00	.4639	.08335	.04812	81.79	1.732
1.55	.8843	.7475	.5578	1.155	1.340	7.00	.4496	.06192	.03555	191.3	1.742
1.60	.8701	.7230	.5314	1.185	1.360	8.00	.4402	.04775	.02732	413.4	1.748
1.65	.8560	.6990	.5067	1.219	1.379	9.00	.4336	.03792	.02164	833.4	1.753
1.70	.8421	.6756	.4835	1.256	1.397	10.00	.4289	.03082	.01756	1582	1.756
						∞	.4083	0	0	∞	1.769

Table 41 Rayleigh Line—One-Dimensional Compressible-Flow Functions for Stagnation-Temperature Change in the Absence of Friction and Area Change

For a perfect gas with constant specific heat and molecular weight
k = 1.67

M	$\frac{T_0}{T_0{}^*}$	$\frac{T}{T^*}$	$\frac{p}{p^*}$	$\frac{p_0}{p_0{}^*}$	$\frac{V}{V^*}$	M	$\frac{T_0}{T_0{}^*}$	$\frac{T}{T^*}$	$\frac{p}{p^*}$	$\frac{p_0}{p_0{}^*}$	$\frac{V}{V^*}$
0	0	0	2.670	1.299	0	1.75	.8862	.5840	.4367	1.235	1.337
0.05	.01325	.01767	2.659	1.297	.00665	1.80	.8779	.5620	.4165	1.266	1.349
.10	.05183	.06896	2.626	1.289	.02626	1.85	.8699	.5410	.3976	1.299	1.360
.15	.11243	.1490	2.573	1.276	.05790	1.90	.8621	.5209	.3799	1.334	1.371
.20	.19020	.2506	2.503	1.259	.10011	1.95	.8546	.5018	.3633	1.370	1.381
.25	.2794	.3653	2.418	1.239	.1511	2.00	.8474	.4835	.3477	1.408	1.391
.30	.3742	.4849	2.321	1.216	.2089	2.05	.8405	.4660	.3330	1.448	1.400
.35	.4693	.6018	2.216	1.192	.2715	2.10	.8338	.4493	.3192	1.490	1.408
.40	.5606	.7103	2.107	1.168	.3371	2.15	.8274	.4334	.3062	1.534	1.415
.45	.6448	.8062	1.995	1.144	.4040	2.20	.8213	.4183	.2940	1.580	1.423
.50	.7201	.8870	1.884	1.1202	.4709	2.25	.8154	.4038	.2824	1.628	1.430
.55	.7853	.9519	1.774	1.0981	.5366	2.30	.8097	.3899	.2715	1.678	1.436
.60	.8402	1.0010	1.667	1.0778	.6003	2.35	.8043	.3767	.2612	1.729	1.442
.65	.8853	1.0354	1.565	1.0597	.6614	2.40	.7991	.3641	.2514	1.783	1.448
.70	.9213	1.0565	1.468	1.0438	.7195	2.45	.7941	.3521	.2422	1.839	1.454
.75	.9491	1.0662	1.377	1.0303	.7744	2.50	.7893	.3406	.2334	1.897	1.459
.80	.9697	1.0660	1.291	1.0193	.8260	2.55	.7847	.3296	.2251	1.956	1.464
.85	.9842	1.0578	1.210	1.0108	.8742	2.60	.7803	.3191	.2173	2.018	1.469
.90	.9935	1.0432	1.135	1.0048	.9192	2.65	.7761	.3090	.2098	2.082	1.473
.95	.9985	1.0235	1.0649	1.0012	.9611	2.70	.7721	.2994	.2027	2.148	1.477
1.00	1.0000	1.0000	1.0000	1.0000	1.0000	2.75	.7682	.2902	.1959	2.216	1.481
1.05	.9987	.9736	.9398	1.0012	1.0361	2.80	.7644	.2814	.1895	2.287	1.485
1.10	.9952	.9454	.8839	1.0046	1.0695	2.85	.7608	.2730	.1834	2.360	1.489
1.15	.9899	.9158	.8321	1.0103	1.1005	2.90	.7574	.2649	.1775	2.435	1.493
1.20	.9833	.8855	.7842	1.0181	1.1292	2.95	.7541	.2571	.1719	2.512	1.496
1.25	.9757	.8550	.7397	1.0280	1.156	3.00	.7509	.2497	.1666	2.587	1.499
1.30	.9674	.8246	.6985	1.0400	1.181	3.50	.7251	.1897	.1244	3.521	1.524
1.35	.9586	.7946	.6603	1.0540	1.204	4.00	.7072	.1484	.09632	4.716	1.541
1.40	.9495	.7652	.6249	1.0700	1.225	4.50	.6943	.1191	.07669	6.213	1.553
1.45	.9403	.7365	.5919	1.0880	1.245	5.00	.6848	.0975	.06246	8.044	1.561
1.50	.9310	.7087	.5612	1.108	1.263	6.00	.6721	.06870	.04368	12.86	1.573
1.55	.9217	.6818	.5327	1.130	1.280	7.00	.6642	.05092	.03224	19.44	1.580
1.60	.9125	.6559	.5062	1.154	1.296	8.00	.6590	.03920	.02475	28.07	1.584
1.65	.9035	.6309	.4814	1.179	1.311	9.00	.6553	.03110	.01959	39.05	1.587
1.70	.8947	.6069	.4583	1.206	1.324	10.00	.6527	.02526	.01589	52.66	1.589
						∞	.6414	0	0	∞	1.599

Table 42 Fanno Line—One-Dimensional Compressible-Flow Functions for Adiabatic Flow at Constant Area with Friction

For a perfect gas with constant specific heat and molecular weight
k = 1.4

M	$\dfrac{T}{T^*}$	$\dfrac{p}{p^*}$	$\dfrac{p_0}{p_0{}^*}$	$\dfrac{V}{V^*}$	$\dfrac{F}{F^*}$	$4\dfrac{fL_{max}}{D}$	M	$\dfrac{T}{T^*}$	$\dfrac{p}{p^*}$	$\dfrac{p_0}{p_0{}^*}$	$\dfrac{V}{V^*}$	$\dfrac{F}{F^*}$	$4\dfrac{fL_{max}}{D}$
0	1.2000	∞	∞	0	∞	∞	0.40	1.1628	2.6958	1.5901	.43133	1.3749	2.3085
0.01	1.2000	109.544	57.874	.01095	45.650	7134.40	.41	1.1610	2.6280	1.5587	.44177	1.3527	2.1344
.02	1.1999	54.770	28.942	.02191	22.834	1778.45	.42	1.1591	2.5634	1.5289	.45218	1.3318	1.9744
.03	1.1998	36.511	19.300	.03286	15.232	787.08	.43	1.1572	2.5017	1.5007	.46257	1.3122	1.8272
.04	1.1996	27.382	14.482	.04381	11.435	440.35	.44	1.1553	2.4428	1.4739	.47293	1.2937	1.6915
.05	1.1994	21.903	11.5914	.05476	9.1584	280.02	.45	1.1533	2.3865	1.4486	.48326	1.2763	1.5664
.06	1.1991	18.251	9.6659	.06570	7.6428	193.03	.46	1.1513	2.3326	1.4246	.49357	1.2598	1.4509
.07	1.1988	15.642	8.2915	.07664	6.5620	140.66	.47	1.1492	2.2809	1.4018	.50385	1.2443	1.3442
.08	1.1985	13.684	7.2616	.08758	5.7529	106.72	.48	1.1471	2.2314	1.3801	.51410	1.2296	1.2453
.09	1.1981	12.162	6.4614	.09851	5.1249	83.496	.49	1.1450	2.1838	1.3595	.52433	1.2158	1.1539
.10	1.1976	10.9435	5.8218	.10943	4.6236	66.922	.50	1.1429	2.1381	1.3399	.53453	1.2027	1.06908
.11	1.1971	9.9465	5.2992	.12035	4.2146	54.688	.51	1.1407	2.0942	1.3212	.54469	1.1903	.99042
.12	1.1966	9.1156	4.8643	.13126	3.8747	45.408	.52	1.1384	2.0519	1.3034	.55482	1.1786	.91741
.13	1.1960	8.4123	4.4968	.14216	3.5880	38.207	.53	1.1362	2.0112	1.2864	.56493	1.1675	.84963
.14	1.1953	7.8093	4.1824	.15306	3.3432	32.511	.54	1.1339	1.9719	1.2702	.57501	1.1571	.78662
.15	1.1946	7.2866	3.9103	.16395	3.1317	27.932	.55	1.1315	1.9341	1.2549	.58506	1.1472	.72805
.16	1.1939	6.8291	3.6727	.17482	2.9474	24.198	.56	1.1292	1.8976	1.2403	.59507	1.1378	.67357
.17	1.1931	6.4252	3.4635	.18568	2.7855	21.115	.57	1.1268	1.8623	1.2263	.60505	1.1289	.62286
.18	1.1923	6.0662	3.2779	.19654	2.6422	18.543	.58	1.1244	1.8282	1.2130	.61500	1.1205	.57568
.19	1.1914	5.7448	3.1123	.20739	2.5146	16.375	.59	1.1219	1.7952	1.2003	.62492	1.1126	.53174
.20	1.1905	5.4555	2.9635	.21822	2.4004	14.533	.60	1.1194	1.7634	1.1882	.63481	1.10504	.49081
.21	1.1895	5.1936	2.8293	.22904	2.2976	12.956	.61	1.1169	1.7325	1.1766	.64467	1.09793	.45270
.22	1.1885	4.9554	2.7076	.23984	2.2046	11.596	.62	1.1144	1.7026	1.1656	.65449	1.09120	.41720
.23	1.1874	4.7378	2.5968	.25063	2.1203	10.416	.63	1.1118	1.6737	1.1551	.66427	1.08485	.38411
.24	1.1863	4.5383	2.4956	.26141	2.0434	9.3865	.64	1.1091	1.6456	1.1451	.67402	1.07883	.35330
.25	1.1852	4.3546	2.4027	.27217	1.9732	8.4834	.65	1.10650	1.6183	1.1356	.68374	1.07314	.32460
.26	1.1840	4.1850	2.3173	.28291	1.9088	7.6876	.66	1.10383	1.5919	1.1265	.69342	1.06777	.29785
.27	1.1828	4.0280	2.2385	.29364	1.8496	6.9832	.67	1.10114	1.5662	1.1179	.70306	1.06271	.27295
.28	1.1815	3.8820	2.1656	.30435	1.7950	6.3572	.68	1.09842	1.5413	1.1097	.71267	1.05792	.24978
.29	1.1802	3.7460	2.0979	.31504	1.7446	5.7989	.69	1.09567	1.5170	1.1018	.72225	1.05340	.22821
.30	1.1788	3.6190	2.0351	.32572	1.6979	5.2992	.70	1.09290	1.4934	1.09436	.73179	1.04915	.20814
.31	1.1774	3.5002	1.9765	.33637	1.6546	4.8507	.71	1.09010	1.4705	1.08729	.74129	1.04514	.18949
.32	1.1759	3.3888	1.9219	.34700	1.6144	4.4468	.72	1.08727	1.4482	1.08057	.75076	1.04137	.17215
.33	1.1744	3.2840	1.8708	.35762	1.5769	4.0821	.73	1.08442	1.4265	1.07419	.76019	1.03783	.15606
.34	1.1729	3.1853	1.8229	.36822	1.5420	3.7520	.74	1.08155	1.4054	1.06815	.76958	1.03450	.14113
.35	1.1713	3.0922	1.7780	.37880	1.5094	3.4525	.75	1.07865	1.3848	1.06242	.77893	1.03137	.12728
.36	1.1697	3.0042	1.7358	.38935	1.4789	3.1801	.76	1.07573	1.3647	1.05700	.78825	1.02844	.11446
.37	1.1680	2.9209	1.6961	.39988	1.4503	2.9320	.77	1.07279	1.3451	1.05188	.79753	1.02570	.10262
.38	1.1663	2.8420	1.6587	.41039	1.4236	2.7055	.78	1.06982	1.3260	1.04705	.80677	1.02314	.09167
.39	1.1646	2.7671	1.6234	.42087	1.3985	2.4983	.79	1.06684	1.3074	1.04250	.81598	1.02075	.08159

See page 121 for definitions of symbols.

Table 42   Fanno Line—One-Dimensional Compressible-Flow Functions for Adiabatic Flow at Constant Area with Friction

For a perfect gas with constant specific heat and molecular weight
k = 1.4

M	$\frac{T}{T^*}$	$\frac{p}{p^*}$	$\frac{p_0}{p_0{}^*}$	$\frac{V}{V^*}$	$\frac{F}{F^*}$	$4\frac{fL_{max}}{D}$	M	$\frac{T}{T^*}$	$\frac{p}{p^*}$	$\frac{p_0}{p_0{}^*}$	$\frac{V}{V^*}$	$\frac{F}{F^*}$	$4\frac{fL_{max}}{D}$
0.80	1.06383	1.2892	1.03823	.82514	1.01853	.07229	1.20	.93168	.80436	1.03044	1.1583	1.01082	.03364
.81	1.06080	1.2715	1.03422	.83426	1.01646	.06375	1.21	.92820	.79623	1.03344	1.1658	1.01178	.03650
.82	1.05775	1.2542	1.03047	.84334	1.01455	.05593	1.22	.92473	.78822	1.03657	1.1732	1.01278	.03942
.83	1.05468	1.2373	1.02696	.85239	1.01278	.04878	1.23	.92125	.78034	1.03983	1.1806	1.01381	.04241
.84	1.05160	1.2208	1.02370	.86140	1.01115	.04226	1.24	.91777	.77258	1.04323	1.1879	1.01486	.04547
.85	1.04849	1.2047	1.02067	.87037	1.00966	.03632	1.25	.91429	.76495	1.04676	1.1952	1.01594	.04858
.86	1.04537	1.1889	1.01787	.87929	1.00829	.03097	1.26	.91080	.75743	1.05041	1.2025	1.01705	.05174
.87	1.04223	1.1735	1.01529	.88818	1.00704	.02613	1.27	.90732	.75003	1.05419	1.2097	1.01818	.05494
.88	1.03907	1.1584	1.01294	.89703	1.00591	.02180	1.28	.90383	.74274	1.05809	1.2169	1.01933	.05820
.89	1.03589	1.1436	1.01080	.90583	1.00490	.01793	1.29	.90035	.73556	1.06213	1.2240	1.02050	.06150
.90	1.03270	1.12913	1.00887	.91459	1.00399	.014513	1.30	.89686	.72848	1.06630	1.2311	1.02169	.06483
.91	1.02950	1.11500	1.00714	.92332	1.00318	.011519	1.31	.89338	.72152	1.07060	1.2382	1.02291	.06820
.92	1.02627	1.10114	1.00560	.93201	1.00248	.008916	1.32	.88989	.71465	1.07502	1.2452	1.02415	.07161
.93	1.02304	1.08758	1.00426	.94065	1.00188	.006694	1.33	.88641	.70789	1.07957	1.2522	1.02540	.07504
.94	1.01978	1.07430	1.00311	.94925	1.00136	.004815	1.34	.88292	.70123	1.08424	1.2591	1.02666	.07850
.95	1.01652	1.06129	1.00215	.95782	1.00093	.003280	1.35	.87944	.69466	1.08904	1.2660	1.02794	.08199
.96	1.01324	1.04854	1.00137	.96634	1.00059	.002056	1.36	.87596	.68818	1.09397	1.2729	1.02924	.08550
.97	1.00995	1.03605	1.00076	.97481	1.00033	.001135	1.37	.87249	.68180	1.09902	1.2797	1.03056	.08904
.98	1.00664	1.02379	1.00033	.98324	1.00014	.000493	1.38	.86901	.67551	1.10419	1.2864	1.03189	.09259
.99	1.00333	1.01178	1.00008	.99164	1.00003	.000120	1.39	.86554	.66931	1.10948	1.2932	1.03323	.09616
1.00	1.00000	1.00000	1.00000	1.00000	1.00000	0	1.40	.86207	.66320	1.1149	1.2999	1.03458	.09974
1.01	.99666	.98844	1.00008	1.00831	1.00003	.000114	1.41	.85860	.65717	1.1205	1.3065	1.03595	.10333
1.02	.99331	.97711	1.00033	1.01658	1.00013	.000458	1.42	.85514	.65122	1.1262	1.3131	1.03733	.10694
1.03	.98995	.96598	1.00073	1.02481	1.00030	.001013	1.43	.85168	.64536	1.1320	1.3197	1.03872	.11056
1.04	.98658	.95506	1.00130	1.03300	1.00053	.001771	1.44	.84822	.63958	1.1379	1.3262	1.04012	.11419
1.05	.98320	.94435	1.00203	1.04115	1.00082	.002712	1.45	.84477	.63387	1.1440	1.3327	1.04153	.11782
1.06	.97982	.93383	1.00291	1.04925	1.00116	.003837	1.46	.84133	.62824	1.1502	1.3392	1.04295	.12146
1.07	.97642	.92350	1.00394	1.05731	1.00155	.005129	1.47	.83788	.62269	1.1565	1.3456	1.04438	.12510
1.08	.97302	.91335	1.00512	1.06533	1.00200	.006582	1.48	.83445	.61722	1.1629	1.3520	1.04581	.12875
1.09	.96960	.90338	1.00645	1.07331	1.00250	.008185	1.49	.83101	.61181	1.1695	1.3583	1.04725	.13240
1.10	.96618	.89359	1.00793	1.08124	1.00305	.009933	1.50	.82759	.60648	1.1762	1.3646	1.04870	.13605
1.11	.96276	.88397	1.00955	1.08913	1.00365	.011813	1.51	.82416	.60122	1.1830	1.3708	1.05016	.13970
1.12	.95933	.87451	1.01131	1.09698	1.00429	.013824	1.52	.82075	.59602	1.1899	1.3770	1.05162	.14335
1.13	.95589	.86522	1.01322	1.10479	1.00497	.015949	1.53	.81734	.59089	1.1970	1.3832	1.05309	.14699
1.14	.95244	.85608	1.01527	1.11256	1.00569	.018187	1.54	.81394	.58583	1.2043	1.3894	1.05456	.15063
1.15	.94899	.84710	1.01746	1.1203	1.00646	.02053	1.55	.81054	.58084	1.2116	1.3955	1.05604	.15427
1.16	.94554	.83827	1.01978	1.1280	1.00726	.02298	1.56	.80715	.57591	1.2190	1.4015	1.05752	.15790
1.17	.94208	.82958	1.02224	1.1356	1.00810	.02552	1.57	.80376	.57104	1.2266	1.4075	1.05900	.16152
1.18	.93862	.82104	1.02484	1.1432	1.00897	.02814	1.58	.80038	.56623	1.2343	1.4135	1.06049	.16514
1.19	.93515	.81263	1.02757	1.1508	1.00988	.03085	1.59	.79701	.56148	1.2422	1.4195	1.06198	.16876

Table 42 Fanno Line—One-Dimensional Compressible-Flow Functions for Adiabatic Flow at Constant Area with Friction

For a perfect gas with constant specific heat and molecular weight
k = 1.4

M	$\dfrac{T}{T^*}$	$\dfrac{p}{p^*}$	$\dfrac{p_0}{p_0^*}$	$\dfrac{V}{V^*}$	$\dfrac{F}{F^*}$	$4\dfrac{fL_{max}}{D}$	M	$\dfrac{T}{T^*}$	$\dfrac{p}{p^*}$	$\dfrac{p_0}{p_0^*}$	$\dfrac{V}{V^*}$	$\dfrac{F}{F^*}$	$4\dfrac{fL_{max}}{D}$
1.60	.79365	.55679	1.2502	1.4254	1.06348	.17236	2.00	.66667	.40825	1.6875	1.6330	1.1227	.30499
1.61	.79030	.55216	1.2583	1.4313	1.06498	.17595	2.01	.66371	.40532	1.7017	1.6375	1.1241	.30796
1.62	.78695	.54759	1.2666	1.4371	1.06648	.17953	2.02	.66076	.40241	1.7160	1.6420	1.1255	.31091
1.63	.78361	.54308	1.2750	1.4429	1.06798	.18311	2.03	.65783	.39954	1.7305	1.6465	1.1269	.31384
1.64	.78028	.53862	1.2835	1.4487	1.06948	.18667	2.04	.65491	.39670	1.7452	1.6509	1.1283	.31675
1.65	.77695	.53421	1.2922	1.4544	1.07098	.19022	2.05	.65200	.39389	1.7600	1.6553	1.1297	.31965
1.66	.77363	.52986	1.3010	1.4601	1.07249	.19376	2.06	.64910	.39110	1.7750	1.6597	1.1311	.32253
1.67	.77033	.52556	1.3099	1.4657	1.07399	.19729	2.07	.64621	.38834	1.7902	1.6640	1.1325	.32538
1.68	.76703	.52131	1.3190	1.4713	1.07550	.20081	2.08	.64333	.38562	1.8056	1.6683	1.1339	.32822
1.69	.76374	.51711	1.3282	1.4769	1.07701	.20431	2.09	.64047	.38292	1.8212	1.6726	1.1352	.33104
1.70	.76046	.51297	1.3376	1.4825	1.07851	.20780	2.10	.63762	.38024	1.8369	1.6769	1.1366	.33385
1.71	.75718	.50887	1.3471	1.4880	1.08002	.21128	2.11	.63478	.37760	1.8528	1.6811	1.1380	.33664
1.72	.75392	.50482	1.3567	1.4935	1.08152	.21474	2.12	.63195	.37498	1.8690	1.6853	1.1393	.33940
1.73	.75067	.50082	1.3665	1.4989	1.08302	.21819	2.13	.62914	.37239	1.8853	1.6895	1.1407	.34215
1.74	.74742	.49686	1.3764	1.5043	1.08453	.22162	2.14	.62633	.36982	1.9018	1.6936	1.1420	.34488
1.75	.74419	.49295	1.3865	1.5097	1.08603	.22504	2.15	.62354	.36728	1.9185	1.6977	1.1434	.34760
1.76	.74096	.48909	1.3967	1.5150	1.08753	.22844	2.16	.62076	.36476	1.9354	1.7018	1.1447	.35030
1.77	.73774	.48527	1.4070	1.5203	1.08903	.23183	2.17	.61799	.36227	1.9525	1.7059	1.1460	.35298
1.78	.73453	.48149	1.4175	1.5256	1.09053	.23520	2.18	.61523	.35980	1.9698	1.7099	1.1474	.35564
1.79	.73134	.47776	1.4282	1.5308	1.09202	.23855	2.19	.61249	.35736	1.9873	1.7139	1.1487	.35828
1.80	.72816	.47407	1.4390	1.5360	1.09352	.24189	2.20	.60976	.35494	2.0050	1.7179	1.1500	.36091
1.81	.72498	.47042	1.4499	1.5412	1.09500	.24521	2.21	.60704	.35254	2.0228	1.7219	1.1513	.36352
1.82	.72181	.46681	1.4610	1.5463	1.09649	.24851	2.22	.60433	.35017	2.0409	1.7258	1.1526	.36611
1.83	.71865	.46324	1.4723	1.5514	1.09798	.25180	2.23	.60163	.34782	2.0592	1.7297	1.1539	.36868
1.84	.71551	.45972	1.4837	1.5564	1.09946	.25507	2.24	.59895	.34550	2.0777	1.7336	1.1552	.37124
1.85	.71238	.45623	1.4952	1.5614	1.1009	.25832	2.25	.59627	.34319	2.0964	1.7374	1.1565	.37378
1.86	.70925	.45278	1.5069	1.5664	1.1024	.26156	2.26	.59361	.34091	2.1154	1.7412	1.1578	.37630
1.87	.70614	.44937	1.5188	1.5714	1.1039	.26478	2.27	.59096	.33865	2.1345	1.7450	1.1590	.37881
1.88	.70304	.44600	1.5308	1.5763	1.1054	.26798	2.28	.58833	.33641	2.1538	1.7488	1.1603	.38130
1.89	.69995	.44266	1.5429	1.5812	1.1068	.27116	2.29	.58570	.33420	2.1733	1.7526	1.1616	.38377
1.90	.69686	.43936	1.5552	1.5861	1.1083	.27433	2.30	.58309	.33200	2.1931	1.7563	1.1629	.38623
1.91	.69379	.43610	1.5677	1.5909	1.1097	.27748	2.31	.58049	.32983	2.2131	1.7600	1.1641	.38867
1.92	.69074	.43287	1.5804	1.5957	1.1112	.28061	2.32	.57790	.32767	2.2333	1.7637	1.1653	.39109
1.93	.68769	.42967	1.5932	1.6005	1.1126	.28372	2.33	.57532	.32554	2.2537	1.7673	1.1666	.39350
1.94	.68465	.42651	1.6062	1.6052	1.1141	.28681	2.34	.57276	.32342	2.2744	1.7709	1.1678	.39589
1.95	.68162	.42339	1.6193	1.6099	1.1155	.28989	2.35	.57021	.32133	2.2953	1.7745	1.1690	.39826
1.96	.67861	.42030	1.6326	1.6146	1.1170	.29295	2.36	.56767	.31925	2.3164	1.7781	1.1703	.40062
1.97	.67561	.41724	1.6461	1.6193	1.1184	.29599	2.37	.56514	.31720	2.3377	1.7817	1.1715	.40296
1.98	.67262	.41421	1.6597	1.6239	1.1198	.29901	2.38	.56262	.31516	2.3593	1.7852	1.1727	.40528
1.99	.66964	.41121	1.6735	1.6284	1.1213	.30201	2.39	.56011	.31314	2.3811	1.7887	1.1739	.40760

Table 42   Fanno Line—One-Dimensional Compressible-Flow Functions for Adiabatic Flow at Constant Area with Friction

For a perfect gas with constant specific heat and molecular weight
k = 1.4

M	$\dfrac{T}{T^*}$	$\dfrac{p}{p^*}$	$\dfrac{p_0}{p_0{}^*}$	$\dfrac{V}{V^*}$	$\dfrac{F}{F^*}$	$4\dfrac{fL_{max}}{D}$	M	$\dfrac{T}{T^*}$	$\dfrac{p}{p^*}$	$\dfrac{p_0}{p_0{}^*}$	$\dfrac{V}{V^*}$	$\dfrac{F}{F^*}$	$4\dfrac{fL_{max}}{D}$
2.40	.55762	.31114	2.4031	1.7922	1.1751	.40989	2.80	.46729	.24414	3.5001	1.9140	1.2182	.48976
2.41	.55514	.30916	2.4254	1.7956	1.1763	.41216	2.81	.46526	.24274	3.5336	1.9167	1.2192	.49148
2.42	.55267	.30720	2.4479	1.7991	1.1775	.41442	2.82	.46324	.24135	3.5674	1.9193	1.2202	.49321
2.43	.55021	.30525	2.4706	1.8025	1.1786	.41667	2.83	.46122	.23997	3.6015	1.9220	1.2211	.49491
2.44	.54776	.30332	2.4936	1.8059	1.1798	.41891	2.84	.45922	.23861	3.6359	1.9246	1.2221	.49660
2.45	.54533	.30141	2.5168	1.8092	1.1810	.42113	2.85	.45723	.23726	3.6707	1.9271	1.2230	.49828
2.46	.54291	.29952	2.5403	1.8126	1.1821	.42333	2.86	.45525	.23592	3.7058	1.9297	1.2240	.49995
2.47	.54050	.29765	2.5640	1.8159	1.1833	.42551	2.87	.45328	.23458	3.7413	1.9322	1.2249	.50161
2.48	.53810	.29579	2.5880	1.8192	1.1844	.42768	2.88	.45132	.23326	3.7771	1.9348	1.2258	.50326
2.49	.53571	.29395	2.6122	1.8225	1.1856	.42983	2.89	.44937	.23196	3.8133	1.9373	1.2268	.50489
2.50	.53333	.29212	2.6367	1.8257	1.1867	.43197	2.90	.44743	.23066	3.8498	1.9398	1.2277	.50651
2.51	.53097	.29031	2.6615	1.8290	1.1879	.43410	2.91	.44550	.22937	3.8866	1.9423	1.2286	.50812
2.52	.52862	.28852	2.6865	1.8322	1.1890	.43621	2.92	.44358	.22809	3.9238	1.9448	1.2295	.50973
2.53	.52627	.28674	2.7117	1.8354	1.1910	.43831	2.93	.44167	.22682	3.9614	1.9472	1.2304	.51133
2.54	.52394	.28498	2.7372	1.8386	1.1912	.44040	2.94	.43977	.22556	3.9993	1.9497	1.2313	.51291
2.55	.52163	.28323	2.7630	1.8417	1.1923	.44247	2.95	.43788	.22431	4.0376	1.9521	1.2322	.51447
2.56	.51932	.28150	2.7891	1.8448	1.1934	.44452	2.96	.43600	.22307	4.0763	1.9545	1.2331	.51603
2.57	.51702	.27978	2.8154	1.8479	1.1945	.44655	2.97	.43413	.22185	4.1153	1.9569	1.2340	.51758
2.58	.51474	.27808	2.8420	1.8510	1.1956	.44857	2.98	.43226	.22063	4.1547	1.9592	1.2348	.51912
2.59	.51247	.27640	2.8689	1.8541	1.1967	.45059	2.99	.43041	.21942	4.1944	1.9616	1.2357	.52064
2.60	.51020	.27473	2.8960	1.8571	1.1978	.45259	3.0	.42857	.21822	4.2346	1.9640	1.2366	.52216
2.61	.50795	.27307	2.9234	1.8602	1.1989	.45457	3.5	.34783	.16850	6.7896	2.0642	1.2743	.58643
2.62	.50571	.27143	2.9511	1.8632	1.2000	.45654	4.0	.28571	.13363	10.719	2.1381	1.3029	.63306
2.63	.50349	.26980	2.9791	1.8662	1.2011	.45850	4.5	.23762	.10833	16.562	2.1936	1.3247	.66764
2.64	.50127	.26818	3.0074	1.8691	1.2021	.46044	5.0	.20000	.08944	25.000	2.2361	1.3416	.69380
2.65	.49906	.26658	3.0359	1.8721	1.2031	.46237	6.0	.14634	.06376	53.180	2.2953	1.3655	.72987
2.66	.49687	.26499	3.0647	1.8750	1.2042	.46429	7.0	.11111	.04762	104.14	2.3333	1.3810	.75280
2.67	.49469	.26342	3.0938	1.8779	1.2052	.46619	8.0	.08696	.03686	190.11	2.3591	1.3915	.76819
2.68	.49251	.26186	3.1234	1.8808	1.2062	.46807	9.0	.06977	.02935	327.19	2.3772	1.3989	.77898
2.69	.49035	.26032	3.1530	1.8837	1.2073	.46996	10.0	.05714	.02390	535.94	2.3905	1.4044	.78683
2.70	.48820	.25878	3.1830	1.8865	1.2083	.47182	∞	0	0	∞	2.4495	1.4289	.82153
2.71	.48606	.25726	3.2133	1.8894	1.2093	.47367							
2.72	.48393	.25575	3.2440	1.8922	1.2103	.47551							
2.73	.48182	.25426	3.2749	1.8950	1.2113	.47734							
2.74	.47971	.25278	3.3061	1.8978	1.2123	.47915							
2.75	.47761	.25131	3.3376	1.9005	1.2133	.48095							
2.76	.47553	.24985	3.3695	1.9032	1.2143	.48274							
2.77	.47346	.24840	3.4017	1.9060	1.2153	.48452							
2.78	.47139	.24697	3.4342	1.9087	1.2163	.48628							
2.79	.46933	.24555	3.4670	1.9114	1.2173	.48803							

Table 43 Fanno Line—One-Dimensional Compressible-Flow Functions for Adiabatic Flow at Constant Area with Friction

For a perfect gas with constant specific heat and molecular weight
k = 1.0

M	$\frac{T}{T^*}$	$\frac{p}{p^*}$	$\frac{p_0}{p_0^*}$	$\frac{V}{V^*}$	$\frac{F}{F^*}$	$4\frac{fL_{max}}{D}$	M	$\frac{T}{T^*}$	$\frac{p}{p^*}$	$\frac{p_0}{p_0^*}$	$\frac{V}{V^*}$	$\frac{F}{F^*}$	$4\frac{fL_{max}}{D}$
0	1.000	∞	∞	0	∞	∞	1.75	1.000	.5714	1.603	1.750	1.161	.4458
0.05		20.000	12.146	.0500	10.025	393.01	1.80		.5556	1.703	1.800	1.178	.4842
.10		10.000	6.096	.1000	5.050	94.39	1.85		.5406	1.815	1.850	1.195	.5225
.15		6.667	4.089	.1500	3.408	39.65	1.90		.5263	1.941	1.900	1.213	.5607
.20		5.000	3.094	.2000	2.600	20.78	1.95		.5128	2.082	1.950	1.231	.5986
.25		4.000	2.503	.2500	2.125	12.227	2.00		.5000	2.241	2.000	1.250	.6363
.30		3.333	2.115	.3000	1.817	7.703	2.05		.4878	2.419	2.050	1.269	.6736
.35		2.857	1.842	.3500	1.604	5.064	2.10		.4762	2.620	2.100	1.288	.7106
.40		2.500	1.643	.4000	1.450	3.417	2.15		.4651	2.846	2.150	1.308	.7472
.45		2.222	1.492	.4500	1.336	2.341	2.20		.4545	3.100	2.200	1.327	.7835
.50		2.000	1.375	.5000	1.250	1.614	2.25		.4444	3.388	2.250	1.347	.8194
.55		1.818	1.283	.5500	1.184	1.110	2.30		.4348	3.714	2.300	1.367	.8549
.60		1.667	1.210	.6000	1.133	.7561	2.35		.4256	4.083	2.350	1.388	.8900
.65		1.539	1.153	.6500	1.0942	.5053	2.40		.4167	4.502	2.400	1.408	.9246
.70		1.429	1.107	.7000	1.0643	.3275	2.45		.4082	4.979	2.450	1.429	.9588
.75		1.333	1.0714	.7500	1.0417	.2024	2.50		.4000	5.522	2.500	1.450	.9926
.80		1.250	1.0441	.8000	1.0250	.1162	2.55		.3922	6.142	2.550	1.471	1.0260
.85		1.176	1.0240	.8500	1.0132	.05904	2.60		.3847	6.852	2.600	1.492	1.0590
.90		1.111	1.0104	.9000	1.0056	.02385	2.65		.3774	7.665	2.650	1.514	1.0916
.95		1.0526	1.0026	.9500	1.0013	.00545	2.70		.3704	8.600	2.700	1.535	1.1237
1.00		1.0000	1.0000	1.0000	1.0000	0	2.75		.3636	9.676	2.750	1.557	1.155
1.05		.9524	1.0025	1.0500	1.0012	.00461	2.80		.3571	10.92	2.800	1.579	1.187
1.10		.9091	1.0097	1.100	1.0045	.01707	2.85		.3509	12.35	2.850	1.600	1.218
1.15		.8695	1.0217	1.150	1.0098	.03567	2.90		.3449	14.02	2.900	1.622	1.248
1.20		.8333	1.0384	1.200	1.0167	.05909	2.95		.3390	15.95	2.950	1.644	1.279
1.25		.8000	1.0598	1.250	1.0250	.08629	3.00		.3333	18.20	3.000	1.667	1.308
1.30		.7692	1.0862	1.300	1.0346	.1164	3.50		.2857	79.22	3.500	1.893	1.587
1.35		.7407	1.118	1.350	1.0453	.1489	4.00		.2500	452.01	4.000	2.125	1.835
1.40		.7143	1.154	1.400	1.0571	.1831	4.50		.2222	3364	4.500	2.361	2.058
1.45		.6897	1.196	1.450	1.0698	.2188	5.00		.2000	32550	5.000	2.600	2.259
1.50		.6667	1.245	1.500	1.0833	.2554	6.00		.1667	$664(10)^4$	6.000	3.083	2.611
1.55		.6452	1.300	1.550	1.0976	.2927	7.00		.1429	$378(10)^7$	7.000	3.571	2.912
1.60		.6250	1.363	1.600	1.112	.3306	8.00		.1250	$599(10)^{10}$	8.000	4.062	3.174
1.65		.6061	1.434	1.650	1.128	.3689	9.00		.1111	$262(10)^{14}$	9.000	4.556	3.407
1.70		.5882	1.514	1.700	1.144	.4073	10.00		.1000	$314(10)^{18}$	10.000	5.050	3.615
							∞	1.000	0	∞	∞	∞	∞

Table 44 Fanno Line—One-Dimensional Compressible-Flow Functions for Adiabatic Flow at Constant Area with Friction

For a perfect gas with constant specific heat and molecular weight
k = 1.1

M	$\frac{T}{T^*}$	$\frac{p}{p^*}$	$\frac{p_0}{p_0^*}$	$\frac{V}{V^*}$	$\frac{F}{F^*}$	$4\frac{fL_{max}}{D}$	M	$\frac{T}{T^*}$	$\frac{p}{p^*}$	$\frac{p_0}{p_0^*}$	$\frac{V}{V^*}$	$\frac{F}{F^*}$	$4\frac{fL_{max}}{D}$
0	1.0500	∞	∞	0	∞	∞	1.75	.9105	.5453	1.528	1.670	1.134	.3667
0.05	1.0499	20.493	11.999	.05123	9.785	357.05	1.80	.9036	.5281	1.610	1.711	1.148	.3969
.10	1.0495	10.244	6.023	.1024	4.932	85.65	1.85	.8966	.5118	1.701	1.752	1.161	.4268
.15	1.0488	6.828	4.042	.1536	3.332	35.92	1.90	.8895	.4964	1.801	1.792	1.175	.4563
.20	1.0479	5.118	3.059	.2047	2.545	18.79	1.95	.8823	.4817	1.911	1.832	1.189	.4854
.25	1.0467	4.092	2.476	.2558	2.083	11.03	2.00	.8750	.4677	2.032	1.871	1.203	.5140
.30	1.0453	3.408	2.094	.3067	1.784	6.936	2.05	.8677	.4544	2.165	1.910	1.217	.5422
.35	1.0436	2.919	1.825	.3575	1.577	4.549	2.10	.8603	.4417	2.312	1.948	1.231	.5698
.40	1.0417	2.552	1.628	.4082	1.429	3.062	2.15	.8529	.4295	2.473	1.986	1.245	.5970
.45	1.0395	2.266	1.480	.4588	1.319	2.093	2.20	.8454	.4179	2.651	2.023	1.259	.6237
.50	1.0370	2.037	1.365	.5092	1.237	1.439	2.25	.8379	.4068	2.846	2.060	1.273	.6498
.55	1.0343	1.849	1.275	.5594	1.174	.9871	2.30	.8304	.3962	3.061	2.096	1.286	.6754
.60	1.0314	1.693	1.204	.6094	1.125	.6705	2.35	.8228	.3860	3.299	2.132	1.300	.7005
.65	1.0283	1.560	1.148	.6591	1.0882	.4468	2.40	.8152	.3762	3.560	2.167	1.314	.7251
.70	1.0249	1.446	1.104	.7086	1.0599	.2887	2.45	.8076	.3668	3.848	2.202	1.328	.7491
.75	1.0213	1.347	1.0689	.7579	1.0386	.1780	2.50	.8000	.3578	4.165	2.236	1.342	.7726
.80	1.0174	1.261	1.0425	.8069	1.0231	.1019	2.55	.7924	.3491	4.515	2.270	1.355	.7957
.85	1.0133	1.184	1.0231	.8557	1.0122	.05160	2.60	.7848	.3407	4.902	2.303	1.369	.8182
.90	1.0091	1.116	1.0100	.9041	1.0051	.02078	2.65	.7771	.3327	5.328	2.336	1.382	.8402
.95	1.0047	1.0551	1.0024	.9522	1.0012	.00472	2.70	.7695	.3249	5.799	2.368	1.395	.8617
1.00	1.0000	1.0000	1.0000	1.0000	1.0000	0	2.75	.7619	.3174	6.320	2.400	1.409	.8828
1.05	.9951	.9501	1.0023	1.0474	1.0011	.00398	2.80	.7543	.3102	6.895	2.432	1.422	.9034
1.10	.9901	.9046	1.0092	1.0945	1.0041	.01468	2.85	.7467	.3032	7.532	2.463	1.434	.9235
1.15	.9849	.8630	1.0204	1.1412	1.0087	.03058	2.90	.7392	.2965	8.237	2.493	1.447	.9432
1.20	.9795	.8247	1.0360	1.1876	1.0148	.05050	2.95	.7316	.2900	9.016	2.523	1.460	.9624
1.25	.9739	.7895	1.0559	1.234	1.0221	.07350	3.00	.7241	.2837	9.880	2.553	1.472	.9812
1.30	.9682	.7569	1.0801	1.279	1.0304	.09885	3.50	.6512	.2305	25.83	2.824	1.589	1.147
1.35	.9623	.7266	1.109	1.324	1.0397	.1260	4.00	.5833	.1909	71.74	3.055	1.691	1.280
1.40	.9563	.6985	1.142	1.369	1.0498	.1544	4.50	.5217	.1605	205.7	3.250	1.779	1.386
1.45	.9501	.6722	1.180	1.413	1.0605	.1838	5.00	.4667	.1366	597.7	3.416	1.854	1.472
1.50	.9438	.6476	1.223	1.457	1.0717	.2138	6.00	.3750	.1021	4949	3.674	1.973	1.601
1.55	.9374	.6246	1.272	1.501	1.0835	.2443	7.00	.3043	.07881	37976	3.862	2.060	1.689
1.60	.9309	.6030	1.326	1.544	1.0958	.2749	8.00	.2500	.06250	$262(10)^3$	4.000	2.125	1.752
1.65	.9242	.5826	1.387	1.586	1.108	.3056	9.00	.2079	.05067	$161(10)^4$	4.104	2.174	1.798
1.70	.9174	.5634	1.454	1.628	1.121	.3362	10.00	.1750	.04183	$887(10)^4$	4.183	2.211	1.832
							∞	0	0	∞	4.583	2.400	1.997

Table 45 Fanno Line—One-Dimensional Compressible-Flow Functions for Adiabatic Flow at Constant Area with Friction

For a perfect gas with constant specific heat and molecular weight
k = 1.2

M	$\frac{T}{T^*}$	$\frac{p}{p^*}$	$\frac{p_0}{p_0^*}$	$\frac{V}{V^*}$	$\frac{F}{F^*}$	$4\frac{fL_{max}}{D}$	M	$\frac{T}{T^*}$	$\frac{p}{p^*}$	$\frac{p_0}{p_0^*}$	$\frac{V}{V^*}$	$\frac{F}{F^*}$	$4\frac{fL_{max}}{D}$
0	1.1000	∞	∞	0	∞	∞	1.75	.8421	.5244	1.471	1.606	1.114	.3072
0.05	1.0997	20.974	11.857	.05243	9.562	327.09	1.80	.8308	.5064	1.540	1.641	1.125	.3316
.10	1.0989	10.483	5.953	.1048	4.822	78.36	1.85	.8195	.4894	1.615	1.675	1.136	.3556
.15	1.0975	6.984	3.996	.1571	3.260	32.81	1.90	.8082	.4732	1.697	1.708	1.147	.3791
.20	1.0956	5.234	3.026	.2093	2.493	17.13	1.95	.7970	.4578	1.787	1.741	1.158	.4021
.25	1.0932	4.182	2.451	.2614	2.044	10.04	2.00	.7857	.4432	1.884	1.773	1.168	.4247
.30	1.0902	3.480	2.073	.3133	1.753	6.298	2.05	.7745	.4293	1.989	1.804	1.179	.4468
.35	1.0867	2.978	1.809	.3649	1.553	4.121	2.10	.7634	.4160	2.103	1.835	1.190	.4684
.40	1.0827	2.601	1.615	.4162	1.409	2.768	2.15	.7523	.4034	2.226	1.865	1.201	.4894
.45	1.0782	2.307	1.469	.4672	1.304	1.887	2.20	.7413	.3913	2.359	1.894	1.211	.5099
.50	1.0732	2.072	1.356	.5179	1.224	1.294	2.25	.7303	.3798	2.504	1.923	1.221	.5299
.55	1.0677	1.879	1.268	.5683	1.164	.8855	2.30	.7194	.3688	2.660	1.951	1.232	.5493
.60	1.0618	1.717	1.199	.6183	1.118	.5999	2.35	.7086	.3582	2.829	1.978	1.242	.5683
.65	1.0554	1.581	1.144	.6678	1.0826	.3987	2.40	.6980	.3481	3.011	2.005	1.252	.5868
.70	1.0486	1.463	1.100	.7168	1.0561	.2570	2.45	.6874	.3384	3.208	2.031	1.262	.6047
.75	1.0414	1.361	1.0666	.7654	1.0360	.1579	2.50	.6769	.3291	3.420	2.057	1.272	.6222
.80	1.0338	1.271	1.0410	.8134	1.0214	.09016	2.55	.6665	.3202	3.650	2.082	1.281	.6392
.85	1.0259	1.192	1.0222	.8609	1.0112	.04554	2.60	.6563	.3116	3.898	2.106	1.291	.6557
.90	1.0176	1.121	1.0096	.9078	1.0047	.01829	2.65	.6462	.3033	4.166	2.130	1.300	.6718
.95	1.0089	1.0573	1.0023	.9542	1.0011	.00414	2.70	.6362	.2954	4.455	2.154	1.309	.6874
1.00	1.0000	1.0000	1.0000	1.0000	1.0000	0	2.75	.6263	.2878	4.767	2.176	1.318	.7026
1.05	.9908	.9480	1.0022	1.0451	1.0010	.00347	2.80	.6166	.2804	5.103	2.199	1.327	.7173
1.10	.9813	.9005	1.0087	1.0896	1.0037	.01277	2.85	.6070	.2733	5.466	2.220	1.335	.7316
1.15	.9715	.8571	1.0194	1.134	1.0079	.02657	2.90	.5975	.2665	5.858	2.242	1.344	.7456
1.20	.9615	.8172	1.0340	1.177	1.0134	.04368	2.95	.5882	.2600	6.280	2.263	1.352	.7592
1.25	.9514	.7803	1.0525	1.219	1.0197	.06338	3.00	.5789	.2536	6.735	2.283	1.360	.7724
1.30	.9410	.7462	1.0749	1.261	1.0270	.08500	3.50	.4944	.2009	13.76	2.461	1.434	.8857
1.35	.9304	.7145	1.101	1.302	1.0351	.1080	4.00	.4231	.1626	28.35	2.602	1.493	.9718
1.40	.9197	.6850	1.132	1.342	1.0437	.1320	4.50	.3636	.1340	57.96	2.714	1.541	1.0380
1.45	.9089	.6575	1.166	1.382	1.0529	.1567	5.00	.3143	.1121	116.34	2.803	1.580	1.0896
1.50	.8980	.6317	1.205	1.421	1.0625	.1817	6.00	.2391	.08150	435.9	2.934	1.637	1.163
1.55	.8869	.6076	1.248	1.459	1.0724	.2069	7.00	.1864	.06168	1469	3.023	1.677	1.212
1.60	.8758	.5849	1.296	1.497	1.0826	.2323	8.00	.1486	.04819	4467	3.084	1.704	1.245
1.65	.8646	.5635	1.349	1.534	1.0930	.2575	9.00	.1209	.03863	12383	3.129	1.724	1.268
1.70	.8534	.5434	1.407	1.570	1.1036	.2825	10.00	.1000	.03162	31623	3.162	1.739	1.286
							∞	0	0	∞	3.317	1.809	1.365

Table 46 Fanno Line—One-Dimensional Compressible-Flow Functions for Adiabatic Flow at Constant Area with Friction

For a perfect gas with constant specific heat and molecular weight
k = 1.3

M	$\frac{T}{T^*}$	$\frac{p}{p^*}$	$\frac{p_0}{p_0^*}$	$\frac{V}{V^*}$	$\frac{F}{F^*}$	$4\frac{fL_{max}}{D}$	M	$\frac{T}{T^*}$	$\frac{p}{p^*}$	$\frac{p_0}{p_0^*}$	$\frac{V}{V^*}$	$\frac{F}{F^*}$	$4\frac{fL_{max}}{D}$
0	1.150	∞	∞	0	∞	∞	1.75	.7880	.5073	1.424	1.554	1.0986	.2613
0.05	1.149	21.444	11.721	.05361	9.354	301.74	1.80	.7739	.4887	1.484	1.584	1.108	.2814
.10	1.148	10.716	5.885	.1072	4.720	72.20	1.85	.7599	.4712	1.549	1.613	1.116	.3010
.15	1.146	7.137	3.952	.1606	3.194	30.18	1.90	.7460	.4546	1.618	1.641	1.125	.3202
.20	1.143	5.346	2.994	.2138	2.445	15.73	1.95	.7323	.4388	1.693	1.669	1.134	.3390
.25	1.139	4.270	2.426	.2668	2.007	9.201	2.00	.7188	.4239	1.773	1.696	1.143	.3573
.30	1.134	3.551	2.054	.3195	1.724	5.759	2.05	.7054	.4097	1.859	1.722	1.151	.3751
.35	1.129	3.036	1.793	.3719	1.530	3.760	2.10	.6922	.3962	1.951	1.747	1.160	.3924
.40	1.123	2.649	1.602	.4239	1.391	2.520	2.15	.6791	.3833	2.050	1.772	1.168	.4092
.45	1.116	2.348	1.459	.4754	1.289	1.714	2.20	.6662	.3710	2.156	1.796	1.176	.4255
.50	1.1084	2.106	1.348	.5264	1.213	1.172	2.25	.6536	.3593	2.268	1.819	1.184	.4413
.55	1.1001	1.907	1.261	.5769	1.155	.8004	2.30	.6412	.3482	2.388	1.842	1.192	.4566
.60	1.0911	1.741	1.193	.6267	1.111	.5409	2.35	.6290	.3375	2.517	1.864	1.200	.4715
.65	1.0815	1.600	1.140	.6759	1.0777	.3586	2.40	.6170	.3273	2.654	1.885	1.208	.4860
.70	1.0713	1.479	1.0972	.7245	1.0524	.2305	2.45	.6051	.3175	2.800	1.906	1.215	.5000
.75	1.0605	1.373	1.0644	.7724	1.0336	.14131	2.50	.5935	.3082	2.954	1.926	1.223	.5136
.80	1.0493	1.280	1.0395	.8195	1.0199	.08044	2.55	.5822	.2992	3.119	1.946	1.230	.5267
.85	1.0376	1.198	1.0214	.8658	1.0104	.04053	2.60	.5711	.2906	3.295	1.965	1.237	.5394
.90	1.0254	1.125	1.0092	.9113	1.0043	.01623	2.65	.5601	.2824	3.482	1.983	1.244	.5517
.95	1.0129	1.0594	1.0022	.9561	1.0010	.00367	2.70	.5493	.2745	3.681	2.001	1.250	.5636
1.00	1.0000	1.0000	1.0000	1.0000	1.0000	0	2.75	.5388	.2669	3.892	2.019	1.257	.5752
1.05	.9868	.9461	1.0021	1.0430	1.0009	.00305	2.80	.5285	.2596	4.116	2.036	1.263	.5864
1.10	.9733	.8969	1.0083	1.0852	1.0033	.01122	2.85	.5184	.2526	4.354	2.052	1.270	.5972
1.15	.9596	.8518	1.0183	1.1266	1.0071	.02324	2.90	.5085	.2459	4.607	2.068	1.276	.6077
1.20	.9457	.8104	1.0321	1.1670	1.0120	.03820	2.95	.4988	.2394	4.875	2.084	1.282	.6179
1.25	.9316	.7722	1.0495	1.206	1.0177	.05524	3.00	.4894	.2332	5.160	2.099	1.288	.6277
1.30	.9174	.7368	1.0704	1.245	1.0241	.07388	3.50	.4053	.1819	9.110	2.228	1.338	.7110
1.35	.9031	.7039	1.0948	1.283	1.0312	.09365	4.00	.3382	.1454	15.94	2.326	1.378	.7726
1.40	.8887	.6734	1.1227	1.320	1.0388	.11417	4.50	.2848	.1186	27.39	2.402	1.409	.8189
1.45	.8743	.6448	1.1543	1.356	1.0467	.13513	5.00	.2421	.09841	45.95	2.460	1.433	.8543
1.50	.8598	.6182	1.189	1.391	1.0549	.1564	6.00	.1797	.07065	120.1	2.543	1.468	.9037
1.55	.8454	.5932	1.228	1.425	1.0634	.1777	7.00	.1377	.05302	285.3	2.598	1.491	.9355
1.60	.8309	.5697	1.271	1.458	1.0721	.1989	8.00	.1085	.04117	623.1	2.635	1.507	.9570
1.65	.8165	.5477	1.318	1.491	1.0808	.2200	9.00	.08745	.03286	1266	2.662	1.519	.9722
1.70	.8022	.5269	1.369	1.523	1.0897	.2408	10.00	.07188	.02681	2416	2.681	1.527	.9832
							∞	0	0	∞	2.769	1.565	1.0326

Table 47 Fanno Line—One-Dimensional Compressible-Flow Functions for Adiabatic Flow at Constant Area with Friction

For a perfect gas with constant specific heat and molecular weight
k = 1.67

M	$\frac{T}{T^*}$	$\frac{p}{p^*}$	$\frac{p_0}{p_0^*}$	$\frac{V}{V^*}$	$\frac{F}{F^*}$	$4\frac{fL_{max}}{D}$	M	$\frac{T}{T^*}$	$\frac{p}{p^*}$	$\frac{p_0}{p_0^*}$	$\frac{V}{V^*}$	$\frac{F}{F^*}$	$4\frac{fL_{max}}{D}$
0	1.335	∞	∞	0	∞	∞	1.75	.6590	.4639	1.312	1.421	1.0623	.1580
0.05	1.334	23.099	11.265	.05775	8.687	234.36	1.80	.6402	.4445	1.351	1.440	1.0673	.1692
.10	1.331	11.535	5.661	.1154	4.392	55.83	1.85	.6219	.4263	1.392	1.459	1.0722	.1800
.15	1.325	7.674	3.805	.1727	2.982	23.21	1.90	.6042	.4091	1.436	1.477	1.0770	.1905
.20	1.317	5.739	2.887	.2296	2.293	12.11	1.95	.5871	.3929	1.482	1.494	1.0817	.2007
.25	1.308	4.574	2.344	.2859	1.892	6.980	2.00	.5705	.3776	1.530	1.510	1.0863	.2105
.30	1.296	3.795	1.989	.3415	1.635	4.337	2.05	.5544	.3632	1.580	1.526	1.0908	.2199
.35	1.282	3.235	1.741	.3963	1.460	2.810	2.10	.5388	.3496	1.632	1.541	1.0952	.2290
.40	1.267	2.814	1.560	.4502	1.336	1.868	2.15	.5238	.3367	1.687	1.556	1.0994	.2377
.45	1.250	2.485	1.424	.5031	1.245	1.260	2.20	.5093	.3244	1.744	1.570	1.1035	.2461
.50	1.232	2.220	1.320	.5549	1.178	.8549	2.25	.4952	.3128	1.803	1.583	1.107	.2542
.55	1.212	2.002	1.239	.6056	1.128	.5787	2.30	.4816	.3017	1.865	1.596	1.111	.2620
.60	1.191	1.819	1.176	.6548	1.0909	.3877	2.35	.4684	.2912	1.929	1.608	1.115	.2694
.65	1.169	1.664	1.126	.7029	1.0628	.2548	2.40	.4557	.2813	1.995	1.620	1.119	.2766
.70	1.146	1.530	1.0874	.7496	1.0418	.1625	2.45	.4434	.2718	2.064	1.631	1.122	.2835
.75	1.1233	1.413	1.0576	.7949	1.0265	.09870	2.50	.4315	.2628	2.135	1.642	1.126	.2901
.80	1.0993	1.311	1.0351	.8388	1.0155	.05576	2.55	.4200	.2542	2.209	1.653	1.129	.2965
.85	1.0748	1.220	1.0189	.8812	1.0080	.02780	2.60	.4089	.2460	2.285	1.663	1.132	.3026
.90	1.0501	1.139	1.0081	.9222	1.0033	.01106	2.65	.3982	.2381	2.364	1.672	1.135	.3085
.95	1.0251	1.0657	1.0019	.9618	1.0008	.00248	2.70	.3878	.2306	2.445	1.682	1.138	.3141
1.00	1.0000	1.0000	1.0000	1.0000	1.0000	0	2.75	.3778	.2235	2.529	1.691	1.141	.3196
1.05	.9749	.9404	1.0018	1.0368	1.0006	.00203	2.80	.3681	.2167	2.616	1.699	1.144	.3248
1.10	.9499	.8860	1.0070	1.0721	1.0024	.00740	2.85	.3587	.2102	2.705	1.707	1.146	.3299
1.15	.9251	.8364	1.0154	1.1061	1.0051	.01522	2.90	.3497	.2039	2.797	1.715	1.149	.3348
1.20	.9006	.7908	1.0266	1.1388	1.0084	.02481	2.95	.3410	.1979	2.892	1.723	1.152	.3395
1.25	.8763	.7489	1.0406	1.170	1.0124	.03564	3.00	.3325	.1922	2.990	1.730	1.154	.3440
1.30	.8524	.7102	1.0573	1.200	1.0167	.04733	3.50	.2616	.1461	4.134	1.790	1.174	.3810
1.35	.8289	.6744	1.0765	1.229	1.0213	.05957	4.00	.2099	.1145	5.608	1.833	1.189	.4071
1.40	.8059	.6412	1.0981	1.257	1.0262	.07212	4.50	.1715	.09203	7.456	1.864	1.200	.4261
1.45	.7833	.6104	1.1220	1.284	1.0313	.08481	5.00	.1424	.07547	9.721	1.887	1.208	.4402
1.50	.7612	.5817	1.148	1.309	1.0364	.09749	6.00	.10222	.05329	15.68	1.918	1.220	.4594
1.55	.7397	.5549	1.176	1.333	1.0416	.1101	7.00	.07666	.03955	23.85	1.938	1.227	.4714
1.60	.7187	.5298	1.207	1.356	1.0468	.1225	8.00	.05949	.03049	34.58	1.951	1.232	.4793
1.65	.6982	.5064	1.240	1.378	1.0520	.1346	9.00	.04745	.02420	48.24	1.960	1.235	.4849
1.70	.6783	.4845	1.275	1.400	1.0572	.1465	10.00	.03870	.01967	65.18	1.967	1.238	.4889
							∞	0	0	∞	1.996	1.249	.5064

Table 48 One-Dimensional Normal-Shock Functions

For a perfect gas with constant specific heat and molecular weight
$k = 1.4$

M_x	M_y	$\dfrac{p_y}{p_x}$	$\dfrac{\rho_y}{\rho_x}$	$\dfrac{T_y}{T_x}$	$\dfrac{p_{0y}}{p_{0x}}$	$\dfrac{p_{0y}}{p_x}$	M_x	M_y	$\dfrac{p_y}{p_x}$	$\dfrac{\rho_y}{\rho_x}$	$\dfrac{T_y}{T_x}$	$\dfrac{p_{0y}}{p_{0x}}$	$\dfrac{p_{0y}}{p_x}$
1.00	1.00000	1.00000	1.00000	1.00000	1.00000	1.8929	1.40	.73971	2.1200	1.6896	1.2547	.95819	3.0493
1.01	.99013	1.02345	1.01669	1.00665	.99999	1.9152	1.41	.73554	2.1528	1.7070	1.2612	.95566	3.0844
1.02	.98052	1.04713	1.03344	1.01325	.99998	1.9379	1.42	.73144	2.1858	1.7243	1.2676	.95306	3.1198
1.03	.97115	1.07105	1.05024	1.01981	.99997	1.9610	1.43	.72741	2.2190	1.7416	1.2742	.95039	3.1555
1.04	.96202	1.09520	1.06709	1.02634	.99994	1.9845	1.44	.72345	2.2525	1.7589	1.2807	.94765	3.1915
1.05	.95312	1.1196	1.08398	1.03284	.99987	2.0083	1.45	.71956	2.2862	1.7761	1.2872	.94483	3.2278
1.06	.94444	1.1442	1.10092	1.03931	.99976	2.0325	1.46	.71574	2.3202	1.7934	1.2938	.94196	3.2643
1.07	.93598	1.1690	1.11790	1.04575	.99962	2.0570	1.47	.71198	2.3544	1.8106	1.3004	.93901	3.3011
1.08	.92772	1.1941	1.13492	1.05217	.99944	2.0819	1.48	.70829	2.3888	1.8278	1.3070	.93600	3.3382
1.09	.91965	1.2194	1.15199	1.05856	.99921	2.1072	1.49	.70466	2.4234	1.8449	1.3136	.93292	3.3756
1.10	.91177	1.2450	1.1691	1.06494	.99892	2.1328	1.50	.70109	2.4583	1.8621	1.3202	.92978	3.4133
1.11	.90408	1.2708	1.1862	1.07130	.99858	2.1588	1.51	.69758	2.4934	1.8792	1.3269	.92658	3.4512
1.12	.89656	1.2968	1.2034	1.07764	.99820	2.1851	1.52	.69413	2.5288	1.8962	1.3336	.92331	3.4894
1.13	.88922	1.3230	1.2206	1.08396	.99776	2.2118	1.53	.69073	2.5644	1.9133	1.3403	.91999	3.5279
1.14	.88204	1.3495	1.2378	1.09027	.99726	2.2388	1.54	.68739	2.6003	1.9303	1.3470	.91662	3.5667
1.15	.87502	1.3762	1.2550	1.09657	.99669	2.2661	1.55	.68410	2.6363	1.9473	1.3538	.91319	3.6058
1.16	.86816	1.4032	1.2723	1.10287	.99605	2.2937	1.56	.68086	2.6725	1.9643	1.3606	.90970	3.6451
1.17	.86145	1.4304	1.2896	1.10916	.99534	2.3217	1.57	.67768	2.7090	1.9812	1.3674	.90615	3.6847
1.18	.85488	1.4578	1.3069	1.11544	.99455	2.3499	1.58	.67455	2.7458	1.9981	1.3742	.90255	3.7245
1.19	.84846	1.4854	1.3243	1.12172	.99371	2.3786	1.59	.67147	2.7828	2.0149	1.3811	.89889	3.7645
1.20	.84217	1.5133	1.3416	1.1280	.99280	2.4075	1.60	.66844	2.8201	2.0317	1.3880	.89520	3.8049
1.21	.83601	1.5414	1.3590	1.1343	.99180	2.4367	1.61	.66545	2.8575	2.0485	1.3949	.89144	3.8456
1.22	.82998	1.5698	1.3764	1.1405	.99073	2.4662	1.62	.66251	2.8951	2.0652	1.4018	.88764	3.8866
1.23	.82408	1.5984	1.3938	1.1468	.98957	2.4961	1.63	.65962	2.9330	2.0820	1.4088	.88380	3.9278
1.24	.81830	1.6272	1.4112	1.1531	.98835	2.5263	1.64	.65677	2.9712	2.0986	1.4158	.87992	3.9693
1.25	.81264	1.6562	1.4286	1.1594	.98706	2.5568	1.65	.65396	3.0096	2.1152	1.4228	.87598	4.0111
1.26	.80709	1.6855	1.4460	1.1657	.98568	2.5876	1.66	.65119	3.0482	2.1318	1.4298	.87201	4.0531
1.27	.80165	1.7150	1.4634	1.1720	.98422	2.6187	1.67	.64847	3.0870	2.1484	1.4369	.86800	4.0954
1.28	.79631	1.7448	1.4808	1.1782	.98268	2.6500	1.68	.64579	3.1261	2.1649	1.4440	.86396	4.1379
1.29	.79108	1.7748	1.4983	1.1846	.98106	2.6816	1.69	.64315	3.1654	2.1813	1.4512	.85987	4.1807
1.30	.78596	1.8050	1.5157	1.1909	.97935	2.7135	1.70	.64055	3.2050	2.1977	1.4583	.85573	4.2238
1.31	.78093	1.8354	1.5331	1.1972	.97758	2.7457	1.71	.63798	3.2448	2.2141	1.4655	.85155	4.2672
1.32	.77600	1.8661	1.5505	1.2035	.97574	2.7783	1.72	.63545	3.2848	2.2304	1.4727	.84735	4.3108
1.33	.77116	1.8970	1.5680	1.2099	.97382	2.8112	1.73	.63296	3.3250	2.2467	1.4800	.84312	4.3547
1.34	.76641	1.9282	1.5854	1.2162	.97181	2.8444	1.74	.63051	3.3655	2.2629	1.4873	.83886	4.3989
1.35	.76175	1.9596	1.6028	1.2226	.96972	2.8778	1.75	.62809	3.4062	2.2791	1.4946	.83456	4.4433
1.36	.75718	1.9912	1.6202	1.2290	.96756	2.9115	1.76	.62570	3.4472	2.2952	1.5019	.83024	4.4880
1.37	.75269	2.0230	1.6376	1.2354	.96534	2.9455	1.77	.62335	3.4884	2.3113	1.5093	.82589	4.5330
1.38	.74828	2.0551	1.6550	1.2418	.96304	2.9798	1.78	.62104	3.5298	2.3273	1.5167	.82152	4.5783
1.39	.74396	2.0874	1.6723	1.2482	.96065	3.0144	1.79	.61875	3.5714	2.3433	1.5241	.81711	4.6238

See page 121 for definitions of symbols.

Table 48   One-Dimensional Normal-Shock Functions

For a perfect gas with constant specific heat and molecular weight
k = 1.4

M_x	M_y	$\dfrac{p_y}{p_x}$	$\dfrac{\rho_y}{\rho_x}$	$\dfrac{T_y}{T_x}$	$\dfrac{p_{0y}}{p_{0x}}$	$\dfrac{p_{0y}}{p_x}$	M_x	M_y	$\dfrac{p_y}{p_x}$	$\dfrac{\rho_y}{\rho_x}$	$\dfrac{T_y}{T_x}$	$\dfrac{p_{0y}}{p_{0x}}$	$\dfrac{p_{0y}}{p_x}$
1.80	.61650	3.6133	2.3592	1.5316	.81268	4.6695	2.20	.54706	5.4800	2.9512	1.8569	.62812	6.7163
1.81	.61428	3.6554	2.3751	1.5391	.80823	4.7155	2.21	.54572	5.5314	2.9648	1.8657	.62358	6.7730
1.82	.61209	3.6978	2.3909	1.5466	.80376	4.7618	2.22	.54440	5.5831	2.9783	1.8746	.61905	6.8299
1.83	.60993	3.7404	2.4067	1.5542	.79926	4.8083	2.23	.54310	5.6350	2.9918	1.8835	.61453	6.8869
1.84	.60780	3.7832	2.4224	1.5617	.79474	4.8551	2.24	.54182	5.6872	3.0052	1.8924	.61002	6.9442
1.85	.60570	3.8262	2.4381	1.5694	.79021	4.9022	2.25	.54055	5.7396	3.0186	1.9014	.60554	7.0018
1.86	.60363	3.8695	2.4537	1.5770	.78567	4.9498	2.26	.53929	5.7922	3.0319	1.9104	.60106	7.0597
1.87	.60159	3.9130	2.4693	1.5847	.78112	4.9974	2.27	.53805	5.8451	3.0452	1.9194	.59659	7.1178
1.88	.59957	3.9568	2.4848	1.5924	.77656	5.0453	2.28	.53683	5.8982	3.0584	1.9285	.59214	7.1762
1.89	.59758	4.0008	2.5003	1.6001	.77197	5.0934	2.29	.53561	5.9515	3.0715	1.9376	.58772	7.2348
1.90	.59562	4.0450	2.5157	1.6079	.76735	5.1417	2.30	.53441	6.0050	3.0846	1.9468	.58331	7.2937
1.91	.59368	4.0894	2.5310	1.6157	.76273	5.1904	2.31	.53322	6.0588	3.0976	1.9560	.57891	7.3529
1.92	.59177	4.1341	2.5463	1.6236	.75812	5.2394	2.32	.53205	6.1128	3.1105	1.9652	.57452	7.4123
1.93	.58988	4.1790	2.5615	1.6314	.75347	5.2886	2.33	.53089	6.1670	3.1234	1.9745	.57015	7.4720
1.94	.58802	4.2242	2.5767	1.6394	.74883	5.3381	2.34	.52974	6.2215	3.1362	1.9838	.56580	7.5319
1.95	.58618	4.2696	2.5919	1.6473	.74418	5.3878	2.35	.52861	6.2762	3.1490	1.9931	.56148	7.5920
1.96	.58437	4.3152	2.6070	1.6553	.73954	5.4378	2.36	.52749	6.3312	3.1617	2.0025	.55717	7.6524
1.97	.58258	4.3610	2.6220	1.6633	.73487	5.4880	2.37	.52638	6.3864	3.1743	2.0119	.55288	7.7131
1.98	.58081	4.4071	2.6369	1.6713	.73021	5.5385	2.38	.52528	6.4418	3.1869	2.0213	.54862	7.7741
1.99	.57907	4.4534	2.6518	1.6794	.72554	5.5894	2.39	.52419	6.4974	3.1994	2.0308	.54438	7.8354
2.00	.57735	4.5000	2.6666	1.6875	.72088	5.6405	2.40	.52312	6.5533	3.2119	2.0403	.54015	7.8969
2.01	.57565	4.5468	2.6814	1.6956	.71619	5.6918	2.41	.52206	6.6094	3.2243	2.0499	.53594	7.9587
2.02	.57397	4.5938	2.6962	1.7038	.71152	5.7434	2.42	.52100	6.6658	3.2366	2.0595	.53175	8.0207
2.03	.57231	4.6411	2.7109	1.7120	.70686	5.7952	2.43	.51996	6.7224	3.2489	2.0691	.52758	8.0830
2.04	.57068	4.6886	2.7255	1.7203	.70218	5.8473	2.44	.51894	6.7792	3.2611	2.0788	.52344	8.1455
2.05	.56907	4.7363	2.7400	1.7286	.69752	5.8997	2.45	.51792	6.8362	3.2733	2.0885	.51932	8.2083
2.06	.56747	4.7842	2.7545	1.7369	.69284	5.9523	2.46	.51691	6.8935	3.2854	2.0982	.51521	8.2714
2.07	.56589	4.8324	2.7690	1.7452	.68817	6.0052	2.47	.51592	6.9510	3.2975	2.1080	.51112	8.3347
2.08	.56433	4.8808	2.7834	1.7536	.68351	6.0584	2.48	.51493	7.0088	3.3095	2.1178	.50706	8.3983
2.09	.56280	4.9295	2.7977	1.7620	.67886	6.1118	2.49	.51395	7.0668	3.3214	2.1276	.50303	8.4622
2.10	.56128	4.9784	2.8119	1.7704	.67422	6.1655	2.50	.51299	7.1250	3.3333	2.1375	.49902	8.5262
2.11	.55978	5.0275	2.8216	1.7789	.66957	6.2194	2.51	.51204	7.1834	3.3451	2.1474	.49502	8.5904
2.12	.55830	5.0768	2.8402	1.7874	.66492	6.2736	2.52	.51109	7.2421	3.3569	2.1574	.49104	8.6549
2.13	.55683	5.1264	2.8543	1.7960	.66029	6.3280	2.53	.51015	7.3010	3.3686	2.1674	.48709	8.7198
2.14	.55538	5.1762	2.8683	1.8046	.65567	6.3827	2.54	.50923	7.3602	3.3802	2.1774	.48317	8.7850
2.15	.55395	5.2262	2.8823	1.8132	.65105	6.4377	2.55	.50831	7.4196	3.3918	2.1875	.47927	8.8505
2.16	.55254	5.2765	2.8962	1.8219	.64644	6.4929	2.56	.50740	7.4792	3.4034	2.1976	.47540	8.9162
2.17	.55114	5.3270	2.9100	1.8306	.64185	6.5484	2.57	.50651	7.5391	3.4149	2.2077	.47155	8.9821
2.18	.54976	5.3778	2.9238	1.8393	.63728	6.6042	2.58	.50562	7.5992	3.4263	2.2179	.46772	9.0482
2.19	.54841	5.4288	2.9376	1.8481	.63270	6.6602	2.59	.50474	7.6595	3.4376	2.2281	.46391	9.1146

Table 48   One-Dimensional Normal-Shock Functions

For a perfect gas with constant specific heat and molecular weight
k = 1.4

M_x	M_y	$\dfrac{p_y}{p_x}$	$\dfrac{\rho_y}{\rho_x}$	$\dfrac{T_y}{T_x}$	$\dfrac{p_{0y}}{p_{0x}}$	$\dfrac{p_{0y}}{p_x}$
2.60	.50387	7.7200	3.4489	2.2383	.46012	9.1813
2.61	.50301	7.7808	3.4602	2.2486	.45636	9.2481
2.62	.50216	7.8418	3.4714	2.2589	.45262	9.3154
2.63	.50132	7.9030	3.4825	2.2693	.44891	9.3829
2.64	.50048	7.9645	3.4936	2.2797	.44522	9.4507
2.65	.49965	8.0262	3.5047	2.2901	.44155	9.5187
2.66	.49883	8.0882	3.5157	2.3006	.43791	9.5869
2.67	.49802	8.1504	3.5266	2.3111	.43429	9.6553
2.68	.49722	8.2128	3.5374	2.3217	.43070	9.7241
2.69	.49642	8.2754	3.5482	2.3323	.42713	9.7932
2.70	.49563	8.3383	3.5590	2.3429	.42359	9.8625
2.71	.49485	8.4014	3.5697	2.3536	.42007	9.9320
2.72	.49408	8.4648	3.5803	2.3643	.41657	10.002
2.73	.49332	8.5284	3.5909	2.3750	.41310	10.072
2.74	.49256	8.5922	3.6014	2.3858	.40965	10.142
2.75	.49181	8.6562	3.6119	2.3966	.40622	10.212
2.76	.49107	8.7205	3.6224	2.4074	.40282	10.283
2.77	.49033	8.7850	3.6328	2.4183	.39945	10.354
2.78	.48960	8.8497	3.6431	2.4292	.39610	10.426
2.79	.48888	8.9147	3.6533	2.4402	.39276	10.498
2.80	.48817	8.9800	3.6635	2.4512	.38946	10.569
2.81	.48746	9.0454	3.6737	2.4622	.38618	10.641
2.82	.48676	9.1111	3.6838	2.4733	.38293	10.714
2.83	.48607	9.1770	3.6939	2.4844	.37970	10.787
2.84	.48538	9.2432	3.7039	2.4955	.37649	10.860
2.85	.48470	9.3096	3.7139	2.5067	.37330	10.933
2.86	.48402	9.3762	3.7238	2.5179	.37013	11.006
2.87	.48334	9.4431	3.7336	2.5292	.36700	11.080
2.88	.48268	9.5102	3.7434	2.5405	.36389	11.154
2.89	.48203	9.5775	3.7532	2.5518	.36080	11.228
2.90	.48138	9.6450	3.7629	2.5632	.35773	11.302
2.91	.48074	9.7127	3.7725	2.5746	.35469	11.377
2.92	.48010	9.7808	3.7821	2.5860	.35167	11.452
2.93	.47946	9.8491	3.7917	2.5975	.34867	11.527
2.94	.47883	9.9176	3.8012	2.6090	.34570	11.603
2.95	.47821	9.986	3.8106	2.6206	.34275	11.679
2.96	.47760	10.055	3.8200	2.6322	.33982	11.755
2.97	.47699	10.124	3.8294	2.6438	.33692	11.831
2.98	.47638	10.194	3.8387	2.6555	.33404	11.907
2.99	.47578	10.263	3.8479	2.6672	.33118	11.984
3.00	.47519	10.333	3.8571	2.6790	.32834	12.061
3.50	.45115	14.125	4.2608	3.3150	.21295	16.242
4.00	.43496	18.500	4.5714	4.0469	.13876	21.068
4.50	.42355	23.458	4.8119	4.8751	.09170	26.539
5.00	.41523	29.000	5.0000	5.8000	.06172	32.654
6.00	.40416	41.833	5.2683	7.941	.02965	46.815
7.00	.39736	57.000	5.4444	10.469	.01535	63.552
8.00	.39289	74.500	5.5652	13.387	.00849	82.865
9.00	.38980	94.333	5.6512	16.693	.00496	104.753
10.00	.38757	116.50	5.7143	20.388	.00304	129.217
∞	.37796	∞	6.000	∞	0	∞

(160)

Table 49 One-Dimensional Normal-Shock Functions

For a perfect gas with constant specific heat and molecular weight
k = 1.0

M_x	M_y	$\dfrac{p_y}{p_x} = \dfrac{\rho_y}{\rho_x}$	$\dfrac{T_y}{T_x}$	$\dfrac{p_{0y}}{p_{0x}}$	$\dfrac{p_{0y}}{p_x}$
1.00	1.0000	1.000	1.000	1.0000	1.649
1.05	.9524	1.103		.9998	1.735
1.10	.9091	1.210		.9988	1.829
1.15	.8696	1.322		.9964	1.930
1.20	.8333	1.440		.9919	2.038
1.25	.8000	1.563		.9851	2.152
1.30	.7692	1.690		.9759	2.272
1.35	.7407	1.822		.9640	2.398
1.40	.7143	1.960		.9494	2.530
1.45	.6897	2.103		.9321	2.667
1.50	.6667	2.250		.9122	2.810
1.55	.6452	2.402		.8899	2.958
1.60	.6250	2.560		.8653	3.112
1.65	.6061	2.723		.8386	3.271
1.70	.5882	2.890		.8100	3.436
1.75	.5714	3.062		.7798	3.606
1.80	.5556	3.240		.7482	3.781
1.85	.5406	3.423		.7155	3.961
1.90	.5263	3.610		.6819	4.146
1.95	.5128	3.802		.6478	4.337
2.00	.5000	4.000		.6134	4.532
2.05	.4878	4.203		.5789	4.733
2.10	.4762	4.410		.5446	4.940
2.15	.4651	4.622		.5106	5.151
2.20	.4545	4.840		.4772	5.367
2.25	.4444	5.063		.4446	5.588
2.30	.4347	5.290		.4129	5.814
2.35	.4255	5.522		.3822	6.045
2.40	.4167	5.760		.3527	6.282
2.45	.4082	6.003		.3244	6.524
2.50	.4000	6.250		.2975	6.771
2.55	.3921	6.502		.2720	7.022
2.60	.3846	6.760		.2479	7.279
2.65	.3774	7.023		.2252	7.541
2.70	.3704	7.290		.2040	7.807
2.75	.3636	7.562		.1842	8.079
2.80	.3571	7.840		.1658	8.356
2.85	.3508	8.123		.1488	8.638
2.90	.3448	8.410		.1332	8.925
2.95	.3390	8.702		.1188	9.217
3.00	.3333	9.000		$1055(10)^{-4}$	9.514
3.50	.2857	12.25		$2791(10)^{-5}$	12.76
4.00	.2500	16.00		$554(10)^{-5}$	16.51
4.50	.2222	20.25		$832(10)^{-6}$	20.76
5.00	.2000	25.00		$951(10)^{-7}$	25.51
6.00	.1667	36.00		$556(10)^{-9}$	36.50
7.00	.1429	49.00		$113(10)^{-11}$	49.50
8.00	.1250	64.00		$817(10)^{-15}$	64.50
9.00	.1111	81.00		$210(10)^{-18}$	81.50
10.00	.1000	100.00		$194(10)^{-22}$	100.50
∞	0	∞	1.000	0	∞

(161)

Table 50 One-Dimensional Normal-Shock Functions

For a perfect gas with constant specific heat and molecular weight
k = 1.1

M_x	M_y	$\dfrac{p_y}{p_x}$	$\dfrac{\rho_y}{\rho_x}$	$\dfrac{T_y}{T_x}$	$\dfrac{p_{oy}}{p_{ox}}$	$\dfrac{p_{oy}}{p_x}$
1.00	1.0000	1.0000	1.000	1.0000	1.0000	1.710
1.05	.9526	1.107	1.097	1.0093	.9998	1.804
1.10	.9099	1.220	1.198	1.0183	.9988	1.906
1.15	.8712	1.338	1.303	1.0271	.9965	2.015
1.20	.8360	1.461	1.410	1.0358	.9921	2.132
1.25	.8038	1.589	1.521	1.0444	.9856	2.255
1.30	.7743	1.723	1.636	1.0529	.9769	2.384
1.35	.7471	1.862	1.754	1.0615	.9657	2.520
1.40	.7221	2.006	1.875	1.0701	.9519	2.662
1.45	.6989	2.155	1.998	1.0788	.9358	2.810
1.50	.6773	2.309	2.124	1.0876	.9174	2.964
1.55	.6573	2.469	2.252	1.0965	.8969	3.124
1.60	.6386	2.634	2.383	1.1055	.8744	3.289
1.65	.6211	2.804	2.516	1.1146	.8501	3.460
1.70	.6048	2.980	2.651	1.1239	.8242	3.637
1.75	.5895	3.161	2.789	1.133	.7970	3.820
1.80	.5751	3.347	2.928	1.143	.7686	4.008
1.85	.5615	3.538	3.069	1.153	.7393	4.202
1.90	.5487	3.734	3.211	1.163	.7093	4.401
1.95	.5366	3.936	3.355	1.173	.6789	4.606
2.00	.5252	4.143	3.500	1.184	.6483	4.817
2.05	.5144	4.355	3.646	1.194	.6175	5.033
2.10	.5042	4.572	3.793	1.205	.5869	5.254
2.15	.4945	4.795	3.942	1.216	.5566	5.481
2.20	.4853	5.023	4.092	1.228	.5267	5.713
2.25	.4765	5.256	4.242	1.239	.4974	5.951
2.30	.4682	5.494	4.393	1.251	.4687	6.194
2.35	.4603	5.738	4.544	1.263	.4408	6.443
2.40	.4527	5.987	4.696	1.275	.4138	6.697
2.45	.4454	6.241	4.848	1.287	.3878	6.957
2.50	.4385	6.500	5.000	1.300	.3627	7.222
2.55	.4319	6.764	5.152	1.313	.3387	7.492
2.60	.4256	7.034	5.304	1.326	.3157	7.768
2.65	.4196	7.309	5.457	1.339	.2938	8.049
2.70	.4138	7.589	5.610	1.353	.2730	8.335
2.75	.4082	7.875	5.762	1.367	.2533	8.627
2.80	.4029	8.166	5.914	1.381	.2347	8.925
2.85	.3978	8.462	6.065	1.395	.2172	9.228
2.90	.3929	8.763	6.216	1.410	.2007	9.536
2.95	.3882	9.069	6.367	1.424	.1852	9.850
3.00	.3837	9.381	6.517	1.439	.17070	10.17
3.50	.3466	12.786	7.977	1.603	.07126	13.66
4.00	.3203	16.714	9.333	1.791	.02750	17.68
4.50	.3009	21.167	10.565	2.003	.01014	22.24
5.00	.2863	26.143	11.667	2.241	.00366	27.35
6.00	.2661	37.67	13.50	2.790	$472(10)^{-6}$	39.16
7.00	.2531	51.29	14.91	3.439	$645(10)^{-7}$	53.12
8.00	.2443	67.00	16.00	4.188	$965(10)^{-8}$	69.23
9.00	.2381	84.81	16.84	5.036	$161(10)^{-8}$	87.49
10.00	.2336	104.71	17.50	5.984	$297(10)^{-9}$	107.90
∞	.2132	∞	21.00	∞	0	∞

Table 51   One-Dimensional Normal-Shock Functions

For a perfect gas with constant specific heat and molecular weight
k = 1.2

M_x	M_y	$\dfrac{p_y}{p_x}$	$\dfrac{\rho_y}{\rho_x}$	$\dfrac{T_y}{T_x}$	$\dfrac{p_{0y}}{p_{0x}}$	$\dfrac{p_{0y}}{p_x}$
1.00	1.0000	1.000	1.000	1.0000	1.0000	1.772
1.05	.9528	1.112	1.092	1.0178	.9998	1.873
1.10	.9106	1.229	1.187	1.0351	.9989	1.982
1.15	.8726	1.352	1.285	1.0521	.9965	2.099
1.20	.8383	1.480	1.385	1.0689	.9923	2.224
1.25	.8071	1.614	1.486	1.0855	.9861	2.356
1.30	.7787	1.753	1.590	1.1022	.9777	2.495
1.35	.7527	1.897	1.696	1.1189	.9671	2.641
1.40	.7288	2.047	1.803	1.1357	.9542	2.793
1.45	.7067	2.203	1.911	1.1527	.9391	2.951
1.50	.6864	2.364	2.020	1.170	.9220	3.115
1.55	.6676	2.530	2.131	1.187	.9030	3.286
1.60	.6501	2.702	2.242	1.205	.8822	3.463
1.65	.6338	2.879	2.354	1.223	.8599	3.646
1.70	.6186	3.062	2.466	1.241	.8362	3.836
1.75	.6044	3.250	2.579	1.260	.8114	4.031
1.80	.5912	3.444	2.692	1.279	.7856	4.232
1.85	.5788	3.643	2.805	1.299	.7591	4.439
1.90	.5671	3.847	2.918	1.319	.7320	4.652
1.95	.5561	4.057	3.031	1.339	.7045	4.871
2.00	.5458	4.273	3.143	1.360	.6768	5.096
2.05	.5360	4.494	3.255	1.381	.6490	5.326
2.10	.5268	4.720	3.366	1.402	.6213	5.562
2.15	.5181	4.952	3.477	1.424	.5938	5.805
2.20	.5099	5.189	3.587	1.446	.5667	6.053
2.25	.5021	5.432	3.697	1.469	.5400	6.307
2.30	.4947	5.680	3.806	1.492	.5139	6.567
2.35	.4877	5.934	3.914	1.516	.4884	6.832
2.40	.4810	6.193	4.020	1.540	.4636	7.104
2.45	.4746	6.457	4.126	1.565	.4397	7.383
2.50	.4686	6.727	4.231	1.590	.4162	7.664
2.55	.4629	7.003	4.335	1.616	.3937	7.952
2.60	.4574	7.284	4.437	1.642	.3721	8.247
2.65	.4521	7.570	4.538	1.668	.3513	8.547
2.70	.4471	7.862	4.638	1.695	.3314	8.853
2.75	.4424	8.159	4.737	1.723	.3124	9.165
2.80	.4378	8.462	4.834	1.751	.2942	9.483
2.85	.4334	8.770	4.930	1.779	.2768	9.806
2.90	.4292	9.084	5.025	1.808	.2603	10.135
2.95	.4252	9.403	5.118	1.837	.2446	10.470
3.00	.4214	9.727	5.211	1.867	.22980	10.81
3.50	.3904	13.273	6.056	2.192	.11978	14.53
4.00	.3690	17.364	6.769	2.565	.06096	18.83
4.50	.3536	22.000	7.364	2.988	.03093	23.70
5.00	.3421	27.182	7.857	3.459	.01586	29.15
6.00	.3267	39.18	8.609	4.551	$441(10)^{-5}$	41.76
7.00	.3170	53.36	9.136	5.841	$134(10)^{-5}$	56.66
8.00	.3106	69.73	9.513	7.329	$450(10)^{-6}$	73.86
9.00	.3061	88.27	9.791	9.016	$164(10)^{-6}$	93.35
10.00	.3029	109.00	10.000	10.900	$650(10)^{-7}$	115.14
∞	.2887	∞	11.00	∞	0	∞

Table 52   One-Dimensional Normal-Shock Functions

For a perfect gas with constant specific heat and molecular weight
k = 1.3

M_x	M_y	$\dfrac{p_y}{p_x}$	$\dfrac{\rho_y}{\rho_x}$	$\dfrac{T_y}{T_x}$	$\dfrac{p_{0y}}{p_{0x}}$	$\dfrac{p_{0y}}{p_x}$
1.00	1.0000	1.000	1.000	1.0000	1.0000	1.832
1.05	.9530	1.116	1.088	1.0257	.9998	1.941
1.10	.9112	1.237	1.178	1.0507	.9989	2.058
1.15	.8739	1.364	1.269	1.0752	.9966	2.183
1.20	.8403	1.497	1.362	1.0995	.9925	2.316
1.25	.8100	1.636	1.456	1.124	.9866	2.457
1.30	.7825	1.780	1.551	1.148	.9786	2.605
1.35	.7575	1.930	1.646	1.172	.9684	2.760
1.40	.7346	2.085	1.742	1.197	.9562	2.922
1.45	.7136	2.246	1.838	1.222	.9421	3.090
1.50	.6942	2.413	1.935	1.247	.9261	3.265
1.55	.6764	2.585	2.031	1.273	.9084	3.447
1.60	.6599	2.763	2.127	1.299	.8891	3.635
1.65	.6446	2.947	2.223	1.326	.8684	3.830
1.70	.6304	3.137	2.318	1.353	.8466	4.031
1.75	.6172	3.332	2.413	1.380	.8238	4.238
1.80	.6048	3.532	2.507	1.408	.8001	4.452
1.85	.5933	3.738	2.601	1.437	.7758	4.672
1.90	.5825	3.950	2.694	1.467	.7510	4.898
1.95	.5724	4.168	2.785	1.497	.7259	5.131
2.00	.5629	4.391	2.875	1.527	.7006	5.370
2.05	.5539	4.620	2.964	1.558	.6752	5.615
2.10	.5455	4.855	3.052	1.590	.6499	5.866
2.15	.5376	5.095	3.139	1.623	.6248	6.123
2.20	.5301	5.341	3.225	1.656	.6000	6.387
2.25	.5230	5.592	3.309	1.690	.5755	6.657
2.30	.5163	5.849	3.392	1.725	.5515	6.933
2.35	.5100	6.112	3.474	1.760	.5280	7.215
2.40	.5040	6.381	3.554	1.796	.5050	7.503
2.45	.4983	6.655	3.633	1.832	.4827	7.798
2.50	.4929	6.935	3.710	1.869	.4610	8.098
2.55	.4878	7.220	3.786	1.907	.4400	8.405
2.60	.4829	7.511	3.860	1.946	.4196	8.718
2.65	.4782	7.808	3.933	1.985	.3999	9.037
2.70	.4738	8.110	4.005	2.025	.3810	9.362
2.75	.4696	8.418	4.075	2.066	.3628	9.693
2.80	.4655	8.732	4.144	2.108	.3452	10.030
2.85	.4616	9.052	4.211	2.150	.3284	10.373
2.90	.4579	9.377	4.277	2.193	.3123	10.723
2.95	.4544	9.708	4.341	2.236	.2969	11.079
3.00	.4511	10.04	4.404	2.280	.28216	11.44
3.50	.4241	13.72	4.964	2.763	.16774	15.39
4.00	.4058	17.96	5.412	3.318	.09933	19.96
4.50	.3927	22.76	5.768	3.946	.05939	25.13
5.00	.3832	28.13	6.053	4.648	.03613	30.92
6.00	.3704	40.57	6.469	6.271	$1422(10)^{-5}$	44.31
7.00	.3625	55.26	6.749	8.189	$610(10)^{-5}$	60.14
8.00	.3573	72.22	6.943	10.401	$283(10)^{-5}$	78.40
9.00	.3536	91.43	7.084	12.908	$140(10)^{-5}$	99.10
10.00	.3510	112.91	7.188	15.710	$740(10)^{-6}$	122.24
∞	.3397	∞	7.667	∞	0	∞

(164)

Table 53   One-Dimensional Normal-Shock Functions

For a perfect gas with constant specific heat and molecular weight
k = 1.67

M_x	M_y	$\dfrac{p_y}{p_x}$	$\dfrac{\rho_y}{\rho_x}$	$\dfrac{T_y}{T_x}$	$\dfrac{p_{0y}}{p_{0x}}$	$\dfrac{p_{0y}}{p_x}$
1.00	1.0000	1.000	1.000	1.0000	1.0000	2.055
1.05	.9535	1.128	1.075	1.0496	.9998	2.189
1.10	.9131	1.262	1.149	1.0985	.9990	2.333
1.15	.8776	1.403	1.223	1.1471	.9969	2.486
1.20	.8463	1.550	1.297	1.195	.9934	2.649
1.25	.8184	1.703	1.370	1.244	.9883	2.821
1.30	.7935	1.863	1.441	1.293	.9813	3.002
1.35	.7711	2.029	1.511	1.343	.9728	3.191
1.40	.7509	2.201	1.580	1.394	.9627	3.388
1.45	.7325	2.379	1.647	1.445	.9511	3.592
1.50	.7158	2.563	1.713	1.497	.9381	3.804
1.55	.7006	2.754	1.777	1.550	.9239	4.025
1.60	.6866	2.951	1.840	1.604	.9086	4.254
1.65	.6738	3.154	1.901	1.659	.8924	4.490
1.70	.6620	3.364	1.960	1.716	.8754	4.733
1.75	.6511	3.580	2.018	1.774	.8577	4.984
1.80	.6410	3.802	2.074	1.833	.8395	5.243
1.85	.6316	4.030	2.128	1.893	.8209	5.510
1.90	.6229	4.265	2.181	1.955	.8019	5.784
1.95	.6148	4.506	2.232	2.018	.7827	6.065
2.00	.6073	4.753	2.282	2.083	.7634	6.354
2.05	.6002	5.006	2.330	2.149	.7441	6.650
2.10	.5936	5.266	2.376	2.216	.7248	6.954
2.15	.5875	5.532	2.421	2.284	.7056	7.266
2.20	.5817	5.804	2.465	2.354	.6866	7.585
2.25	.5762	6.082	2.507	2.426	.6678	7.911
2.30	.5711	6.366	2.548	2.499	.6493	8.244
2.35	.5663	6.657	2.587	2.574	.6310	8.585
2.40	.5617	6.954	2.625	2.650	.6130	8.934
2.45	.5574	7.257	2.662	2.727	.5954	9.290
2.50	.5534	7.567	2.697	2.806	.5783	9.65
2.55	.5495	7.883	2.731	2.886	.5615	10.02
2.60	.5459	8.205	2.764	2.968	.5450	10.40
2.65	.5425	8.533	2.796	3.052	.5289	10.79
2.70	.5392	8.868	2.827	3.137	.5133	11.18
2.75	.5361	9.209	2.857	3.223	.4981	11.58
2.80	.5331	9.556	2.886	3.311	.4833	11.99
2.85	.5303	9.909	2.914	3.401	.4689	12.41
2.90	.5276	10.269	2.941	3.492	.4550	12.83
2.95	.5251	10.635	2.967	3.584	.4415	13.26
3.00	.5227	11.01	2.992	3.678	.4283	13.69
3.50	.5036	15.07	3.204	4.704	.3177	18.47
4.00	.4910	19.76	3.358	5.885	.2384	23.99
4.50	.4822	25.08	3.473	7.221	.1816	30.24
5.00	.4758	31.02	3.560	8.714	.1406	37.23
6.00	.4674	44.78	3.680	12.17	.08831	53.40
7.00	.4623	61.04	3.756	16.25	.05854	72.53
8.00	.4589	79.81	3.807	20.96	.04059	94.59
9.00	.4566	101.08	3.843	26.30	.02920	119.6
10.00	.4550	124.84	3.870	32.26	.02167	147.6
∞	.4479	∞	3.985	∞	0	∞

Table 54 Two-Dimensional Isentropic Compressible-Flow Functions for Method of Characteristics

For a perfect gas with constant specific heat and molecular weight
k = 1.4

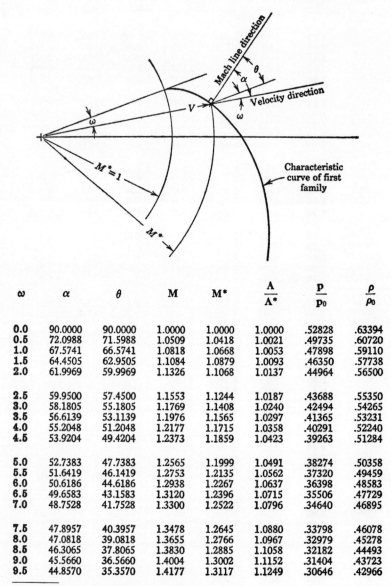

ω	α	θ	M	M*	$\dfrac{A}{A^*}$	$\dfrac{p}{p_0}$	$\dfrac{\rho}{\rho_0}$
0.0	90.0000	90.0000	1.0000	1.0000	1.0000	.52828	.63394
0.5	72.0988	71.5988	1.0509	1.0418	1.0021	.49735	.60720
1.0	67.5741	66.5741	1.0818	1.0668	1.0053	.47898	.59110
1.5	64.4505	62.9505	1.1084	1.0879	1.0093	.46350	.57738
2.0	61.9969	59.9969	1.1326	1.1068	1.0137	.44964	.56500
2.5	59.9500	57.4500	1.1553	1.1244	1.0187	.43688	.55350
3.0	58.1805	55.1805	1.1769	1.1408	1.0240	.42494	.54265
3.5	56.6139	53.1139	1.1976	1.1565	1.0297	.41365	.53231
4.0	55.2048	51.2048	1.2177	1.1715	1.0358	.40291	.52240
4.5	53.9204	49.4204	1.2373	1.1859	1.0423	.39263	.51284
5.0	52.7383	47.7383	1.2565	1.1999	1.0491	.38274	.50358
5.5	51.6419	46.1419	1.2753	1.2135	1.0562	.37320	.49459
6.0	50.6186	44.6186	1.2938	1.2267	1.0637	.36398	.48583
6.5	49.6583	43.1583	1.3120	1.2396	1.0715	.35506	.47729
7.0	48.7528	41.7528	1.3300	1.2522	1.0796	.34640	.46895
7.5	47.8957	40.3957	1.3478	1.2645	1.0880	.33798	.46078
8.0	47.0818	39.0818	1.3655	1.2766	1.0967	.32979	.45278
8.5	46.3065	37.8065	1.3830	1.2885	1.1058	.32182	.44493
9.0	45.5660	36.5660	1.4004	1.3002	1.1152	.31404	.43723
9.5	44.8570	35.3570	1.4177	1.3117	1.1249	.30646	.42966

See page 121 for definitions of symbols.

Table 54   Method of Characteristics

For a perfect gas with constant specific heat and molecular weight
k = 1.4

ω	α	θ	M	M*	$\dfrac{A}{A^*}$	$\dfrac{p}{p_0}$	$\dfrac{\rho}{\rho_0}$
10.0	44.1770	34.1770	1.4349	1.3230	1.1349	.29906	.42222
10.5	43.5233	33.0233	1.4521	1.3341	1.1453	.29184	.41491
11.0	42.8940	31.8940	1.4692	1.3451	1.1560	.28478	.40772
11.5	42.2869	30.7869	1.4862	1.3559	1.1670	.27788	.40064
12.0	41.7007	29.7007	1.5032	1.3666	1.1783	.27114	.39367
12.5	41.1338	28.6338	1.5202	1.3772	1.1900	.26454	.38680
13.0	40.5849	27.5849	1.5371	1.3876	1.2021	.25809	.38004
13.5	40.0529	26.5529	1.5540	1.3979	1.2145	.25178	.37338
14.0	39.5366	25.5366	1.5709	1.4081	1.2273	.24560	.36681
14.5	39.0350	24.5350	1.5878	1.4182	1.2405	.23955	.36034
15.0	38.5474	23.5474	1.6047	1.4282	1.2541	.23363	.35396
15.5	38.0730	22.5730	1.6216	1.4380	1.2680	.22783	.34766
16.0	37.6108	21.6108	1.6385	1.4478	1.2823	.22216	.34145
16.5	37.1605	20.6605	1.6555	1.4575	1.2970	.21661	.33533
17.0	36.7212	19.7212	1.6725	1.4671	1.3121	.21117	.32929
17.5	36.2925	18.7925	1.6895	1.4766	1.3277	.20584	.32334
18.0	35.8739	17.8739	1.7065	1.4860	1.3437	.20062	.31747
18.5	35.4648	16.9648	1.7235	1.4953	1.3602	.19551	.31168
19.0	35.0648	16.0648	1.7406	1.5046	1.3771	.19051	.30596
19.5	34.6735	15.1735	1.7578	1.5138	1.3945	.18562	.30032
20.0	34.2904	14.2904	1.7750	1.5229	1.4123	.18082	.29475
20.5	33.9153	13.4153	1.7922	1.5319	1.4306	.17612	.28926
21.0	33.5479	12.5479	1.8095	1.5409	1.4494	.17152	.28385
21.5	33.1877	11.6877	1.8269	1.5498	1.4687	.16702	.27851
22.0	32.8344	10.8344	1.8443	1.5586	1.4886	.16261	.27324
22.5	32.4879	9.9879	1.8618	1.5673	1.5090	.15830	.26804
23.0	32.1478	9.1478	1.8793	1.5760	1.5300	.15408	.26291
23.5	31.8138	8.3138	1.8969	1.5846	1.5515	.14995	.25786
24.0	31.4859	7.4859	1.9146	1.5932	1.5736	.14590	.25287
24.5	31.1637	6.6637	1.9324	1.6017	1.5963	.14194	.24795
25.0	30.8469	5.8469	1.9503	1.6101	1.6197	.13806	.24310
25.5	30.5355	5.0355	1.9682	1.6184	1.6437	.13427	.23831
26.0	30.2293	4.2293	1.9862	1.6267	1.6683	.13057	.23359
26.5	29.9281	3.4281	2.0044	1.6350	1.6936	.12694	.22894
27.0	29.6316	2.6316	2.0226	1.6432	1.7196	.12339	.22435
27.5	29.3397	1.8397	2.0409	1.6513	1.7464	.11992	.21982
28.0	29.0524	1.0524	2.0593	1.6594	1.7739	.11653	.21536
28.5	28.7694	0.2694	2.0778	1.6674	1.8022	.11321	.21097
29.0	28.4906	−0.5094	2.0964	1.6753	1.8312	.10997	.20664
29.5	28.2158	−1.2842	2.1151	1.6832	1.8611	.10680	.20237

Table 54 Method of Characteristics

For a perfect gas with constant specific heat and molecular weight
k = 1.4

ω	α	θ	M	M*	$\frac{A}{A^*}$	$\frac{p}{p_0}$	$\frac{\rho}{\rho_0}$
30.0	27.9451	−2.0549	2.1339	1.6911	1.8918	.10370	.19816
30.5	27.6782	−2.8218	2.1528	1.6989	1.9233	.10068	.19401
31.0	27.4149	−3.5851	2.1718	1.7066	1.9557	.09773	.18992
31.5	27.1552	−4.3448	2.1910	1.7143	1.9891	.09484	.18590
32.0	26.8991	−5.1009	2.2103	1.7220	2.0235	.09202	.18194
32.5	26.6464	−5.8536	2.2297	1.7296	2.0588	.08927	.17804
33.0	26.3970	−6.6030	2.2492	1.7371	2.0951	.08658	.17419
33.5	26.1507	−7.3493	2.2689	1.7446	2.1324	.08395	.17040
34.0	25.9076	−8.0924	2.2887	1.7521	2.1709	.08139	.16667
34.5	25.6675	−8.8325	2.3086	1.7595	2.2105	.07889	.16300
35.0	25.4304	−9.5696	2.3287	1.7669	2.2512	.07646	.15938
35.5	25.1962	−10.3038	2.3489	1.7742	2.2931	.07408	.15582
36.0	24.9648	−11.0352	2.3693	1.7814	2.3363	.07176	.15232
36.5	24.7361	−11.7639	2.3898	1.7886	2.3807	.06950	.14887
37.0	24.5101	−12.4899	2.4105	1.7958	2.4264	.06729	.14548
37.5	24.2866	−13.2134	2.4313	1.8029	2.4736	.06514	.14215
38.0	24.0657	−13.9343	2.4523	1.8100	2.5222	.06304	.13886
38.5	23.8473	−14.6527	2.4734	1.8170	2.5722	.06100	.13563
39.0	23.6313	−15.3687	2.4947	1.8240	2.6237	.05901	.13246
39.5	23.4176	−16.0824	2.5162	1.8310	2.6768	.05707	.12934
40.0	23.2061	−16.7939	2.5378	1.8379	2.7316	.05519	.12627
40.5	22.9969	−17.5031	2.5596	1.8447	2.7881	.05335	.12325
41.0	22.7900	−18.2100	2.5816	1.8515	2.8463	.05156	.12029
41.5	22.5852	−18.9148	2.6038	1.8583	2.9063	.04982	.11738
42.0	22.3824	−19.6176	2.6261	1.8650	2.9682	.04813	.11452
42.5	22.1816	−20.3184	2.6487	1.8717	3.0321	.04648	.11170
43.0	21.9828	−21.0172	2.6714	1.8783	3.0981	.04488	.10894
43.5	21.7860	−21.7140	2.6944	1.8849	3.1662	.04332	.10622
44.0	21.5911	−22.4089	2.7176	1.8915	3.2364	.04181	.10356
44.5	21.3980	−23.1020	2.7409	1.8980	3.3089	.04034	.10094
45.0	21.2068	−23.7932	2.7644	1.9045	3.3838	.03891	.09837
45.5	21.0174	−24.4826	2.7882	1.9109	3.4611	.03752	.09585
46.0	20.8297	−25.1703	2.8122	1.9173	3.5410	.03617	.09338
46.5	20.6437	−25.8563	2.8364	1.9236	3.6236	.03486	.09095
47.0	20.4594	−26.5406	2.8609	1.9299	3.7089	.03359	.08856
47.5	20.2767	−27.2233	2.8856	1.9362	3.7971	.03235	.08622
48.0	20.0956	−27.9044	2.9105	1.9424	3.8883	.03115	.08393
48.5	19.9160	−28.5840	2.9356	1.9486	3.9827	.02999	.08168
49.0	19.7380	−29.2620	2.9610	1.9547	4.0803	.02886	.07948
49.5	19.5615	−29.9385	2.9867	1.9608	4.1812	.02777	.07732

Table 54   Method of Characteristics

For a perfect gas with constant specific heat and molecular weight
k = 1.4

ω	α	θ	M	M*	$\frac{A}{A^*}$	$\frac{p}{p_0}$	$\frac{\rho}{\rho_0}$
50.0	19.3865	−30.6135	3.0126	1.9669	4.2857	.02671	.07520
50.5	19.2129	−31.2871	3.0388	1.9729	4.3938	.02568	.07313
51.0	19.0408	−31.9592	3.0652	1.9789	4.5058	.02469	.07110
51.5	18.8700	−32.6300	3.0919	1.9848	4.6218	.02373	.06911
52.0	18.7005	−33.2995	3.1189	1.9907	4.7419	.02280	.06716
52.5	18.5324	−33.9676	3.1462	1.9966	4.8663	.02190	.06525
53.0	18.3657	−34.6343	3.1738	2.0024	4.9953	.02103	.06338
53.5	18.2002	−35.2998	3.2016	2.0082	5.1290	.02018	.06155
54.0	18.0360	−35.9640	3.2298	2.0139	5.2676	.01936	.05976
54.5	17.8730	−36.6270	3.2583	2.0196	5.4114	.01857	.05801
55.0	17.7112	−37.2888	3.2871	2.0253	5.5606	.01781	.05629
55.5	17.5506	−37.9494	3.3162	2.0309	5.7154	.01707	.05461
56.0	17.3911	−38.6089	3.3457	2.0365	5.8761	.01636	.05297
56.5	17.2328	−39.2672	3.3755	2.0421	6.0429	.01567	.05137
57.0	17.0757	−39.9243	3.4056	2.0476	6.2162	.01500	.04981
57.5	16.9196	−40.5804	3.4361	2.0531	6.3963	.01436	.04828
58.0	16.7646	−41.2354	3.4669	2.0585	6.5834	.01374	.04678
58.5	16.6107	−41.8893	3.4981	2.0639	6.7778	.01314	.04532
59.0	16.4579	−42.5421	3.5297	2.0692	6.9799	.01257	.04389
59.5	16.3061	−43.1939	3.5616	2.0745	7.1902	.01202	.04250
60.0	16.1552	−43.8448	3.5940	2.0798	7.4090	.01148	.04114
60.5	16.0053	−44.4947	3.6268	2.0850	7.6368	.01096	.03981
61.0	15.8564	−45.1436	3.6600	2.0902	7.8739	.01047	.03851
61.5	15.7085	−45.7915	3.6936	2.0954	8.1208	.00999	.03725
62.0	15.5615	−46.4385	3.7276	2.1005	8.3780	.00953	.03602
62.5	15.4154	−47.0846	3.7620	2.1056	8.6460	.00909	.03482
63.0	15.2703	−47.7297	3.7969	2.1107	8.9254	.00867	.03365
63.5	15.1260	−48.3740	3.8323	2.1157	9.2168	.00826	.03251
64.0	14.9826	−49.0174	3.8681	2.1207	9.5208	.00786	.03140
64.5	14.8400	−49.6600	3.9044	2.1256	9.8380	.00749	.03032
65.0	14.6983	−50.3017	3.9412	2.1305	10.169	.00712	.02926
65.5	14.5574	−50.9426	3.9785	2.1353	10.515	.00678	.02823
66.0	14.4174	−51.5826	4.0163	2.1401	10.876	.00644	.02723
66.5	14.2781	−52.2219	4.0547	2.1449	11.253	.00612	.02626
67.0	14.1396	−52.8604	4.0936	2.1497	11.648	.00582	.02532
67.5	14.0019	−53.4981	4.1330	2.1544	12.061	.00552	.02440
68.0	13.8650	−54.1350	4.1730	2.1591	12.493	.00524	.02350
68.5	13.7288	−54.7712	4.2136	2.1637	12.945	.00497	.02263
69.0	13.5934	−55.4066	4.2548	2.1683	13.418	.00472	.02179
69.5	13.4587	−56.0413	4.2966	2.1728	13.913	.00447	.02097

Table 54   Method of Characteristics

For a perfect gas with constant specific heat and molecular weight
k = 1.4

ω	α	θ	M	M*	$\frac{A}{A^*}$	$\frac{p}{p_0}$	$\frac{\rho}{\rho_0}$
70.0	13.3247	−56.6753	4.3390	2.1773	14.433	.00423	.02017
70.5	13.1913	−57.3087	4.3821	2.1818	14.978	.00401	.01940
71.0	13.0587	−57.9413	4.4258	2.1863	15.549	.00379	.01865
71.5	12.9268	−58.5732	4.4702	2.1907	16.148	.00359	.01792
72.0	12.7955	−59.2045	4.5152	2.1951	16.777	.00339	.01721
72.5	12.6649	−59.8351	4.5610	2.1994	17.438	.00320	.01652
73.0	12.5349	−60.4651	4.6076	2.2037	18.132	.00302	.01586
73.5	12.4055	−61.0945	4.6549	2.2080	18.862	.00285	.01522
74.0	12.2768	−61.7232	4.7029	2.2122	19.630	.00269	.01460
74.5	12.1487	−62.3513	4.7517	2.2164	20.438	.00254	.01400
75.0	12.0212	−62.9788	4.8014	2.2205	21.288	.00239	.01342
75.5	11.8943	−63.6057	4.8519	2.2246	22.183	.00225	.01285
76.0	11.7680	−64.2320	4.9032	2.2287	23.126	.00212	.01230
76.5	11.6422	−64.8578	4.9554	2.2327	24.121	.00199	.01177
77.0	11.5170	−65.4830	5.0085	2.2367	25.171	.00187	.01126
77.5	11.3924	−66.1076	5.0626	2.2407	26.279	.00176	.01077
78.0	11.2683	−66.7317	5.1176	2.2446	27.449	.00165	.01029
78.5	11.1447	−67.3553	5.1736	2.2485	28.685	.00155	.00983
79.0	11.0217	−67.9783	5.2306	2.2523	29.992	.00145	.00939
79.5	10.8992	−68.6008	5.2887	2.2561	31.374	.00136	.00896
80.0	10.7772	−69.2228	5.3479	2.2599	32.837	.00127	.00854
80.5	10.6558	−69.8442	5.4081	2.2636	34.387	.00119	.00814
81.0	10.5348	−70.4652	5.4694	2.2673	36.029	.00111	.00776
81.5	10.4143	−71.0857	5.5320	2.2710	37.770	.00104	.00739
82.0	10.2942	−71.7058	5.5959	2.2746	39.618	.00097	.00703
82.5	10.1746	−72.3254	5.6610	2.2782	41.580	.00090	.00669
83.0	10.0555	−72.9445	5.7274	2.2818	43.665	.00084	.00636
83.5	9.9369	−73.5631	5.7950	2.2853	45.882	.00078	.00604
84.0	9.8187	−74.1813	5.8640	2.2888	48.240	.00073	.00574
84.5	9.7010	−74.7990	5.9345	2.2922	50.750	.00068	.00545
85.0	9.5837	−75.4163	6.0064	2.2956	53.424	.00063	.00517
85.5	9.4668	−76.0332	6.0799	2.2990	56.276	.00058	.00490
86.0	9.3503	−76.6497	6.1550	2.3023	59.320	.00054	.00464
86.5	9.2342	−77.2658	6.2317	2.3056	62.570	.00050	.00439
87.0	9.1185	−77.8815	6.3101	2.3088	66.043	.00046	.00416
87.5	9.0032	−78.4968	6.3902	2.3120	69.758	.00043	.00393
88.0	8.8884	−79.1116	6.4720	2.3152	73.734	.00040	.00371
88.5	8.7740	−79.7260	6.5558	2.3183	77.994	.00037	.00351
89.0	8.6599	−80.3401	6.6415	2.3214	82.561	.00034	.00331
89.5	8.5462	−80.9538	6.7292	2.3245	87.463	.00031	.00312
90.0	8.4328	−81.5672	6.8190	2.3275	92.730	.00029	.00294

Table 55   Wedge Angle for Sonic Flow Downstream of and Maximum Wedge Angle for Two-Dimensional Shock

For a perfect gas with constant specific heat and molecular weight
k = 1.4

M_x	ω_a'	ω_m'	M_x	ω_a'	ω_m'	M_x	ω_a'	ω_m'	M_x	ω_a'	ω_m'
1.01	0.05°	0.05°	1.36	7.94°	8.32°	1.71	16.86°	17.24°	2.06	23.73°	23.98°
1.02	0.13	0.15	1.37	8.21	8.60	1.72	17.09	17.46	2.07	23.90	24.14
1.03	0.24	0.26	1.38	8.48	8.88	1.73	17.31	17.68	2.08	24.06	24.30
1.04	0.37	0.40	1.39	8.76	9.15	1.74	17.54	17.90	2.09	24.22	24.46
1.05	0.52	0.56	1.40	9.03	9.43	1.75	17.76	18.12	2.10	24.38	24.61
1.06	0.67	0.73	1.41	9.30	9.70	1.76	17.98	18.34	2.11	24.54	24.77
1.07	0.84	0.91	1.42	9.57	9.97	1.77	18.20	18.55	2.12	24.70	24.92
1.08	1.02	1.10	1.43	9.84	10.25	1.78	18.41	18.76	2.13	24.85	25.08
1.09	1.21	1.30	1.44	10.10	10.52	1.79	18.63	18.97	2.14	25.01	25.23
1.10	1.41	1.51	1.45	10.37	10.79	1.80	18.84	19.18	2.15	25.16	25.38
1.11	1.61	1.73	1.46	10.64	11.05	1.81	19.05	19.39	2.16	25.31	25.52
1.12	1.82	1.96	1.47	10.90	11.32	1.82	19.26	19.59	2.17	25.46	25.67
1.13	2.04	2.19	1.48	11.17	11.59	1.83	19.46	19.80	2.18	25.61	25.82
1.14	2.26	2.43	1.49	11.43	11.85	1.84	19.67	20.00	2.19	25.76	25.96
1.15	2.49	2.67	1.50	11.69	12.11	1.85	19.87	20.20	2.20	25.90	26.10
1.16	2.73	2.92	1.51	11.96	12.37	1.86	20.07	20.40	2.21	26.05	26.24
1.17	2.97	3.17	1.52	12.21	12.63	1.87	20.27	20.59	2.22	26.19	26.38
1.18	3.21	3.42	1.53	12.47	12.89	1.88	20.47	20.78	2.23	26.33	26.52
1.19	3.45	3.68	1.54	12.73	13.15	1.89	20.67	20.98	2.24	26.47	26.66
1.20	3.70	3.94	1.55	12.99	13.40	1.90	20.86	21.17	2.25	26.61	26.80
1.21	3.95	4.21	1.56	13.24	13.66	1.91	21.05	21.36	2.3	27.28	27.45
1.22	4.21	4.48	1.57	13.49	13.91	1.92	21.24	21.54	2.4	28.53	28.68
1.23	4.46	4.74	1.58	13.74	14.16	1.93	21.43	21.73	2.5	29.67	29.80
1.24	4.72	5.01	1.59	13.99	14.41	1.94	21.62	21.91	2.6	30.70	30.81
1.25	4.99	5.29	1.60	14.24	14.65	1.95	21.81	22.09			
									2.7	31.64	31.74
1.26	5.25	5.56	1.61	14.49	14.90	1.96	21.99	22.27	2.8	32.50	32.59
1.27	5.52	5.83	1.62	14.73	15.14	1.97	22.17	22.45	2.9	33.29	33.36
1.28	5.78	6.11	1.63	14.98	15.38	1.98	22.35	22.63	3.0	34.01	34.07
1.29	6.05	6.39	1.64	15.22	15.62	1.99	22.53	22.80			
1.30	6.32	6.66	1.65	15.46	15.86	2.00	22.71	22.97	4	38.75	38.77
									5	41.11	41.12
1.31	6.59	6.94	1.66	15.70	16.09	2.01	22.88	23.14	6	42.44	42.44
1.32	6.86	7.22	1.67	15.93	16.32	2.02	23.05	23.31	8	43.79	43.79
1.33	7.13	7.49	1.68	16.17	16.55	2.03	23.23	23.48			
1.34	7.40	7.77	1.69	16.40	16.78	2.04	23.40	23.65	10	44.43	44.43
1.35	7.67	8.05	1.70	16.63	17.01	2.05	23.56	23.81	15	45.07	45.07
									20	45.29	45.29
									∞	45.58	45.58

See page 121 for definitions of symbols.

Table 56 Upstream Mach Number, M_x, for Two-Dimensional Shock

For a perfect gas with constant specific heat and molecular weight
$k = 1.4$

$\alpha' \rightarrow$ $\omega' \downarrow$	15°	16°	17°	18°	19°	20°	21°	22°	23°	24°	25°	26°	27°	28°	29°	30°	31°
1°	4.036	3.779	3.555	3.358	3.182	3.023	2.882	2.754	2.637	2.531	2.433	2.344	2.262	2.186	2.116	2.050	1.989
2°	4.232	3.950	3.706	3.491	3.301	3.132	2.981	2.844	2.721	2.608	2.505	2.410	2.325	2.244	2.171	2.102	2.039
3°	4.457	4.145	3.876	3.643	3.436	3.252	3.089	2.943	2.811	2.691	2.582	2.482	2.391	2.307	2.230	2.158	2.091
4°	4.717	4.369	4.070	3.811	3.586	3.386	3.211	3.051	2.910	2.782	2.666	2.560	2.463	2.374	2.293	2.217	2.147
5°	5.032	4.632	4.294	4.005	3.755	3.536	3.344	3.173	3.019	2.881	2.757	2.644	2.541	2.447	2.360	2.280	2.207
6°	5.414	4.946	4.559	4.232	3.949	3.708	3.495	3.307	3.139	2.990	2.856	2.735	2.625	2.525	2.433	2.349	2.271
7°	5.897	5.332	4.876	4.498	4.177	3.904	3.666	3.459	3.275	3.112	2.966	2.835	2.717	2.609	2.511	2.421	2.339
8°	6.531	5.823	5.267	4.818	4.445	4.131	3.864	3.630	3.427	3.249	3.089	2.946	2.818	2.702	2.597	2.500	2.413
9°	7.425	6.475	5.767	5.215	4.771	4.403	4.094	3.830	3.601	3.403	3.226	3.069	2.930	2.804	2.690	2.587	2.492
10°	8.823	7.407	6.442	5.729	5.177	4.734	4.371	4.065	3.805	3.579	3.381	3.209	3.055	2.916	2.793	2.681	2.579
11°	11.486	8.902	7.419	6.427	5.706	5.151	4.709	4.346	4.042	3.783	3.560	3.366	3.195	3.043	2.907	2.785	2.675
12°	20.743	11.921	9.034	7.459	6.433	5.698	5.137	4.691	4.329	4.027	3.769	3.547	3.356	3.185	3.035	2.901	2.780
13°		26.089	12.533	9.223	7.532	6.460	5.703	5.134	4.684	4.321	4.017	3.761	3.540	3.348	3.180	3.030	2.898
14°			45.340	13.410	9.485	7.638	6.507	5.725	5.142	4.686	4.319	4.015	3.757	3.537	3.346	3.178	3.030
15°					14.693	9.830	7.782	6.579	5.761	5.161	4.698	4.324	4.017	3.759	3.538	3.348	3.180
16°						16.729	10.295	7.970	6.672	5.815	5.193	4.717	4.338	4.027	3.767	3.546	3.353
17°							20.472	10.911	8.213	6.794	5.885	5.237	4.748	4.359	4.044	3.781	3.557
18°								30.452	11.765	8.519	6.949	5.974	5.295	4.787	4.390	4.067	3.800
19°										12.994	8.915	7.139	6.085	5.368	4.839	4.428	4.098
20°											14.923	9.422	7.377	6.222	5.457	4.901	4.475
21°												18.474	10.104	7.669	6.388	5.564	4.978
22°													28.833	11.044	9.039	6.590	5.694
23°															12.431	8.512	6.835
24°																14.720	9.133
25°																	19.514
26°																	
27°																	
28°																	
29°																	
30°																	

$\alpha' \rightarrow$ $\omega' \downarrow$	48°	49°	50°	51°	52°	53°	54°	55°	56°	57°	58°	59°	60°	61°	62°	63°	64°
1°	1.375	1.354	1.334	1.314	1.296	1.280	1.263	1.249	1.234	1.220	1.207	1.195	1.183	1.172	1.162	1.152	1.142
2°	1.404	1.383	1.362	1.343	1.325	1.308	1.291	1.276	1.262	1.248	1.235	1.223	1.211	1.201	1.191	1.181	1.172
3°	1.434	1.412	1.391	1.372	1.354	1.337	1.320	1.305	1.290	1.277	1.264	1.252	1.240	1.230	1.220	1.211	1.203
4°	1.465	1.443	1.422	1.403	1.383	1.366	1.349	1.334	1.319	1.305	1.293	1.281	1.269	1.260	1.250	1.241	1.234
5°	1.497	1.474	1.453	1.433	1.414	1.396	1.379	1.363	1.349	1.335	1.322	1.310	1.300	1.290	1.280	1.272	1.265
6°	1.530	1.508	1.485	1.465	1.445	1.427	1.410	1.394	1.379	1.366	1.353	1.341	1.330	1.320	1.312	1.304	1.296
7°	1.565	1.541	1.519	1.498	1.478	1.459	1.441	1.426	1.411	1.397	1.384	1.372	1.361	1.352	1.343	1.335	1.328
8°	1.601	1.576	1.553	1.531	1.511	1.492	1.474	1.458	1.443	1.429	1.416	1.404	1.393	1.383	1.375	1.368	1.362
9°	1.638	1.612	1.589	1.566	1.546	1.526	1.508	1.491	1.476	1.462	1.449	1.436	1.426	1.416	1.408	1.401	1.395
10°	1.677	1.651	1.625	1.603	1.581	1.561	1.543	1.526	1.510	1.496	1.482	1.471	1.460	1.450	1.442	1.435	1.430
11°	1.717	1.689	1.664	1.641	1.618	1.598	1.579	1.561	1.545	1.530	1.517	1.505	1.495	1.485	1.477	1.470	1.465
12°	1.760	1.731	1.704	1.680	1.656	1.636	1.616	1.598	1.581	1.567	1.553	1.541	1.530	1.521	1.513	1.506	1.502
13°	1.804	1.774	1.747	1.721	1.697	1.675	1.655	1.636	1.620	1.604	1.590	1.578	1.567	1.557	1.550	1.544	1.539
14°	1.851	1.819	1.790	1.764	1.739	1.716	1.695	1.676	1.659	1.643	1.629	1.616	1.605	1.596	1.588	1.582	1.577
15°	1.900	1.867	1.837	1.809	1.783	1.760	1.738	1.718	1.699	1.683	1.669	1.656	1.645	1.635	1.628	1.621	1.617
16°	1.952	1.917	1.886	1.857	1.830	1.805	1.782	1.762	1.743	1.725	1.710	1.698	1.686	1.676	1.668	1.662	1.658
17°	2.007	1.971	1.937	1.907	1.878	1.853	1.829	1.807	1.787	1.770	1.754	1.740	1.729	1.719	1.711	1.705	1.701
18°	2.067	2.029	1.992	1.960	1.930	1.902	1.877	1.855	1.835	1.816	1.800	1.786	1.774	1.764	1.756	1.750	1.746
19°	2.131	2.089	2.051	2.016	1.985	1.955	1.930	1.906	1.884	1.865	1.848	1.833	1.820	1.811	1.802	1.796	1.792
20°	2.199	2.154	2.114	2.077	2.043	2.012	1.985	1.959	1.937	1.916	1.898	1.883	1.870	1.859	1.851	1.845	1.841
21°	2.273	2.225	2.181	2.142	2.105	2.073	2.042	2.016	1.992	1.970	1.952	1.935	1.922	1.911	1.902	1.896	1.892
22°	2.353	2.302	2.254	2.211	2.172	2.137	2.105	2.076	2.051	2.028	2.008	1.991	1.977	1.965	1.955	1.949	1.946
23°	2.441	2.385	2.333	2.287	2.244	2.207	2.172	2.142	2.114	2.090	2.069	2.050	2.035	2.022	2.013	2.006	2.003
24°	2.538	2.476	2.419	2.369	2.323	2.281	2.244	2.211	2.182	2.155	2.133	2.113	2.096	2.083	2.073	2.066	2.062
25°	2.646	2.577	2.515	2.459	2.408	2.363	2.322	2.287	2.254	2.226	2.201	2.180	2.163	2.148	2.137	2.130	2.126
26°	2.767	2.689	2.620	2.557	2.502	2.452	2.408	2.368	2.333	2.302	2.275	2.253	2.233	2.218	2.206	2.198	2.193
27°	2.903	2.815	2.738	2.668	2.606	2.551	2.501	2.458	2.419	2.385	2.355	2.330	2.310	2.293	2.280	2.271	2.266
28°	3.061	2.960	2.871	2.791	2.722	2.660	2.605	2.556	2.513	2.476	2.443	2.416	2.393	2.374	2.359	2.350	2.344
29°	3.244	3.126	3.023	2.932	2.852	2.782	2.720	2.665	2.617	2.575	2.539	2.509	2.483	2.462	2.446	2.435	2.429
30°	3.462	3.321	3.199	3.094	3.001	2.920	2.849	2.788	2.733	2.686	2.646	2.610	2.582	2.558	2.541	2.528	2.521

See page 121 for definitions of symbols.

Table 56 Upstream Mach Number, M$_x$, for Two-Dimensional Shock

For a perfect gas with constant specific heat and molecular weight
k = 1.4

α′→ ω′↓	32°	33°	34°	35°	36°	37°	38°	39°	40°	41°	42°	43°	44°	45°	46°	47°
1°	1.932	1.879	1.830	1.783	1.740	1.698	1.660	1.623	1.589	1.557	1.527	1.498	1.471	1.445	1.420	1.397
2°	1.979	1.925	1.872	1.824	1.779	1.737	1.697	1.660	1.624	1.591	1.559	1.530	1.502	1.475	1.450	1.426
3°	2.029	1.972	1.919	1.869	1.821	1.777	1.736	1.698	1.661	1.627	1.594	1.563	1.534	1.507	1.481	1.457
4°	2.082	2.023	1.966	1.914	1.865	1.819	1.776	1.736	1.698	1.663	1.629	1.598	1.568	1.540	1.514	1.489
5°	2.138	2.075	2.017	1.962	1.911	1.864	1.819	1.777	1.738	1.701	1.667	1.634	1.604	1.575	1.547	1.522
6°	2.198	2.133	2.071	2.013	1.959	1.910	1.863	1.820	1.779	1.740	1.705	1.671	1.640	1.610	1.582	1.556
7°	2.263	2.192	2.128	2.067	2.012	1.959	1.910	1.865	1.822	1.783	1.745	1.710	1.678	1.647	1.618	1.591
8°	2.331	2.258	2.199	2.125	2.065	2.011	1.960	1.913	1.868	1.827	1.788	1.751	1.717	1.685	1.655	1.627
9°	2.406	2.327	2.253	2.187	2.124	2.066	2.012	1.963	1.916	1.872	1.832	1.794	1.758	1.725	1.694	1.665
10°	2.487	2.402	2.324	2.253	2.187	2.125	2.068	2.016	1.966	1.921	1.879	1.839	1.802	1.767	1.735	1.705
11°	2.575	2.483	2.400	2.324	2.253	2.188	2.128	2.072	2.021	1.973	1.928	1.886	1.848	1.812	1.778	1.746
12°	2.672	2.573	2.482	2.401	2.325	2.256	2.192	2.133	2.078	2.027	1.981	1.937	1.896	1.858	1.823	1.790
13°	2.779	2.671	2.573	2.484	2.404	2.330	2.260	2.197	2.140	2.086	2.036	1.990	1.947	1.907	1.871	1.836
14°	2.898	2.780	2.673	2.576	2.488	2.408	2.335	2.268	2.205	2.148	2.095	2.046	2.002	1.959	1.920	1.884
15°	3.033	2.902	2.784	2.678	2.583	2.495	2.416	2.344	2.276	2.215	2.159	2.107	2.059	2.014	1.973	1.935
16°	3.187	3.039	2.909	2.792	2.686	2.591	2.505	2.426	2.354	2.288	2.228	2.172	2.121	2.073	2.030	1.990
17°	3.365	3.197	3.049	2.919	2.802	2.697	2.602	2.516	2.439	2.367	2.302	2.242	2.167	2.137	2.090	2.047
18°	3.574	3.380	3.211	3.063	2.932	2.815	2.711	2.617	2.531	2.454	2.382	2.318	2.259	2.205	2.155	2.109
19°	3.826	3.596	3.399	3.230	3.080	2.949	2.832	2.728	2.634	2.549	2.471	2.401	2.337	2.278	2.224	2.176
20°	4.136	3.857	3.623	3.424	3.253	3.103	2.971	2.853	2.748	2.654	2.569	2.492	2.422	2.359	2.300	2.248
21°	4.534	4.182	3.896	3.656	3.454	3.280	3.128	2.995	2.877	2.771	2.678	2.592	2.516	2.446	2.383	2.326
22°	5.069	4.602	4.236	3.942	3.697	3.490	3.313	3.159	3.023	2.905	2.799	2.704	2.619	2.543	2.473	2.411
23°	5.846	5.176	4.683	4.301	3.995	3.743	3.532	3.350	3.195	3.057	2.937	2.830	2.735	2.650	2.573	2.504
24°	7.137	6.031	5.303	4.778	4.377	4.058	3.798	3.579	3.394	3.234	3.096	2.974	2.866	2.770	2.684	2.607
25°	9.973	7.516	6.254	5.454	4.890	4.465	4.131	3.860	3.635	3.445	3.281	3.140	3.016	2.906	2.810	2.723
26°	43.967	11.198	7.993	6.523	5.632	5.021	4.568	4.216	3.931	3.698	3.502	3.334	3.189	3.063	2.952	2.854
27°		13.162	8.624	6.854	5.847	5.173	4.687	4.313	4.015	3.770	3.567	3.394	3.246	3.116	3.003	
28°			17.083	9.495	7.272	6.104	5.365	4.825	4.424	4.108	3.853	3.641	3.462	3.309	3.176	
29°				32.120	10.771	7.809	6.420	5.571	4.988	4.554	4.218	3.948	3.725	3.539	3.380	
30°					12.895	8.531	6.812	5.831	5.178	4.705	4.344	4.056	3.822	3.628		

α′→ ω′↓	65°	66°	67°	68°	69°	70°	71°	72°	73°	74°	75°	76°	77°	78°	79°	80°
1°	1.134	1.125	1.118	1.111	1.105	1.099	1.094	1.089	1.085	1.081	1.078	1.077	1.075	1.075	1.075	1.077
2°	1.164	1.156	1.150	1.144	1.139	1.134	1.130	1.126	1.124	1.122	1.122	1.122	1.124			
3°	1.195	1.188	1.182	1.177	1.173	1.169	1.166	1.164	1.163	1.164	1.165	1.168				
4°	1.227	1.220	1.215	1.210	1.207	1.204	1.203	1.202	1.203	1.205	1.208	1.214				
5°	1.258	1.252	1.248	1.244	1.242	1.240	1.239	1.240	1.243	1.247	1.252	1.260				
6°	1.291	1.285	1.281	1.278	1.276	1.276	1.277	1.279	1.283	1.289	1.297	1.307				
7°	1.323	1.319	1.315	1.313	1.312	1.313	1.314	1.318	1.324	1.331	1.341	1.354				
8°	1.357	1.352	1.350	1.348	1.348	1.350	1.353	1.358	1.365	1.375	1.386	1.402				
9°	1.390	1.387	1.385	1.384	1.385	1.388	1.392	1.398	1.407	1.418	1.433	1.450				
10°	1.426	1.423	1.421	1.421	1.423	1.426	1.432	1.440	1.450	1.463	1.479	1.500				
11°	1.461	1.459	1.458	1.459	1.462	1.466	1.473	1.482	1.494	1.509	1.528	1.551				
12°	1.498	1.496	1.496	1.497	1.501	1.507	1.514	1.525	1.539	1.556	1.577	1.603				
13°	1.535	1.535	1.535	1.537	1.541	1.548	1.557	1.569	1.584	1.604	1.627	1.657				
14°	1.575	1.574	1.575	1.578	1.583	1.591	1.601	1.615	1.632	1.653	1.679	1.712				
15°	1.615	1.614	1.616	1.620	1.626	1.635	1.647	1.662	1.681	1.704	1.733	1.768				
16°	1.656	1.657	1.659	1.663	1.670	1.680	1.694	1.710	1.731	1.757	1.788	1.827				
17°	1.700	1.700	1.703	1.708	1.716	1.727	1.742	1.760	1.783	1.811	1.846	1.888				
18°	1.745	1.745	1.749	1.755	1.764	1.776	1.792	1.812	1.837	1.868	1.906	1.952				
19°	1.791	1.792	1.796	1.804	1.813	1.827	1.844	1.866	1.893	1.927	1.968	2.019				
20°	1.840	1.842	1.846	1.854	1.865	1.879	1.899	1.923	1.953	1.989	2.033	2.088				
21°	1.891	1.893	1.898	1.907	1.919	1.935	1.956	1.982	2.014	2.054	2.102	2.161				
22°	1.945	1.948	1.953	1.962	1.975	1.993	2.016	2.043	2.079	2.122	2.174	2.239				
23°	2.002	2.005	2.011	2.021	2.035	2.054	2.079	2.109	2.147	2.193	2.251	2.321				
24°	2.062	2.064	2.071	2.083	2.098	2.119	2.145	2.178	2.219	2.269	2.332	2.410				
25°	2.125	2.129	2.136	2.148	2.165	2.187	2.216	2.251	2.296	2.351	2.419	2.504				
26°	2.193	2.197	2.205	2.218	2.236	2.260	2.290	2.329	2.378	2.437	2.512	2.604				
27°	2.266	2.270	2.278	2.292	2.311	2.337	2.370	2.413	2.466	2.531	2.612	2.715				
28°	2.344	2.348	2.357	2.372	2.393	2.420	2.457	2.503	2.560	2.631	2.720	2.833				
29°	2.428	2.432	2.441	2.457	2.481	2.511	2.549	2.599	2.662	2.740	2.839	2.965				
30°	2.520	2.523	2.533	2.551	2.575	2.608	2.651	2.705	2.774	2.860	2.970	3.110				

Table 57 Downstream Mach Number, M_y, for Two-Dimensional Shock

For a perfect gas with constant specific heat and molecular weight
k = 1.4

$\alpha' \to$ / $\omega' \downarrow$	24°	25°	26°	27°	28°	.29°	30°	31°	32°	33°	34°	35°	36°	37°	38°
1°	2.487	2.392	2.303	2.222	2.147	2.077	2.014	1.953	1.897	1.844	1.795	1.749	1.705	1.665	1.626
2°		2.420	2.329	2.244	2.167	2.095	2.028	1.966	1.908	1.854	1.804	1.756	1.711	1.669	1.630
3°			2.357	2.270	2.189	2.114	2.045	1.981	1.921	1.865	1.812	1.763	1.718	1.675	1.633
4°				2.298	2.214	2.136	2.065	1.998	1.936	1.878	1.825	1.774	1.726	1.682	1.640
5°					2.242	2.161	2.087	2.018	1.954	1.894	1.837	1.785	1.736	1.690	1.646
6°						2.189	2.111	2.039	1.973	1.910	1.852	1.798	1.748	1.700	1.655
7°							2.139	2.063	1.994	1.930	1.869	1.813	1.760	1.711	1.665
8°								2.090	2.018	1.951	1.888	1.830	1.776	1.724	1.676
9°									2.045	1.974	1.910	1.848	1.792	1.739	1.689
10°										2.001	1.933	1.869	1.810	1.755	1.704
11°											1.959	1.893	1.831	1.774	1.720
12°												1.918	1.854	1.794	1.738
13°													1.879	1.817	1.759
14°														1.842	1.781
15°															1.806
16°															
17°															
18°															
19°															
20°															
21°															
22°															
23°															
24°															
25°															
26°															
27°															
28°															
29°															
30°															

$\alpha' \to$ / $\omega' \downarrow$	54°	55°	56°	57°	58°	59°	60°	61°	62°	63°	64°	65°	66°	67°	68°
1°	1.225	1.209	1.194	1.179	1.165	1.152	1.139	1.127	1.115	1.104	1.093	1.083	1.074	1.064	1.055
2°	1.215	1.199	1.183	1.167	1.153	1.138	1.125	1.112	1.099	1.087	1.076	1.064	1.054	1.043	1.033
3°	1.206	1.189	1.173	1.156	1.141	1.126	1.112	1.098	1.085	1.072	1.060	1.048	1.036	1.025	1.014
4°	1.199	1.181	1.164	1.147	1.130	1.115	1.100	1.086	1.072	1.058	1.045	1.032	1.020	1.008	.9958
5°	1.192	1.173	1.155	1.138	1.121	1.105	1.089	1.074	1.060	1.046	1.032	1.018	1.005	.9920	.9794
6°	1.186	1.167	1.148	1.130	1.113	1.096	1.080	1.064	1.049	1.034	1.020	1.005	.9915	.9778	.9644
7°	1.181	1.161	1.142	1.123	1.105	1.088	1.071	1.055	1.039	1.024	1.009	.9938	.9790	.9650	.9507
8°	1.177	1.156	1.136	1.117	1.099	1.081	1.064	1.047	1.030	1.014	.9984	.9829	.9681	.9528	.9382
9°	1.174	1.153	1.132	1.112	1.093	1.075	1.056	1.039	1.022	1.005	.9891	.9735	.9575	.9422	.9268
10°	1.171	1.150	1.128	1.108	1.088	1.069	1.051	1.032	1.015	.9978	.9809	.9644	.9483	.9322	.9162
11°	1.170	1.147	1.125	1.104	1.084	1.064	1.045	1.027	1.009	.9911	.9736	.9566	.9397	.9233	.9070
12°	1.169	1.146	1.123	1.102	1.081	1.060	1.041	1.022	1.003	.9849	.9669	.9495	.9321	.9150	.8984
13°	1.169	1.145	1.122	1.100	1.078	1.057	1.037	1.018	.9982	.9794	.9612	.9433	.9252	.9077	.8904
14°	1.170	1.145	1.121	1.099	1.076	1.055	1.034	1.014	.9943	.9749	.9561	.9374	.9192	.9013	.8834
15°	1.171	1.146	1.122	1.098	1.075	1.053	1.032	1.011	.9907	.9710	.9517	.9327	.9138	.8953	.8771
16°	1.174	1.148	1.123	1.099	1.075	1.052	1.030	1.009	.9881	.9678	.9479	.9284	.9091	.8902	.8714
17°	1.177	1.150	1.124	1.100	1.075	1.052	1.030	1.007	.9860	.9653	.9448	.9248	.9053	.8857	.8666
18°	1.182	1.154	1.127	1.101	1.077	1.052	1.029	1.007	.9846	.9632	.9424	.9219	.9018	.8819	.8622
19°	1.186	1.158	1.130	1.104	1.078	1.054	1.030	1.006	.9840	.9622	.9407	.9196	.8991	.8787	.8584
20°	1.192	1.163	1.134	1.107	1.081	1.056	1.031	1.007	.9840	.9614	.9395	.9179	.8968	.8760	.8554
21°	1.200	1.169	1.140	1.112	1.084	1.058	1.033	1.008	.9846	.9615	.9389	.9169	.8953	.8740	.8530
22°	1.207	1.176	1.145	1.116	1.088	1.061	1.036	1.010	.9861	.9622	.9391	.9164	.8943	.8725	.8511
23°	1.216	1.183	1.152	1.122	1.093	1.066	1.039	1.013	.9878	.9636	.9398	.9166	.8939	.8716	.9497
24°	1.226	1.192	1.160	1.129	1.099	1.071	1.043	1.016	.9906	.9655	.9413	.9174	.8943	.8714	.8489
25°	1.237	1.202	1.169	1.137	1.106	1.076	1.048	1.020	.9939	.9682	.9432	.9188	.8949	.8717	.8487
26°	1.249	1.213	1.178	1.145	1.113	1.083	1.053	1.025	.9979	.9716	.9459	.9209	.8963	.8725	.8490
27°	1.263	1.225	1.189	1.155	1.122	1.090	1.060	1.031	1.003	.9756	.9491	.9234	.8983	.8738	.8499
28°	1.278	1.238	1.201	1.165	1.131	1.099	1.067	1.037	1.008	.9802	.9531	.9267	.9010	.8759	.8513
29°	1.294	1.253	1.214	1.177	1.142	1.108	1.075	1.044	1.015	.9856	.9578	.9306	.9042	.8786	.8534
30°	1.312	1.269	1.228	1.190	1.153	1.118	1.085	1.053	1.022	.9919	.9631	.9352	.9082	.8818	.8559

Table 57 Downstream Mach Number, M_y, for Two-Dimensional Shock

For a perfect gas with constant specific heat and molecular weight
k = 1.4

$\alpha' \rightarrow$ $\omega' \downarrow$	39°	40°	41°	42°	43°	44°	45°	46°	47°	48°	49°	50°	51°	52°	53°
1°	1.590	1.556	1.523	1.492	1.463	1.436	1.409	1.385	1.361	1.339	1.317	1.297	1.278	1.260	1.242
2°	1.592	1.557	1.524	1.492	1.461	1.433	1.407	1.381	1.357	1.333	1.311	1.290	1.271	1.251	1.232
3°	1.595	1.558	1.524	1.492	1.461	1.432	1.404	1.378	1.353	1.329	1.306	1.284	1.263	1.243	1.224
4°	1.600	1.562	1.527	1.494	1.461	1.431	1.403	1.376	1.350	1.325	1.302	1.279	1.257	1.237	1.217
5°	1.605	1.567	1.531	1.496	1.463	1.432	1.403	1.375	1.348	1.323	1.298	1.275	1.253	1.231	1.211
6°	1.613	1.573	1.535	1.500	1.466	1.434	1.403	1.375	1.347	1.321	1.296	1.272	1.249	1.227	1.206
7°	1.621	1.580	1.541	1.504	1.470	1.436	1.405	1.375	1.347	1.320	1.294	1.270	1.246	1.224	1.202
8°	1.631	1.588	1.548	1.510	1.474	1.441	1.408	1.378	1.348	1.320	1.294	1.268	1.244	1.221	1.198
9°	1.642	1.599	1.557	1.517	1.480	1.445	1.412	1.380	1.350	1.322	1.294	1.268	1.243	1.219	1.196
10°	1.655	1.610	1.566	1.526	1.488	1.451	1.417	1.384	1.353	1.323	1.295	1.269	1.242	1.218	1.194
11°	1.670	1.622	1.578	1.536	1.496	1.459	1.423	1.389	1.357	1.327	1.298	1.270	1.243	1.218	1.193
12°	1.686	1.637	1.591	1.547	1.506	1.467	1.430	1.396	1.362	1.331	1.301	1.272	1.245	1.219	1.193
13°	1.704	1.653	1.605	1.560	1.517	1.477	1.439	1.402	1.368	1.336	1.305	1.275	1.247	1.220	1.194
14°	1.724	1.671	1.621	1.574	1.530	1.488	1.448	1.411	1.376	1.342	1.310	1.279	1.250	1.222	1.195
15°	1.746	1.691	1.638	1.589	1.543	1.500	1.459	1.421	1.384	1.349	1.316	1.284	1.254	1.225	1.198
16°	1.770	1.712	1.658	1.607	1.559	1.514	1.472	1.431	1.393	1.358	1.323	1.290	1.259	1.230	1.201
17°		1.736	1.680	1.626	1.576	1.529	1.485	1.444	1.404	1.367	1.331	1.298	1.265	1.235	1.205
18°			1.703	1.648	1.595	1.546	1.500	1.457	1.416	1.377	1.340	1.306	1.272	1.241	1.211
19°			1.729	1.671	1.617	1.565	1.517	1.472	1.429	1.389	1.351	1.315	1.281	1.248	1.217
20°			1.759	1.697	1.640	1.586	1.536	1.489	1.444	1.402	1.363	1.325	1.290	1.256	1.224
21°			1.791	1.726	1.665	1.609	1.556	1.507	1.461	1.417	1.376	1.337	1.300	1.265	1.231
22°			1.826	1.757	1.694	1.634	1.579	1.527	1.479	1.433	1.390	1.350	1.311	1.275	1.240
23°			1.865	1.792	1.725	1.662	1.604	1.550	1.499	1.451	1.406	1.364	1.324	1.286	1.250
24°			1.909	1.831	1.759	1.693	1.631	1.574	1.521	1.471	1.424	1.380	1.338	1.299	1.262
25°			1.957	1.873	1.797	1.726	1.662	1.601	1.545	1.493	1.444	1.397	1.354	1.313	1.274
26°			2.011	1.921	1.839	1.764	1.695	1.631	1.572	1.516	1.465	1.417	1.371	1.328	1.288
27°			2.071	1.974	1.885	1.805	1.731	1.664	1.601	1.543	1.488	1.438	1.390	1.345	1.303
28°			2.139	2.033	1.938	1.851	1.772	1.700	1.633	1.572	1.514	1.461	1.411	1.364	1.320
29°			2.217	2.100	1.996	1.902	1.818	1.740	1.669	1.603	1.542	1.486	1.434	1.384	1.338
30°			2.306	2.177	2.062	1.960	1.868	1.785	1.708	1.638	1.574	1.514	1.459	1.407	1.358

$\alpha' \rightarrow$ $\omega' \downarrow$	69°	70°	71°	72°	73°	74°	75°	76°	77°	78°	79°	80°
1°	1.046	1.037	1.030	1.021	1.014	1.007	.9992	.9916	.9844	.9771	.9697	
2°	1.023	1.014	1.004	.9949	.9857	.9764	.9671	.9577				
3°	1.003	.9920	.9815	.9709	.9603	.9493	.9384	.9274				
4°	.9837	.9722	.9604	.9490	.9369	.9253	.9130	.9004				
5°	.9665	.9542	.9416	.9291	.9160	.9031	.8900	.8761				
6°	.9511	.9376	.9244	.9109	.8974	.8833	.8689	.8542				
7°	.9367	.9225	.9087	.8944	.8800	.8654	.8501	.8344				
8°	.9235	.9088	.8940	.8793	.8642	.8487	.8330	.8167				
9°	.9115	.8962	.8810	.8655	.8498	.8338	.8172	.8004				
10°	.9006	.8848	.8688	.8529	.8364	.8200	.8030	.7853				
11°	.8905	.8741	.8577	.8412	.8245	.8074	.7899	.7718				
12°	.8813	.8645	.8477	.8307	.8135	.7959	.7777	.7593				
13°	.8732	.8559	.8384	.8211	.8034	.7853	.7669	.7478				
14°	.8656	.8479	.8301	.8122	.7942	.7758	.7567	.7373				
15°	.8588	.8407	.8225	.8041	.7856	.7669	.7475	.7279				
16°	.8528	.8343	.8155	.7969	.7780	.7588	.7393	.7190				
17°	.8475	.8285	.8094	.7903	.7711	.7514	.7315	.7110				
18°	.8426	.8232	.8040	.7844	.7648	.7449	.7245	.7038				
19°	.8386	.8187	.7990	.7791	.7591	.7388	.7182	.6971				
20°	.8350	.8149	.7946	.7743	.7540	.7334	.7125	.6911				
21°	.8321	.8116	.7909	.7703	.7496	.7286	.7074	.6857				
22°	.8299	.8087	.7876	.7668	.7455	.7243	.7028	.6808				
23°	.8279	.8065	.7851	.7637	.7421	.7206	.6987	.6764				
24°	.8268	.8048	.7830	.7612	.7394	.7174	.6951	.6725				
25°	.8260	.8037	.7813	.7592	.7370	.7146	.6920	.6692				
26°	.8258	.8030	.7803	.7577	.7351	.7124	.6895	.6663				
27°	.8262	.8029	.7797	.7566	.7336	.7106	.6874	.6638				
28°	.8271	.8033	.7797	.7562	.7327	.7093	.6857	.6619				
29°	.8285	.8041	.7801	.7562	.7323	.7085	.6845	.6603				
30°	.8305	.8057	.7810	.7566	.7323	.7081	.6837	.6592				

Table 58 Ratio of Downstream to Upstream Pressure for Two-Dimensional Shock

For a perfect gas with constant specific heat and molecular weight
k = 1.4

ω ↓ \ α' →	1°	2°	3°	4°	5°	6°	7°	8°	9°	10°	11°	12°	13°	14°	15°	16°	17°	18°	19°
1°		2.753	1.780	1.501	1.370	1.293	1.243	1.209	1.182	1.162	1.146	1.134	1.122	1.114	1.107	1.099	1.093	1.089	1.085
2°			5.674	2.754	2.081	1.783	1.614	1.506	1.431	1.376	1.333	1.299	1.273	1.251	1.233	1.216	1.203	1.191	1.181
3°				11.553	4.014	2.763	2.247	1.965	1.790	1.669	1.581	1.514	1.461	1.419	1.386	1.356	1.331	1.311	1.293
4°					29.425	5.706	3.571	2.772	2.353	2.098	1.924	1.799	1.706	1.632	1.575	1.525	1.485	1.452	1.423
5°							8.092	4.548	3.371	2.783	2.432	2.199	2.033	1.908	1.812	1.735	1.672	1.620	1.577
6°								11.722	5.752	4.063	3.264	2.798	2.495	2.282	2.124	2.002	1.906	1.828	1.762
7°									17.907	7.283	4.248	3.804	3.203	2.819	2.551	2.353	2.204	2.087	1.991
8°										30.803	9.275	5.824	4.412	3.647	3.167	2.838	2.600	2.420	2.276
9°											74.645	12.004	6.972	5.111	4.047	3.550	3.150	2.864	2.649
10°													15.944	8.381	5.918	4.697	3.971	3.490	3.147
11°														22.158	10.144	6.859	5.322	4.437	3.859
12°															33.459	12.431	7.972	6.033	4.950
13°																60.163	15.499	9.309	6.850
14°																	204.843	19.868	10.957
15°																		26.528	
16°																			
17°																			
18°																			
19°																			
20°																			
21°																			
22°																			
23°																			
24°																			
25°																			
26°																			
27°																			
28°																			
29°																			
30°																			

ω ↓ \ α' →	39°	40°	41°	42°	43°	44°	45°	46°	47°	48°	49°	50°	51°	52°	53°	54°	55°	56°	57°
1°	1.051	1.051	1.051	1.051	1.051	1.051	1.051	1.051	1.051	1.051	1.051	1.051	1.051	1.051	1.052	1.052	1.054	1.054	1.055
2°	1.107	1.105	1.104	1.104	1.104	1.104	1.102	1.102	1.102	1.104	1.104	1.104	1.104	1.105	1.107	1.107	1.108	1.109	1.111
3°	1.165	1.163	1.162	1.160	1.159	1.159	1.159	1.157	1.157	1.159	1.159	1.159	1.160	1.162	1.163	1.165	1.166	1.168	1.171
4°	1.225	1.224	1.222	1.222	1.219	1.218	1.216	1.216	1.216	1.216	1.216	1.218	1.219	1.219	1.222	1.224	1.225	1.228	1.231
5°	1.293	1.289	1.286	1.284	1.283	1.281	1.280	1.278	1.278	1.278	1.278	1.280	1.281	1.283	1.284	1.286	1.289	1.293	1.296
6°	1.364	1.359	1.354	1.351	1.348	1.347	1.345	1.344	1.344	1.342	1.344	1.344	1.345	1.347	1.348	1.351	1.354	1.359	1.364
7°	1.440	1.434	1.429	1.425	1.420	1.418	1.415	1.414	1.412	1.412	1.412	1.412	1.414	1.415	1.418	1.420	1.425	1.429	1.434
8°	1.524	1.516	1.509	1.503	1.498	1.493	1.490	1.487	1.485	1.485	1.483	1.485	1.485	1.487	1.490	1.493	1.498	1.503	1.509
9°	1.614	1.602	1.594	1.587	1.581	1.574	1.569	1.566	1.564	1.561	1.561	1.561	1.561	1.564	1.566	1.569	1.574	1.581	1.587
10°	1.711	1.697	1.687	1.677	1.669	1.662	1.655	1.650	1.647	1.645	1.644	1.642	1.644	1.645	1.647	1.650	1.655	1.662	1.669
11°	1.818	1.801	1.788	1.776	1.764	1.755	1.749	1.742	1.737	1.733	1.730	1.730	1.730	1.730	1.733	1.737	1.742	1.749	1.755
12°	1.935	1.915	1.897	1.883	1.869	1.856	1.848	1.839	1.833	1.828	1.825	1.821	1.821	1.821	1.825	1.828	1.833	1.839	1.848
13°	2.065	2.040	2.018	1.998	1.982	1.967	1.955	1.945	1.936	1.929	1.924	1.922	1.920	1.920	1.922	1.924	1.929	1.940	1.945
14°	2.210	2.178	2.151	2.126	2.106	2.089	2.072	2.059	2.048	2.040	2.033	2.027	2.026	2.024	2.025	2.027	2.033	2.040	2.048
15°	2.371	2.331	2.298	2.269	2.243	2.220	2.201	2.183	2.170	2.158	2.149	2.143	2.140	2.138	2.138	2.140	2.143	2.149	2.158
16°	2.524	2.505	2.461	2.427	2.393	2.365	2.341	2.321	2.304	2.288	2.276	2.268	2.262	2.259	2.257	2.259	2.262	2.268	2.276
17°	2.759	2.700	2.646	2.602	2.562	2.526	2.497	2.470	2.448	2.430	2.415	2.403	2.395	2.389	2.387	2.387	2.389	2.395	2.403
18°	2.997	2.922	2.856	2.798	2.750	2.706	2.670	2.638	2.610	2.587	2.568	2.551	2.541	2.532	2.526	2.524	2.526	2.532	2.541
19°	3.272	3.177	3.096	3.023	2.961	2.908	2.861	2.821	2.787	2.759	2.733	2.713	2.698	2.687	2.678	2.676	2.676	2.678	2.687
20°	3.594	3.472	3.371	3.281	3.203	3.136	3.079	3.028	2.986	2.949	2.917	2.892	2.872	2.856	2.845	2.841	2.838	2.841	2.845
21°	3.977	3.823	3.690	3.579	3.480	3.396	3.323	3.262	3.208	3.162	3.124	3.091	3.065	3.044	3.030	3.018	3.014	3.014	3.018
22°	4.445	4.240	4.072	3.926	3.801	3.695	3.605	3.526	3.459	3.401	3.353	3.311	3.279	3.252	3.232	3.218	3.208	3.207	3.208
23°	5.020	4.753	4.526	4.340	4.180	4.046	3.929	3.831	3.747	3.674	3.613	3.560	3.519	3.483	3.457	3.437	3.424	3.416	3.416
24°	5.753	5.386	5.086	4.839	4.632	4.457	4.309	4.183	4.075	3.983	3.907	3.840	3.787	3.741	3.706	3.679	3.661	3.650	3.645
25°	6.718	6.204	5.792	5.457	5.182	4.955	4.759	4.600	4.460	4.343	4.246	4.163	4.092	4.034	3.989	3.952	3.926	3.907	3.898
26°	8.045	7.283	6.700	6.238	5.864	5.559	5.305	5.094	4.917	4.766	4.638	4.533	4.442	4.367	4.306	4.261	4.222	4.198	4.183
27°	9.982	8.799	7.928	7.258	6.737	6.320	5.978	5.694	5.460	5.265	5.101	4.965	4.849	4.753	4.674	4.609	4.561	4.526	4.501
28°	13.084	11.057	9.660	8.650	7.889	7.298	6.825	6.442	6.128	5.868	5.655	5.475	5.323	5.200	5.097	5.013	4.948	4.897	4.863
29°	18.876	14.796	12.325	10.667	9.487	8.608	7.928	7.395	6.963	6.613	6.328	6.090	5.889	5.725	5.593	5.483	5.393	5.327	5.276
30°	33.459	22.203	16.906	13.841	11.848	10.455	9.432	8.650	8.045	7.555	7.163	6.839	6.576	6.359	6.178	6.032	5.917	5.824	5.753

Table 58 Ratio of Downstream to Upstream Pressure for Two-Dimensional Shock

For a perfect gas with constant specific heat and molecular weight
k = 1.4

α' → ω ↓	20°	21°	22°	23°	24°	25°	26°	27°	28°	29°	30°	31°	32°	33°	34°	35°	36°	37°	38°
1°	1.081	1.078	1.075	1.072	1.069	1.066	1.065	1.064	1.062	1.061	1.059	1.058	1.056	1.055	1.055	1.054	1.054	1.052	1.052
2°	1.172	1.165	1.157	1.152	1.146	1.141	1.136	1.133	1.128	1.125	1.122	1.120	1.117	1.115	1.112	1.111	1.109	1.108	1.107
3°	1.277	1.263	1.249	1.240	1.231	1.222	1.215	1.208	1.202	1.197	1.267	1.187	1.182	1.180	1.177	1.174	1.171	1.168	1.166
4°	1.398	1.378	1.358	1.342	1.327	1.315	1.302	1.292	1.283	1.275	1.267	1.260	1.254	1.249	1.243	1.239	1.236	1.231	1.228
5°	1.540	1.509	1.482	1.456	1.436	1.417	1.401	1.386	1.373	1.361	1.350	1.341	1.331	1.324	1.318	1.312	1.305	1.301	1.296
6°	1.709	1.662	1.624	1.589	1.559	1.533	1.511	1.490	1.472	1.456	1.442	1.429	1.417	1.407	1.398	1.389	1.381	1.375	1.368
7°	1.913	1.848	1.792	1.743	1.702	1.667	1.635	1.609	1.584	1.563	1.543	1.527	1.511	1.496	1.485	1.474	1.464	1.455	1.447
8°	2.162	2.070	1.991	1.926	1.870	1.821	1.780	1.743	1.711	1.682	1.658	1.635	1.614	1.597	1.581	1.566	1.553	1.541	1.532
9°	2.478	2.345	2.235	2.143	2.068	2.002	1.946	1.897	1.855	1.818	1.785	1.755	1.730	1.708	1.686	1.669	1.652	1.637	1.624
10°	2.891	2.696	2.540	2.410	2.306	2.216	2.141	2.078	2.020	1.973	1.929	1.892	1.860	1.830	1.804	1.781	1.761	1.742	1.725
11°	3.455	3.155	2.915	2.743	2.595	2.475	2.373	2.288	2.214	2.151	2.096	2.048	2.005	1.967	1.935	1.906	1.879	1.856	1.835
12°	4.264	3.786	3.438	3.171	2.963	2.794	2.655	2.541	2.442	2.359	2.288	2.226	2.172	2.124	2.081	2.046	2.013	1.984	1.958
13°	5.527	4.708	4.149	3.741	3.437	3.196	3.004	2.847	2.715	2.606	2.512	2.432	2.363	2.302	2.249	2.202	2.162	2.126	2.092
14°	7.797	6.179	5.200	4.542	4.072	3.720	3.447	3.228	3.051	2.904	2.779	2.674	2.585	2.508	2.440	2.381	2.329	2.284	2.245
15°	13.022	8.905	6.921	5.745	4.977	4.432	4.026	3.714	3.467	3.267	3.103	2.963	2.847	2.748	2.661	2.587	2.522	2.464	2.415
16°	38.025	15.712	10.234	7.763	6.359	5.453	4.823	4.358	4.003	3.725	3.501	3.313	3.160	3.030	2.919	2.825	2.741	2.670	2.608
17°		62.627	19.326	11.848	8.743	7.050	5.982	5.254	4.720	4.318	4.003	3.749	3.542	3.371	3.225	3.103	2.997	2.906	2.827
18°			151.658	24.488	13.841	9.895	7.834	6.576	5.725	5.118	4.658	4.303	4.017	3.787	3.594	3.434	3.299	3.182	3.084
19°					32.419	16.393	11.258	8.737	7.243	6.255	5.551	5.031	4.629	4.309	4.049	3.837	3.658	3.508	3.381
20°						46.239	19.735	12.919	9.787	8.000	6.839	6.032	5.438	4.982	4.622	4.334	4.098	3.901	3.736
21°							76.353	24.382	14.958	11.023	8.862	7.503	6.567	5.885	5.371	4.965	4.642	4.380	4.160
22°								192.864	31.194	17.553	12.499	9.866	8.251	7.163	6.381	5.796	5.342	4.979	4.687
23°										42.209	20.965	14.290	11.032	9.106	7.834	6.934	6.268	5.753	5.349
24°											63.034	25.646	16.523	12.421	10.092	8.595	7.555	6.792	6.212
25°												117.675	32.419	19.381	14.103	11.250	9.473	8.257	7.379
26°													630.364	43.226	23.142	16.166	12.617	10.486	9.060
27°															63.034	28.381	18.770	14.278	11.668
28°																111.845	36.169	22.180	16.307
29°																	415.687	48.859	26.802
30°																			73.365

α' → ω ↓	58°	59°	60°	61°	62°	63°	64°	65°	66°	67°	68°	69°	70°	71°	72°	73°	74°	75°	76°
1°	1.055	1.056	1.058	1.059	1.061	1.062	1.064	1.065	1.066	1.069	1.072	1.075	1.078	1.081	1.085	1.089	1.093	1.099	1.107
2°	1.112	1.115	1.117	1.120	1.122	1.125	1.128	1.133	1.136	1.141	1.146	1.152	1.157	1.165	1.172	1.181	1.191	1.203	1.216
3°	1.174	1.177	1.180	1.182	1.187	1.192	1.197	1.202	1.208	1.215	1.222	1.231	1.240	1.249	1.263	1.277	1.293	1.311	1.331
4°	1.236	1.239	1.243	1.249	1.254	1.260	1.267	1.275	1.283	1.292	1.302	1.315	1.327	1.342	1.358	1.378	1.398	1.423	1.452
5°	1.301	1.305	1.312	1.318	1.324	1.331	1.341	1.350	1.361	1.373	1.386	1.401	1.417	1.436	1.456	1.482	1.509	1.540	1.577
6°	1.368	1.375	1.381	1.389	1.398	1.407	1.417	1.429	1.442	1.456	1.472	1.490	1.511	1.533	1.559	1.589	1.624	1.662	1.709
7°	1.440	1.447	1.455	1.464	1.474	1.485	1.496	1.511	1.527	1.543	1.563	1.584	1.609	1.635	1.667	1.702	1.743	1.792	1.848
8°	1.516	1.524	1.532	1.541	1.553	1.566	1.581	1.597	1.614	1.635	1.658	1.682	1.711	1.743	1.780	1.821	1.870	1.926	1.991
9°	1.594	1.602	1.614	1.624	1.637	1.652	1.669	1.686	1.708	1.730	1.755	1.785	1.818	1.855	1.897	1.946	2.002	2.068	2.143
10°	1.677	1.687	1.697	1.711	1.725	1.742	1.761	1.781	1.804	1.830	1.860	1.892	1.929	1.973	2.020	2.078	2.141	2.216	2.306
11°	1.764	1.776	1.788	1.801	1.818	1.835	1.856	1.879	1.906	1.935	1.967	2.005	2.048	2.096	2.151	2.214	2.288	2.373	2.475
12°	1.856	1.869	1.883	1.897	1.915	1.935	1.958	1.984	2.013	2.046	2.081	2.124	2.172	2.226	2.288	2.359	2.442	2.541	2.655
13°	1.955	1.967	1.982	1.998	2.018	2.040	2.065	2.092	2.126	2.162	2.202	2.249	2.302	2.363	2.432	2.512	2.606	2.715	2.847
14°	2.059	2.072	2.089	2.106	2.126	2.151	2.178	2.210	2.245	2.284	2.329	2.381	2.440	2.508	2.585	2.674	2.779	2.904	3.051
15°	2.170	2.183	2.201	2.220	2.243	2.269	2.298	2.331	2.371	2.415	2.464	2.522	2.587	2.661	2.748	2.847	2.963	3.103	3.267
16°	2.288	2.304	2.321	2.341	2.365	2.393	2.427	2.461	2.505	2.524	2.608	2.670	2.741	2.825	2.919	3.030	3.160	3.313	3.501
17°	2.415	2.430	2.448	2.470	2.497	2.526	2.562	2.602	2.646	2.700	2.759	2.827	2.906	2.997	3.103	3.225	3.371	3.542	3.749
18°	2.551	2.568	2.587	2.610	2.638	2.670	2.706	2.750	2.798	2.856	2.922	2.997	3.084	3.182	3.299	3.434	3.594	3.787	4.017
19°	2.698	2.713	2.737	2.759	2.787	2.821	2.861	2.908	2.963	3.023	3.096	3.177	3.272	3.381	3.508	3.658	3.837	4.049	4.309
20°	2.856	2.872	2.892	2.917	2.949	2.986	3.028	3.079	3.136	3.203	3.281	3.371	3.472	3.594	3.736	3.901	4.098	4.334	4.622
21°	3.030	3.044	3.065	3.091	3.124	3.162	3.208	3.262	3.323	3.396	3.480	3.579	3.690	3.824	3.977	4.160	4.380	4.642	4.965
22°	3.218	3.232	3.252	3.252	3.279	3.311	3.353	3.401	3.459	3.526	3.605	3.695	3.801	3.926	4.240	4.445	4.687	4.979	5.342
23°	3.424	3.437	3.457	3.483	3.519	3.560	3.613	3.674	3.747	3.831	3.929	4.046	4.180	4.340	4.526	4.753	5.020	5.349	5.753
24°	3.650	3.661	3.679	3.706	3.741	3.787	3.840	3.907	3.983	4.075	4.183	4.309	4.457	4.632	4.839	5.086	5.386	5.753	6.212
25°	3.898	3.907	3.926	3.952	3.989	4.034	4.092	4.163	4.246	4.343	4.460	4.600	4.759	4.955	5.182	5.457	5.792	6.204	6.718
26°	4.177	4.183	4.198	4.222	4.261	4.306	4.367	4.442	4.533	4.638	4.766	4.917	5.094	5.305	5.559	5.864	6.238	6.700	7.283
27°	4.488	4.488	4.501	4.526	4.561	4.609	4.674	4.753	4.849	4.965	5.101	5.265	5.460	5.694	5.978	6.320	6.737	7.258	7.928
28°	4.843	4.836	4.843	4.863	4.897	4.948	5.013	5.097	5.200	5.323	5.475	5.655	5.868	6.128	6.442	6.825	7.298	7.889	8.650
29°	5.243	5.229	5.229	5.243	5.276	5.327	5.393	5.483	5.593	5.725	5.889	6.090	6.328	6.613	6.963	7.395	7.928	8.608	9.487
30°	5.706	5.674	5.667	5.674	5.706	5.753	5.824	5.917	6.032	6.178	6.359	6.576	6.839	7.163	7.555	8.045	8.650	9.432	10.455

Table 59 Ratio of Downstream to Upstream Density for Two-Dimensional Shock

For a perfect gas with constant specific heat and molecular weight
k = 1.4

α' → / ω' ↓	1°	2°	3°	4°	5°	6°	7°	8°	9°	10°	11°	12°	13°	14°	15°	16°	17°	18°	19°
1°		2.001	1.501	1.334	1.251	1.201	1.168	1.145	1.127	1.113	1.102	1.094	1.086	1.080	1.075	1.070	1.066	1.063	1.060
2°			3.002	2.002	1.669	1.503	1.403	1.337	1.290	1.255	1.227	1.205	1.188	1.173	1.161	1.150	1.141	1.133	1.126
3°				4.006	2.505	2.006	1.756	1.606	1.507	1.436	1.383	1.342	1.309	1.283	1.261	1.242	1.226	1.213	1.201
4°					5.012	3.010	2.343	2.010	1.810	1.678	1.583	1.512	1.458	1.414	1.378	1.349	1.324	1.303	1.285
5°						6.021	3.516	2.682	2.265	2.015	1.849	1.731	1.643	1.574	1.520	1.475	1.438	1.407	1.381
6°							7.034	4.025	3.022	2.522	2.222	2.022	1.880	1.774	1.692	1.626	1.573	1.529	1.491
7°								8.052	4.536	3.365	2.780	2.430	2.197	2.031	1.907	1.810	1.734	1.672	1.620
8°									9.074	5.049	3.709	3.040	2.639	2.372	2.182	2.040	1.930	1.843	1.771
9°										10.102	5.566	4.056	3.302	2.850	2.549	2.335	2.175	2.051	1.953
10°											11.136	6.087	4.405	3.566	3.063	2.728	2.490	2.312	2.174
11°												12.177	6.611	4.757	3.832	3.278	2.909	2.646	2.450
12°													13.226	7.140	5.113	4.101	3.495	3.091	2.804
13°															7.673	5.471	4.372	3.714	3.276
14°																8.211	5.834	4.647	3.936
15°																	8.755	6.200	4.924
16°																		9.304	6.570
17°																			9.860
18°																			
19°																			
20°																			
21°																			
22°																			
23°																			
24°																			
25°																			
26°																			
27°																			
28°																			
29°																			
30°																			

α' → / ω' ↓	39°	40°	41°	42°	43°	44°	45°	46°	47°	48°	49°	50°	51°	52°	53°	54°	55°	56°	57°
1°	1.036	1.036	1.036	1.036	1.036	1.036	1.036	1.036	1.036	1.036	1.036	1.036	1.036	1.036	1.037	1.037	1.038	1.038	1.039
2°	1.075	1.074	1.073	1.073	1.073	1.073	1.072	1.072	1.072	1.073	1.073	1.073	1.073	1.074	1.075	1.075	1.076	1.077	1.078
3°	1.115	1.114	1.113	1.112	1.111	1.111	1.111	1.110	1.110	1.111	1.111	1.111	1.112	1.113	1.114	1.115	1.116	1.117	1.119
4°	1.156	1.155	1.154	1.152	1.152	1.151	1.150	1.150	1.150	1.150	1.150	1.151	1.152	1.152	1.154	1.155	1.156	1.158	1.160
5°	1.201	1.198	1.196	1.195	1.194	1.193	1.192	1.191	1.191	1.191	1.191	1.192	1.193	1.194	1.195	1.196	1.198	1.201	1.203
6°	1.247	1.244	1.241	1.239	1.237	1.236	1.235	1.234	1.234	1.233	1.234	1.234	1.235	1.236	1.237	1.239	1.241	1.244	1.247
7°	1.296	1.292	1.289	1.286	1.283	1.282	1.280	1.279	1.278	1.278	1.278	1.278	1.279	1.280	1.282	1.283	1.286	1.289	1.292
8°	1.348	1.343	1.339	1.335	1.332	1.329	1.327	1.325	1.324	1.324	1.323	1.324	1.324	1.325	1.327	1.329	1.332	1.335	1.339
9°	1.403	1.396	1.391	1.387	1.383	1.379	1.376	1.374	1.373	1.371	1.371	1.371	1.371	1.373	1.374	1.376	1.379	1.383	1.387
10°	1.461	1.453	1.447	1.441	1.436	1.432	1.428	1.425	1.423	1.422	1.421	1.420	1.421	1.422	1.423	1.425	1.428	1.432	1.436
11°	1.523	1.514	1.506	1.499	1.492	1.487	1.483	1.479	1.476	1.474	1.472	1.472	1.472	1.472	1.474	1.476	1.479	1.483	1.487
12°	1.589	1.578	1.568	1.560	1.552	1.545	1.540	1.535	1.532	1.529	1.527	1.525	1.525	1.525	1.527	1.529	1.532	1.535	1.540
13°	1.660	1.647	1.635	1.624	1.615	1.607	1.600	1.595	1.590	1.586	1.583	1.582	1.581	1.581	1.582	1.583	1.586	1.590	1.595
14°	1.737	1.720	1.706	1.693	1.682	1.673	1.664	1.657	1.651	1.647	1.643	1.640	1.639	1.638	1.639	1.640	1.643	1.647	1.651
15°	1.819	1.799	1.782	1.767	1.754	1.742	1.732	1.723	1.716	1.710	1.705	1.702	1.700	1.699	1.699	1.700	1.702	1.705	1.710
16°	1.908	1.885	1.864	1.846	1.830	1.816	1.804	1.794	1.785	1.777	1.771	1.767	1.764	1.762	1.761	1.762	1.764	1.767	1.771
17°	2.004	1.977	1.952	1.931	1.912	1.895	1.881	1.868	1.857	1.848	1.841	1.835	1.831	1.828	1.827	1.827	1.828	1.831	1.835
18°	2.110	2.077	2.048	2.022	2.000	1.980	1.963	1.948	1.935	1.924	1.915	1.907	1.902	1.898	1.895	1.894	1.895	1.898	1.902
19°	2.225	2.186	2.152	2.121	2.094	2.071	2.050	2.032	2.017	2.004	1.992	1.983	1.976	1.971	1.967	1.966	1.966	1.967	1.971
20°	2.352	2.305	2.265	2.229	2.197	2.169	2.145	2.123	2.105	2.089	2.075	2.064	2.055	2.048	2.043	2.041	2.040	2.041	2.043
21°	2.492	2.437	2.388	2.346	2.308	2.275	2.246	2.221	2.199	2.180	2.164	2.150	2.139	2.130	2.124	2.119	2.117	2.117	2.119
22°	2.649	2.582	2.525	2.474	2.429	2.390	2.356	2.326	2.300	2.277	2.258	2.241	2.228	2.217	2.209	2.203	2.199	2.198	2.199
23°	2.824	2.745	2.675	2.615	2.562	2.516	2.475	2.440	2.409	2.382	2.359	2.339	2.323	2.309	2.299	2.291	2.286	2.283	2.283
24°	3.022	2.926	2.843	2.771	2.708	2.653	2.605	2.563	2.526	2.494	2.467	2.443	2.424	2.407	2.394	2.384	2.377	2.373	2.371
25°	3.248	3.132	3.032	2.945	2.870	2.805	2.747	2.698	2.654	2.616	2.584	2.556	2.532	2.512	2.496	2.483	2.474	2.467	2.464
26°	3.508	3.365	3.244	3.140	3.050	2.972	2.904	2.845	2.794	2.749	2.710	2.677	2.648	2.624	2.604	2.589	2.576	2.568	2.563
27°	3.810	3.635	3.487	3.360	3.252	3.159	3.078	3.007	2.946	2.893	2.847	2.808	2.774	2.745	2.721	2.701	2.686	2.675	2.667
28°	4.166	3.948	3.765	3.611	3.480	3.368	3.271	3.187	3.114	3.051	2.997	2.950	2.909	2.875	2.846	2.822	2.803	2.788	2.778
29°	4.593	4.317	4.090	3.900	3.740	3.604	3.487	3.387	3.300	3.225	3.161	3.105	3.056	3.015	2.981	2.952	2.928	2.910	2.896
30°	5.113	4.759	4.472	4.236	4.039	3.873	3.732	3.611	3.508	3.418	3.341	3.274	3.217	3.168	3.126	3.091	3.063	3.040	3.022

Table 59 Ratio of Downstream to Upstream Density for Two-Dimensional Shock

For a perfect gas with constant specific heat and molecular weight
k = 1.4

α' → ω' ↓	20°	21°	22°	23°	24°	25°	26°	27°	28°	29°	30°	31°	32°	33°	34°	35°	36°	37°	38°
1°	1.057	1.055	1.053	1.051	1.049	1.047	1.046	1.045	1.044	1.043	1.042	1.041	1.040	1.039	1.039	1.038	1.038	1.037	1.037
2°	1.120	1.115	1.110	1.106	1.102	1.099	1.095	1.093	1.090	1.088	1.086	1.084	1.082	1.081	1.079	1.078	1.077	1.076	1.075
3°	1.190	1.181	1.173	1.166	1.160	1.154	1.149	1.144	1.140	1.137	1.133	1.130	1.127	1.125	1.123	1.121	1.119	1.117	1.116
4°	1.269	1.256	1.243	1.233	1.223	1.215	1.207	1.200	1.194	1.189	1.184	1.179	1.175	1.172	1.168	1.165	1.163	1.160	1.158
5°	1.358	1.339	1.322	1.306	1.293	1.281	1.271	1.261	1.253	1.245	1.238	1.232	1.226	1.221	1.217	1.213	1.209	1.206	1.203
6°	1.460	1.433	1.409	1.388	1.370	1.354	1.340	1.327	1.316	1.306	1.297	1.289	1.281	1.275	1.269	1.263	1.258	1.254	1.250
7°	1.577	1.540	1.508	1.480	1.456	1.435	1.416	1.400	1.385	1.372	1.360	1.350	1.340	1.331	1.324	1.317	1.311	1.305	1.300
8°	1.712	1.663	1.620	1.584	1.553	1.525	1.501	1.480	1.461	1.444	1.429	1.416	1.403	1.393	1.383	1.374	1.366	1.359	1.353
9°	1.872	1.806	1.750	1.702	1.662	1.626	1.595	1.568	1.544	1.523	1.504	1.487	1.472	1.459	1.446	1.436	1.426	1.417	1.409
10°	2.064	1.975	1.901	1.839	1.786	1.740	1.701	1.667	1.636	1.610	1.586	1.565	1.547	1.530	1.515	1.502	1.490	1.479	1.469
11°	2.298	2.177	2.079	1.997	1.928	1.870	1.820	1.777	1.739	1.706	1.677	1.651	1.628	1.607	1.589	1.573	1.558	1.545	1.533
12°	2.590	2.424	2.291	2.184	2.095	2.020	1.956	1.902	1.854	1.813	1.777	1.745	1.717	1.692	1.669	1.650	1.632	1.616	1.602
13°	2.964	2.731	2.551	2.407	2.291	2.194	2.113	2.044	1.984	1.933	1.888	1.849	1.815	1.784	1.757	1.733	1.712	1.693	1.675
14°	3.463	3.126	2.875	2.680	2.525	2.399	2.295	2.207	2.133	2.069	2.013	1.965	1.923	1.886	1.853	1.824	1.798	1.775	1.755
15°	4.160	3.652	3.291	3.020	2.811	2.645	2.509	2.397	2.303	2.223	2.155	2.095	2.044	1.999	1.959	1.924	1.893	1.865	1.841
16°	5.205	4.388	3.844	3.457	3.168	2.944	2.766	2.621	2.501	2.401	2.316	2.242	2.179	2.124	2.076	2.034	1.996	1.963	1.934
17°	6.945	5.490	4.618	4.039	3.626	3.318	3.079	2.890	2.735	2.608	2.501	2.410	2.332	2.265	2.206	2.155	2.110	2.070	2.035
18°	10.423	7.325	5.778	4.852	4.236	3.798	3.470	3.217	3.015	2.852	2.716	2.603	2.506	2.424	2.352	2.290	2.236	2.188	2.147
19°			7.709	6.070	5.089	4.437	3.972	3.625	3.357	3.144	2.970	2.827	2.707	2.605	2.517	2.442	2.376	2.319	2.269
20°				8.099	6.367	5.330	4.640	4.150	3.783	3.500	3.274	3.091	2.940	2.813	2.705	2.613	2.534	2.465	2.405
21°					8.495	6.669	5.575	4.848	4.330	3.944	3.645	3.408	3.215	3.055	2.922	2.808	2.711	2.628	2.555
22°						8.898	6.975	5.824	5.059	4.514	4.108	3.794	3.544	3.341	3.173	3.033	2.914	2.812	2.725
23°							9.306	7.287	6.077	5.274	4.702	4.275	3.945	3.683	3.470	3.294	3.147	3.022	2.916
24°								7.604	6.336		5.493	4.894	4.446	4.100	3.825	3.602	3.418	3.264	3.134
25°										7.927	6.599	5.717	5.089	4.621	4.259	3.971	3.738	3.545	3.384
26°											8.256	6.868	5.945	5.289	4.799	4.421	4.120	3.877	3.676
27°												8.593	7.142	6.179	5.493	4.982	4.587	4.274	4.019
28°														7.423	6.418	5.703	5.170	4.758	4.431
29°																6.662	5.917	5.362	4.933
30°																	6.913	6.137	5.559

α' → ω' ↓	58°	59°	60°	61°	62°	63°	64°	65°	66°	67°	68°	69°	70°	71°	72°	73°	74°	75°	76°
1°	1.039	1.040	1.041	1.042	1.043	1.044	1.045	1.046	1.047	1.049	1.051	1.053	1.055	1.057	1.060	1.063	1.066	1.070	1.075
2°	1.079	1.081	1.082	1.084	1.086	1.088	1.090	1.093	1.095	1.099	1.102	1.106	1.110	1.115	1.120	1.126	1.133	1.141	1.150
3°	1.121	1.123	1.125	1.127	1.130	1.133	1.137	1.140	1.144	1.149	1.154	1.160	1.166	1.173	1.181	1.190	1.201	1.213	1.226
4°	1.163	1.165	1.168	1.172	1.175	1.179	1.184	1.189	1.194	1.200	1.207	1.215	1.223	1.233	1.243	1.256	1.269	1.285	1.303
5°	1.206	1.209	1.213	1.217	1.221	1.226	1.232	1.238	1.245	1.253	1.261	1.271	1.281	1.293	1.306	1.322	1.339	1.358	1.381
6°	1.250	1.254	1.258	1.263	1.269	1.275	1.281	1.289	1.297	1.306	1.316	1.327	1.340	1.354	1.370	1.388	1.409	1.433	1.460
7°	1.296	1.300	1.305	1.311	1.317	1.324	1.331	1.340	1.350	1.360	1.372	1.385	1.400	1.416	1.435	1.456	1.480	1.508	1.540
8°	1.343	1.348	1.353	1.359	1.366	1.374	1.383	1.393	1.403	1.416	1.429	1.444	1.461	1.480	1.501	1.525	1.553	1.584	1.620
9°	1.391	1.396	1.403	1.409	1.417	1.426	1.436	1.446	1.459	1.472	1.487	1.504	1.523	1.544	1.568	1.595	1.626	1.662	1.702
10°	1.441	1.447	1.453	1.461	1.469	1.479	1.490	1.502	1.515	1.530	1.547	1.565	1.586	1.610	1.636	1.667	1.701	1.740	1.786
11°	1.492	1.499	1.506	1.514	1.523	1.533	1.545	1.558	1.573	1.589	1.607	1.628	1.651	1.677	1.706	1.739	1.777	1.820	1.870
12°	1.545	1.552	1.560	1.568	1.578	1.589	1.602	1.616	1.632	1.650	1.669	1.692	1.717	1.745	1.777	1.813	1.854	1.902	1.956
13°	1.600	1.607	1.615	1.624	1.635	1.647	1.660	1.675	1.693	1.712	1.733	1.757	1.784	1.815	1.849	1.888	1.933	1.984	2.044
14°	1.657	1.664	1.673	1.682	1.693	1.706	1.720	1.737	1.755	1.775	1.798	1.824	1.853	1.886	1.923	1.965	2.013	2.069	2.133
15°	1.716	1.723	1.732	1.742	1.754	1.767	1.782	1.799	1.819	1.841	1.865	1.893	1.924	1.959	1.999	2.044	2.095	2.155	2.223
16°	1.777	1.785	1.794	1.804	1.816	1.830	1.846	1.864	1.885	1.908	1.934	1.963	1.996	2.034	2.076	2.124	2.179	2.242	2.316
17°	1.841	1.848	1.857	1.868	1.881	1.895	1.912	1.931	1.952	1.977	2.004	2.035	2.070	2.110	2.155	2.206	2.265	2.332	2.410
18°	1.907	1.915	1.924	1.935	1.948	1.963	1.980	2.000	2.022	2.048	2.077	2.110	2.147	2.188	2.236	2.290	2.352	2.424	2.506
19°	1.976	1.983	1.992	2.004	2.017	2.032	2.050	2.071	2.094	2.121	2.152	2.186	2.225	2.269	2.319	2.376	2.442	2.517	2.605
20°	2.048	2.055	2.064	2.075	2.089	2.105	2.123	2.145	2.169	2.197	2.229	2.265	2.305	2.352	2.405	2.465	2.534	2.613	2.705
21°	2.124	2.130	2.139	2.150	2.164	2.180	2.199	2.221	2.246	2.275	2.308	2.346	2.388	2.437	2.492	2.555	2.628	2.711	2.803
22°	2.203	2.209	2.217	2.228	2.241	2.258	2.277	2.300	2.326	2.356	2.390	2.429	2.474	2.525	2.582	2.649	2.725	2.812	2.914
23°	2.286	2.291	2.299	2.309	2.323	2.339	2.359	2.382	2.409	2.440	2.475	2.516	2.562	2.615	2.675	2.745	2.824	2.916	3.022
24°	2.373	2.377	2.384	2.394	2.407	2.424	2.443	2.467	2.494	2.526	2.563	2.605	2.653	2.708	2.771	2.843	2.926	3.022	3.134
25°	2.464	2.467	2.474	2.483	2.496	2.512	2.532	2.556	2.584	2.616	2.654	2.698	2.747	2.805	2.870	2.945	3.032	3.132	3.248
26°	2.561	2.563	2.568	2.576	2.589	2.604	2.624	2.648	2.677	2.710	2.749	2.794	2.845	2.904	2.972	3.050	3.140	3.244	3.365
27°	2.663	2.663	2.667	2.675	2.686	2.701	2.721	2.745	2.774	2.808	2.847	2.893	2.946	3.007	3.078	3.159	3.252	3.360	3.487
28°	2.772	2.770	2.772	2.778	2.788	2.803	2.822	2.846	2.875	2.909	2.950	2.997	3.051	3.114	3.187	3.271	3.368	3.480	3.611
29°	2.887	2.883	2.883	2.887	2.896	2.910	2.928	2.952	2.981	3.015	3.056	3.105	3.161	3.225	3.300	3.387	3.487	3.604	3.740
30°	3.010	3.002	3.000	3.002	3.010	3.022	3.040	3.063	3.091	3.126	3.168	3.217	3.274	3.341	3.418	3.508	3.611	3.732	3.873

Table 60 Standard Atmosphere[a]

Altitude ft	Temperature °F	Temperature R	Pressure in. Hg	Pressure lb/ft²	Density lb/ft³	Density ratio, ρ/ρ_0	Speed of Sound mi/hr	Absolute Viscosity lbm/ft sec	Kinematic Viscosity ft²/sec
0	58.7	518.4	29.92	2116.	.002378	1.0000	760.9	3.725×10^{-7}	1.566×10^{-4}
2000	51.5	511.2	27.82	1968.	.002242	.9428	755.7	3.685	1.644
4000	44.4	504.1	25.84	1828.	.002112	.8881	750.4	3.644	1.725
6000	37.3	497.0	23.98	1696.	.001988	.8358	745.1	3.602	1.812
8000	30.2	489.9	22.22	1572.	.001869	.7859	739.7	3.561	1.905
10000	23.0	482.7	20.58	1455.	.001756	.7384	734.3	3.519	2.004
12000	15.9	475.6	19.03	1346.	.001648	.6931	728.8	3.476	2.109
14000	8.8	468.5	17.57	1243.	.001545	.6499	723.4	3.434	2.223
16000	1.6	461.3	16.21	1146.	.001448	.6088	718.7	3.391	2.342
18000	−5.5	454.2	14.94	1056.	.001355	.5698	712.2	3.348	2.471
20000	−12.6	447.1	13.75	972.1	.001267	.5327	706.6	3.305	2.608
22000	−19.8	439.9	12.63	893.3	.001183	.4974	701.1	3.261	2.756
24000	−26.9	432.8	11.59	819.8	.001103	.4640	695.3	3.217	2.916
26000	−34.0	425.7	10.62	751.2	.001028	.4323	689.5	3.173	3.086
28000	−41.2	418.5	9.720	687.4	.000957	.4023	683.7	3.128	3.268
30000	−48.3	411.4	8.880	628.0	.000889	.3740	677.9	3.083	3.468
32000	−55.4	404.3	8.101	572.9	.000826	.3472	672.0	3.038	3.678
34000	−62.5	397.2	7.377	521.7	.000765	.3218	666.0	2.992	3.911
35332	−67.3	392.4	6.926	489.8	.000727	.3058	662.0	2.962	4.073
36000	−67.3	392.4	6.711	474.4	.000705	.2963	662.0	2.962	4.204
38000	−67.3	392.4	6.098	431.1	.000640	.2692	662.0	2.962	4.625
40000	−67.3	392.4	5.544	391.9	.000582	.2448	662.0	2.962	5.089
42000	−67.3	392.4	5.038	356.2	.000529	.2225	662.0	2.962	5.599
44000	−67.3	392.4	4.578	323.7	.000480	.2021	662.0	2.962	6.161
46000	−67.3	392.4	4.162	294.2	.000437	.1838	662.0	2.962	6.778
48000	−67.3	392.4	3.782	267.4	.000397	.1670	662.0	2.962	7.459
50000	−67.3	392.4	3.438	243.1	.000361	.1518	662.0	2.962	8.206
52000	−67.3	392.4	3.124	220.9	.000328	.1379	662.0	2.962	9.028
54000	−67.3	392.4	2.840	200.8	.000298	.1254	662.0	2.962	9.933
56000	−67.3	392.4	2.581	182.5	.000271	.1140	662.0	2.962	10.93
58000	−67.3	392.4	2.346	165.9	.000246	.1036	662.0	2.962	12.02
60000	−67.3	392.4	2.132	150.8	.000224	.09415	662.0	2.962	13.23
62000	−67.3	392.4	1.938	137.1	.000203	.08557	662.0	2.962	14.56
64000	−67.3	392.4	1.761	124.6	.000185	.07777	662.0	2.962	16.02
65000	−67.3	392.4	1.679	118.7	.000176	.07414	662.0	2.962	16.80
70000	−67.3	392.4	1.322	93.53	.000139	.05839	662.0	2.962	21.33
75000	−67.3	392.4	1.042	73.66	.000109	.04599	662.0	2.962	27.09
80000	−67.3	392.4	.8202	58.01	.0000861	.03621	662.0	2.962	34.39
85000	−67.3	392.4	.6460	45.68	.0000678	.02852	662.0	2.962	43.67
90000	−67.3	392.4	.5086	35.97	.0000534	.02246	662.0	2.962	55.45
95000	−67.3	392.4	.4006	28.33	.0000421	.01769	662.0	2.962	70.41
100000	−67.3	392.4	.3156	22.31	.0000331	.01394	662.0	2.962	89.41

[a]Data from reference 40.

Table 61 Physical Constants[a]

Quantity	Symbol	Value	Uncertainty ppm
Speed of light in vacuum	c	299792458 m/sec	0.004
Avogadro number	N	6.022045×10^{23} mole^{-1}	5.1
Faraday constant	F	96484.56 C/mole	2.8
Planck constant	h	6.626176×10^{-34} J/Hz	5.4
Molar gas constant	R	8.31441 J/mole K	31.
Elementary charge	e	$1.6021892 \times 10^{-19}$ C	2.9
Boltzmann constant	k	1.380662×10^{-23} J/K	32.
First radiation constant	$c_1 = 2\pi hc^2$	3.741832×10^{-16} W m^2	5.4
Second radiation constant	$c_2 = hc/k$	0.01438786 m/K	31.
Molar volume, ideal gas at T = 273.15 K, p = 1 atm	V_0	0.02241383 m^3/mole	31.
Gravitational constant	G	6.6720×10^{-11} N m^2/kg^2	615.

[a]Data from Reference 41.

Table 62 Conversion Factors

To convert the numerical value of a property expressed in one of the units in the left-hand column of a table to the numerical value of the same property given in one of the units in the top row of the same table, multiply the former value by the factor in the block common to both units. Numbers followed by an asterisk (*) are definitions of the relation between the two units. SI is the abbreviation of the International System of Units.

Table 62a Length

Units	in.	ft	cm	m(SI)
1 in. =	1	0.0833333	2.54*	0.0254
1 ft =	12*	1	30.48	0.3048
1 cm =	0.3937008	0.0328084	1	0.01*
1 m(SI) =	39.37008	3.280840	100	1

Table 62b Volume

Units	in.^{3}	ft^3	gal	liter	cm^3	m^3(SI)
1 in.^{3} =	1	5.787037×10^{-4}	4.329004×10^{-3}	0.01638706	16.38706*	1.638706×10^{-5}
1 ft^3 =	1728*	1	7.480519	28.31684	28316.84	0.02831684
1 gal =	231*	0.1336806	1	3.785411	3785.411	3.785411×10^{-3}
1 liter =	61.02374	0.03531467	0.2641721	1	1000*	10^{-3}
1 cm^3 =	0.06102374	3.531467×10^{-5}	2.641721×10^{-4}	10^{-3}	1	10^{-6}
1 m^3(SI) =	61023.74	35.31467	264.1721	10^3	10^6	1

Table 62c Mass

Units	lb	g	kg(SI)	ton	metric ton
1 lb =	1	453.59237*	0.45359237	0.0005	4.5359237×10^{-4}
1 g =	2.204623×10^{-3}	1	10^{-3}	1.1023115×10^{-6}	10^{-6}
1 kg(SI) =	2.204623	1000	1	1.1023115×10^{-3}	10^{-3}
1 ton =	2000*	907184.74	907.18474	1	0.9071846
1 metric ton =	2204.623	10^6	1000*	1.1023115	1

Table 62d Density

Units	lb/ft^3	lb/gal	g/cm^3	kg/m^3(SI)
1 lb/ft^3 =	1	0.13368056	0.01601847	16.01847
1 lb/gal =	7.480519	1	0.11982646	119.82646
1 g/cm^3 =	62.42793	8.345402	1	10^3
1 kg/m^3(SI) =	0.06242793	8.345402×10^{-3}	10^{-3}	1

Table 62e Force

Unit	lbf	pdl[a]	dyn	g[b]	N(SI)
1 lbf =	1	32.1740	4.44822×10^5	453.59237	4.448222
1 pdl =	0.0310809	1	1.38255×10^4	14.0981	0.138255
1 dyn =	2.24809×10^{-6}	7.23301×10^{-5}	1	1.01972×10^{-3}	10^{-5}
1 g =	2.20462×10^{-3}	0.070931	980.665	1	9.80665×10^{-3}
1 N(SI) =	0.224809	7.23298	10^5	101.972	1

[a]Poundal.
[b]Gram force.

Table 62f Pressure

Unit	lb/in.^{2}	ft H$_2$O 60°F	atm	kg/cm^2	torr	bar	N/m^2(SI)
1 lb/in.^{2} =	1	2.30897	0.06804596	0.07030669	51.71495	0.06894757	6.894757×10^3
1 ft H$_2$O, 60°F =	0.43309	1	0.0294703	0.0304495	22.3974	0.0298608	2.98608×10^3
1 atm =	14.69595	33.9325	1	1.033227	760*	1.013250	1.013250×10^5
1 kg/cm^2 =	14.22335	32.8413	0.9678411	1	735.5596	0.980665	9.80665×10^4
1 torr =	0.0193368	0.0446480	1.315789×10^{-3}	1.359509×10^{-3}	1	1.333224×10^{-3}	133.3224
1 bar =	14.503775	33.4887	0.9869233	1.019716	750.0617	1	10^5
1 N/m^2(SI) =	1.450378×10^{-4}	3.34887×10^{-4}	9.869233×10^{-6}	1.019716×10^{-5}	7.500617×10^{-3}	10^{-5}	1

Table 62g Energy

Units	Btu$_{IT}$	cal$_{IT}$	ft-lb	ft^3-lb/in.^{2}	hp hr	kg m	liter atm	cal	J(SI)
1 Btu$_{IT}$ =	1	251.9958*	778.1693	5.403953	3.930148×10^{-4}	107.5858	10.41259	252.1644	1055.056
1 cal$_{IT}$ =	3.968320×10^{-3}	1	3.088025	0.02144462	1.559608×10^{-6}	0.4269349	0.04132048	1.000669	4.1868*
1 ft-lb =	1.285067×10^{-3}	0.3238316	1	6.944444×10^{-3}	5.050505×10^{-7}	0.1382550	0.01338088	0.3240482	1.355818
1 ft^3-lb/in.^{2} =	0.1850497	46.63175	144*	1	7.272727×10^{-5}	19.90873	1.926847	46.66294	195.2378
1 hp hr =	2544.433	6.411866×10^5	1980000*	13750	1	2.73745×10^5	2.649414×10^4	6.41615×10^5	2.684520×10^6
1 kg m =	9.294906×10^{-3}	2.342277	7.233009	0.0502292	3.653035×10^{-6}	1	0.0967840	2.343844	9.80665
1 liter atm =	0.09603759	24.20107	74.73351	0.5189827	3.774420×10^{-5}	10.33229	1	24.21726	101.3250
1 cal =	3.965667×10^{-3}	0.9993314	3.085960	2.143028×10^{-2}	1.558566×10^{-6}	0.4266495	0.04129286	1	4.184*
1 J(SI) =	9.47817×10^{-4}	0.2388459	0.7375621	5.121959×10^{-3}	3.725062×10^{-7}	0.1019716	9.869233×10^{-3}	0.2390057	1

Table 62h Power

Units	ft-lb/sec	Btu$_{IT}$/hr	hp	W(SI)
1 ft-lb/sec =	1	4.626243	1.818182×10^{-3}	1.355818
1 Btu$_{IT}$/hr =	0.216158	1	3.930148×10^{-4}	0.2930711
1 hp =	550 *	2544.43	1	745.6999
1 W(SI) =	0.737562	3.412142	1.341022×10^{-3}	1

Table 62i Specific Energy

Units	Btu$_{IT}$/lb	cal$_{IT}$/g	cal/g	J/g	J/kg(SI)
1 Btu$_{IT}$/lb =	1	0.55555556	0.5559272	2.326000	2326
1 cal$_{IT}$/g =	1.8*	1	1.000669	4.1868*	4186.8
1 cal/g =	1.798797	0.9993314	1	4.1840*	4184
1 J/g =	0.4299226	0.2388459	0.2390057	1	1000
1 J/kg(SI) =	4.299226×10^{-4}	2.388459×10^{-4}	2.390057×10^{-4}	10^{-3}	1

Table 62j   Specific Energy per Degree

Units	Btu$_{IT}$/R lb	cal$_{IT}$/K g	cal/K g	J/K g	J/K kg(SI)
1 Btu$_{IT}$/R lb =	1	1	1.000669	4.1868	4186.8
1 cal$_{IT}$/K g =	1	1	1.000669	4.1868*	4186.8
1 cal/K g =	0.9993314	0.9993314	1	4.1840*	4184
1 J/K g =	0.2388459	0.2388459	0.2390057	1	1000
1 J/K kg(SI) =	2.388459×10^{-4}	2.388459×10^{-4}	2.390057×10^{-4}	10^{-3}	1

Table 62k   Absolute Viscosity

Unit	centipoise	lbm/ft sec	lbf sec/ft^2	N sec/m^2(SI)
1 centipoise =	1	6.71971×10^{-4}	2.08855×10^{-5}	10^{-3}*
1 lbm/ft sec =	1488.16	1	0.031081	1.48816
1 lbf sec/ft^2 =	47880.3	32.1741	1	47.8803
1 N sec/m^2(SI) =	10^3	0.671971	0.0208855	1

Table 62l   Kinematic Viscosity

Units	centistoke	ft^2/sec	cm^2/hr	m^2/sec(SI)
1 centistoke =	1	1.076391×10^{-5}	36.0000	10^{-6}*
1 ft^2/sec =	92903.04	1	3344509.	0.09290304*
1 cm^2/hr =	0.02777778	2.989975×10^{-7}	1	2.777778×10^{-8}
1 m^2/sec(SI) =	10^6	10.76391	3.6×10^7	1

Table 62m   Thermal Conductivity

Units	Btu$_{IT}$/ft hr R	cal$_{IT}$/cm sec K	kcal$_{IT}$/m hr K	W/m K(SI)
1 Btu$_{IT}$/ft hr R =	1	0.0041336	1.4881	1.7307
1 cal$_{IT}$/cm sec K =	241.92	1	360	418.68
1 kcal$_{IT}$/m hr K =	0.67199	0.0027778	1	1.1630
1 W/m K(SI) =	0.57781	0.0023885	0.85985	1

Table 62n   Molar Gas Constant

R = 1.98586	cal$_{IT}$/K g-mole
1.98586	Btu$_{IT}$/R lb-mole
10.73150	ft^3/lb-in.^{2} R lb-mole
0.730235	ft^3 atm/R lb-mole
62363.2	cm^3 torr/K g-mole
82.0568	cm^3 atm/K g-mole
0.0820568	liter atm/K g-mole
1.98719	cal/K g-mole
8.31441	J/K g-mole(SI)

(184)

Table 62o Miscellaneous

Temperature

Triple point of water $= 273.16$ K*

$t(°C) = T(K) - 273.15$

$T(R) = 1.8T(K)$

$t(°F) = T(R) - 459.67$*

$t(°F) = 1.8t(°C) + 32$

Force and Mass Conversion

$g_c = 32.1740$ lbm ft/lbf sec^2

$\quad = 1.0$ kg m/N sec^2(SI)

Gravity

$g = 980.665$ cm/sec^2

$\quad = 9.80665$ m/sec^2(SI)

$(PV)_{T=0°C}^{P=0}$

5276.36 ft^3/lb-in.^{2} lb-mole

22.41383 liter atm/g-mole

542.8015 cal/g-mole

2271.082 J/g-mole(SI)

Sources of Data and Calculation Methods

1 GENERAL DESCRIPTION

For decades engineers and scientists have sought to determine with precision the thermodynamic properties of the air around us and the simple gases from which it is composed. A great many experimental data across a broad range of variables have been accumulated, but accurate measurements have proved elusive, perhaps unobtainable, at certain critical extremes of pressure and temperature.

Where experimental techniques and the methods for deriving additional data produced unreliable results, researchers turned to theoretical analysis in an effort to learn even more about the properties of gases. It was well known, for instance, that "real" gases begin to acquire the characteristics attributed to an "ideal" gas when pressures are reduced to critically low levels. The theoretical concept of an ideal gas, therefore, became a vital link in the understanding of gas properties and in the practical application of this knowledge.

1.1 Ideal Gas Properties

As pressure is reduced to zero, the properties of real gases approach those of an ideal gas. The latter is assumed to exhibit two characteristic properties—its molecules occupy no space and they produce no molecular interactions. Consequently, the equation of state for an ideal gas is

$$pv = RT \qquad (1)$$

where p denotes the pressure, v the specific volume, R the gas constant, and T the absolute thermodynamic temperature.[1]*

From (1) the relations between internal energy u and enthalpy h, for unit mass, and specific heat capacity at constant volume c_v and specific heat capacity at constant pressure c_p are

$$u = h - RT \qquad (2)$$

*Superscript numbers refer to items in the Bibliography, which begins on page 216.

and

$$c_v = c_p - R. \qquad (3)$$

It may be shown that h, u, c_p, and c_v are functions of temperature only.[2] Hence, complete presentation of these quantities will consist of a table with a single variable, the temperature.

The entropy, on the other hand, proves to be a function of both temperature and pressure, so that an equally simple presentation is not possible. In an isentropic (constant-entropy) process the ratio of the pressures corresponding to a given pair of temperatures is the same for all isentropics. Therefore the pressure ratio of any two pressures p_a and p_b, corresponding to the given temperatures T_a and T_b of a gas in an isentropic process, can be obtained from the relative pressures p_{ra} and p_{rb} as tabulated for T_a and T_b in the property table under the heading of the given gas. It is also true that the ratio of the volumes corresponding to a given pair of temperatures is the same for all isentropics. These two statements may be proved as shown in the following.

For a homogeneous system in the absence of gravity, electricity, capillarity, magnetism, and chemical change, we may write for changes between equilibrium states

$$T\,ds = dh - v\,dp$$

where s denotes the entropy, h the enthalpy, v the specific volume—all for unit mass—and p the pressure. Since h is a function of T only, we have

$$T\,ds = c_p\,dT - v\,dp \qquad (4)$$

where c_p is the specific heat capacity at constant pressure, $(\partial h/\partial T)_P = dh/dt$. For an infinitesimal step in an isentropic process we have

$$0 = c_p\,dT - v\,dp$$

and, upon substituting from (1) and transposing,

$$\frac{dp}{p} = \frac{c_P}{R} \cdot \frac{dT}{T}.$$

Because c_p, like the enthalpy, is a function of temperature only we get, upon integrating between temperatures T_0 and T,

$$\ln\frac{p}{p_0} = \frac{1}{R}\int_{T_0}^{T}\frac{c_P}{T}\,dT. \qquad (5)$$

When a base temperature T_0 is selected the ratio p/p_0 becomes a single function of the temperature T regardless of the value of the entropy.

The corresponding volume ratio is a single function of the temperature T also and is given by

$$\ln\frac{v}{v_0} = -\frac{1}{R}\int_{T_0}^{T}\frac{c_v}{T}\,dT \qquad (6)$$

or, alternatively,

$$\frac{v}{v_0} = \frac{p_0}{p}\cdot\frac{T}{T_0}. \qquad (7)$$

Solving (4) for ds and substituting from (1), we get

$$ds = \frac{c_P}{T}\,dT - R\frac{dp}{p}.$$

The entropy s at any state reckoned from $T_0 = 0$ and unit pressure is then

$$s = \int_{T_0}^{T}\frac{c_P}{T}\,dT - R\ln p = \phi - R\ln p \qquad (8)$$

in which

$$\phi = \int_{T_0}^{T}\frac{c_P}{T}\,dT. \qquad (9)$$

The change in entropy between states 1 and 2 is then

$$s_2 - s_1 = \phi_2 - \phi_1 - R\ln\frac{p_2}{p_1}. \qquad (10)$$

From (5), it is evident that the ratio of the pressures p_a and p_b, corresponding to the temperatures T_a and T_b, respectively, along a given isentropic, is equal to the ratio of the relative pressures p_{ra} and p_{rb} as tabulated for T_a and T_b, respectively. Thus

$$\left(\frac{p_a}{p_b}\right)_s = \text{constant} = \frac{p_{ra}}{p_{rb}}.$$

Similarly, from (6),

$$\left(\frac{v_a}{v_b}\right)_s = \text{constant} = \frac{v_{ra}}{v_{rb}}.$$

In terms of an ideal gas, thermodynamic properties represented by h, p_r, u, v_r, ϕ, c_p, c_v, and k ($= c_p/c_v$) are presented as functions of temperatures, R and °F. Included in this book are tables for air, fuel combustion products with 400%, 200%, and 100% theoretical air, N_2, O_2, H_2O, CO_2, H_2, CO, and Ar.

Velocity of sound, α, as a function of temperatures was included in all these tables except those for the fuel combustion products. In Table 2, the transport properties for air at low pressures, for example

mass velocity (G), viscosity (μ), thermal conductivity (λ), and Prandtl number (Pr) at selected temperatures were also listed.

The sources of data and methods for evaluating the tabulated properties are described in the remaining sections.

1.2 Calculation of Ideal Gas Thermodynamic Properties

The thermodynamic properties like c_p, s, h, and so on may be calculated for simple gases in the ideal gaseous state over a wide range of temperatures with high reliability by applying the method of statistical mechanics to basic empirical data.[3] The basic data are the energy levels of various molecular motions, which include both external motions (translation of the molecules) and internal motions (rotational, vibrational, etc.). Based on such energy levels (ε_i) and their respective statistical weights (g_i), the molecular partition function

$$q = \sum_i g_i e^{-\varepsilon_i/kt}$$

is obtained for each of the $3n$ degrees of freedom for a gaseous molecule having n atoms. From these partition functions, the thermodynamic properties of the given substance in the ideal gaseous state at one atmosphere pressure can be evaluated by use of the standard method of statistical mechanics.[3]

The internal energy of a gaseous molecule is composed of rotational kinetic energy ε_{rot}, which is produced by collisions of the gaseous molecules, and interatomic energies or vibrational energy ε_{vib} of the molecule. The atoms within the molecule possess kinetic energy due to vibration, as well as potential energy arising from forces acting between the atoms. In complex molecules certain groups of atoms can rotate against the remaining part of the molecule to produce "internal rotation." This kind of motion inside the molecule is treated as a special kind of vibration in statistical calculations.

For monatomic substances, such as argon, neon, and so on, there are no rotational and vibrational degrees of freedom. The total energy of such substances contains only translational and electronic energies. The total energy of diatomic and polyatomic molecules, on the other hand, includes energy contributions from $3n$ degrees of freedom as shown in Table A. For polyatomic substances the contributions from electronic energy are usually neglected

Table A Distribution of Energy in Gaseous Molecule[a]

Type	Translation	Rotation	Vibration	Internal Rotation	Total
Monatomic	3	0	0	0	3
Diatomic	3	2	1	0	6
Triatomic					
linear	3	2	4	0	9
nonlinear	3	3	3	0	9
Polyatomic					
linear	3	2	$3n-a-5$	a	$3n$
nonlinear	3	3	$3n-b-6$	b	$3n$

[a] In degrees of freedom.

since the electronic state of the molecule is undisturbed under ordinary conditions (i.e., it remains at the electronic ground state). Similarly the nuclear energy inside the atoms of the gaseous molecules is not considered in statistical calculations. Some simple molecules ($CClO$, $TiCl_3$, etc.) have low-lying electronic levels, the contributions to the ideal gas thermodynamic properties from these electronic energy levels are included.

In Table B are summarized the partition functions that can be employed for evaluation of thermodynamic properties for chemical substances in the ideal gaseous state at given temperatures. The derivation of these partition functions may be found in standard textbooks on statistical mechanics.[4] In this table: V denotes volume of the given gas; m, mass of the molecule; k, Boltzmann's constant; T, absolute thermodynamic temperature; h, Planck's constant; I, moment of inertia of the linear molecule; σ, symmetry number; $I_x I_y I_z$, product of the principal moments of inertia of a nonlinear molecule; ω_i, wavenumber of ith fundamental vibration; g_0, statistical weight of electronic ground state; g_i, statistical weight of electronic energy level ε_i; I_r, reduced moment of inertia of the rotating top; and σ_{IR}, symmetry number of the top.

The thermodynamic properties of a gaseous substance can be calculated by substituting each of the partition functions listed in Table B into the equations shown in Table C, where R denotes the gas constant; q', $T(\partial q/\partial T)_v$; E_0°, energy of the substance at absolute zero temperature; and N, Avogadro's number. The calculated contributions from different molecular motions for each of the thermodynamic properties presented in Table C are added to yield the final property values at the given temperature T.

Table C Equations for Calculating Thermodynamic Properties

$$E^\circ - E_0^\circ = RT^2\left(\frac{\partial \ln q}{\partial T}\right)_v = RT\left(\frac{q'}{q}\right)$$

$$H^\circ - E_0^\circ = E^\circ - E_0^\circ + RT = RT^2\left(\frac{\partial \ln q}{\partial T}\right)_P$$

$$C_P^\circ = \left(\frac{\partial E}{\partial T}\right)_v + R = \frac{R}{T^2}\left[\frac{\partial^2 \ln q}{\partial(1/T)^2}\right]_P$$

$$C_v^\circ = C_P^\circ - R = \frac{R}{T^2}\left[\frac{\partial^2 \ln q}{\partial(1/T)^2}\right]_v$$

$$S^\circ = \frac{H^\circ - E_0^\circ}{T} + R\ln\frac{q}{N}$$

Table B Partition Functions of Molecular Motion

Motion	Partition Function	Degrees of Freedom Included
Translation	$q_{\text{trans}} = V(2\pi mkT)^{3/2}h^{-3}$	3
Rotation	$q_{\text{rot}} = \dfrac{8\pi^2 IkT}{\sigma h^2}$ (linear molecule)	2
	$q_{\text{rot}} = \dfrac{(8\pi^2 kT)^{3/2}(\pi I_x I_y I_z)^{1/2}}{\sigma h^3}$ (nonlinear molecule)	3
Vibration	$q_{\text{vib}} = (1 - e^{-h\omega_i/kT})^{-1}$	1
Electronic	$q_{\text{el}} = g_0 + \sum\limits_i g_i e^{-\varepsilon_i/kT}$	—
Free internal rotation	$q_f = (8\pi^3 I_r kT)^{1/2}(h\sigma_{IR})^{-1}$	1

(189)

1.3 Equations for Calculating Ideal Gas Thermodynamic Properties

In Table D a summary of equations for calculation of thermodynamic properties for monatomic substances in the ideal gaseous state at one atmosphere pressure is presented. The input data needed for evaluation of translational contributions to thermodynamic properties are molecular weight (M) of the given substance and the absolute temperature in degrees Kelvin (K) at which the thermodynamic properties are required. For evaluation of thermodynamic properties due to electronic contributions, the values of electronic energy levels ε_i and the corresponding statistical weight g_i for each of the i energy levels are employed. These equations were used for calculating the ideal gas thermodynamic properties for Ar over the temperature range from 100 to 6000°R and at one atmosphere pressure.

In calculating the thermodynamic properties for diatomic gases, for example N_2, O_2, and CO, in the ideal gaseous state at one atmosphere pressure, the equations listed in Table E were employed. These equations were derived based on a nonrigid-rotor and anharmonic-oscillator molecular model.[3] In Table E u denotes $(\omega_e - 2\omega_e X_e)hc/kT$, where ω_e denotes vibrational constant; $\omega_e X_e$, anharmonicity constant; h, Planck's constant; c, speed of light; k, Boltzmann's constant; T, absolute temperature in K; I, moment of inertia of the given molecule; χ, $\omega_e X_e/\omega_e$; and δ, α_e/B_e, where α_e = vibrational-rotational coupling constant, B_e = rotational constant; and $\gamma = B_e/\omega_e$.

As mentioned in Section 1.2, the total thermodynamic property G of a diatomic substance is obtained as

$$G = G_{\text{trans}} + G_{\text{rot}} + G_{\text{vib}} + G_{\text{el}} + G_{\text{anh}}$$

where G_{trans} denotes contributions from 3 translational degrees of freedom; G_{rot}, contributions from 2 rotational degrees of freedom; G_{vib}, vibrational contribution; G_{el}, electronic contribution; and G_{anh}, contributions from anharmonicity corrections.

The input data, such as moment of inertia I, may be derived from molecular structure of the given substance, which can be determined by microwave spectroscopy or X-ray diffraction. The vibrational constants are usually obtained from infrared and Raman spectroscopy.[5] The thermodynamic properties for H_2, H_2O, and CO_2 were adopted from literature.[6] These have been the best values available. Based on the calculated values of h, u, p_r, v_r, and ϕ, and c_p, c_v, and k ($= c_p/c_v$) for N_2, O_2, and Ar, the corresponding properties for air at low pressures (for one pound) were evaluated at close temperature ranges. The thermodynamic properties at the intermediate temperatures can be obtained by linear interpolation of the tabulated property values without significant error.

2 SOURCES AND METHODS FOR INDIVIDUAL TABLES

2.1 Tables 1 and 2—Air at Low Pressures (for one pound)

In principle, a mathematical model of the equation of state for air could be developed based upon the experimental P–V–T measurements. By combining

---

Table D Equations for Calculating Ideal Gas Thermodynamic Properties for Monatomic Molecules at a Pressure of 1 atm[a]

Contribution	Property	Equation
Translation	C_p°	4.967975
	$H^\circ - H_0^\circ$	$4.967975\,T$
	S°	$6.863511 \log M + 11.439185 \log T - 2.314820$
Electronic	C_p°	$\dfrac{4.113692}{T^2}\left[\dfrac{\sum \varepsilon_i^2 g_i e^{-1.438786\varepsilon_i/T}}{\sum g_i e^{-1.438786\varepsilon_i/T}} - \left(\dfrac{\sum \varepsilon_i g_i e^{-1.438786\varepsilon_i/T}}{\sum g_i e^{-1.438786\varepsilon_i/T}}\right)^2\right]$
	$H^\circ - H_0^\circ$	$2.859141\left(\dfrac{\sum \varepsilon_i g_i e^{-1.438786\varepsilon_i/T}}{\sum g_i e^{-1.438786\varepsilon_i/T}}\right)$
	S°	$\dfrac{2.859141}{T}\left(\dfrac{\sum \varepsilon_i g_i e^{-1.438786\varepsilon_i/T}}{\sum g_i e^{-1.438786\varepsilon_i/T}}\right) + 4.575674 \log \sum g_i e^{-1.438786\varepsilon_i/T}$

[a]Units: cal/mole for $H^\circ - H_0^\circ$ and cal/K mole for the remaining quantities.

Table E Equations for Calculating Ideal Gas Thermodynamic Properties for Diatomic Molecules at a Pressure of 1 atm[a]

Contribution	Property	Equation
Translation	C_p°	4.967975
	$H^\circ - H_0^\circ$	$4.967975\,T$
	S°	$6.863511 \log M + 11.439185 \log T - 2.314820$
Rotation	C_p°	1.98719
	$H^\circ - H_0^\circ$	$1.98719\,T$
	S°	$4.575674 \log[(IT \times 10^{39})/\sigma] - 4.349171$
Vibration	C_p°	$1.98719 u^2 e^{-u}/(1 - e^{-u})^2$
	$H^\circ - H_0^\circ$	$1.98719\,Tue^{-u}/(1 - e^{-u})$
	S°	$1.98719 ue^{-u}/(1 - e^{-u}) - 4.575674 \log(1 - e^{-u})$
Electronic	C_p°	$\dfrac{4.113692}{T^2}\left[\dfrac{\sum \varepsilon_i^2 g_i e^{-1.438786\varepsilon_i/T}}{\sum g_i e^{-1.438786\varepsilon_i/T}} - \left(\dfrac{\sum \varepsilon_i g_i e^{-1.438786\varepsilon_i/T}}{\sum g_i e^{-1.438786\varepsilon_i/T}}\right)^2\right]$
	$H^\circ - H_0^\circ$	$2.859141\left(\dfrac{\sum \varepsilon_i g_i e^{-1.438786\varepsilon_i/T}}{\sum g_i e^{-1.438786\varepsilon_i/T}}\right)$
	S°	$\dfrac{2.859141}{T}\left(\dfrac{\sum \varepsilon_i g_i e^{-1.438786\varepsilon_i/T}}{\sum g_i e^{-1.438786\varepsilon_i/T}}\right) + 4.575674 \log \sum g_i e^{-1.438786\varepsilon_i/T}$
Anharmonicity Corrections	C_p°	$1.98719\left[\dfrac{16\gamma}{u} - \dfrac{\delta u^2 e^u}{(e^u-1)^2} + \dfrac{u^2 e^u(2\delta e^u - 4Xu - 8X)}{(e^u-1)^3} + \dfrac{12Xu^3 e^{2u}}{(e^u-1)^4}\right]$
	$H^\circ - H_0^\circ$	$1.98719\,T\left[\dfrac{8\gamma}{u} + \dfrac{u(\delta e^u - 2X)}{(e^u-1)^2} + \dfrac{4Xu^2 e^u}{(e^u-1)^3}\right]$
	S°	$1.98719\left[\dfrac{16\gamma}{u} + \dfrac{\delta}{(e^u-1)} + \dfrac{\delta u e^u}{(e^u-1)^2} + \dfrac{4Xu^2 e^u}{(e^u-1)^3}\right]$

[a]Units: cal/mole for $H^\circ - H_0^\circ$ and cal/K mole for the remaining quantities.

this model with the standard thermodynamic relationships, it appears likely that certain properties, such as heat capacity, enthalpy, entropy, and internal energy, might be derived as a result. In fact, the accuracy of the experimentally measured property values is less than those obtained by theoretical calculations, because an adequate amount of high-precision, experimental data is lacking.

More than 100 sources of information on the specific heat of air at selected pressures are found in the literature. These results were obtained by theoretical calculations, correlations, calorimetry, adiabatic expansion, Joule-Thomson experiments, the velocity of sound measurements, and heat transfer measurements.[8] Several sets of extensive values were derived by use of the method of statistical mechanics.[6,9,10] These results are more reliable than those evaluated from correlations, earlier statistical calculations, and extrapolated values to zero pressures for $P-V-T$ measurements.

This work employed the modern physical constants,[11] atomic weights,[12] and the recent molecular and spectroscopic constants for nitrogen,[13] oxygen,[13] and argon[14] for reevaluation of the thermodynamic properties of air by using statistical mechanical method.[3-5] In these calculations a molecular model with a nonrigid rotor and anharmonic oscillator was adopted for both nitrogen and oxygen molecules. As mentioned previously, the calculated results, as given in Table 1, are adequate for industrial applications, where the pressure of air is low and the temperature is moderate.

For computation of the thermodynamic properties of air, the composition of air was assumed to be as follows:

Gas	Molecular Weight	Percentage by Volume
Nitrogen	28.0134	78.03
Oxygen	31.9988	20.99
Argon	39.948	0.98

The molecular weight and the gas constant for air are calculated as 28.9669 and 53.348 ft-lb/R lb or 0.068556 Btu/R lb, respectively.

The enthalpy, entropy, and relative pressure of air were derived from the corresponding values of the three constituents in air in accordance with the Gibbs-Dalton law as follows:

$$h = \sum_i x_i h_i \qquad (11)$$

$$\phi = \sum_i x_i \phi_i \qquad (12)$$

$$\ln P_r = \sum_i x_i \ln P_{ri} \qquad (13)$$

where x_i denotes the mole fraction of gas i, and the summation takes place over all the constituent gases. The values of h, P_r, u, V_r, and ϕ, evaluated over the temperature range from 100 to 6500R (-359.67 to 6040.33 F) at low pressures for one pound of air are presented in Table 1.

The internal energy and relative volume for air were calculated from the following two equations:

$$u = h - 0.0685561 T \qquad (14)$$

and

$$v_r = 0.370475 \left(\frac{T}{P_r} \right). \qquad (15)$$

Each value of ϕ listed in Table 1 is the value computed from (12) minus unity.

In Table 2, the specific heat at constant volume and the velocity of sound for air were computed from the relations

$$c_v = c_p - R \qquad (16)$$

and

$$\alpha = \sqrt{gkRT}. \qquad (17)$$

The maximum flow per unit area is the maximum value for isentropic expansion from temperature T, unit pressure, and zero velocity. Its value for any other pressure is that for unit pressure multiplied by the pressure in pounds per square inch absolute. The values given in Table 2 were obtained from Table 1 by trial for the maximum value of the quantity

$$\frac{G_x}{p_i} = \frac{p_{rx}\sqrt{2g(h_i - h_x)}}{p_{ri}RT_x} \qquad (18)$$

where subscript i refers to the initial state of the expansion and temperature T in Table 2, and subscript x refers to any state at the same entropy as state i but at lower pressures.

More than fifty sets of experimental measurements on viscosity of gaseous air have been reported.[6,17,18] Most of them were made close to room temperature. Touloukian et al.[18] critically evaluated the reported experimental data and obtained a consistent set of air viscosities in the temperature range from 80 to 2000 K or 144 to 3600 R. Their recommended values were adopted to recalculate the η values for air, as shown in Table 2. The uncertainty of the calculated results is estimated as $\pm 2\%$ in the whole temperature range.

Numerous thermal conductivity (λ) measurements for air were made, only few covered extensive ranges of temperatures.[6,7] The listed values of λ in Table 2 were based on the thermal conductivity data selected for air in the temperature range from 50 to 1500 K or 90 to 2700 R by Touloukian et al.[19] They obtained these data from large scale graphs of different sets of measurements and checked by differencing. Below 750 R the recommended λ values should be accurate to within about 1%; the uncertainty then increased to about 5% at 2700 R. The Prandtl numbers, Pr, for air were computed in accordance with the following relationship: $\text{Pr} = c_p \eta / \lambda$ where the effect of pressure on η and λ is less than the uncertainty in the given values for pressures up to 200 psia.

Real gas thermodynamic properties of air have been derived by several authors.[6,15,16] Hilsenrath et al.[6] divided the range of the calculated thermodynamic properties for air into two regions. In the region below 1500 K or 2700 R, the composition of air was considered fixed and the corrections for gas imperfection were significant. Above 1500 K, the corrections for gas imperfection were small and the predominant influence on the thermodynamic properties was the effect of the dissociation of the constituents of air at high temperatures. In this region, the properties of air were based on the contributions from each of the molecular and atomic species present in the equilibrium composition at each temperature and pressure. The properties of ideal gas for each of the molecular and atomic species in air were obtained by use of the method of statistical mechanic calculations.

A comparison of calculated values of h, c_p, and ϕ with those derived for air in the real gaseous state by Hilsenrath et al.,[6] at temperatures 540, 720, 1080, 1800 R and pressures 0.147, 14.70, 147.0, 588.0, 1470 psia, is shown in Table F. In general, the average

Table F   Comparison of Calculated Ideal Gas Properties With Real Gas Properties[a]

| | T | | | P, psia | | |
	R	0.147	14.70	147.0	588.0	1470.0
$\Delta(h - h_0)$						
	540	0.01	−0.06	−0.74	−3.02	−7.33
	720	0.01	−0.02	−0.27	−1.08	−2.51
	1080	0.02	0.01	0.00	−0.06	−0.08
	1800	0.04	0.04	0.08	0.26	0.60
ΔC_p						
	540	0.04	0.17	1.56	5.98	13.52
	720	0.00	0.08	0.78	3.02	6.82
	1080	0.04	0.08	0.36	1.22	2.75
	1800	0.14	0.14	0.24	0.49	0.96
$\Delta\phi$						
	540	2.16	2.38	2.44	2.20	1.60
	720	2.09	2.29	2.38	2.30	2.06
	1080	1.99	2.02	2.27	2.28	2.23
	1800	1.88	2.02	2.10	2.15	2.16

[a]Percent = (NBS value − calculated value) × 100/NBS value.

deviation in enthalpy is less than five percent at these conditions. The agreement is better at lower pressures and higher temperatures. The differences in ϕ between the real gas and ideal gas for air are 2 to 3%, in the temperature range 500 to 2000 R, and at pressures up to 1500 psia.

2.2   Tables 3 to 8 – Products of Combustion with 400, 200, and 100% of Theoretical Air

Tables 3, 5, and 7 are for 1 lb-mole of products of combustion of a hydrocarbon fuel of composition $(CH_2)_n$ with 400, 200, and 100% of theoretical air, respectively. The composition of products of combustion employed for evaluation of their thermodynamic properties are presented in Table G. The molecular weights of these mixtures are 28.9512, 28.9360, 28.9072; and their gas constants are 53.3773 ft-lb/lb R or 0.068593 Btu/lb R, 53.4053 ft-lb/lb R or 0.068629 Btu/lb R, and 53.4586 ft-lb/lb R or 0.068698 Btu/lb R, respectively.

The molar enthalpy, entropy ϕ, and relative pressure were computed by combining the corresponding values of the constituents as shown in Table G in accordance with equations (11), (12), and (13). The molar internal energy was derived from the equation

$$\bar{u} = \bar{h} - 1.98586\,T, \qquad (19)$$

where the bar over the symbol indicates that the quantity is per pound-mole.

The relative volume was calculated by the equation

$$v_r = 10.7315\,\frac{T}{p_r}. \qquad (20)$$

Although the thermodynamic properties given in Tables 3, 5, and 7 were calculated for hydrocarbon fuel of composition $(CH_2)_n$, it has been shown[20] that they represent with high precision the properties of fuel combustion products over a wide range of composition with the same percent of theoretical air.

Table G   Molecular Weight and Composition of Combustion Gases

| Gases in the Combustion Products | Molecular Weight | Percentage by Volume | | |
		400% Theoretical Air	200% Theoretical Air	100% Theoretical Air
Nitrogen (N_2)	28.0134	76.6886	75.3925	72.9275
Oxygen (O_2)	31.9988	15.4719	10.1403	0.0000
Carbon dioxide (CO_2)	44.0098	3.4382	6.7602	13.0783
Water (H_2O)	18.0152	3.4382	6.7602	13.0783
Argon (Ar)	39.948	0.9631	0.9468	0.9159

Evidence to this effect appear in Tables 4, 6, and 8, which show specific heats and ratios of specific heats for three different compositions of hydrocarbon fuel, namely, $(CH)_n$, $(CH_2)_n$, and $(CH_3)_n$. In tables 4, 6, and 8, the molar specific heats at constant pressure were calculated by means of the Gibbs-Dalton law from the composition of combustion products and from the specific heats of the individual component gases.

The molar specific heat at constant volume was then computed from the equation

$$\bar{c}_v = \bar{c}_p - \bar{R}$$

The ratio of the specific heats was obtained from

$$k = \frac{\bar{c}_p}{\bar{c}_v}.$$

2.3 Use of Tables 1, 3, 5, and 7 for Other Mixtures

Table 1 for air may be converted to a table for 1 lb-mole merely by multiplication of the enthalpy, internal energy, and $(\phi + 1)$ by the molecular weight of air. The resultant table for air is also a molar products table for infinite percent of theoretical air or zero percent of theoretical fuel. It has been shown[20] that Table 1 converted to the molar form and the molar Tables 3, 5, and 7 have utility for mixtures other than those for which they were calculated. They represent combustion products of hydrocarbon fuels ranging from $(CH)_n$ to $(CH_3)_n$ for the percentages of theoretical air as indicated. They also represent certain mixtures of air and octane vapor and of air and water vapor in accordance with Table H.

Problems involving mixtures intermediate between those indicated in Table H may be solved by linear interpolation. These extensions of the tables yield a precision in general better than one part in five hundred.

2.4 Tables 9 and 10

In Table 9 the molecular weight and gas constant are listed for the combustion products of hydrocarbon fuels containing $(CH_x)_n$ and air. The values of x, atomic ratio of H/C, were chosen in the range 0.715 to 4.051, which corresponds to 0.06 to 0.34 lb of H/lb of C in the fuel. The percent of theoretical fuel for combustion increases from 0 (i.e., pure air) to 100 in 10 steps with increment of 10% for each step.

Table 10 presents the enthalpy of combustion of three classes of pure hydrocarbons related to liquid and gaseous fuels. These values were taken from the extensive tabulations of the American Petroleum Institute Research Project 44.[21] The enthalpy of combustion of hydrocarbons other than those listed in Table 10 can be calculated from the chemical equation of combustion reaction with the enthalpy of formation values for all the reactants and products known at the given temperature. In complete combustion at 77°F, the products formed are CO_2 (gas) and H_2O (liquid), of which the enthalpy of formation is reported[22] to be -393.51 ± 0.13 kJ/g-mole or $-169180. \pm 56.$ Btu/lb-mole and -285.830 ± 0.042 kJ/g-mole or 122885 ± 18 Btu/lb-mole, respectively. Also available are the enthalpies of formation at 77°F for hundreds of hydrocarbons related to petroleum and coal products.[21]

The enthalpy of formation of higher members of an homologous series can be estimated by the addition of appropriate numbers of methylene (CH_2) increments to known values for a lower homolog. The n-alkyl chain of the lower homolog must be long enough so that the increment for the addition of another CH_2 group has approached the limiting value for the long chains.[23]

Table H   Application of Evaluated Combustion Product Properties for Other Gas Mixtures

Table Number	Products		Reactants	Air and Water Vapor
	% Theoretical Air	% Theoretical Fuel	% Theoretical Fuel	Mass % Water
1, 2	∞	0	0	0
3, 4	400	25	14	6.7
5, 6	200	50	28	13
7, 8	100	100	54	26

2.5 Tables 11 to 23—N_2, O_2, H_2O, CO_2, H_2, CO, and Ar

Tables 11 to 23 are for 1 lb-mole of gas. The ideal gas thermodynamic properties for N_2, O_2, H_2O, CO_2, H_2, and CO are listed for each compound in two consecutive tables. The values of $\bar{h}$, p_r, $\bar{u}$, v_r, and $\bar{\phi}$ are given in the first table. Those of $\bar{c}_p$, $\bar{c}_v$, $\bar{c}_p/\bar{c}_v$, and α are presented in the following one. For a monatomic species such as argon, only the first table is included since $\bar{c}_p$ and $\bar{c}_v$ are constants and independent of temperature. Using the statistical mechanical method, the molecular and spectroscopic constants reported by Rosen[13] were employed in computing the thermodynamic properties for diatomic molecules such as N_2, O_2, and CO.

The thermodynamic properties for argon in the ideal gaseous state were derived from statistical mechanical calculation. The required spectroscopic data for argon were obtained from C. E. Moore.[14] In calculating the required thermodynamic properties of atomic argon at each given temperature 265 electronic energy levels (up to 127970 cm^{-1}) were used. The ideal gas thermodynamic properties of CO_2 (gas) were taken from those reported by J. Hilsenrath et al.[6] The properties for hydrogen (75% ortho and 25% para) were adopted from those reported by Woolley et al.[24] and those for H_2O (gas) were adopted from a recent work recommended by L. Haar,[25] where the thermodynamic properties were computed by statistical mechanical method, employing direct summation of electronic energy levels. The above calculated results are considered to be the most reliable ones.

Appropriate nonlinear interpolations were applied to the reported heat capacity at constant pressure, entropy (ϕ), and enthalpy data in converting the adopted ideal gas properties of hydrogen, carbon dioxide, and water vapor into engineering units at close temperatures as shown in Tables 15 to 20.

The formulas employed to derive the other related thermodynamic quantities as given in Tables 11 to 23 are similar to those described in Section 2.1 for air. For ease of presentation, the tabulated values of the relative pressure have been decreased from the calculated values by a constant factor of 10^{10} for N_2, O_2, H_2O, and CO; 10^{12} for CO_2; 10^{15} for H_2; and 10^7 for Ar. The low critical temperatures of N_2, O_2, H_2,

CO, and Ar insure that the equation, $p\bar{v} = \bar{R}T$, is a good approximation to the true equation of state in each instance over a wide range of pressures. Therefore, Tables 11, 13, 19, 21, and 23, have validity over a range comparable to that of Table 1.

The critical temperature of CO_2 is close to room temperature, far below the critical point encountered in power plants. Consequently the range of validity for Table 17 should compare favorably with that of Table 1, although it is doubtless slightly less extensive. The critical temperature of water being the highest of all the gases considered here, the range of validity found in Table 15 is at the low end of the scale. Nevertheless, the range is considerable, owing largely to the high critical pressure of water. The pressures of combustion products in powerplants seldom exceed 200 psia. For 200% of theoretical air, the corresponding partial pressures are: nitrogen 15, oxygen 20, carbon dioxide 13.5, and argon 2 psi. It follows from this example that the present tables are reliable for pressures several times as large as those now employed in power plants.

Presented in Table I is a comparison of our calculated results with those reported in the literature for molar enthalpy, heat capacity at constant pressure, and entropy, at 298.15, 1000, and 3000 K (or 536.67, 1800, and 5400 R) and 1 atm (or 14.696 psia). The agreement between them is excellent. Some slight discrepancies are caused by the use of different values of physical, molecular, and spectroscopic constants for calculations. Nevertheless, in most cases, these discrepancies are well within their respective assigned uncertainties. The thermodynamic properties of nitrogen, oxygen, water, carbon dioxide, hydrogen, carbon monoxide, and argon in the real gaseous state have been critically evaluated by J. Hilsenrath et al.[6] The reported values are internally consistent, very reliable, and highly recommended for use. Comprehensive reports on argon,[29] carbon dioxide,[30] and air components[36] are available. V. J. Johnson has published a review[31] of recent data compiled on hydrogen, argon, nitrogen, and oxygen. A survey of existing data on thermodynamic properties of nitrogen and carbon dioxide was reported by G. M. Wilson et al.[32] F. Din surveyed the existing data, calculated the thermodynamic functions, and constructed the thermodynamic diagrams for CO_2,[33] CO,[33] Ar,[34] and N_2.[35]

Table I Comparison of the Ideal Gas Thermodynamic Data at 1 atm and 298.15 K

	$\bar{h}$, cal/mole			$\bar{c}_p$, cal/K mole			$\bar{\phi}$, cal/K mole		
	298.15 K	1000 K	3000 K	298.15 K	1000 K	3000 K	298.15 K	1000 K	3000 K
Nitrogen (N_2)									
CODATA,[22] 1975	2072						45.770		
JANAF,[26] 1977	2072	7202	24231	6.961	7.815	8.850	45.770	54.508	63.762
TRCDP,[27] 1971	2072	7202	24233	6.961	7.815	8.852	45.761	54.500	63.755
Gurvich et al.,[28] 1962	2072	7202	24236				45.771	54.510	63.766
NBS 564,[6] 1955		7202	24234		7.815	8.852		54.501	63.756
This work	2072	7202	24225	6.961	7.814	8.842	45.761	54.500	63.751
Oxygen (O_2)									
CODATA,[22] 1975	2075						49.005		
JANAF,[26] 1977	2075	7501	25501	7.021	8.334	9.528	49.005	58.190	67.963
TRCDP,[27] 1968	2075	7501	25519	7.020	8.336	9.551	49.003	58.191	67.971
Gurvich et al.,[28] 1962	2075	7502	25525				49.006	58.193	67.976
NBS 564,[6] 1955		7501	25521		8.336	9.551		58.200	67.981
This work	2075	7501	25424	7.021	8.334	9.496	48.994	58.180	67.924
Argon (Ar)									
CODATA,[22] 1975	1481						36.982		
JANAF,[26] 1977	1481	4968	14904	4.968	4.968	4.968	36.983	42.995	48.453
Gurvich et al.[28] 1962	1481	4968	14905				36.983	42.995	48.454
NBS 564,[6] 1955		4968	14904		4.968	4.968		42.997	48.454
This work	1481	4968	14904	4.968	4.968	4.968	36.983	42.995	48.452
Water (H_2O)									
CODATA,[22] 1975	2368						45.106		
JANAF,[26] 1961	2367	8576	32568	8.025	9.851	13.304	45.106	55.592	68.421
TRCDP,[27] 1969	2367	8576	32567	8.025	9.850	13.303	45.103	55.589	68.418
Gurvich et al.,[28] 1962	2368	8591	32910				45.108	55.614	68.594
NBS 564,[6] 1955		8576	32567		9.850	13.303		55.590	68.419
This work	2367	8624	32773	8.028	9.864	13.331	45.104	55.598	68.446
Carbon Dioxide (CO_2)									
CODATA,[22] 1975	2238						51.070		
JANAF,[26] 1965	2238	10222	38773	8.874	12.980	14.873	51.072	64.344	79.848
TRCDP,[27] 1966	2238	10221	38762	8.874	12.980	14.855	51.070	64.337	79.837
Gurvich et al.,[28] 1962	2239	10220	38781				51.071	64.333	79.838
NBS 564,[6] 1955		10221	38770		12.980	14.872		64.337	79.841
This work	2238	10221	38771	8.874	12.980	14.872	51.070	64.337	79.841
Hydrogen (H_2)									
CODATA,[22] 1975	2024						31.207		
JANAF,[26] 1977	2024	6967	23233	6.892	7.219	8.864	31.207	39.700	48.466
TRCDP,[27] 1968	2024	6966	23231	6.889	7.219	8.859	31.206	39.701	48.465
Gurvich et al.,[28] 1962	2019	6967	23232				31.195	39.702	48.467
NBS 564,[6] 1955	2037	6967	23232	6.894	7.219	8.859	31.251	39.700	48.466
This work	2024	6966	23231	6.890	7.220	8.860	31.251	39.700	48.466
Carbon Monoxide (CO)									
CODATA,[22] 1975	2073						47.217		
JANAF,[26] 1965	2072	7255	24429	6.965	7.931	8.895	47.214	56.028	65.370
TRCDP,[27] 1966	2073	7256	24423	6.965	7.931	8.886	47.219	56.033	65.372
Gurvich et al.,[28] 1962	2073	7256	24429				47.217	56.032	65.373
NBS 564,[6] 1955	2085	7256	24430	6.965	7.931	8.895	47.259	56.031	65.371
This work	2072	7256	24418	6.965	7.929	8.883	47.208	56.022	65.358

2.6 Tables 30 to 53—One-Dimensional Compressible-Flow Functions

Developments in high-speed propulsion have led to a better understanding of the flow of compressible fluids. Generalized[37]* treatments of one-dimensional flow are available in both analytical and numerical formulations. Tables 30 to 53 represent a portion of the numerical formulations prepared by A. H. Shapiro and G. M. Edelman. Some of this material has been published in the *Journal of Applied Mechanics*.[38] These tables contain the functions useful in many engineering problems in the one-dimensional flow of a perfect gas with constant specific heat and molecular weight.

The isentropic, which is represented by the line *abx* in Figure 1, is the locus of states for a frictionless adiabatic process. Segments *ab* and *bx* represent subsonic and supersonic flows, respectively. At point *b* the Mach number is unity, and this condition is denoted by a superscript asterisk.

The following relations hold along the isentropic:

$$T_0 = \text{constant} = T_0^* = T_a$$

$$T^* = \text{constant} = T_b$$

$$p_0 = \text{constant} = p_0^* = p_a$$

$$p^* = \text{constant} = p_b$$

$$M^* = \frac{V}{V^*} = M \sqrt{\frac{k+1}{2\left(1 + \frac{k-1}{2} M^2\right)}}$$

$$\frac{A}{A^*} = \frac{1}{M} \left[\frac{2\left(1 + \frac{k-1}{2} M^2\right)}{k+1} \right]^{\frac{k+1}{2(k-1)}}$$

$$\frac{T}{T^*} = \frac{k+1}{2\left(1 + \frac{k-1}{2} M^2\right)}$$

$$\frac{\rho}{\rho^*} = \left[\frac{k+1}{2\left(1 + \frac{k-1}{2} M^2\right)} \right]^{\frac{1}{k-1}}$$

$$\frac{p}{p^*} = \left[\frac{k+1}{2\left(1 + \frac{k-1}{2} M^2\right)} \right]^{\frac{k}{k-1}}$$

*References to earlier papers are given in reference 37.

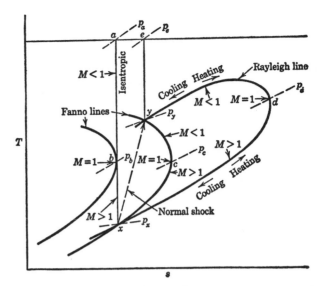

Figure 1

$$\frac{F}{F^*} = \frac{1 + kM^2}{M\sqrt{2(k+1)\left(1 + \frac{k-1}{2} M^2\right)}} .$$

2.7 Tables 36 to 41—Rayleigh Line—One-Dimensional Compressible-Flow Functions for Stagnation-Temperature Change in the Absence of Friction and Area Change

The Rayleigh line, which is represented by the curve *ydx* of Figure 1, is the locus of states for a process at constant flow per unit area—that is, constant *A*—and constant impulse *F*. The process being reversible, increasing entropy corresponds to heat flow to the stream, decreasing entropy to heat flow from the stream. It follows that the stagnation temperature increases with increasing entropy and decreases with decreasing entropy. At point *d* both the entropy and the stagnation temperature are at a maximum and the Mach number is unity. This condition is denoted by a superscript asterisk.

The following relations hold along the Rayleigh line:

$$A = A^* = \text{constant}$$
$$F = F^* = \text{constant}$$

$$\frac{T_0}{T_0^*} = \frac{2(k+1)M^2\left(1 + \frac{k-1}{2} M^2\right)}{(1 + kM^2)^2}$$

$$\frac{T}{T^*} = \frac{(k+1)^2 M^2}{(1+kM^2)^2}$$

$$\frac{p_0}{p_0^*} = \frac{k+1}{1+kM^2}\left[\frac{2\left(1+\frac{k-1}{2}M^2\right)}{k+1}\right]^{\frac{k}{k-1}}$$

$$\frac{p}{p^*} = \frac{k+1}{1+kM^2}$$

$$\frac{\rho}{\rho^*} = \frac{V^*}{V} = \frac{1}{M^*} = \frac{1+kM^2}{(k+1)M^2}.$$

2.8   Tables 42 to 47—Fanno Line—One Dimensional Compressible-Flow Functions for Adiabatic Flow at Constant Area with Friction

The Fanno line, which is represented by the curve *ycx* in Figure 1, is the locus of states for an adiabatic process at constant flow per unit area, that is, constant A. A second Fanno line, passing through the point b in Figure 1, is for a higher flow per unit area. At points b and c the Mach number is unity, and this condition is denoted by a superscript asterisk. At higher points on the line the velocities are subsonic; at lower points supersonic.

It should be noted that the superscript asterisk always refers to a state where the Mach number is unity, but that this state corresponding to a given state x (Figure 1) is state b for the isentropic, state c for the Fanno line, and state d for the Rayleigh line.

The following relations hold along the Fanno line:

$$A = A^*$$
$$T_0 = T_0^* = T_a$$

$$\frac{T}{T^*} = \frac{k+1}{2\left(1+\frac{k-1}{2}M^2\right)}$$

$$\frac{p_0}{p_0^*} = \frac{1}{M}\left[\frac{2\left(1+\frac{k-1}{2}M^2\right)}{k+1}\right]^{\frac{k+1}{2(k-1)}}$$

$$\frac{p}{p^*} = \frac{1}{M}\sqrt{\frac{k+1}{2\left(1+\frac{k-1}{2}M^2\right)}}$$

$$\frac{\rho}{\rho^*} = \frac{V^*}{V} = \frac{1}{M^*} = \frac{1}{M}\sqrt{\frac{2\left(1+\frac{k-1}{2}M^2\right)}{k+1}}$$

$$\frac{F}{F^*} = \frac{1+kM^2}{M\sqrt{2(k+1)\left(1+\frac{k-1}{2}M^2\right)}}$$

$$\frac{4fL_{\max}}{D} = \frac{1-M^2}{kM^2} + \frac{k+1}{2k}\log_e\frac{(k+1)M^2}{2\left(1+\frac{k-1}{2}M^2\right)}.$$

2.9   Tables 48 to 53—One-Dimensional Normal Shock Functions

States x and y in Figure 1, which have the same flow per unit area, the same stagnation temperature and the same impulse, represent the initial and final states of a normal shock.

The following relations hold for the normal shock:

$$T_{0x} = T_{0y}$$

$$M_y^2 = \frac{M_x^2 + \frac{2}{k-1}}{\frac{2k}{k-1}M_x^2 - 1}$$

$$\frac{T_y}{T_x} = \frac{\left(1+\frac{k-1}{2}M_x^2\right)\left(\frac{2k}{k-1}M_x^2 - 1\right)}{\frac{(k+1)^2}{2(k-1)}M_x^2}$$

$$\frac{p_y}{p_x} = \frac{2k}{k+1}M_x^2 - \frac{k-1}{k+1}$$

$$\frac{\rho_y}{\rho_x} = \left(\frac{p_y}{p_x}\right)\left(\frac{T_x}{T_y}\right)$$

$$\frac{p_{0y}}{p_{0x}} = \left[\frac{\frac{k+1}{2}M_x^2}{1+\frac{k-1}{2}M_x^2}\right]^{\frac{k}{k-1}}\left[\frac{2k}{k+1}M_x^2 - \frac{k-1}{k+1}\right]^{\frac{1}{1-k}}$$

$$\frac{p_{0y}}{p_x} = \left[\frac{k+1}{2}M_x^2\right]^{\frac{k}{k-1}}\left[\frac{2k}{k+1}M_x^2 - \frac{k-1}{k+1}\right]^{\frac{1}{1-k}}$$

2.10   Tables 54 to 59—Two-Dimensional Compressible-Flow Functions

Table 54, taken from the data of Shapiro and Edelman,[38] presents functions useful in the application of the method of characteristics to two-dimensional, isentropic, supersonic-flow problems. Under these conditions the following relations hold:

$$\frac{p}{p_0} = \left(1 + \frac{k-1}{2} M^2\right)^{\frac{k}{1-k}}$$

$$\frac{\rho}{\rho_0} = \left(1 + \frac{k-1}{2} M^2\right)^{\frac{1}{1-k}}$$

$$M^* = \frac{V}{V^*} = M \sqrt{\frac{k+1}{2\left(1 + \frac{k-1}{2} M^2\right)}}$$

$$\frac{A}{A^*} = \frac{1}{M} \left[\frac{2\left(1 + \frac{k-1}{2} M^2\right)}{k+1}\right]^{\frac{k+1}{2(k-1)}}$$

$$\alpha = \arcsin \frac{1}{M}.$$

Tables 55 to 59, taken from the data of Edmonson, Murnaghan, and Snow,[39] present functions useful in applications of the theory of two-dimensional shocks, or oblique shock waves. For these the following relations hold:

$$\frac{\rho_y}{\rho_x} = X = \frac{\tan \alpha}{\tan(\alpha - \omega)}$$

$$M_x = \left(\frac{5X}{6-X}\right)^{1/2} \csc \alpha$$

$$M_y = \left(\frac{5}{6X-1}\right)^{1/2} \csc (\alpha - \omega)$$

$$\frac{p_y}{p_x} = \frac{6X-1}{6-X}.$$

2.11 Tables 60 to 62 — Standard Atmosphere, Physical Constants, and Conversion Factors

The contents in Table 60 represent a combination of two tables found in the handbook, *Smithsonian Physical Tables*.[40] The values of physical constants in Table 61 are those of the 1973 International Physical Constants recommended by CODATA-ICSU.[41] As the most reliable values presently available, these constants were employed throughout this work in computing the ideal gas thermodynamic properties of chemical compounds by means of statistical mechanics.

The conversion factor tables demonstrate the numerical relationships among English, cgs, and SI units for such quantities as length, volume, mass, density, force, pressure, energy, power, specific energy, specific energy per degree, absolute viscosity, kinematic viscosity, thermal conductivity, and molar gas constants. Of course, the trend of the future is towards international conformity in the use of SI units.[42]

EXAMPLES

Illustrating the Use of the Tables

Example 1   Compression of Air in Steady Flow

Air at a pressure of 1 atm abs and a temperature of 520 R is compressed in steady flow to a pressure of 6 atm abs. Find the work of compression and the temperature after compression for (a) 100% efficiency of compression and (b) 60% efficiency of compression. The efficiency of compression is here defined as the ratio of the isentropic work of compression to the actual work of compression.

Solution.   (a) From Table 1 we get for $T_1 = 520$ R,

$$p_{r1} = 1.2094, \qquad h_1 = 124.36 \text{ Btu/lb},$$

where subscript 1 refers to the state at the compressor inlet. To determine the properties at the compressor outlet for isentropic compression we compute the relative pressure there

$$p_{r2} = \tfrac{6}{1} \times 1.2094 = 7.256.$$

Entering Table 1 with this value of p_r, we find, for h_{2s} and T_{2s}, the enthalpy and temperature at the compressor outlet for isentropic compression

$$h_{2s} = 207.69 \text{ Btu/lb}, \qquad T_{2s} = 864.57 \text{ R}$$

The work of compression for 100% efficiency is then

$$h_{2s} - h_1 = 83.33 \text{ Btu/lb}.$$

(b) Since the efficiency of compression η is defined by the equation $\eta = (h_{2s} - h_1)/\text{work per lb}$, we have for 60% efficiency

$$\text{work per pound} = \frac{83.33}{0.60} = 138.88 \text{ Btu/lb}.$$

For the enthalpy at state 2, the state at the compressor outlet, we have

$$h_2 = h_1 + \text{work per pound}$$

$$= 124.36 + 138.88 = 263.24 \text{ Btu/lb}.$$

Entering Table 1 with this value of the enthalpy, we get the temperature at the compressor outlet

$$T_2 = 1088.6 \text{ R}.$$

(The value of p_{r2} is irrelevant because the process is not isentropic.)

If in this problem the definition of the efficiency is altered to be the ratio of the reversible isothermal work of compression to the actual work of compression, Table 1 is not necessary to the solution, for it is readily shown that the work of reversible isothermal compression in steady flow is given by

$$RT \ln \frac{p_2}{p_1},$$

provided only that

$$pv = RT.$$

Example 2   Change in Entropy

Find the increase in entropy of each pound of air in Example 1(b).

Solution.

$$s_2 - s_1 = \phi_2 - \phi_1 - R \ln \frac{p_2}{p_1}.$$

From Table 1,

$$T_1 = 520 \text{ R}, \qquad p_1 = 1 \text{ atm}, \qquad \phi_1 = 0.59160,$$
$$T_2 = 1088.6 \text{ R}, \qquad p_2 = 6 \text{ atm}, \qquad \phi_2 = 0.77155.$$

From Table 3,

$$- R \ln \frac{p_2}{p_1} = -0.12282.$$

Hence

$$s_2 - s_1 = 0.77155 - 0.59160 - 0.12282$$

$$= .05713 \text{ Btu/lb R}.$$

(200)

An alternative method consists of determining the increase in entropy between T_{2s} and T_2 at p_2. Then

$$s_2 - s_1 = \phi_2 - \phi_{2s}$$
$$= 0.77155 - 0.71443$$
$$= 0.05712 \text{ Btu/lb R}.$$

Example 3   The Pressure in a Turbine Stage

A steady stream of air approaches a ring of convergent nozzles at a pressure of 50 psia, a temperature of 1500 R, and at a rate of flow per unit area of 80 lb/sec ft^2 of nozzle exit area. Assuming a discharge coefficient of unity, find the pressure in the exit plane of the nozzle.

Solution. The alternative methods (*a*) and (*b*) given below are based on Tables 1 and 29 respectively.

(*a*) From the equation

$$G = \frac{w}{A} = \frac{V}{v} = \frac{p\sqrt{2g(h_0 - h)}}{RT},$$

where h denotes exit enthalpy and h_0, entrance enthalpy (see page 121 for definition of other symbols) and from Table 1, the exit pressure is determined by a trial-and-error procedure to be 40.2 psia.

(*b*) The value of k at 1500 R from Table 2 is 1.35. The rate of flow per unit area is given by the equation (page 122)

$$G = \frac{p_0\sqrt{\bar{m}}}{\sqrt{T_0}} \sqrt{\frac{2g}{\bar{R}} \frac{n}{n-1}} \; r^{\frac{1}{n}} \sqrt{1 - r^{\frac{n-1}{n}}}$$

where $n = k$. Substituting given values and obtaining the value for $\sqrt{(2g/\bar{R})[n/(n-1)]}$ from Table 24, we get

$$80 = \frac{50 \times 144\sqrt{28.970}}{\sqrt{1500}} (0.40039) r^{\frac{1}{n}} \sqrt{1 - r^{\frac{n-1}{n}}}$$

or

$$0.1997 = r^{\frac{1}{n}} \sqrt{1 - r^{\frac{n-1}{n}}}$$

From Table 29, $r = 0.804$. It follows that the exit pressure is 40.2 psia.

Example 4   Reversible Adiabatic Steady Flow

A steady stream of air expands reversibly and adiabatically in a nozzle passage from a pressure of 100 psia, a temperature of 510 R, and negligible velocity. Find the specific volume, velocity, Mach number, and mass velocity at the cross sections of the stream where the values of the pressure are 20, 90, and 99.5 psia respectively.

Solution. A solution may be found from Table 1 upon noting that

$$\frac{v}{v_0} = \frac{v_r}{v_{r0}},$$

$$V = \sqrt{2g(h_0 - h)}$$

(where the subscript 0 refers to the section where the velocity is zero),

$$M = \frac{V}{a}$$

(where a denotes the velocity of sound as given in Table 2), and

$$G = \frac{V}{v}.$$

Since the ratio of the specific heats k is constant at a value of 1.4 throughout this expansion, according to the data of Table 2, a second solution may be found from Table 25 with the aid of the relations given on page 122.

For small changes in pressure, a third method of solution yields precise results. Noting that

$$\left(\frac{\partial h}{\partial p}\right)_s = v,$$

we get

$$h_0 - h \simeq \frac{v_0 + v}{2}(p_0 - p),$$

in which v_0 and v may be obtained from Table 1.

The results of these three methods are summarized in Table J. The precision of Table 1, as indicated by the value of Δh_s when p/p_0 is 0.995, becomes inadequate for small changes in pressure. The precision of the method of Table 25 is good in all three instances. It deteriorates as the pressure change is reduced still further. For large pressure changes it is precise provided that a mean value of k is used and the variation in k is not large. The method employing $\bar{v}\Delta p$ is good for small changes in pressure and is distinguished from the other methods in that it improves in precision as the pressure change approaches zero.

Table J

$\frac{p}{p_0}$	Method	$\frac{T}{R}$	$\frac{\Delta h_s}{\text{Btu/lb}}$	$\frac{v}{\text{ft}^3/\text{lb}}$	$\frac{V}{\text{ft/sec}}$	M	$\frac{G}{\text{lb/sec ft}^2}$
	Table 1	321.11	45.09	5.957	1502	1.709	252.2
0.20	Table 25	322.01	45.11	5.963	1503	1.709	252.0
	$\bar{v}\Delta p$	321.71	58.09	5.957	1705	1.94	286.3
	Table 1	494.86	3.63	2.036	426.3	.3909	209.4
0.90	Table 25	494.88	3.629	2.036	426.2	.3909	209.3
	$\bar{v}\Delta p$	494.86	3.633	2.036	426.5	.3910	209.4
	Table 1	509.27	0.18	1.896	94.9	.0859	50.08
0.995	Table 25	509.27	0.1751	1.896	93.64	.0847	49.40
	$\bar{v}\Delta p$	509.27	0.1751	1.896	93.63	.0847	49.39

If the value of k were different from 1.4, Tables 26 to 29 would be used instead of Table 25 in the second method.

Example 5   Polytropic Process

Air at 100 psia and 3000 R expands according to the relation

$$pv^{1.3} = \text{constant}$$

to a pressure of 5 psia. Find (a) the temperature at the end of the expansion, (b) the change in entropy, (c) the work done on a piston by a pound of air expanding slowly, (d) the heat flow to a pound of air expanding slowly, (e) the work delivered to a turbine shaft per pound of air entering the turbine for reversible expansion in steady flow, and the corresponding flow of heat.

Solution.   (a)

$$\frac{T_2}{T_1} = \left(\frac{p_2}{p_1}\right)^{\frac{n-1}{n}} = 0.499, \text{ according to Table 27.}$$

$$T_2 = 0.499 \times 3000 = 1497 \text{ R.}$$

(b) Referring to Tables 1 and 3, we get

$$s_2 - s_1 = \phi_2 - \phi_1 + R \ln \frac{p_1}{p_2}$$

$$= 0.85351 - 1.04755 + 0.20535$$

$$= 0.0113 \text{ Btu/lb R.}$$

(c)

$$W = \int_1^2 p \, dv = \frac{RT_1}{n-1}\left[1 - \left(\frac{p_2}{p_1}\right)^{\frac{n-1}{n}}\right]$$

$$= \frac{53.342 \times 3000}{0.3}[1 - 0.499] = 267,243 \text{ ft-lb/lb.}$$

(d) $Q = u_2 - u_1 + W$, which, upon substitution of values from Table 1 and from above, becomes

$$Q = 265.83 - 584.83 + \frac{267,243}{778.16} = 24.4 \text{ Btu/lb.}$$

(e) For an infinitesimal step between states of zero velocity in a steady-flow process, the shaft work dW_x is given by

$$dW_x = -dh + dQ,$$

which for reversibility becomes

$$dW_x = -dh + T \, ds = -v \, dp,$$

and, therefore,

$$W_x = -\int_1^2 v \, dp = \frac{nR}{n-1}(T_1 - T_2) = 347,417 \text{ ft-lb/lb.}$$

$$Q = h_2 - h_1 + W_x,$$

which, upon substitution of values from Table 1 and from above, becomes

$$Q = 368.46 - 790.49 + \frac{347,417}{778.16}$$

$$= 24.4 \text{ Btu/lb,}$$

a value identical with that found in (d) above, both being equal to the integral of $T \, ds$.

Example 6   Compression of Air and Water Vapor

A mixture of air and water vapor at 560 R and 15.0 psia has a specific humidity of 0.030 pound of water per pound of dry air. The mixture is compressed isentropically in steady flow to 60 psia. Calculate the work of compression per pound of mixture.

Solution.   The molal products table for 400% of theoretical air corresponds to a mixture of air and

6.70% by mass of water vapor (see Table H). Linear interpolation is necessary to obtain the solution.

*From Table 3: Air with 6.70%
water vapor*

$T_1 = 560$ R
$p_{r1} = 1.5576$
$h_1 = 3911.2$ Btu/lb-mole
$p_{r2} = 4 \times p_{r1} = 6.2304$
$T_2 = 824.30$ R
$h_2 = 5794.6$ Btu/lb-mole
$(h_2 - h_1)_s = 1883.4$ Btu/lb-mole

*From Table 1: Air with 0%
water vapor*

$T_1 = 560$ R
$p_{r1} = 1.5675$
$h_1 = 3880.8$
$p_{r2} = 6.2700$
$T_2 = 829.89$ R
$h_2 = 5770.9$ Btu/lb-mole
$(h_2 - h_1)_s = 1890.1$ Btu/lb-mole

Since the mixture is $(0.03/1.03) \times 100$ or 2.91% by mass of water vapor, linear interpolation with respect to the percentage of water vapor yields the following expression for the work of compression per pound of mixture:

$$W = \frac{\frac{2.91}{6.70}(1883.4 - 1890.1) + 1890.1}{28.466} = 66.30 \text{ Btu/lb},$$

where 28.466 is the molecular weight of the mixture.

Example 7 Compression of Air and Octane Vapor

A mixture of air and octane vapor, corresponding to 25% of theoretical fuel, is compressed isentropically in steady flow from 500 R and 15 psia to 90 psia. Calculate the work of compression per pound of reactants.

Solution. The molal products tables for 200 and 400% of theoretical air correspond to a mixture of air and octane vapor for 28 and 14% of theoretical fuel, respectively (see Table H). Linear interpolation is necessary to obtain the solution.

*From Table 5: Reactants for 28%
theoretical fuel*

$T_1 = 500$ R, $p_{r1} = 1.0299$
$h_1 = 3512.4$ Btu/lb-mole
$p_{r2} = 6 \times p_{r1} = 6.179$
$T_2 = 818.68$ R Btu/lb-mole
$h_2 = 5814.9$ Btu/lb-mole
$(h_2 - h_1)_s = 2302.5$ Btu/lb-mole

*From Table 3: Reactants for 14%
theoretical fuel*

$T_1 = 500$ R, $p_{r1} = 1.0419$
$h_1 = 3488.4$ Btu/lb-mole
$p_{r2} = 6 \times p_{r1} = 6.251$
$T_2 = 825.04$ R
$h_2 = 5800$ Btu/lb-mole
$(h_2 - h_1)_s = 2311.6$ Btu/lb-mole

Interpolating linearly with respect to the percentage of theoretical fuel, we get for the work of compression per pound of reactants

$$W = \frac{\frac{0.25 - 0.14}{0.28 - 0.14}(2302.5 - 2311.6) + 2311.6}{29.326}$$

$$= 78.58 \text{ Btu/lb},$$

where 29.326 is the molecular weight of the mixture.

Example 8 Expansion of Products of Combustion of Benzene

The products of combustion of benzene with 200% of theoretical air expand in steady flow in a turbine from an initial temperature of 1500 R and an initial pressure of 10 atm to an exit pressure of 1 atm. The efficiency of the turbine is 80% based on the isentropic work of expansion. Calculate the work to the turbine shaft per pound of products.

Solution. The composition of benzene is C_6H_6. It has been shown[20] that a molal products table based on a fuel composition of $(CH_2)_n$ will yield precise results for the products of combustion of a hydrocarbon with the same composition as benzene, provided that the percentage of theoretical air is the same for the benzene as for the $(CH_2)_n$. Using Table 5 for 200% of theoretical air, we obtain for the

(203)

isentropic expansion

$T_1 = 1500$ R, $h_1 = 11050.5$ Btu/lb-mole,

$p_{r1} = 63.66$ $p_{r2} = 63.66 \times \frac{1}{10} = 6.366$

$T_{2s} = 825.29$ R, $h_{2s} = 5863.4$ Btu/lb-mole,

where subscript 1 refers to the inlet of the turbine and subscript $2s$ refers to the state at the exit pressure for isentropic expansion.

The work per pound of products is given by

$$W = 0.8 \frac{h_1 - h_{2s}}{\overline{m}} = 0.8 \frac{11050.5 - 5863.4}{29.445}$$

$$= 140.93 \text{ Btu/lb},$$

where $\overline{m}$ is the molecular weight of the products of combustion for benzene with 200% of theoretical air.

Example 9   Heat Transfer to Products of Combustion of Benzene

The products of combustion of benzene with 300% of theoretical air flow steadily through a heat exchanger and drop in temperature from 1200 to 800 R at a constant pressure of 1 atm. Calculate the heat transfer per pound of products.

Solution. The molal products table based on a fuel composition of $(CH_2)_n$ may be used for this mixture (see Table H). Linear interpolation between Tables 3 and 5 is necessary to obtain the answer for 300% of theoretical air.

From Table 5: Products for 50%
theoretical fuel
$T_1 = 1200$ R
$h_1 = 8685.0$ Btu/lb-mole
$T_2 = 800$ R
$h_2 = 5677.9$ Btu/lb-mole
$h_1 - h_2 = 3007.1$ Btu/lb-mole

From Table 3: Products for 25%
theoretical fuel
$T_1 = 1200$ R
$h_1 = 8565.3$ Btu/lb-mole
$T_2 = 800$ R
$h_2 = 5619.6$ Btu-lb-mole
$h_1 - h_2 = 2945.7$ Btu/lb-mole

Interpolating linearly with respect to percentage of theoretical fuel for 33.33% of theoretical fuel we

get for the heat transfer per pound of mixture

$$Q = \frac{\dfrac{33.33 - 25.0}{50.0 - 25.0}(3007.1 - 2945.7) + 2945.7}{29.289}$$

$$= 101.27 \text{ Btu/lb},$$

where 29.289 is the molecular weight of the mixture.

Example 10   Adiabatic Combustion at Constant Pressure

Liquid octane originally at 60 F is burned at constant pressure in steady flow in 200% of theoretical air originally at 400 F. Find the flame temperature for complete adiabatic combustion.

Solution. Application of the First Law indicates equality of the enthalpy of reactants, H_R, and that of products, H_P; thus

$$H_R = H_P,$$

where H_R and H_P are rederived from the same base state. Expressing this in terms of the enthalpy of combustion of liquid octane to gaseous products at 25 C (which is given in Table 10), we have

$$[H_R - H_R'] = H_P - H_P' + (H_P' - H_R')$$
$$= [H_P - H_P'] + H_{RP}'. \qquad [a]$$

In general,

$$Q = H_P - H_R = (H_P - H_P') - (H_R - H_R') + (H_P' - H_R')$$
$$= (H_P - H_P') - (H_R - H_R') + H_{RP}'$$

where the primed symbols refer to a temperature of 25 C (536.7 R) and H_{RP} denotes enthalpy of combustion. The quantities in brackets, being differences in enthalpy between two states of the same chemical aggregation, are independent of the base state selected and therefore may be taken from any tables of reactants and products.

The enthalpy of the reactants may be considered to be the sum of the enthalpies of liquid octane and of air. The former is given satisfactorily by the equation

$$h_f = 0.5T - 287$$

where h_f is the enthalpy in Btu per pound of liquid octane and T is the absolute temperature in degrees Rankine. (The state of zero enthalpy, which is irrelevant in this analysis, is octane vapor at 0 R.) The enthalpy of air may be taken from Table 1 and that of products from Table 5. The enthalpy of combustion at 25 C is taken from Table 10.

For 200% of theoretical air the chemical equation is

$$C_8H_{18} + 25O_2 + 94.10N_2$$

$$= 8CO_2 + 9H_2O + 12.5O_2 + 94.10N_2,$$

where N_2 represents nitrogen and argon. Per mole of liquid octane we have

$$n_A = 119.10$$

$$m_A = 119.10 \times 28.970 = 3450.3 \text{ lb}$$

$$n_P = 123.60,$$

where n_A and n_P denote numbers of moles of air and products, respectively, and m_A denotes the number of pounds of air.

Expressing equation [a] in terms of specific enthalpies and solving for the enthalpy per mole of products, we have

$$\bar{h}_P = \frac{1}{n_P} \left[\bar{m}_f(h_f - h_f' - h_{RP}') + m_A(h_A - h_A') \right] + \bar{h}_P',$$

where $\bar{m}_f$ denotes the molecular weight of octane. Substituting numbers, we get

$$\bar{h}_P = \frac{1}{123.60} \{ 114.22 [.5(519.7 - 536.7) + 19,100]$$

$$+ 3450.3(206.57 - 128.37) \} + 3774.3$$

$$= 23599.9 \text{ Btu/lb-mole.}$$

From Table 5 we get the flame temperature

$$T_P = 2958.1 \text{ R.}$$

It is evident from equation [a] that a simplification results if the enthalpy of combustion at 0 R, H_{RP}^0, is used, for we get

$$H_P = H_R - H_{RP}^0,$$

or, in terms of the specific enthalpies and the masses of air and fuel,

$$(m_A + m_F)h_P = (m_A + m_F)h_R - m_F h_{RP}^0$$

and

$$h_P = h_R - \frac{m_F}{m_A + m_F} h_{RP}^0. \qquad [b]$$

Equation [b] is often employed in combustion calculations.

Example 11   Adiabatic Combustion at Constant Volume

A mixture of octane vapor and 200% of theoretical air at a pressure of 1 atm and a temperature of 600 R is burned adiabatically and completely at constant volume. Find the temperature and pressure after combustion.

Solution. Application of the First Law indicates equality of the internal energy of reactants, U_R, and that of products, U_P; thus

$$U_R = U_P.$$

Expressing this in terms of the internal energy of combustion U_{RP} at 25 C, we have

$$[U_R - U_R'] = U_P - U_P' + (U_P' - U_R')$$

$$= [U_P - U_P'] + U_{RP}', \qquad [a]$$

where the primed symbols refer to a temperature of 25 C (536.7 R). The quantities in brackets, being differences in internal energy between two states of the same chemical aggregation, are independent of the base state selected. That for reactants may be taken from a table for reactants, and that for products from a table for products.

The values of the properties of a mixture of octane vapor and 200% of theoretical air (or 50% of theoretical fuel) are obtained by linear extrapolation, on the basis of percentage of theoretical fuel, with the aid of Tables 3 and 5, and with the data given in Table H on page 194.

Values of the properties of the products may be obtained directly from Table 5.

The value of U_{RP}' is obtained from H_{RP}' through the relation

$$U_{RP} = H_{RP} + [(pV)_R - (pV)_P],$$

or

$$U_{RP} = H_{RP} + \bar{R}T(n_R - n_P),$$

where T denotes the absolute temperature and n_R and n_P the number of moles of reactants and products, respectively. The value of H_{RP} at 25 C is given in Table 10.

For 200% of theoretical air the chemical equation is

$$C_8H_{18} + 25O_2 + 25\frac{0.7901}{0.2099}N_2$$

$$= 8CO_2 + 9H_2O + 12.5O_2 + 94.10N_2,$$

where N_2 represents nitrogen and argon.

Per mole of gaseous octane we have

$$n_R = 120.10$$

and

$$n_P = 123.60,$$

so that

$$U_R = 120.10 \bar{u}_R,$$

$$U_P = 123.60 \bar{u}_P,$$

and, at 25 C or 536.7 R,

$$U_{RP} = H_{RP} + 1.98586 \times 536.7 [120.10 - 123.60]$$

$$= -19,253 \times 114.22 - 3730 = -2,199,078 - 3730$$

$$= -2,202,808 \text{ Btu/lb-mole},$$

and per pound of octane

$$u_{RP} = -\frac{2,202,808}{114.22} = -19,286 \text{ Btu/lb}.$$

The value of $\bar{u}_R$ at 600 R is obtained as follows: From Table 5, for 28% of theoretical fuel,

$$\bar{u}_R = 3036.2 \text{ Btu/lb-mole}.$$

From Table 3, for 14% of theoretical fuel,

$$\bar{u}_R = 3002.3 \text{ Btu/lb-mole}.$$

Hence, for the mixture of air and octane vapor having 50% of theoretical fuel, we have by extrapolation

$$\bar{u}_R = \frac{50 - 28}{28 - 14}(3036.2 - 3002.3) + 3036.2$$

$$= 3089.5 \text{ Btu/lb-mole}.$$

Similarly,

$$\bar{u}_R' = 2751.5 \text{ Btu/lb-mole}.$$

Introducing these numbers and a value for $\bar{u}_P'$ from Table 5 into equation [a] and solving for $\bar{u}_P$, we get

$$\bar{u}_P = \frac{1}{123.10}\left[120.10(\bar{u}_R - \bar{u}_R') - U_{RP}'\right] + \bar{u}_P'$$

$$= \frac{1}{123.60}\left[120.10(3089.5 - 2751.5)\right.$$

$$\left. + 2,203,150\right] + 2708.4$$

$$= 20,858.9 \text{ Btu/lb-mole}.$$

The last entry in Table 5 lists an internal energy of 17985.7 Btu/lb-mole corresponding to a temperature of 2995 R. Thus we must use Table 6 in conjunction with Table 5. The change in internal energy

from state 1 to state 2 may be expressed as

$$u_2 - u_1 = \int_{T_1}^{T_2} c_v \, dT.$$

Letting $u_1 = 17985.7$ Btu/lb-mole, $T_1 = 2995$ R, and $u_2 = 20858.9$ Btu/lb-mole and using Table 6 we get

$$20858.9 - 1798.7 = 7.051(3100 - 2995)$$

$$+ 7.121(3300 - 3100) + 7.184(T_p - 3300)$$

or

$$T_p = 3399 \text{ R}.$$

The ratio of the pressure of the products to that of the reactants is given by

$$\frac{p_P}{p_R} = \frac{n_P T_P}{n_R T_R}$$

since

$$V_P = V_R.$$

Thus

$$p_P = 1 \times \frac{123.60}{120.10} \times \frac{3399}{600} = 5.83 \text{ atm}.$$

Example 12   Enthalpy of Combustion of n-Octane

The enthalpy of combustion at 25 C of gaseous n-octane is given in Table 10 as 19,253 Btu per pound of octane for gaseous products of combustion. Correct this value to 0 R.

Solution. The chemical equation may be given in the form

$$C_8H_{18}(g) + 12.5O_2(g) + D \rightarrow 8CO_2(g) + 9H_2O(g) + D,$$

where D denotes diluent gases which are unaffected by the reaction.

For a given temperature T the enthalpy of combustion per mole of octane is given by

$$\bar{h}_{RP} = -\bar{h}_A - 12.5\bar{h}_B + 8\bar{h}_M + 9\bar{h}_N,$$

where $\bar{h}_A$, $\bar{h}_B$, $\bar{h}_M$ and $\bar{h}_N$ denote the enthalpy per mole of C_8H_{18}, O_2, CO_2, and H_2O, respectively, all at the temperature T and all reckoned from the same base (for example, zero enthalpy for each element at 0 R).

The relation between the enthalpies of combustion per pound of octane at two temperatures T' and T'' is then given by

$$h_{RP}' = h_{RP}'' + \frac{1}{\bar{m}_A}\left[-(\bar{h}' - \bar{h}'')_A - 12.5(\bar{h}' - \bar{h}'')_B\right.$$

$$\left. + 8(\bar{h}' - \bar{h}'')_M + 9(\bar{h}' - \bar{h}'')_N\right],$$

where the superscript refers to the corresponding temperature, and $\bar{m}_A$ denotes the molecular weight of octane. In this equation the enthalpies for each substance appear only in the difference between two values for that substance, and the base selected is, therefore, of no consequence. Values for O_2, CO_2, and H_2O may be taken from Tables 13, 17, and 15, respectively. Values for octane must be found elsewhere and are here taken from Table 3u-E (part I) of "Tables of Selected Values of Chemical Thermodynamic Properties" (December 31, 1944), published by the National Bureau of Standards. The numerical solution for 0 R in terms of values at that temperature and at 25 C is as follows.

$$h_{RP} \text{ at } 0 \text{ R} = -19{,}256 + \frac{1}{114.22}\big[-(0-114.22$$

$$\times 137.58) - 12.5(0-3733.3)$$

$$+ 8(0-4026.4) + 9(0-4258.1)\big]$$

$$= -19{,}253 - 71 = -19{,}324 \text{ Btu/lb of octane.}$$

Example 13 Discharge of Hydrogen from a Closed Container

Hydrogen in a tank having a volume of 10 ft^3 is initially at a pressure of 1000 psia and at a temperature of 700 R. The hydrogen discharges to a pressure of 1 atm through a small nozzle until one half the original mass remains in the tank. No heat is exchanged between the hydrogen and the walls of the tank. Find the final pressure and temperature in the tank.

Solution. The process experienced by the hydrogen which finally remains in the tank is an isentropic expansion. Using subscripts 1 and 2 to denote, respectively, the initial and final states of hydrogen

inside the tank, we have from Table 19

$$T_1 = 700 \text{ R}, \qquad v_{r1} = 448.9, \qquad p_{r1} = 16.735.$$

Moreover,

$$v_2 = 2v_1$$

and, therefore,

$$v_{r2} = 2v_{r1} = 897.8.$$

From Table 19,

$$T_2 = 530.1, \quad p_{r2} = 6.337$$

$$p_2 = p_1 \times \frac{p_{r2}}{p_{r1}} = 378.7 \text{ psia.}$$

Example 14 Gas Turbine

Air is compressed isentropically in steady flow from 1 atm and 520 R to 5 atm. Liquid octane at 520 R is introduced into the stream of compressed air at such a rate that the resultant mixture contains 300% of theoretical air. The octane burns completely at constant pressure, and the products of combustion expand isentropically to 1 atm. Find the efficiency.

Solution. Compression: Let subscripts 1 and 2 refer to states at inlet and outlet of the compressor, respectively. Then from Table 1 we have

$$T_1 = 520, \qquad h_1 = 124.36, \qquad p_{r1} = 1.2094$$

$$p_{r2} = \tfrac{5}{1} \times 1.2094 = 6.047, \qquad T_2 = 821.5, \qquad h_2 = 197.15.$$

Mixing: Application of the First Law to the adiabatic process of mixing liquid octane and air shows that the enthalpy of the resultant air-fuel mixture is equal to the sum of the enthalpies of the air and fuel before mixing. To evaluate the temperature of the mixture of air and octane vapor, a common base state is selected from which the en-

Table K. States in the Gas Turbine, Figure 2

State	Table Number	T R	h Btu/lb	$\bar{h}$ Btu/lb-mole	p_r	p atm
1	1	520	124.36	3,602.7[a]	1.2094	1
2	1	821.5	197.15	5,711.4	6.047	5
3	3, 5	794.8	192.31	5,662.6		5
4	3, 5	2,309.5	609.72	17,610.5	364.0	5
5	3, 5	1,568.5	397.37	11,477.1	72.8	1

[a]Base state is a fuel-air mixture. Where values of these properties are not so marked the base state is a mixture of air and products of complete combustion.

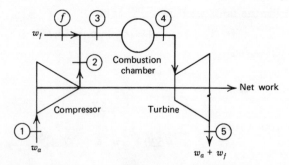

Figure 1

thalpies of the mixture and of the components may be reckoned.

In Table 1 the enthalpy of air is zero at 0 R. Using an analogous convention, we may fix the enthalpy of octane vapor at zero at 0 R. Data relative to this base have been published by the Bureau of Standards for hydrocarbons. The enthalpy of liquid octane as obtained from these data is given with satisfactory precision over a range of temperatures centering on 500 R by the equation

$$h_f = 0.5T - 287 \text{ Btu/lb},$$

where h_f denotes the enthalpy of a pound of liquid octane at the temperature T R. The effect of pressure on enthalpy may be safely ignored in problems involving combustion.

The theoretical fuel-air ratio for octane (C_8H_{18}) is 0.06621, based on the composition of air used in constructing Table 1. Then for 300% of theoretical air, each pound of fuel-air mixture contains 0.0216 lb of fuel and 0.9784 lb of air. The enthalpy of each pound of liquid fuel at 520 R, based on vapor at 0 R, is

$$h_f = 0.5 \times 520 - 287 = -27 \text{ Btu/lb}.$$

The enthalpy of the fuel-air mixture based on a state having the same chemical aggregation is then

$$h_3 = 0.0216(-27) + 0.9784 \times 197.15$$
$$= 192.31 \text{ Btu/lb},$$

and

$$\bar{h}_3 = 192.31 \times 29.445 = 5662.6 \text{ Btu/lb-mole},$$

where subscript 3 refers to the state immediately after mixing and 29.445 is the molecular weight of the mixture. The temperature of this mixture is obtained by linear extrapolation, on the basis of percentage of theoretical fuel, with the aid of Tables 3 and 5. From Table H on page 194, a mixture of air and octane vapor for 28% of theoretical fuel corre-

sponds closely to a products table for 50% of theoretical fuel. From the value of h_3 we get:

From Table 5, for 28% of theoretical fuel, $T_3 = 797.9$.

From Table 3, for 14% of theoretical fuel, $T_3 = 806.0$.

Hence, for 33.33% of theoretical fuel, we have by extrapolation

$$T_3 = 797.9 + \frac{33.33 - 28}{28 - 14}(797.9 - 806.0) = 794.8 \text{ R}.$$

Combustion: Application of the First Law to the process of adiabatic combustion between states 3 and 4 (see Figure 2) results in the equation

$$h_3 = h_4,$$

where h_3 and h_4 represent the enthalpy per pound for reactants and products of combustion, respectively. The base states for h_3 and h_4 must, of course, be the same as explained on page 204.

The values of enthalpy used above in the calculation of the mixing process are based on an enthalpy of zero for gaseous reactants at 0 R. The values of enthalpy given in Tables 3 and 5, on the other hand, are based on an enthalpy of zero for gaseous products of combustion at 0 R. If the enthalpy of the reactants is to be reckoned from this latter or products base, the values on the reactants base must be augmented by the increase in enthalpy when products at 0 R are changed to reactants at 0 R. This quantity is the negative of the *enthalpy of combustion* at 0 R and is equal to the constant-pressure "heat of combustion" at 0 R.

For octane, the enthalpy of combustion per pound of fuel at 0 R, $h_{RP}{}^0$ (see Example 12) is constant over the range from 0 to 100% of theoretical fuel and is given by

$$h_{RP}{}^0 = h_P{}^0 - h_R{}^0 = -19{,}324 \text{ Btu/lb of octane},$$

where $h_R{}^0$ and $h_P{}^0$ denote the enthalpy at 0 R of the reactants and of the products, respectively, per pound of octane.

Values of the enthalpy *per pound of reactants* on the reactants base must therefore be augmented by the negative of $h_{RP}{}^0$ multiplied by the fraction of a pound of octane in each pound of reactants; that is, by

$$-\frac{w_f}{w_a - w_f} h_{RP}{}^0,$$

where w_f and w_a denote the mass rate of flow of fuel and air, respectively. Hence the enthalpy of the fuel-air mixture before combustion at state 3, reckoned from the products base, is given by

$$h_3 = 192.31 + 0.0216 \times 19{,}324 = 609.72 \text{ Btu/lb.}$$

Since the enthalpies at states 3 and 4 are equal,

$$h_4 = 609.72 \text{ Btu/lb,}$$

where h_4 is reckoned from the products base state. Then

$$\bar{h}_4 = 609.72 \times 28.883 = 17{,}610.5 \text{ Btu/lb-mole,}$$

where 28.883 is the molecular weight of the mixture found by interpolation from Table 9.

This value of h_4 and a table of enthalpies for products of combustion for 300% of theoretical air will yield the temperature at state 4 at the exit of the combustion chamber, because there is no change in chemical aggregation for such a mixture between the products base state and state 4. Since, however, a table for products of combustion for 300% of theoretical air or 33.33% of theoretical fuel is not available, linear interpolation, with respect to the percentage of theoretical fuel, between Tables 3 and 5 yields a value for the temperature as follows:

From Table 3, for 25% of theoretical fuel, $T_4 = 2323.1$.

From Table 5, for 50% of theoretical fuel, $T_4 = 2282.3$.

Hence, for 33.33% of theoretical fuel, we get

$$T_4 = 2323.1 + \frac{33.33 - 25}{50 - 25}(2282.3 - 2323.1) = 2309.5 \text{ R.}$$

Similarly, we get for the relative pressure at state 4,

$$p_{r4} = 359.3 + \frac{33.33 - 25}{50 - 25}(372.8 - 359.3) = 363.8.$$

Expansion: For isentropic expansion from 5 atm to 1 atm, we find the relative pressure at state 5 from that at state 4.

$$p_{r5} = p_{r4} \times \tfrac{1}{5} = 72.76.$$

The temperature and enthalpy at state 5 are found by linear interpolation, with respect to the percentage of theoretical fuel, between Tables 3 and 5. Thus for a value of p_{r5} of 72.76, we get:

From Table 3, for 25% of theoretical fuel, $T_5 = 1577.6$.

From Table 5, for 50% of theoretical fuel, $T_5 = 1550.2$.

Hence, for 33.33% of theoretical fuel,

$$T_5 = 1577.6 + \frac{33.33 - 25}{50 - 25}(1550.2 - 1577.6) = 1568.5.$$

A similar calculation for h_5 yields

$$h_5 = 397.37 \text{ Btu/lb.}$$

The shaft work of the turbine per pound of products is now

$$W_t = 609.72 - 397.37 = 212.35 \text{ Btu/lb.}$$

Performance: The shaft work of the compressor per pound of products is (from Table K)

$$W_c = \frac{w_a}{w_a + w_f}(h_2 - h_1) = 0.9784 \times 72.79 = 71.22 \text{ Btu/lb.}$$

The net work of the gas turbine per pound of products is

$$W_n = 212.4 - 71.2 = 141.2 \text{ Btu/lb,}$$

since the work of compressing the liquid fuel is negligible. The heating value of the fuel supplied may be taken to be

$$-\frac{w_f}{w_a + w_f} h_{RP} = -0.0216 \times (-19{,}253) = 415.9 \text{ Btu/lb,}$$

where 19,256 is an arbitrary selected heat of combustion for octane at 25 C. The efficiency of the gas turbine is then

$$\eta = \frac{141.2}{415.9} = 0.339.$$

The mixing and combustion processes may be combined and analyzed as a single process between sections 1 and 2 on the one hand, and section 4 on the other. Then the values for section 3 need not be determined.

If the products of combustion were treated as air in this example, the calculations for combustion and expansion would have been greatly simplified. Table 1 would have been used in place of Tables 3 and 5; no conversions to molal quantities and no interpolation beteen these tables would have been necessary. The temperature found at state 4 would have been 2374 R, an error of 64 degrees, and the efficiency would have been 0.349, an error of about 3%.

Example 15   Diesel Engine

Air in a cylinder at 1 atm and 600 R is compressed isentropically to one-tenth its initial volume. As the piston moves outward, liquid octane at 520 R is introduced and burned at such a rate as to maintain constant pressure. The total mass of octane introduced is one third the theoretical mass for combustion in the air in the cylinder. The products

(209)

Table L   States in the Diesel Engine, Figure 2

State	Table Number	T R	h Btu/lb	h̄ Btu/lb-mole	p_r	u Btu/lb
1	1	600	143.57	4,159.2	1.996	102.44
2	1	1,444	354.51	10,270.2	48.04	255.52
3	3, 5	2,822.7	763.68	22,057.4	873.2	569.60
4	3, 5	1,724.1	440.71	12,729.0	106.74	322.17
5	3, 5	1,315.3	328.52	9,488.6	36.28	238.09

State	Table Number	ū Btu/lb-mole	v_r	p atm	v̂ ft³/lb	Mass lb
1	1	2,967.7	111.35	1	15.12	1
2	1	7,402.4	11.135	24.07	1.512	1
3	3, 5	16,451.9	34.72	24.07	3.029	1.0221
4	3, 5	9,305.3	173.35	2.942	15.12	1.0221
5	3, 5	6876.7	389.1	1	33.94	1.0221

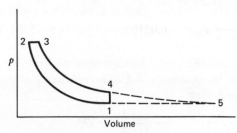

Figure 2

are expanded isentropically to the initial volume, at which point the exhaust valve opens and the pressure falls to 1 atm. Find the states before and after combustion, at the opening of the exhaust valve, and at the return to atmospheric pressure.

Solution.

From Table 1 we obtain the data for state 1 given in Table L. To identify state 2, for which the entropy is the same as for state 1, we note that for state 2 the relative volume must be one tenth as large as for state 1.

For the adiabatic constant-pressure process between states 2 and 3 the enthalpy at state 3 is identical with the sum of the enthalpies of air in state 2 and fuel at 520 R. For each pound of air at 2, one third of the theoretical fuel or 0.0221 lb of fuel is introduced. Thus, each pound of products at 3 is formed from 0.0216 lb of fuel and 0.9784 lb of air. The enthalpy of this combination at 3 is then

$$h_3 = 0.9784(354.51) + 0.0216(19,324 - 27)$$
$$= 763.68 \text{ Btu/lb,}$$

as reckoned from the products base (see Example 14).

The temperature at state 3 is obtained from the value of h_3 by means of linear interpolation between Tables 3 and 5 (see Example 14), as follows:

From Table 3, for 25% of theoretical fuel, $T_3 = 2840.8$.

From Table 5, for 50% of theoretical fuel, $T_3 = 2786.4$.

Hence, for 33.33% of theoretical fuel, we get

$$T_3 = 2840.8 + \frac{33.33 - 25}{50 - 25}(2786.4 - 2840.8) = 2822.7.$$

The internal energy, relative pressure, and relative volume at state 4 are calculated similarly by interpolation between Tables 3 and 5. The molecular weight of this mixture, which is 28.883 and is obtained from Table 9, is then used to calculate the molal enthalpy and internal energy. The specific volume is determined with the aid of the equation of state

$$pv = RT.$$

The values are summarized in Table L.

Since states 3 and 4 lie on the same isentropic, state 4 may be identified from the relative volume, thus

$$\frac{v_{r4}}{v_{r3}} = \frac{\hat{v}_4}{\hat{v}_3},$$

where $\hat{v}$ denotes the volume per pound of air compressed and $\hat{v}_4$ is identical with $\hat{v}_1$.

The temperature T_4 is obtained from v_{r4} by means of linear interpolation between Tables 3 and 5, as for T_3 above.

To identify state 5, which exists in the cylinder when atmospheric pressure is restored, we assume that the expansion of residual gases in the cylinder during exhaust is isentropic. The relative pressure at state 5 is found from the relation

$$\frac{p_{r5}}{p_{r3}} = \frac{p_5}{p_3}.$$

From Tables 3 and 5 we obtain by interpolation the temperature and relative volume and, through the equation of state, the specific volume.

The net indicated work is given by the algebraic sum of the areas under the curves 12, 23, and 34, thus

$$W = u_1 - u_2 + \frac{14.696 \times 144}{778.16} p_2 (\hat{v}_3 - \hat{v}_2) + \frac{m_3}{m_2}(u_3 - u_4).$$

Substituting values from Table L, we get

$$W = 102.44 - 255.52 + 2.720 \times 24.06(3.029 - 1.512)$$

$$+ 1.0221(569.60 - 322.17)$$

$$= 199.1 \text{ Btu/lb of air}$$

and (as in Example 14)

$$\eta = \frac{199.1}{0.0221 \times 19{,}253} = 0.468.$$

If the products of combustion were treated as air in this example, the temperature found at state 4 would have been 1739 R, an error of 15 degrees, and the efficiency would have been 0.478, an error of about 2%.

Example 16   Turbojet

A turbojet is to be designed for a speed of 400 mph at an altitude of 30,000 ft. (See Figure 3.) Air is diffused isentropically from a relative velocity of 400 mph at section 1, the entrance to the diffuser, to zero velocity at section 2, the exit of the diffuser. The air is then compressed isentropically in steady flow through a fourfold increase in pressure from section 2 to section 3. Liquid octane at 500 R is introduced into the stream of compressed air at such a rate that the resultant homogeneous mixture at section 4 contains 25% of theoretical fuel. The octane burns completely at constant pressure, and the products of combustion leave the combustion chamber at section 5. Next, the products expand isentropically through a turbine to that pressure at section 6 which results in equality of power output from the turbine and power input to the compressor. Finally, the products expand reversibly and adiabatically through a nozzle to atmospheric pressure at section 7.

Calculate the efficiency of the power plant.

Solution.

Diffusion: The air enters the diffuser with a velocity of 400 mph or 586.7 ft/sec, a temperature of 412 R, and a pressure of 8.88 in. Hg. For zero exit velocity, application of the First Law results in

$$h_2 = h_1 + \frac{V_1^2}{2g} = 98.42 + \frac{(586.7)^2}{(2)(32.174)(778.16)}$$

$$= 105.29 \text{ Btu/lb},$$

where the numerical value of h_1 is obtained from Table 1 at T_1. Table 1 at h_2 now gives $T_2 = 440.42$ R, $p_{r2} = 0.6768$, and $p_2 = 11.24$ psia.

These values are summarized in Table M.

Compression: The work of isentropic compression in steady flow is the increase in enthalpy across the compressor if the velocities are small. State 3 at the

Table M   States in the Turbojet, Figure 3

State	Table Number	T R	h Btu/lb	$\bar{h}$ Btu/lb-mole	p_r	p in. Hg	V ft/sec
1	1, 60	411.7	98.42	2,851.2	0.5348	8.88	586.7
2	1	440.42	105.29	3,050.3	0.6768	11.24	0
3	1	654.4	156.67	4,538.7	2.707	44.95	0
4	3, 5	639	153.51	4,501.8		44.95	0
5	3	1,831.9	468.11	13,530.3	133.14	44.95	0
6	3	1,650.7	417.57	12,069.3	87.24	29.45	0
7	3	1,215.2	300.30	8,680.0	26.31	8.88	2423.2

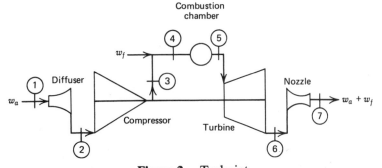

Figure 3.   Turbojet

(211)

exit of the compressor is identified through the relative pressure (Table 1),

$$p_{r3} = 4p_{r2} = 4 \times 0.6768 = 2.707.$$

Hence T_3 is 654.40 R and h_3 is 156.67 Btu/lb of air. The work of compression is given by

$$W_c = h_3 - h_2 = 51.38 \text{ Btu/lb of air.}$$

Mixing and combustion: The same technique is used here as in Example 14, on the gas turbine. The results are given in Table M.

Turbine expansion: The equality of compressor power and turbine power is given by

$$h_3 - h_2 = \left(1 + \frac{w_f}{w_a}\right)(h_5 - h_6),$$

where w_f and w_a denote the mass rate of flow of fuel and air, respectively, and h_6 is the only unknown.

Nozzle expansion: For steady flow through the nozzle,

$$V_7{}^2 = 2g(h_6 - h_7).$$

Since the nozzle expansion is isentropic, the relative pressure (from Table 3) at the exit of the nozzle determines state 7 as follows:

$$p_{r7} = p_{r6} \frac{p_7}{p_6} = 87.24 \frac{8.88}{29.45} = 26.31$$

$$T_7 = 1215.2 \text{ R}, \qquad h_7 = 300.30 \text{ Btu/lb.}$$

Hence the velocity V_7 leaving the nozzle is

$$V_7 = \sqrt{50{,}073(417.57 - 300.30)} = 2423.2 \text{ ft/sec.}$$

Performance: The thrust for a mass flow of one pound of air per second is given by the momentum equation in the form

$$\frac{T}{w_a} = \frac{\left(1 + \dfrac{w_f}{w_a}\right)V_7 - V_1}{g}.$$

The propulsive work for a mass flow of one pound of air per second is

$$\frac{W}{w_a} = \frac{T}{w_a} V_1.$$

The propulsive efficiency, the ratio of the propulsive work to the heating value of the fuel burned in the same period of time, is given by

$$\eta = \frac{\dfrac{W}{w_a}}{\dfrac{w_f}{w_a}(-h_{RP})} = \frac{V_1\left(1 + \dfrac{w_f}{w_a}\right)V_7 - V_1}{-g\dfrac{w_f}{w_a}h_{RP}},$$

where h_{RP} is the enthalpy of combustion at 25 C. Hence

$$\eta = \frac{586.7\left(\dfrac{2423.2}{0.9837} - 586.7\right)}{32.174\dfrac{0.01628}{0.9837}(19{,}253)(778.16)}$$

$$\eta = 0.138.$$

Example 17   Isentropic Flow

Consider the steady, isentropic, one-dimensional flow of air at supersonic velocities. At a given section of the stream, the Mach number is 2.5, the stagnation temperature is 560 R, and the static pressure is 0.5 atm.

Find stagnation pressure, static pressure, temperature, density, velocity, and mass velocity at sections of the stream where the Mach number is 2.5, 1.2, and 1.0.

Solution.   From Table 30, for a Mach number of 2.5, we get

$$\frac{p}{p_0} = 0.05853, \quad \frac{T}{T_0} = 0.44444, \quad \frac{\rho}{\rho_0} = 0.13169,$$

$$M^* = 1.8258.$$

From these and given values for p and T_0, values for p_0 and T are found. Either density, ρ or ρ_0, may be found from the general relation

$$\rho = \frac{p}{RT},$$

and the other found from the value of ρ/ρ_0 above.

The velocity of sound is

$$a = \sqrt{gkRT} = 49.02\sqrt{T} = 773.3 \text{ ft/sec.}$$

The velocity is

$$V = Ma$$

and the mass velocity

$$G = \rho V.$$

Since Table 30 gives values of p/p_0, the pressure at any other Mach number M may be found from p/p_0 and the previously calculated value of p_0.

The resultant values are summarized in Table N.

M	2.5	1.2	1.0
p (atm)	0.50	3.523	4.513
T (R)	248.9	434.8	466.7
p_0 (atm)	8.543	8.543	8.543
T_0 (R)	560.0	560.0	560.0
ρ (lb/ft³)	0.07970	0.3215	0.3837
V (ft/sec)	1933.3	1226.5	1058.9
G (lb/sec ft²)	154.1	394.3	406.2

Example 18   Rayleigh Line

Consider the steady, one-dimensional supersonic flow of air which is heated in the absence of friction and of area change. At one section of the stream, the Mach number is 2.5, the stagnation temperature is 560 R, and the static pressure is 0.5 atm.

Find stagnation pressure, stagnation temperature, static pressure, temperature, density, velocity, and mass velocity at sections of the stream where the Mach number is 1.2 and 1.0.

Solution. (*a*) The values of the quantities called for are given in Table N of Example 17 for a Mach number of 2.5. Table 36 gives the pressure, temperature, density, and velocity, each as a ratio to the corresponding quantity at a Mach number of unity on the same Rayleigh line. The values for a Mach number of unity can therefore be calculated. Thus,

$$p^* = \frac{p}{\dfrac{p}{p^*}} = \frac{0.5}{0.24616} = 2.031,$$

and, similarly,

$$T^* = 657.2 \text{ R}, \quad \rho^* = 0.1226 \text{ lb/ft}^3, \quad V^* = 1256.6 \text{ ft/sec.}$$

For other values of the Mach number the ratios p/p^*, etc., now permit calculation of p, etc. The results are presented in Table O.

Table O

M	2.5	1.2	1.0
p (atm)	0.5	1.616	2.031
T (R)	248.9	599.3	657.2
p_0 (atm)	8.543	3.920	3.845
T_0 (R)	560.0	771.9	788.7
ρ (lb/ft³)	0.07970	0.1070	0.1226
V (ft/sec)	1933.3	1439.9	1256.6
G (lb/sec ft²)	154.1	154.1	154.1

Example 19   Fanno Line

Consider the steady, one-dimensional, adiabatic flow of air at constant area with friction. At a given section of the stream, the Mach number is 2.5, the stagnation temperature is 560 R, and the static pressure is 0.5 atm.

(*a*) Find stagnation pressure, stagnation temperature, static pressure, temperature, density, velocity, and mass velocity at sections of the stream where the Mach number is 1.2 and 1.0.

(*b*) Find the distance between two sections of Mach number 2.5 and 1.2, respectively, if the flow is through a circular duct with a diameter of 1 in. and with an average friction coefficient of 0.003.

Solution. (*a*) The procedure parallels that of the preceding example. Values are first found for a Mach number of unity on the same Fanno line (p^*, T_0^*, etc.) from the values for a Mach number of 2.5 and the ratios of Table 42. These with the values from Table 42 for any other given value of the Mach number yield the desired quantities. The results are presented in Table P.

(*b*) Table 42 gives values of the quantity $4fL_{max}/D$, where L_{max} denotes the length of duct from Mach number M to Mach number unity. The difference between the values of L_{max} for two different values of Mach number is, therefore, the length of duct between the two sections corresponding to the two values of Mach number. Thus, for constant friction coefficient f,

$$L = \frac{D}{4f}\left[\left(\frac{4fL_{max}}{D}\right)'' - \left(\frac{4fL_{max}}{D}\right)'\right].$$

Thus, for a constant friction coefficient f of 0.003, and a diameter of 1 inch the distance between two sections of Mach number 2.5 and 1.2, respectively, is given by

$$L = \frac{D}{4f}\left[\left(\frac{4fL_{max}}{D}\right)_{2.5} - \left(\frac{4fL_{max}}{D}\right)_{1.2}\right]$$

$$= \frac{1}{4 \times 0.003}(0.43197 - 0.03364) = 33.2 \text{ in.}$$

Table P

M	2.5	1.2	1.0
p (atm)	0.5	1.377	1.712
T (R)	248.9	434.8	466.7
p_0 (atm)	8.543	3.339	3.240
T_0 (R)	560.0	560.0	560.0
ρ (lb/ft³)	0.07970	0.1256	0.1455
V (ft/sec)	1933.3	1226.5	1058.9
G (lb/sec ft²)	154.1	154.1	154.1

Example 20 Normal Shock

A steady flow of air passes through a one-dimensional normal shock. At the upstream section, the Mach number is 2.5, the stagnation temperature is 560 R, and the static pressure is 0.5 atm.

Find the properties of the air at the upstream and downstream sections of the shock.

Solution. The properties at a Mach number of 2.5 are given in Table N of Example 17. These, together with the quantities of Table 48, by obvious methods, give the results presented in Table Q.

Table Q

Section	Upstream	Downstream
M	2.50	0.51299
p (atm)	0.5	3.563
T (R)	248.9	532.0
p_0 (atm)	8.543	4.263
T_0 (R)	560.0	560.0
ρ (lb/ft^3)	0.07970	0.2657
V (ft/sec)	1933.3	580.0
G (lb/sec ft^2)	154.1	154.1

Example 21 Two-Dimensional Shock

A steady parallel stream of air approaches a sharp corner with an angle of 20°, as shown in Figure 4. The air has a pressure of 20 psia, a temperature of 500 R, and a Mach number of 2.0. For both weak and strong shocks, find the downstream Mach number and pressure, the ratio of downstream to upstream mass velocity, and the angle the shock makes with the original direction of flow.

Solution. Two solutions are possible, depending on the nature of the shock; both are given below. Let x and y refer to conditions upstream and downstream of the shock, respectively.

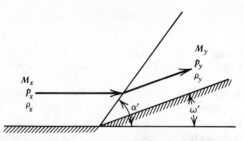

Figure 4. Two-dimensional shock

From Table 56, for $\omega' = 20°$ and $M_x = 2.0$, we get two values of α'.

From Table 57, for $\omega' = 20°$ and the two values of α', we get two values of M_y.

From Tables 58 and 59, we get the corresponding values for p_y/p_x and ρ_y/ρ_x.

The ratio of mass velocities downstream to upstream is given by

$$\frac{G_y}{G_x} = \frac{\rho_y}{\rho_x} \cdot \frac{V_y}{V_x} = \frac{\rho_y}{\rho_x} \cdot \frac{M_y}{M_x} \cdot \sqrt{\frac{T_y}{T_x}} \, ,$$

where

$$\frac{T_y}{T_x} = \frac{p_y/p_x}{\rho_y/\rho_x} .$$

The resulting values are given in Table R.

Table R

	Upstream Section	Downstream Section Weak Shock	Strong Shock
M	2.0	1.210	0.7282
p (psia)	20	56.86	83.14
T (R)	500	696.1	813.9
ρ (lb/ft^3)	0.1080	0.2205	0.2758
α' (deg)	0	53.44	74.25
G/Gx	1	1.458	1.186
V (ft/sec)	2192	1565	1018

Acknowledgments

Tables 30 to 53 were reprinted from Meteor Report 14, entitled "The Mechanics and Thermodynamics of Steady One-Dimensional Gas Flow with Tables for Numerical Solutions," by A. H. Shapiro, W. R. Hawthorne, and G. M. Edelman, December 1, 1947. This report was prepared in the Gas Turbine Laboratory of the Massachusetts Institute of Technology for Project Meteor under contract with the Office of Naval Research of the United States Navy, and was reprinted by permission of the director of the project and of the authors. For this edition, the data in these tables have been rechecked and corrected where necessary.

Tables 55 to 59 are reprinted from Bumblebee Report 26, entitled "The Theory and Practice of Two-Dimensional Supersonic Pressure Calculations," by N. Edmonson, F. D. Murnaghan, and R. M. Snow, December 1945. This report was prepared in the Applied Physics Laboratory, The Johns Hopkins University, under contract with the Bureau of Ordnance of the United States Navy, and is reprinted here by permission of the director of the laboratory.

Personnel of the Research Division of the United Aircraft Corporation provided assistance in the preparation of Tables 26 through 29.

Bibliography

1. J. O. Hirschfelder, C. F. Curtiss, and R. B. Bird, *Molecular Theory of Gases and Liquids*, John Wiley & Sons, Inc., New York, 1954.

2. J. H. Keenan, *Thermodynamics*, John Wiley & Sons, Inc., New York, 1941, p. 96.

3. J. E. Mayer and M. G. Mayer, *Statistical Mechanics*, John Wiley & Sons, Inc., New York, 1940; F. D. Rossini *Chemical Thermodynamics*, John Wiley & Sons, Inc., New York, 1950; and many others.

4. F. C. Andrews, *Equilibrium Statistical Mechanics*, John Wiley & Sons, Inc., New York, 1975.

5. G. Herzberg, *Spectra of Diatomic Molecules*, D. Van Nostrand Co., Inc., New York, 1950.

6. J. Hilsenrath, C. W. Beckett, W. S. Benedict, L. Fano, H. J. Hoge, J. F. Masi, R. L. Nuttall, Y. S. Touloukian, and H. W. Wolley, NBS Circular 564, National Bureau of Standards, Washington, D.C., 1955.

7. B. LeNeindre and B. Vodar, *Experimental Thermodynamics*, Vol. II, Butterworth and Co., Ltd., London, 1975.

8. Y. S. Touloukian and T. Makita, *Thermophysical Properties of Matter*, Vol. 6, IFI/Plenum Data Corporation, New York, 1970.

9. R. L. Dommett, RAE-TN-GW 429, 1–39, 1956 (AD 115386).

10. H. K. Kallman, RM 442, 1–44, 1950 (AD 103216).

11. E. R. Cohen and B. N. Taylor, *J. Phys. Chem. Ref. Data*, **2**, 663 (1973).

12. N. N. Greenwood, *Pure Appl. Chem.*, **37**, 4 (1974).

13. B. Rosen, *Spectroscopic Data of Diatomic Molecules*, Pergamon Press, London, 1970.

14. C. E. Moore, *Atomic Energy Levels*, Vol. I, National Bureau of Standards, Washington, D.C., 1949.

15. F. Din, *Thermodynamic Functions of Gases*, Vol. 2, Butterworths Scientific Publications, London, 1956.

16. N. B. Vargaftik, *Tables on the Thermophysical Properties of Liquids and Gases*, 2nd Ed., John Wiley & Sons, Inc., New York, 1975.

17. G. C. Maitland and E. B. Smith, *J. Chem. Eng. Data*, **17**, 150 (1972).

18. Y. S. Touloukian, S. C. Saxena, and P. Hestermans, *Thermophysical Properties of Matter*, Vol. 11, IFI/Plenum Data Corporation, New York, 1975.

19. Y. S. Touloukian, T. E. Liley, and S. C. Saxena, *Thermophysical Properties of Matter*, Vol. 3, IFI/Plenum Data Corporation, New York, 1970.

20. J. Kaye, paper presented at Annual Meeting of A.S.M.E., December 1947.

21. "Selected Values of Properties of Hydrocarbons and Related Compounds," American Petroleum Institute Research Project 44, Thermodynamics Research Center, Texas A & M University, College Station, Texas 77843 (loose-leaf data sheets, extant 1977).

22. Report of the CODATA Task Group on Key Values for Thermodynamics, J. D. Cox, Chairman, CODATA Bulletin No. 17, January 1976.

23. D. W. Scott, *J. Chem. Phys.*, **60**, 3144 (1974).

24. H. W. Woolley, R. B. Scott, and F. G. Brickwedde, *J. Res. Natl. Bur. Std.*, **41** 379 (1948).

25. L. Haar, Private communication, 1978.

26. D. R. Stull and H. Prophet, ed., *JANAF Thermochemical Tables*, 2nd Edition, NSRDS-NBS 37, U.S. Government Printing Office, Washington, D.C., 1971; M. W. Chase, private communication, the Dow Chemical Company, Midland, Michigan, 1977.

27. "Selected Values of Properties of Chemical Compounds," Thermodynamics Research Center Data Project, Thermodynamics Research Center, Texas A & M University, College Station, Texas 77843 (loose-leaf data sheets, extant 1978).

28. L. V. Gurvich, G. A. Khachkuruzov, V. A. Medvedev, I. V. Veyts, G. A. Bergman, V. S. Yungman, N. P. Rtishcheva, L. F. Kuratova, G. N. Yurkov, A. A. Kane, B. F. Yudin, B. I. Brounshteyn, V. F. Baybuz, V. A. Kvlividze, Ye. A Prozorovskiy, B. A. Vorobyev, *Thermodynamic Properties of Chemical Substances*, Academy of Sciences, U.S.S.R., Moscow, 1962.

29. S. Angus and B. Armstrong, *International Thermodynamic Tables of the Fluid State*, Vol. 1, Pergamon Press, New York, 1971.

30. S. Angus, B. Armstrong, and K. M. deReuck, *International Thermodynamic Tables of the Fluid State*, Vol. 3, Pergamon Press, New York, 1976.

31. V. J. Johnson, "Cryogens and Gases," Publication No. STP 537, American Society for Testing and Materials, Philadelphia, Pa., 1973.

32. G. M. Wilson, R. G. Clark, and F. L. Hyman, *Applied Thermodynamics*, American Chemical Society, Washington, D.C., 1968.

33. F. Din, *Thermodynamic Functions of Gases*, Vol. 1, Butterworths Scientific Publications, London, 1956.

34. F. Din, *Thermodynamic Functions of Gases*, Vol. 2, Butterworths Scientific Publications, London, 1956.

35. F. Din, *Thermodynamic Functions of Gases*, Vol. 3, Butterworths Scientific Publications, London, 1961.

36. A. A. Vasserman, Ya. Z. Kazavchinskii, and V. A. Rabinovich, *Thermophysical Properties of Air and Air Components*, Israel Program for Scientific Translation, Jerusalem, 1971 (translated from Russian).

37. A. H. Shapiro and W. R. Hawthorne, *J. Applied Mech.*, **14**, A317–337 (1947).

38. A. H. Shapiro and G. M. Edelman, *J. Applied Mech.*, **14** A154–162 (1947).

39. N. Edmonson, F. D. Murnaghan, and R. M. Snow, Bumblebee Report 26 (December 1945).

40. W. E. Forsythe, Ed., *Smithsonian Physical Tables*, Smithsonian Institution, Washington, D.C., 1969, 347.

41. E. R. Cohen and B. N. Taylor, *J. Phys. Chem. Ref. Data*, **2**, 663 (1973).

42. "The International System of Units (SI)," U.S. National Bureau of Standards Special Publication 330, U.S. Government Printing Office, Washington, D.C., 1977.